S0-BLY-859

ALLE ZEIT WACH
1842

G. Gierz K. H. Hofmann K. Keimel
J. D. Lawson M. Mislove D. S. Scott

A Compendium of Continuous Lattices

Springer-Verlag
Berlin Heidelberg New York 1980

Gerhard Gierz
Fachbereich Mathematik, Technische Hochschule Darmstadt
Schloßgartenstr. 7
D-6100 Darmstadt/Germany

Karl Heinrich Hofmann
Department of Mathematics, Tulane University
New Orleans, LA 70118/U.S.A.

Klaus Keimel
Fachbereich Mathematik, Technische Hochschule Darmstadt
Schloßgartenstr. 7
D-6100 Darmstadt/Germany

Jimmie D. Lawson
Department of Mathematics
Louisiana State University
Baton Rouge, LA 70803/U.S.A.

Michael W. Mislove
Department of Mathematics
Tulane University
New Orleans, LA 70118/U.S.A.

Dana S. Scott
Merton College
Oxford OX1 4JD/Great Britain

AMS Subject Classification (1980): 06 B30, 06 F30, 54 F05, 54 F20

ISBN 3-540-10111-X Springer-Verlag Berlin Heidelberg New York
ISBN 0-387-10111-X Springer-Verlag New York Heidelberg Berlin

Library of Congress Cataloging in Publication Data. Main entry under title:
A Compendium of continuous lattices. Bibliography: p. Includes indes. 1. Lattices, Continuous. I. Gierz, Gerhard. QA171.5.C65. 511.3'3. 80-19122

Printed in Germany

Printing: Beltz Offsetdruck, Hemsbach/Bergstr. Bookbinding: K. Triltsch, Würzburg
2141/3140-543210

TABLE OF CONTENTS

FOREWORD

A mathematics book with six authors is perhaps a rare enough occurrence to make a reader ask how such a collaboration came about. We begin, therefore, with a few words on how we were brought to the subject over a ten-year period, during part of which time we did not all know each other. We do not intend to write here the history of continuous lattices but rather to explain our own personal involvement. History in a more proper sense is provided by the bibliography and the notes following the sections of the book, as well as by many remarks in the text. A coherent discussion of the content and motivation of the whole study is reserved for the introduction.

In October of 1969 Dana Scott was lead by problems of semantics for computer languages to consider more closely partially ordered structures of function spaces. The idea of using partial orderings to correspond to spaces of partially defined functions and functionals had appeared several times earlier in recursive function theory; however, there had not been very sustained interest in structures of *continuous* functionals. These were the ones Scott saw that he needed. His first insight was to see that – in more modern terminology – the category of algebraic lattices and the (so-called) Scott-continuous functions is cartesian closed. Later during 1969 he incorporated lattices like the reals into the theory and made the first steps to defining continuous lattices as "quotients" of algebraic lattices. It took about a year for the topological ideas to mature in his mind culminating in the paper published as Scott [1972a]. (For historical points we cannot touch on in this book the reader is referred to Scott's papers.) Of course, a large part of Scott's work was devoted to a presentation of models for the *type-free* λ-calculus, but the search for such models was *not* the initial aim of the investigation of partially ordered structures; on the contrary, it was the avoiding of the formal and unmotivated use of λ-calculus that prompted Scott to look more closely at the structures of the functions themselves, and it was only well after he began to see their possibilities that he realized there had to exist non-trivial T_0-spaces homeomorphic to their own function spaces.

Quite separately from this development, Karl Hofmann, Jimmie Lawson, Mike Mislove, and Al Stralka (among others) recognized the importance of compact semilattices as a central ingredient in the structure theory of compact semigroups. In his dissertation [1967], Lawson initiated the study of a class of compact semilattices distinguished by the property that each had enough continuous semilattice morphisms into the unit interval semilattice (in its natural order) to separate points. (Such a program had already been started by Nachbin for partially ordered spaces in [1965].) Lawson characterized this class of compact semilattices as those which admitted a basis of subsemilattice neighborhoods at each point (small subsemilattices): the class proved to be of considerable theoretical interest and attracted the attention of other workers in the field. In fact, it was believed for some time that all compact semilattices were members of the class, partly because the theory was so satisfactory (for example, purely "order-theoretic" characterizations were discovered for the class by Lawson [1973a, b]), and because no natural counterexamples seemed to exist. However, Lawson found the first example of a compact semilattice which was not in the class, one in fact which admitted only constant morphisms into the unit interval [1970] (see Chapter VI).

At about the same time, Klaus Keimel had been working on lattices and lattice-ordered algebras in pursuit of their spectral theory and their representation in sheaves. In his intensive collaboration with Gerhard Gierz on topological representations of non-distributive lattices, a spectral property emerged which turned out to be quite significant for compact semilattices with small subsemilattices.

The explanation for the fact that the topological algebra of Lawson's semilattices had been so satisfactory emerged clearly when Hofmann and Stralka gave a completely lattice-theoretical description of the class [1976]. It was Stralka who first recognized the relation of this class to Scott's continuous lattices, and this observation came about as follows. Two monographs on duality theories for lattices and topological structures emerged in the early seventies: One for topology and lattices by Hofmann and Keimel [1970], and the other for compact zero-dimensional semilattices and lattices by Hofmann, Mislove, and Stralka [1974]. At the lattice theory conference in Houston in 1973, where such dualities were discussed, B. Banaschewski spoke on filters and mentioned Scott's work which was just about to appear in the Proceedings of the Dalhousie Category Theory Conference. Stralka checked out this hint, and while he and Hofmann were working on the algebraic theory of Lawson semilattices [1976], he realized the significance of this work as a link between the topological algebra of compact semilattices and the lattice theory of Scott's continuous lattices. This led to correspondence with Scott and much subsequent activity.

In the summer of 1976, Hofmann and Mislove spent some time collaborating with Keimel and Gierz at the Technische Hochschule in Darmstadt, and together they began a "write-in" seminar called the *Seminar on Continuity in Semilattices*, or SCS for short. The authors formed the core membership of the seminar, but their colleagues and students contributed greatly to the seminar by communicating their results, ideas, and problems. (A list of these seminar reports (SCS Memos) which resulted is provided at the end of this monograph.) The seminar then convened in person for several lively and well-attended workshops. The first was hosted by Tulane University in the spring of 1977, the second by the Technische Hochschule Darmstadt in the summer of 1978, and the third by the University of California at Riverside in the spring of 1979. A fourth workshop was held at the University of Bremen in the fall of 1979. We are very much indebted to all who participated in these seminars and others whose influence on this book is very considerable. In particular we thank H. Bauer, J. H. Carruth, Alan Day (who discovered an independent access to continuous lattices through the filter monad). Marcel Erné, R.-E. Hoffmann, John Isbell (whose also gave very detailed remarks on the present manuscript), Jaime Niño, A. R. Stralka, and O. Wyler.

It was at the Tulane Workshop that the idea of collecting together the results of research – common and individual – was first discussed. A preliminary version of the Compendium worked out primarily by Hofmann, Lawson, and Gierz was circulated among the participants of the Darmstadt Workshop, and many people gave us their useful reactions. For help in typing the earlier versions of this book we would like to thank Frau Salder in Darmstadt and Mrs. Meredith Mickel at Tulane University.

The preparation of the final version of the text, which is reproduced from camera-ready copy, was carried out by and under the direction of Scott at the Xerox Palo Alto Research Center (PARC) in its Computer Science Laboratory (CSL). Scott spent the academic year 1978/79 on sabbatical as a Visiting Scientist at Xerox PARC, and the facilities of CSL, including extra secretarial aid, were very generously put at his disposal. The text was prepared on an Alto computer using the very flexible BRAVO text-editing system and a special computer-controlled printer. The typist, who in the course of the project also became a skilled computer-aided book illustrator and copy editor, was Melinda Maggiani. Without her loyal efforts and concentrated labor the book would never have been put into anywhere near the form seen here; the authors are extremely grateful to her. Special thanks are also due to many members of CSL for their interest and patience in helping Scott learn to use BRAVO, with which he spent long, long hours; he wishes to mention with great warmth in particular Sara Dake, Leo Guibas, Jim Horning, Jeannette Jenkins, Joe Maleson, Jim Morris, HayChan Sargent, and Dan Swinehart. In the very last stages of the book make-up it was necessary to reprogram some printing routines to overcome several most irritating difficulties, and Lyle Ramshaw then stepped in and solved all the programming problems in record time. We take our hats off to him! (See especially in this regard Chapter IV, Section 3.)

The computerized editing system made it possible to produce in a very few months what were in effect two complete sets of galley proofs and two complete sets of page proofs; this is something that would never have been possible in our wildest dreams with the conventional manuscript-typescript-type style of book production. Computer-controlled editing allowed the authors to make, through the fingers of Maggiani and Scott, innumerable substantive corrections and to do extensive rewritting at every stage of the proof reading up to the last day before printing the camera-ready copy. Authors and publishers alike can only hope that such systems will soon become widely available. It was a real privilege to prepare this book at Xerox PARC, and the authors record here their heart-felt thanks to Dr. Robert J. Spinrad, Vice President and Manager of Xerox PARC, and especially to Robert Taylor, Manager of CSL. Aside from the support and cooperation, the remarkably friendly and informal atmosphere of PARC contributed much to the project.

For the support and sponsorship over the years of their research and their workshops, the authors are also happy to express their gratitude to the Alexander von Humboldt Stiftung, the Deutsche Forschungsgemeinschaft, the National Science Foundation, the Simon Guggenheim Foundation, and the Universität Bremen, and to their own institutions, Louisiana State University, Oxford University and Merton College, Technische Hochschule Darmstadt, and Tulane University.

The Authors

January, 1980

INTRODUCTION

Background and Plan of the Work

The purpose of this monograph is to present a fairly complete account of the development of the theory of continuous lattices as it currently exists. An attempt has been made to keep the body of the text expository and reasonably self-contained; somewhat more leeway has been allowed in the exercises. Much of what appears here constitutes basic, foundational or elementary material needed for the theory, but a considerable amount of more advanced exposition is also included.

BACKGROUND AND MOTIVATION

The theory of continuous lattices is of relatively recent origin and has arisen more or less independently in a variety of mathematical contexts. We attempt a brief survey in the following paragraphs in the hope of pointing out some of the motivation behind the current interest in the study of these structures. We first indicate a definition for these lattices and then sketch some ways in which they arise.

A DEFINITION. In the body of the Compendium the reader will find many equivalent characterizations of continuous lattices, but it would perhaps be best to begin with one rather straight-forward definition–though it is not the primary one employed in the main text. Familiarity with *algebraic lattices* will be assumed for the moment, but even if the exact details are vague, the reader is surely familiar with many examples: the lattice of ideals of a ring, the lattice of subgroups of a group.

Abstractly (and up to isomorphism) we can say that an algebraic lattice is a lattice of sets–contained, say, in the lattice of all subsets of a given set A–closed under arbitrary intersection of families of sets and under unions of directed families of sets (e.g. chains of sets). These are important closure properties of the lattice of ideals, for example. If we think of the powerset lattice as a product 2^A of A copies of the two-element lattice $2 = \{0,1\}$, then an algebraic lattice is just a *sublattice* of 2^A with respect to the infinite operations of arbitrary pointwise inf and pointwise sup of *directed* families of lattice elements. (Note, however, that *finite* sups are different in general; so the meaning of the word "sublattice" has to be understood in a suitable sense.)

Let us now replace the discrete lattice 2 with the "continuous" lattice $[0,1]$, the unit interval of real numbers with its natural order and familiar lattice structure. In a power $[0,1]^A$ we can speak of sublattices with respect to arbitrary pointwise inf and pointwise sup of directed families of elements, just as before. *Up to isomorphism, these are exactly the continuous lattices.* Of course this definition gives no hint as to the internal structure of these lattices and is only a dim indicator as to their naturalness and usefulness. But it does show that they are direct generalizations of well-known kinds of lattices and that they have an important element of "continuity".

THEORY OF COMPUTATION. Often in computational schemes one employs some algorithm successively to gain increasingly refined approximations to the desired result. It is convenient to use, formally or informally, topological language–one talks about "how far" the approximation is from the desired result or how good a "fit" has been obtained. An alternate procedure is to specify at each stage a subset in which the desired result lies. The smaller the set, the better the approximation; we could say that the smaller set gives "more information". This approach leads naturally to the use of order-theoretic language in discussing the partial results, and the data generated, in a way related to the containments among the sets.

Let us now abstract this approach somewhat. Let P be a partially ordered set. We think of a "computation" of an element x in P as being a sequence of increasingly larger elements–"larger" meaning "more" in the sense of information–whose supremum is x. (More generally, we could imagine a directed set whose supremum is x.) We wish to regard x as the "limit" of the sequence (or set) of approximations.

What is needed is a precise definition of how well some "stage" of the "computation" approximates the "limit" x. We take an indirect approach to this question, because there is no metric available to tell us immediately how close an approximation is to the desired limit. We define in place of a metric a notion meaning roughly: an element y is a "finite approximation" to the element x. Then, to have a well-behaved system of limits, we *assume* that every element is the sup in the partial ordering of its finite approximations. A given sequence of approximations to x is then "successful" if it eventually encompasses all the finite approximations.

Specifically, we say that an element y, which is less than or equal to x, is a *finite approximation* to x if for any directed set D with supremum x (D represents the stages of a "computation" of x) we have some member of D which is greater than or equal to y. (Hence, all the members from that stage on are greater than or equal to y.) The idea is that if we use y to measure the accuracy of computations of x, then every computation that *achieves* x must eventually be *as least as accurate as* y.

Strictly speaking the outline just given is actually not quite right. We should say that if a directed set D has supremum *greater than or equal to* x, then some member of D is greater than or equal to y. This ensures that if y is a "finite approximation" for x, then it is also one for every element larger than x–a property we would certainly want to require. In the text we use different terminology. The notion of "finite" in "finite approximation" is somewhat vague, because again there is no measure to distinguish finite elements from infinite ones in general; indeed there are lattices where *all* elements except 0 are normally thought of as infinite (as in the lattice [0,1] for instance). This explains our feeling that another terminology was required. We have used the phrase "y is way below x" for topological and order-theoretic reasons cited in the appropriate section of the book.

To recapitulate: we assume that the "finite approximations" for each element are directed and have that element for their supremum. *The complete lattices with this property are the continuous lattices.* It is the theory of these abstracted, order-theoretic structures that we develop in this monograph. It should be pointed out that only the lattice case is treated in the main text; generalizations appear in the exercises.

Owing to limitations of length and time, the theory of computation based on this approach is not developed extensively here. Certain related examples are, however, mentioned in the present text or in the exercises. For instance, consider the set of all partial functions from the natural numbers into itself (or some distinguished subset such as the recursive partial functions). These can be ordered by inclusion (that is, extension). Here again the larger elements give more information. In this example f is a "finite approximation" for g if and only if g is an extension of f and the domain of f is finite. In many examples such as this the "approximating" property can be interpreted directly as a finiteness condition, since there are finite functions in the set (functions with a finite domain). This circumstance relates directly to the theory of algebraic lattices, a theme which we do cover here in great detail.

GENERAL TOPOLOGY. Continuous lattices have also appeared (frequently in cleverly disguised forms) in *general topology*. Often the context is that of the category of all topological spaces or of topological spaces where one assumes only a T_0-separation axiom. Such spaces have been the objects of renewed interest with the emergence of spectral theory.

In fact, a continuous lattice can be endowed in a natural way with a T_0-topology which is defined from the lattice structure; in this book we call this T_0-topology the *Scott topology*. It is shown in Section 3 of Chapter II that these spaces are exactly the "injectives" (relative injectives in the categorical sense) or "absolute retracts" in the category of all T_0-spaces and continuous functions; that is if f is a continuous function from a subspace X of a topological space Y into a continuous lattice L (equipped with the Scott topology), then there always exists a continuous extension of f from X to Y with values still in L. This property in fact gives a topological characterization of continuous lattices, since any T_0-topology of such a space is just the Scott topology of a continuous lattice naturally determined from it.

In another direction let us say that an open set U is *relatively compact* in an open set V if every open cover of V has finitely many members which cover U. If X is a topological space, then the lattice of open sets is a continuous lattice iff each open set is the union of the open sets which are relatively compact in it. In this case the "way below" relation is viewed as just the relation of one open set being relatively compact in another. This illustrates some of the versatility of the concept of a continuous lattice.

Spaces for which the lattice of open sets is a continuous lattice prove to be quite interesting. For Hausdorff spaces it is precisely the locally compact spaces which have this property, and in more general spaces analogues of this result remain true. We investigate this situation in some detail in the context of the spectral theory of distributive continuous lattices in Section 5 of Chapter V. It is often the case that theorems concerning locally compact Hausdorff spaces extend to spaces with a continuous lattice of open sets in the category of all topological spaces (see, e.g., Day and Kelly [1970]). Such considerations provide another link between continuous lattices and general topology.

The dual of the lattice of open sets–the lattice of closed sets–has long been an object of interest to topologists. If X is a compact Hausdorff space, then the lattice of closed subsets under the Vietoris topology is also a compact Hausdorff space. In Chapter III we introduce a direct generalization of this topology, called here the Lawson topology, which proves to be compact for all complete lattices and Hausdorff for continuous lattices. This connection allows applications of continuous lattice theory to the topological theory of hyperspaces (cf. Example VI-3.8).

Analysis, algebra. Several applications of continuous lattices arise in analysis and functional analysis. For example, consider the family $\mathbf{C}(X, \mathbb{R})$ of continuous real-valued functions on the locally compact space X. Using the pointwise operations and the natural order from $\mathbb{R}$, the space $\mathbf{C}(X, \mathbb{R})$ is a lattice, but its lattice theory is rather unsatisfactory. For example, it is not even complete; however, if we consider this lattice as a sublattice of $\mathbf{LSC}(X, \mathbb{R}^*)$, the lattice of all lower semicontinuous extended real-valued functions on X, then we do have a complete lattice with which to work. In fact, although this is not at all apparent from the functional analysis viewpoint, the lattice $\mathbf{LSC}(X, \mathbb{R}^*)$ is a continuous lattice. This entails several results, not the least of which is the following: *The lattice* $\mathbf{LSC}(X, \mathbb{R}^*)$ *admits a unique compact Hausdorff function-space topology such that* $(f,g) \to f \wedge g$ *is a continuous operation.* (See I-1.2, II-4.7, and II-4.20.) In light of the fact that $\mathbf{C}(X, \mathbb{R})$ is never compact and that $\mathbf{C}(X, \mathbb{R}^*)$ is compact only if X is finite, this result is somewhat suprising; we do not know of a "classical" proof. Indeed, lower semicontinuity motivates one of the canonical topologies on any continuous lattice, and, if we equip $\mathbb{R}^*$ with this canonical topology, then the continuous functions from X into $\mathbb{R}^*$ so topologized are exactly those extended real-valued functions on X which are lower semicontinuous relative to the usual topology on $\mathbb{R}^*$. In the same vein it emerges that the probability distribution functions of random variables with values in the unit interval form a continuous lattice; compact topologies for this example are, however, familiar from classical analysis (cf. I-2.17).

A second example is quite different and demonstrates an overlap between analysis and algebra. With a ring R one associates a topological space, called its *spectrum,* and while there are many ways of doing this, probably the most wide-spread and best known is the space of prime ideals of a commutative ring endowed with hull-kernel topology. This plays a central rôle in algebraic geometry (where the relevant theory deals with noetherian rings and their spectra), and this construction can also be carried out for Banach algebras, in which case the preferred spectrum is the space of closed primitive ideals (which reduces to the more familiar theory of maximal ideals if the algebra is commutative). These particular ideals are relevant since they are precisely the kernels of irreducible representations of the algebra as an algebra of operators on a Banach space or Hilbert space.

Now, the connection of these spectral theories with the theory of continuous lattices emerges more clearly if we first return to the case of a commutative ring R. In this case, the spectrum is the set of prime ideals viewed as a subset of the algebraic lattice of all ideals of the ring; in fact, the spectrum is exactly the family of prime elements of the distributive algebraic lattice of all radical ideals of R. (Recall that a radical ideal is one which is the intersection of prime ideals.) If we define the spectrum of a distributive lattice as being its family of prime elements, then we have just reduced the spectral theory of commutative rings to the spectral theory of distributive algebraic lattices.

It may not be obvious, but the situation in functional analysis is analogous. Here we consider the lattice of *closed* two-sided ideals of a Banach algebra; while this lattice is not algebraic in general, it *is* a continuous lattice (at least in the case of a C*-algebra). Moreover, in the case the algebra is separable, its spectrum is just the traditional primitive ideal spectrum. Again, we have reduced the spectral theory of separable C*-algebras to the spectral theory of distributive continuous lattices. This approach to the spectral theory of C*-algebras affords an affirmative (and perhaps more systematic) proof of the fact that the primitive spectrum of a separable C*-algebra is a locally quasicompact T_0-space (cf. I-1.20 and V-5.5). Indeed, the central result in the spectral theory of a continuous lattice is that its spectrum is just such a space (cf. V-5.5).

Lastly, we mention another area of functional analysis which relates to our theory. If C is a compact convex set in a locally convex topological space, it is useful to know as much as possible about the space of closed convex subsets of C. This space is a lattice, and its opposite lattice is in fact a continuous lattice whose prime elements are exactly the singleton sets containing extreme points of C. Moreover, if the upper semicontinuous affine functions on C are considered, then once again a continuous lattice is found in much the way we encountered one in the example **LSC**(X, $\mathbb{R}^*$) above. We have yet another instance where a function space naturally arises whose lattice and topological properties are essentially those of a continuous lattice.

CATEGORY THEORY, LOGIC. Another area in which continuous lattices have occured somewhat unexpectedly is the area of *category theory*. Constructions of free objects play an important role in mathematics, e.g., free groups, free semigroups, free modules. A somewhat more sophisticated construction is the construction of the free compact Hausdorff space over a set X. This turns out to be the Stone-Čech compactification of the discrete space X, which can be identified as all ultrafilters on X equipped with a suitable topology.

These constructions can all be set in a suitable categorial context: the theory of *monads* or *triples*. Here one has adjoint functors (which can be thought of as a "free" functor and a "forgetful" functor). It is then possible to define categorially the "quotients" of the free objects, which become the "algebras" of the system. It has been found that in this abstract setting it is sometimes possible to identify free objects *before* knowing what the algebras are. A simple example is the powerset monad–but it is very easy to prove that the algebras are just the complete sup-semilattices.

The question arose of identifying the monad for which the free functor is that which assigns to a set the set of all filters on that set. It is the discovery of Alan Day that the algebras for this monad are precisely the continuous lattices. (See Day [1975] and Wyler [1976]).

In our treatment of continuous lattices we do not completely follow the categorial approach, but this is no reflection on its mathematical merit. But we do prove Day's theorem, however, and we have much to say about categorial properties of many classes of structures related to continuous lattices.

One particulary interesting categorial aspect of our work–at least in the authors' eyes–is the ease with which examples of *cartesian closed categories* can be found. (Natural examples are not so very common in mathematics or even in category theory until one comes to the theory of topoi.) The reason for the occurrence of these cartesian closed categories, as we explain in the text, has to do with the *function-space construction*.

Specifically, in the context of our considerations, we have available a rather natural notion of morphism, namely that morphisms should preserve limits of what we have called "computations". More precisely a morphism is a function between partially ordered sets with the property that the image of the supremum of a directed set is the supremum of the image of the directed set. (Such functions are treated in some detail in Chapter II, Section 2.) There are also several other interesting classes of morphisms with various properties, but this notion (called Scott continuity) works especially well in forming function spaces.

If L and M are continuous lattices, let $[L \to M]$ be the set of all morphisms from L to M. It is shown in Chapter II that, if these functions are given the pointwise ordering, then $[L \to M]$ is also a continuous lattice. Moreover, this construction is a functor adjoint to the formation of cartesian products.

By using this construction and an inverse limit procedure, examples of nontrivial continuous lattices L can be found which are actually isomorphic with their own self-function space $[L \to L]$. (This does not seem possible with any stronger separation property beyond the T_0-axiom.) Constructions of this type are treated in great generality in Sections 3 and 4 of Chapter IV.

The examples just mentioned have the striking property that their members can be interpreted either as elements or as functions, and that every self-function corresponds to some element. This is precisely the set-up that one hypothesizes in the λ-calculus approach to logic. Thus, continuous lattices provided the background for the construction of concrete models for an axiomatic logical system that had long existed without them (see Scott [1973]).

Topological algebra. The final area that we wish to mention is that of *topological algebra*. Among the objects investigated in this field were compact topological semilattices and lattices (that is, semilattices or lattices equipped with a compact topology for which the meet or meet and join operations were continuous). In the course of study of compact semilattices it emerged that those which had a neighborhood basis of subsemilattices lent themselves more easily to mathematical investigation; in addition known examples had this property. Hence, attention was particulary focused on this class.

The amazing result of these investigations was the discovery that such semilattices were (modulo an identity or top element) continuous lattices with respect to their lattice structure! (We derive this result in Chapter VI-3 after first developing some of the most basic theory of topological semilattices in VI-1.) Conversely, if L is a continuous lattice, then a topology can be defined from the order which makes L into a compact topological semilattice with small semilattices. The topology in question is the Lawson topology already mentioned; it is defined and investigated in Chapter III.

This identification between compact semilattices with small subsemilattices and continuous lattices has greatly affected the development of the theory of continuous lattices. Not only do many of the results of topological semilattices transfer wholesale to continuous lattices, but also topological techniques and methods play a prominent role in their study (as opposed to most traditional lattice theory). Conversely, lattice-theoretic methods frequently aid investigations of a topological nature. This interplay is illustrated in Chapters VI and VII.

Before we entirely leave our discussion of the roots of continuous lattices, a postscript concerning algebraic lattices is probably in order. Algebraic lattices provide an important link between the theory of continuous lattices and traditional lattice theory and universal algebra. Indeed they are a special class of continuous lattices; their theory has frequently suggested generalizations and directions of research for continuous lattices. They are introduced in Chapter I-4, but resurface on several occasions. With respect to the Lawson topology, they are precisely those continuous lattices which are totally disconnected (equivalently: zero-dimensional), and so they also occupy a natural place in topological algebra.

PLAN OF THE WORK

Chapter O consists essentially of background material of an order-theoretical nature. The reader may review it to the extent he feels necessary. Some familiarity with the language and notation introduced there will probably be necessary. The formalism of Galois connections explained in O-3 is vital for many things which will follow in the main body of the book.

Chapter I introduces continuous lattices from an order-theoretic point of view. In Section 1 continuous lattices are defined and examples are given. The "way-below" relation is introduced, and is characterized among the auxiliary relations. Section 2 gives an equational characterization of continuous lattices and discusses their variety-like properties. Section 3 introduces prime and irreducible elements and generalizations thereof and shows the plentiful supply of such in continuous lattices. The basic properties of algebraic lattices and some of their relationships with continuous lattices appear in Section 4. Most of this material in this chapter is quite basic (except perhaps the material on auxiliary relations in Section 1).

Chapter II defines the Scott topology and develops its applications to continuous lattices. In Section 1 the Scott topology is defined and convergence in the Scott topology is characterized for continuous lattices. Section 2 gives the definition and characterizations of Scott-continuous functions. In Section 3 it is shown that continuous lattices endowed with their Scott topologies form the "injectives" in the category of T_0 topological spaces. Section 4 is concerned with function spaces (particulary the set of Scott-continuous functions between spaces and/or lattices) and questions of the categorical notion of "cartesian closedness". Of fundamental importance in this chapter are the basic properties of the Scott topology and Scott-continuous functions appearing in Sections 1 and 2 (although they are treated in greater detail than may be of interest to the general reader).

Chapter III introduces the second important topology for continuous lattices, the Lawson topology. Like the Scott topology, it is defined in an order-theoretic fashion. In Section 1 it is shown that the Lawson topology is quasicompact and T_1 for every complete lattice and compact Hausdorff for continuous lattices. Indeed in Section 2 it is shown that for meet-continuous complete lattices the Lawson topology is Hausdorff if and only if the lattice is continuous. Section 3 characterizes convergence of nets in the Lawson topology. Section 4 generalizes the notion of a basis for a topology to continuous lattices and derives properties thereof. Section 1 and 2 are the basic sections of this chapter.

Chapter IV considers various important categories of continuous lattices together with certain categorial constructions. Section 1 is the important one here; it presents important duality theorems for the study of continuous lattices. Section 2 contains the important result that a continuous lattice has sufficiently many semilattice homomorphisms of the right kind into the unit interval to separate points. Since we aspired to great generality in the proof, its details may appear dry. Some of the exercises exemplify applications which were made possible on this level of generality. The last two sections give general categorical constructions for obtaining continuous lattices which are "fixed points" with respect to some self-functor of the category. This process is needed for the construction of set-theoretical models of the λ-calculus.

In Chapter V spectral theory is taken up. An important lemma on the behavior of primes appears in Section 1. In Section 2 it is shown one always has a smallest closed generating (or order generating) set, namely the closure of the irreducible elements. Considering elements in this closure leads to generalizations of the notion of prime and irreducible in Section 3. Section 4 introduces the subject of the spectral theory of lattices in general, and Section 5 considers that of continuous lattices. There a duality is set up between the category of all distributive continuous lattices and all locally quasicompact sober spaces (with appropriate morphisms for each category). Here probably Sections 1, 4 and 5 would be of greatest interest to the general reader.

Chapter VI begins the study of topological algebra per se. The sections of primary interest are 1 and 3. In Section 1 the most basic and useful properties of pospaces and topological semilattices are given. In Section 3 the Fundamental Theorem of Compact Semilattices is stated and proved, establishing the equivalence between the category of compact semilattices with small semilattices and the category of continuous lattices.

The rest of Chapter VI and Chapter VII center on more specialized topics in topological algebra. Section VI-2 gives an order-theoretic description of convergence in an arbitrary compact semilattice. Section VI-4 presents examples of compact semilattices which are not continuous lattices (unfortunately, the construction is quite intricate). Section VI-5 covers the topic of the existence of arc-chains in topological semilattices, a topic of considerable historical interest in the theory of topological semilattices and lattices.

In Section 1 of Chapter VII topological semilattices in which every open set is an upper set are considered. It is shown that under rather mild restrictions this topology must be the Scott topology. Section 2 takes up the topic of topological lattices, lattices in which both operations are continuous. In Section 3 we return to spectral theory. A topology finer than the spectral topology is introduced, called the "patch topology", and the conditions under which it is compact Hausdorff are investigated. In Section 4 a lattice-theoretic characterization of compact topological semilattices is given, and it is shown that in such a setting separate continuity of the meet operation implies joint continuity.

SUGGESTIONS TO THE READER

The reader who wishes simply to acquire a basic working knowledge of continuous lattices should be able to do so by becoming familiar with those sections we have especially singled out in the above sketch of the contents. More specifically we would consider the following material to be of basic importance: Chapter O, Chapter I (where to reader may wish to go somewhat lightly over parts of Section 1 and the latter parts of 3 and 4), Sections 1 and 2 of Chapter II (treat 1.6 – 1.8 according to interest), Sections 1 and 2 of Chapter III, Sections 1 and 2 of Chapter IV (at a first reading only survey the results of Section 2), Sections 1 and 5 of Chapter V (with 4 as necessary background for 5), and finally Sections 1 and 3 of Chapter VI.

A rather extensive set of exercises has been included ranging from easy to hard to supplement the various sections. Many of the solutions have also been included, especially when they were not easily referenced. Some quite important material has been put in the exercises when it did not fit cleanly with the contents of the text.

The notes at the end of each section make some attempt to relate the material to the published literature. Also since individual contributions can frequently be identified via SCS memos and since such a multiplicity of authors is involved in this project, it seemed reasonable to depart from traditional practice and more or less identify some of the major contributions of various authors in the notes.

We have attempted to compile a rather complete bibliography not only of citations, but also in the work in continuous lattices and closely allied topics at the end which we hope will prove to be useful for further reading and continuing research.

A word about our system of references: If a result or definition within the same chapter is needed, we quote only the Arabic numerals (as in 3.7); if, however, a reference to another Chapter is necessary, we affix the Roman numeral for the chapter (as in III-3.9). Often a numbered paragraph has parts or displayed conditions: the reference is then given by suffixing the appropriate numeral in parentheses (as in III-3.7(ii) or III-3.9(3)). Bibliographical references are given by name and year (as in Birkhoff [1967])–sometimes with a letter (a, b, c) when an author has several publications in the same year; the SCS memos, however, which were only informal and irregular publications, are referenced by number (as in C. E. Clark [SCS-22]); their exact dates are listed in the appendix to the bibliography.

THE LOGICAL INTERDEPENDENCE OF THE CHAPTERS AND SECTIONS

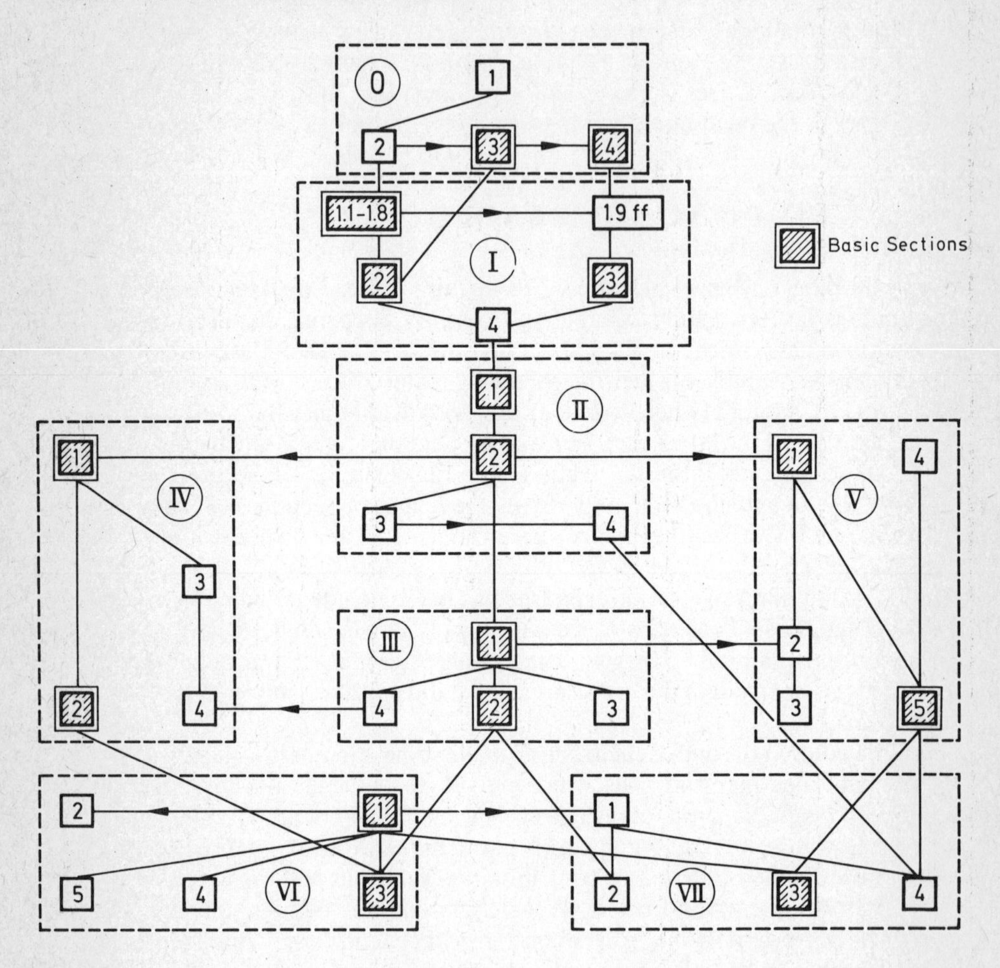

CHAPTER O
A Primer on Complete Lattices

This introductory chapter serves as a convenient source of reference for certain basic aspects of complete lattices needed in the sequel. The experienced reader may wish to skip directly to Chapter I and the beginning of the discussion of the main topic of this book: continuous lattices, a special class of complete lattices.

Section 1 fixes notation, while Section 2 defines complete lattices and lists a number of examples which we shall often encounter. The formalism of Galois connections is presented in Section 3. This is not only a very useful general tool, but it also allows convenient access to the concept of a Heyting algebra. In Section 4 we briefly discuss meet-continuous lattices, of which both continuous lattices and complete Heyting algebras are (overlapping) subclasses. Of course, the more interesting topological aspects of these notions are postponed to later chapters. To aid the student, a few exercises have been included. Brief historical notes and references have been appended, but we have not tried to be exhaustive.

1. GENERALITIES AND NOTATION

Partially ordered sets occur everywhere in mathematics, but it is usually assumed that the partial order is *antisymmetric.* In the discussion of nets and directed limits, however, it is not always so convenient to assume this uniqueness. We begin, therefore, with somewhat more general definitions.

1.1. DEFINITION. Consider a set L equipped with a transitive relation $\leq$. We say that a is a *lower bound* of a set $X \subseteq L$, and b is an *upper bound*, provided that

$$a \leq x \text{ for all } x \in X, \text{ and}$$
$$x \leq b \text{ for all } x \in X, \text{ respectively.}$$

If the set of upper bounds of X has a unique smallest element (that is, the set of upper bounds contains exactly one of its lower bounds), we call this the *least upper bound* and write it as $\bigvee X$ or $\sup X$ (for *supremum*). Similarly the *greatest lower bound* is written as $\bigwedge X$ or inf X (for *infimum*); we will not be dogmatic in our choice of notation. In the case of pairs of elements it is customary to write

$$x \wedge y = \inf \{x,y\},$$
$$x \vee y = \sup \{x,y\}.$$

These operations are also often called *meet* and *join*, and in the case of meet the multiplicative notation xy is common when discussing semigroups.

The set L is *directed* provided every finite subset of L has an upper bound in L. (We also speak of directed subsets of L even when L itself is not directed.) As the empty subset is included in this definition as a finite subset, L must be nonempty. (Aside from non-emptiness, it is sufficient to assume that every *pair* of elements in L has an upper bound in L.) Dually, we call L *filtered* if every finite subset of L has a lower bound in L. □

1.2. DEFINITION. A *net* in a set L is a function $j \mapsto x_j : J \to$ L whose domain is a directed set. (Nets will also be denoted as $(x_j)_{j \in J}$, as (x_j), or even as x_j, if the context is clear.)

If the set L also carries a transitive relation, then the net x_j is called *monotone* (resp., *antitone*), if $i \leq j$ always implies $x_i \leq x_j$ (resp., $x_j \leq x_i$).

If P(x) is a property of the elements $x \in$ L, we say that P(x_j) holds *eventually* in the net if there is a $j_0 \in J$ such that P(x_k) is true whenever $j_0 \leq k$.

The next concept is slightly delicate: if L carries a transitive relation, then the net x_j is a *directed net* provided that for each fixed $i \in J$ one eventually has $x_i \leq x_j$. A *filtered net* is defined dually. □

Every monotone net is directed, but the converse may fail. Exercise 1.12 illustrates pitfalls to avoid in defining directed nets. The next definition gives us some convenient notation connected with upper and lower bounds. Some important special classes of sets are also singled out.

1.3 DEFINITION. Let L have a transitive relation $\leq$. For $X \subseteq$ L and $x \in$ L write:

(i) $\downarrow X = \{y \in \text{L} : y \leq x \text{ for some } x \in X\}$;
(ii) $\uparrow X = \{y \in \text{L} : x \leq y \text{ for some } x \in X\}$;
(iii) $\downarrow x = \downarrow\{x\}$;
(iv) $\uparrow x = \uparrow\{x\}$;
(v) X is a *lower set* iff $X = \downarrow X$;
(vi) X is an *upper set* iff $X = \uparrow X$;
(vii) X is an *ideal* iff it is a directed lower set;
(viii) X is a *filter* iff it is a filtered upper set;
(ix) An ideal is *principal* iff it has a maximum element;
(x) A filter is *principal* iff it has a minimum element;
(xi) Id L (resp., Filt L) is the set of all ideals (resp. filters) of L;
(xii) $\text{Id}_0\text{L} = \text{Id L} \cup \{\emptyset\}$;
(xiii) $\text{Filt}_0\text{L} = \text{Filt L} \cup \{\emptyset\}$. □

In case $\leq$ is reflexive in L, then note that the principal ideals are just the sets $\downarrow x$ for $x \in \mathrm{L}$. The set of lower bounds of a subset $X \subseteq \mathrm{L}$ is equal to the set $\bigcap\{\downarrow x : x \in X\}$, and this will be the same as the set $\downarrow \inf X$ in case $\inf X$ exists. Note, too, that in the case of a reflexive relation we have

$$X \subseteq \downarrow X = \downarrow(\downarrow X),$$

and similarly for $\uparrow X$.

1.4. REMARK. *Let* L *be a set with a transitive, reflexive relation and* $X \subseteq \mathrm{L}$. *Then the following are equivalent:*

(1) X *is directed;*
(2) $\downarrow X$ *is directed;*
(3) $\downarrow X$ *is an ideal.*

Proof. (2) iff (3): By Definition 1.3.

(1) implies (2): If A is a finite subset of $\downarrow X$, then there is a finite subset B of X such that for each $a \in A$ there is a $b \in B$ with $a \leq b$ by 1.3(i). By (1) there is in X an upper bound of B, and this same element must also be an upper bound of A.

(2) implies (1): If A is a finite subset of X, it is also contained in $\downarrow X$; therefore, by (2), there is an upper bound $y \in \downarrow X$ of A. By definition $y \leq x \in X$ for some x, and this x is an upper bound of A. □

1.5. REMARK. *The following conditions are equivalent for* L *and* X *as in 1.4:*

(1) $\sup X$ *exists;*
(2) $\sup \downarrow X$ *exists.*

And if these conditions are satisfied, then $\sup X = \sup \downarrow X$. *Moreover, if every finite subset of* X *has a sup and if* F *denotes the set of all those finite sups, then* F *is directed, and* (1) *and* (2) *are equivalent to:*

(3) $\sup F$ *exists.*

Under these circumstances, $\sup X = \sup F$. *If* X *is nonempty, we need not assume the empty sup belongs to* F.

Proof. Since, by transitivity and reflexivity, the sets X and $\downarrow X$ have the same set of upper bounds, the equivalence of (1) and (2) and the equality of the sups are clear. Now suppose that $\sup A$ exists for every finite $A \subseteq X$ and that F is the set of all these sups. Since $A \subseteq B$ implies $\sup A \leq \sup B$, we know that F is directed. But $X \subseteq F$, and any upper bound of X is an upper bound of $A \subseteq X$; thus, the sets X and F have the same set of upper bounds. The equivalence of (1) and (3) and the equality of the sups is again clear, also in the nonempty case. □

The—rather obvious—theme behind the above remark is that statements about arbitrary sups can often be reduced to statements about finite sups and sups of directed sets. Of course, both 1.4 and 1.5 have straightforward duals.

1.6. DEFINITION. A *partially ordered set*, or *poset* for short, is a nonempty set L equipped with a transitive, reflexive and antisymmetric relation $\leq$. (This last means $x \leq y$ and $y \leq x$ always imply $x = y$.) L is *totally ordered*, or a *chain*, if all elements of L are comparable under $\leq$ (that is, $x \leq y$ or $y \leq x$ for all elements $x, y \in L$). □

We have remarked informally on duality several times already, and the next definition makes duality more precise.

1.7. DEFINITION. For $R \subseteq L \times L$ any binary relation on a set L, we define the *opposite relation* R^{op} (sometimes: the *converse relation*) by the condition that, for all $x, y \in L$, we have $xR^{op}y$ iff yRx.

If in $(L, \leq)$, a set equipped with a transitive relation, the relation is understood, then we write L^{op} as short for $(L, \leq^{op})$. □

The reader should note that if L is a poset or a chain, then so is L^{op}. One should also be aware how the passage from L to L^{op} affects upper and lower bounds. Similar questions of duality are also relevant to the next (standard) definition.

1.8. DEFINITION. A *semilattice* is a poset S in which every nonempty finite subset has an inf. A *sup-semilattice* is a poset S in which every nonempty finite subset has a sup. A poset which is both a semilattice and a sup-semilattice is called a *lattice*.

The empty inf (which, if it exists, is the same as sup S, the maximum element of S) is called the *unit, identity* or *top* element of S and is written as 1 (or, rarely, as $\top$). A semilattice with a unit is called *unital*. The empty sup (which, if it exists, is the same as inf S, the minimum element of S) is called the *zero* or *bottom* element of S and is written 0 (or $\perp$). □

Note that in a semilattice an upper set is a filter iff it is a subsemilattice. A dual remark holds for lower sets and ideals in sup-semilattices. We turn now to the discussion of maps between posets.

1.9. DEFINITION. A function $f: S \to T$ between two posets is called *order preserving* or *monotone* iff $x \leq y$ always implies $f(x) \leq f(y)$. We say that f *preserves*

(i) *finite infs*, or (ii) (*arbitrary*) *infs*, or (iii) *filtered infs*

iff whenever $X \subseteq S$ is

(i) finite, or (ii) arbitrary, or (iii) filtered,

and inf X exists in S, then inf $f(X)$ exists in T and equals $f(\inf X)$. A parallel terminology is applied to the preservation of sups.

A one-one function $f: S \to T$ where both f and f^{-1} are monotone is called an *isomorphism*. □

In the case of (iii) above, the choice of expression may not be quite satisfactory linguistically, but the correct phrase "preserves greatest lower bounds of filtered sets" is too long. For semilattices a map preserving finite infs might be called *multiplicative* or a *homomorphism.* The reader should notice that a function preserving ***all*** finite infs preserves the inf of the empty set; that is, it maps the unit to the unit—provided that units exist. In order to characterize maps f preserving only the non-empty infs (if this is the condition desired), we can employ the usual equation:

$$f(x \wedge y) = f(x) \wedge f(y),$$

for $x,y \in S$. Note that such functions are monotone, and the dual remark also holds for homomorphisms of sup-semilattices.

1.10. REMARK. *Let $f: S \to T$ be a function between posets. The following are equivalent:*

(1) *f preserves directed sups;*
(2) *f preserves sups of ideals.*

Moreover, if S is a sup-semilattice and f preserves finite sups, then (1) *and* (2) *are also equivalent to:*

(3) *f preserves arbitrary sups.*

A dual statement also holds for filtered infs, infs of filters, semilattices and arbitrary infs.

Proof. The equivalence of (1) and (2) is clear from 1.4 and 1.5. Now suppose S is a sup-semilattice and f preserves finite sups. Let $X \subseteq S$ have a sup in S. By the method of 1.5(3), we can replace X by a directed set F having the same sup. Hence, if (1) holds, then $f(\sup X) = \sup f(F)$. But, since f preserves finite sups, it is clear that $f(F)$ is constructed from $f(X)$ in the same way as F was obtained from X. Thus, by another application of 1.5(3), we conclude that $f(\sup X) = \sup f(X)$. That (3) implies (1) is obvious. □

We conclude this section with a few exercises, but no systematic attempt was made to ring all possible changes on the above ideas which, for the most part, are well known.

EXERCISES

1.11. EXERCISE. Let $f: S \to T$ be monotone on sets S and T equipped with a transitive and reflexive relations, and let $X \subseteq S$. Show that $\downarrow f(X) = \downarrow f(\downarrow X)$. □

1.12. EXERCISE. Construct a net $(x_j)_{j \in J}$ with values in a poset such that for all pairs $i,j \in J$ there is a $k \in J$ with $x_i \leq x_k$ and $x_j \leq x_k$ but such that $(x_j)_{j \in J}$ is ***not*** directed.

(HINT: Consider the lattice $2 = \{0,1\}$, let $J = \{0,1,2,...\}$, and let the net be defined so that $x_i=0$ iff i is even.) □

1.13. EXERCISE. Modify 1.10 so that for (3) we have only to assume that f preserves *nonempty* finite sups. □

1.14. EXERCISE. Is the category of sets with transitive and reflexive relations and with monotone maps *equivalent* to the category of posets and monotone maps? In these categories what sort of functor is ^{op} ? □

1.15. EXERCISE. Let L be a set with a transitive relation, and let A and B as well as the A_i for $i \in I$ be ideals of L.

(i) $A \cap B$ is an ideal of L iff $A \cap B \neq \emptyset$, for L a sup-semilattice.

(ii) $\bigcap_i A_i$ is not necessarily an ideal of L, even if $\bigcap_i A_i \neq \emptyset$.

(HINT: Consider the semilattice and ideals in the following figure.)

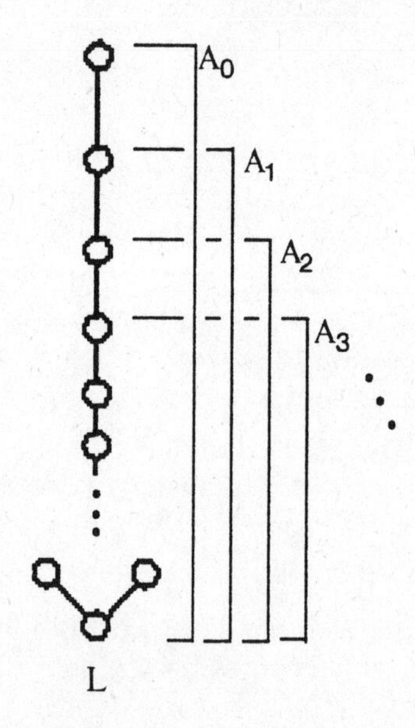

(iii) $A \cup B$ is contained in some ideal of L iff L is directed. (However, even if this is the case, there may not be a smallest ideal containing $A \cup B$.)

(iv) Id L is a sup-semilattice iff L is a sup-semilattice.

(HINT: If L is a sup-semilattice, then $C = \downarrow\{a \vee b : a \in A,\ b \in B\}$ is the sup of the ideals A and B of L. Conversely, if Id L is a sup-semilattice, then we claim there is a unique element $c \in \downarrow a \vee \downarrow b$ with $a, b \leq c$. Indeed, there is at least one since $\downarrow a \vee \downarrow b$ is directed; moreover, if c and c_1 were two such elements, then $\downarrow c$ and $\downarrow c_1$ are two ideals of L both containing a and b and both contained in $\downarrow a \vee \downarrow b$. Hence $\downarrow c = \downarrow c_1 = \downarrow a \vee \downarrow b$.)

(v) Dual statements holds for Filt L, where one assumes L is a semilattice in part (iv). □

1.16. EXERCISE. Let L be a set with a transitive relation, and let $\mathcal{L}$ denote the family of all non-empty lower sets of L.

(i) Id L$\subseteq\mathcal{L}$ and $\mathcal{L}$ is a sup-semilattice.

(ii) If L is a poset, then the map $x \mapsto \downarrow x : \mathrm{L} \rightarrow \mathcal{L}$ is an isomorphism of L onto the family of principal lower sets of L.

(iii) If L is a filtered poset, then $\mathcal{L}$ is a lattice with respect to intersection and union. (However, the converse is false.)

(HINT: For the counterexample to the converse, consider the opposite of the example in the hint for 1.15(ii).)

(iv) Let L and M be semilattices, f: L→M be a function, and $\mathcal{L}$ and $\mathcal{M}$ be the lattices of non-empty lower sets. Let $f_* = (A \mapsto \downarrow f(A)) : \mathcal{L} \rightarrow \mathcal{M}$. Then f is a semilattice morphism iff f_* is a lattice morphism. □

NOTES

The notion of a directed set goes back to the work of Moore and Smith [1922], where they use directed sets and nets to determine topologies. A convenient survey of this theory is provided in Chapter 2 of Kelley [1955]; we shall utilize this approach in our treatment of topologies on lattices, especially in Chapters II and III of this work. The material in this section is basic and elementary; a guide to additional reading—if more background is needed—is provided in the notes for Section 2.

2. COMPLETE LATTICES

No excuse need be given for studying complete lattices, because they arise so frequently in practice. Perhaps the best infinite example (aside from the lattice of all subsets of a set) is the unit interval $\mathbf{I} = [0,1]$. Many more examples will be found in this Compendium—especially involving non-totally ordered lattices.

2.1. DEFINITION. A *complete lattice* is a poset in which *every* subset has a sup and an inf. A totally ordered complete lattice is called a *complete chain.* □

2.2. PROPOSITION. *Let* L *be a poset.*

(i) *For* L *to be a complete lattice it is sufficient to assume the existence of sups (or the existence of infs).*

(ii) *If* L *is a unital semilattice, then for completeness it is sufficient to assume the existence of filtered infs.*

Proof. Assume that all subsets of L have sups. Let $X \subseteq L$ and let

$$B = \bigcap\{\downarrow x : x \in X\}$$

be the set of lower bounds of X. (If X is empty, we take $B = L$.) We wish to show that

$$\sup B = \inf X.$$

If $x \in X$, then x is an upper bound of B; whence, $\sup B \leq x$. This proves that $\sup B \in B$; as it clearly is the maximal element of B, this also proves that X has a greatest lower bound. (There is obviously a dual argument assuming infs exist.) Finally for (ii), since the existence of finite infs is being assumed, the existence of all infs follows from (the dual of) 1.5. □

Many subsets of complete lattices are again complete lattices (with respect to the restricted partial ordering). Obviously, if we assume that $M \subseteq L$ is *closed* under the sups and infs of the complete lattice L, then M is itself a complete lattice. But this is a very strong assumption on M. In view of 2.2, if we assume only that M is closed under the sups of L, then M is a complete lattice (in itself as a poset). The well-worn example is with L equal to *all* subsets of a topological space X and with M the lattice of *open* subsets of X. This example is instructive because in general M is not closed under the infs of L (open sets are not closed under the formation of infinite intersections). Thus the infs of M (as a complete lattice) are *not* the infs of L. (EXERCISE: What is the simple topological definition of the infs of M?) An even more general construction of subsets which form complete lattices is provided by the next theorem from Tarski [1955].

2.3. THEOREM (THE FIXED-POINT THEOREM). *Let* $f: L \to L$ *be a monotone self-map on a complete lattice* L . *Then the set* $M = \{x \in L : x = f(x)\}$ *of fixed points of* f *forms a complete lattice in itself.*

Proof. (At first it is not even clear that M is nonempty let alone a complete lattice. The following argument for completeness, however, also establishes the nonemptiness.) Let $X \subseteq \mathrm{M}$ and let

$$Y = \{y \in \mathrm{L} : x \leq f(y) \leq y \text{ for all } x \in X\}.$$

That is, Y is the set of upper bounds of the set X of fixed points with the special property that $f(y) \leq y$. (For instance, the unit of L has this property.) We wish to show that the inf of Y in L is the sup of X in M. (If f were the identity function, this would be the dual of the proof of 2.2.) Let $z = \inf Y$. If $y \in Y$, then $z \leq y$ and also $f(z) \leq f(y) \leq y$, because f is monotone. It follows that $f(z) \leq z$. But z is an upper bound for X in L; and if $x \in X$, then we have $x = f(x) \leq f(z)$. So $f(z)$ is an upper bound for X. But again by monotonicity, $f(f(z)) \leq f(z)$; so $f(z) \in Y$. This shows us that in fact $z = \inf(Y) \leq f(z)$; therefore, $z = f(z) \in \mathrm{M}$. Clearly, by definition, z is the *least* element of M that is also an upper bound of X. □

If we consider again the topological example with L the power set lattice of the space X, the mapping of a subset to its *interior* is monotone; so the completeness of the lattice of open sets also follows from 2.3. We shall see many other examples of monotone maps. In particular, a function preserving directed sups is monotone (consider directed sets of the trivial form $\{x,y\}$ with $x \leq y$).

2.4. REMARK. *Let $f : \mathrm{L} \to \mathrm{M}$ be a map between complete lattices preserving sups. Then $f(\mathrm{L})$ is closed under sups in* M *and is a complete lattice in itself.*

Proof. Let $Y \subseteq f(\mathrm{L})$ and let $X = f^{-1}(Y)$. Then $f(X) = Y$. Also

$$\sup Y = \sup f(X) = f(\sup X),$$

because f preserves sups. Hence, $\sup Y \in f(\mathrm{L})$. □

The above argument is not sufficient to show that if f preserves directed sups, then its image is closed under directed sups. We have to be satisfied with a special case.

2.5. REMARK. *Let $f : \mathrm{L} \to \mathrm{L}$ be a self-map on a complete lattice preserving directed sups. Provided that the map f is idempotent in the sense that $f = f^2$, then $f(\mathrm{L})$ is closed under directed sups and is a complete lattice.*

Proof. If $D \subseteq f(\mathrm{L})$ is directed, then $f(D) = D$ since $f = f^2$. Thus,

$$\sup(D) = \sup f(D) = f(\sup D),$$

because f preserves directed sups. The reason that $f(\mathrm{L})$ is complete is that the assumption $f = f^2$ means that the range of f is the same as the fixed-point set of f. We can therefore apply 2.3. □

As a very simple example of the application of 2.5, let V be a linear vector space (say, over the reals $\mathbb{R}$) and let L be the lattice of all *subsets* of V. For $x \in L$, define $f(x)$ to be the *convex closure* of the set x (no topology here, only convex linear combinations). The fact that an element of $f(x)$ depends on only *finitely* many elements of x is responsible for f preserving directed unions (sups) of subsets of V. Obviously we have $f(f(x)) = f(x)$. By 2.5, the convex subsets of V form a complete lattice. Note, however, that $x \leq f(x)$ for all $x \in L$. This special property of the function f gives a special property to $f(L)$, as we shall see in Chapter I. In particular, with this property, the set of fixed points of f is closed under infs—which is a simpler reason why $f(L)$ is a complete lattice. And, of course, this can all be verified directly for convex sets.

The next definition introduces some classical kinds of complete lattices that we shall often refer to in the sequel; however, it should be noted that they only partly overlap with the class of continuous lattices.

2.6. DEFINITION. A *complete Boolean algebra* (cBa for short) is a complete lattice which is *distributive* in the sense that, for all elements x,y,z,

$$\text{(D)} \quad x \wedge (y \vee z) = (x \wedge y) \vee (x \wedge z),$$

and where every element x has a *complement* y in the sense that

$$\text{(C)} \quad x \wedge y = 0 \text{ and } x \vee y = 1.$$

A *complete Heyting algebra* (cHa) is a complete lattice which satisfies the following infinite distributive law:

$$\text{(ID)} \quad x \wedge \bigvee Y = \bigvee \{x \wedge y : y \in Y\},$$

for all elements x and all subsets Y. □

A Boolean algebra (*without* completeness) is a lattice (with 0 and 1) satisfying (D) and (C) above. It is well known that (D) implies its dual, and that indeed every Boolean algebra is *isomorphic* to its opposite. Also well known is the fact that complements are *unique*.

The proper definition of a Heyting algebra *without* completeness will emerge in the next section. From the above definition it is not immediately obvious that every cBa is a cHa, but this is the case. We return to these ideas in Exercise 3.19 and Definition 3.20.

We turn now to a list of complete lattices that, so to speak, "occur in nature". This list is far from exhaustive, and many more examples are contained in the remainder of this work. The reader may take these assertions as exercises.

2.7. EXAMPLES. (1) We have already often referred to the *set of all subsets*, or *powerset*, of a set X. We employ the notation 2^X and of course regard this as a lattice under inclusion with union and intersection as sup and inf. It is a cBa but a rather special one. (It is atomic, for instance; and all atomic cBa's are of this form. Here, *atomic* means that every non-zero element contains a minimal non-zero element—an *atom*; for cBa's this is the same as saying that every element is the sup of atoms.)

(2) Generalizing (1), we can form the direct power L^X of any complete lattice L; this is just the poset of *all* functions $f: X \to L$ under the point-wise ordering. Similarly, we can form direct products $\mathsf{X}_{j \in J} L_j$ of any family of complete lattices in the well-known way. Such product lattices have many complicated sublattices.

(3) If X is a topological space, our notation for the *topology*, or *set of open subsets*, of X is $\mathcal{O}(X)$. It is a sublattice of 2^X closed under finite intersections and under arbitrary unions. It is clear then that $\mathcal{O}(X)$ is a cHa since we know the truth of 2.6(ID) for the set-theoretical operations. In general $\mathcal{O}(X)$ is *not* closed under arbitrary intersections, and its opposite is *not* a cHa. (Consider the case of $X = \mathbb{R}$, the real line.)

The opposite of $\mathcal{O}(X)$ is a complete lattice and is obviously isomorphic to the lattice $\Gamma(X)$ of *closed* subsets of X. The isomorphism between $\mathcal{O}(X)^{op}$ and $\Gamma(X)$ is by complements: $U \mapsto X \setminus U$.

Contained in $\mathcal{O}(X)$ is a very interesting complete lattice $\mathcal{O}_{reg}(X)$ of *regular open sets*, that is, those sets equal to the interiors of their closures. The sup is *not* the union of the regular open sets but the *interior of the closure of the union*. The inf is the *interior of the intersection* (which is the same as the inf in $\mathcal{O}(X)$). Remarkably, $\mathcal{O}_{reg}(X)$ is a cBa where the lattice complement of a $U \in \mathcal{O}_{reg}(X)$ is int $(X \setminus U)$. Actually this construction of a cBa can be done abstractly in any cHa, and we return to it in the next section (see Exercise 3.21).

For much more on Boolean algebras and the proof that *every* cBa is isomorphic to $\mathcal{O}_{reg}(X)$ for some space X, the reader is referred to Halmos [1963]. (It is interesting to note that $\mathcal{O}_{reg}(\mathbb{R})$ is an *atomless* cBa. That is to say, there are no minimal non-zero elements.)

(4) Let $\mathcal{A}$ be an abstract algebra with any number of operations. The poset (Cong $\mathcal{A}$, $\subseteq$) of all *congruence relations* under inclusion (of the graphs of the relations) forms a complete lattice, because congruence relations are closed under arbitrary intersections. This example includes numerous special cases:

(i) If $\mathcal{A}$ is a *group*, then the Cong $\mathcal{A}$ can be identified with the lattice of all *normal* subgroups in the usual way, and if $\mathcal{A}$ is an *abelian group* (or a module or a vector space), with the lattice of *all* subgroups (submodules, sub-vector spaces). In general this lattice is **not** distributive.

(ii) If $\mathcal{A}$ is a *ring*, then Cong $\mathcal{A}$ is canonically isomorphic to the lattice of all two-sided ideals. If $\mathcal{A}$ is a *lattice-ordered group* (lattice-ordered ring), then Cong $\mathcal{A}$ can be identified with the lattice of all order-convex ideals which are also sublattices. In general the ideals of a ring **do not** form a distributive lattice.

(iii) If $\mathcal{A}$ is a *lattice*, then Cong $\mathcal{A}$ cannot generally be identified either with the ideals or filters of $\mathcal{A}$, but it **does** form a cHa. (EXERCISE: Prove the distributivity.) If $\mathcal{A}$ is a Boolean algebra, then identification with the lattice of ideals is possible.

Note that in the case of algebras with finitary operations, Cong $\mathcal{A}$ is closed under directed unions. The significance of this remark will become clear in Section I-4.

(5) If $\mathcal{A}$ is an abstract algebra, then (Sub $\mathcal{A}$, $\subseteq$), the structure of all *subalgebras* of $\mathcal{A}$ under inclusion also becomes a complete lattice. The reader can supply special cases easily. In the case of vector spaces, the lattice of subspaces has complements but not unique ones owing to the failure of the distributive law.

(6) Let $\mathcal{A}$ be a compact topological algebra. Then the set $\mathrm{Cong}^{-}\mathcal{A}$ of *closed congruences* (congruences $R\subseteq\mathcal{A}\times\mathcal{A}$ closed in the product space) also forms a complete lattice. The relevance of this example is that these congruences correspond precisely to compact (i.e., Hausdorff) quotient algebras.

(7) Let $\mathcal{A}$ be a Hausdorff topological ring, then the set $\mathrm{Id}^{-}\mathcal{A}$ of *closed two-sided ideals* forms a complete lattice. Again the interest lies in the fact that the quotient rings are Hausdorff.

(8) Let $\mathcal{H}$ be a Hilbert space. Then $\mathrm{Sub}^{-}\mathcal{H}$, the *closed subspaces* of $\mathcal{H}$, forms a complete lattice. This generalizes to the lattice of projections in any von Neumann algebra.

(9) Every nonempty compact interval of real numbers in its natural order is a complete lattice, and all non-singleton intervals are isomorphic to $\mathbf{I} = [0,1]$ and to the infinite interval

$$\mathbb{R}^* = \mathbb{R}\cup\{-\infty,+\infty\} = [-\infty,+\infty].$$

As complete lattices are closed under direct products (cf. (2) above), we can form $\mathbf{I}^X$, where X is an arbitrary set. Such lattices are called *cubes.* In Exercise 2.10, we note what can be said if X is a topological space and only *certain* functions are admitted; this connects with the ideas of semicontinuous functions and real-valued random variables, to which we return in I-1.21. An easy example of a restricted function space which is a complete lattice would be the subspace $M\subseteq\mathbf{I}^{\mathbf{I}}$ of all *monotone* functions from $\mathbf{I}$ into itself.

(10) Let $\mathcal{F}$ be the set of all *partial* functions from the set $\mathbf{N}$ of natural numbers into itself (this could be generalized to any other set besides the set $\mathbf{N}$). Thus, if the function $f\in\mathcal{F}$, then its *domain*, dom f, is a subset of $\mathbf{N}$ and f: dom $f\to\mathbf{N}$. The empty function $\emptyset : \emptyset\to\mathbf{N}$ is allowed. We define $f\leq g$ to mean that

$$\mathrm{dom}\, f\subseteq \mathrm{dom}\, g \text{ and } f = g\,|\,\mathrm{dom}\, f,$$

that is, whenever f is defined, then g is defined and they have the same value.

This definition makes $\mathcal{F}$ into a poset with directed sups and arbitrary *nonempty* infs: it fails to be a lattice only in lacking a top.

In Exercise 2.12 we show how to adjoin a top to such structures. Another repair would be to expand $\mathbf{N}$ to $\mathbf{N}^* = \mathbf{N} \cup \{\bot, \top\}$, which is a poset under the ordering where for $x, y \in \mathbf{N}^*$ we have

$$x \leq y \text{ iff } x = \bot \text{ or } x = y \text{ or } y = \top.$$

Then $\mathcal{F}$ can be regarded as a subset of $(\mathbf{N}^*)^{\mathbf{N}}$ under the pointwise ordering (we define $f(x) = \bot$ if $x \notin \operatorname{dom} f$). (Note that this ordering has nothing to do with the natural ordering of $\mathbf{N}$.) Now $(\mathbf{N}^*)^{\mathbf{N}}$ is a complete lattice, but it is *much larger* than $\mathcal{F} \cup \{\top\}$, because for $f \in (\mathbf{N}^*)^{\mathbf{N}}$ the values taken in $\{\bot, \top\}$ and in $\mathbf{N}$ can be very mixed.

For applications to the theory of computation this proliferation of top elements is most inconvenient. If we read $f \leq g$ as an "information ordering" (roughly, f and g are consistent but g has possibly more information than f), then the only interpretation of $\top$ is to consider it as the *inconsistent* element. (The words "over defined" for $\top$ and "under defined" for $\bot$ have also been used.) As we generally try to keep our values "consistent" as much as possible, it seems natural to avoid $\top$. Because of the importance of the applications to computability, we should keep in mind the need to cover examples like this in our general theory. $\square$

The following paragraph also deals with examples, but they play such a very prominent role in the sequel that we separate them out.

2.8. EXAMPLES. Let L be a poset.

(1) The family of all *lower sets* of L and the family of all *upper sets* are both complete lattices under $\subseteq$; indeed, both of these families are closed under arbitrary intersections and unions in 2^L.

(2) If L is a semilattice, then $\mathrm{Filt}_0 L$ is a complete lattice; if L is also unital, then Filt L is complete. Both lattices of sets are closed under arbitrary intersections in 2^L. In a semilattice the ideals only form a semilattice, since in 2^L both $\mathrm{Id}_0 L$ and Id L are only closed under finite intersections.

(3) In a lattice, both $\mathrm{Filt}_0 L$ and $\mathrm{Id}_0 L$ are complete lattices; and if L has a top and bottom, then Filt L and Id L are complete lattices.

(4) The function $x \mapsto \downarrow x : L \to \mathrm{Id}\, L$ is an embedding preserving arbitrary infs and finite sups; it is called the *principal ideal embedding.* (There is a dual principal filter embedding.) The example $L = \mathbf{N} \cup \{\infty\}$ (with its natural ordering) shows that the principal ideal embedding need *not* preserve arbitrary (or even directed) sups.

(5) If L is a Boolean algebra, we can construe it as an algebra of "propositions" (0 is *false* and 1 is *true*, $\wedge$ and $\vee$ are *conjunction* and *disjunction*, complementation is *negation*). Filt L can be thought of as the lattice of *theories.* Any subset $A \subseteq L$ can be taken as a set of "axioms"

generating the following "theory", which is just a filter and corresponds to the propositions "implied" by the axioms:

$$\{x \in L : \exists\, a_0,\ldots,a_{n-1} \in A.\ a_0 \wedge \ldots \wedge a_{n-1} \leq x\}.$$

The "inconsistent" theory is L, that is, the top filter generated by $\{0\}$. If we eliminate L, then Filt L\{L} is closed under arbitrary nonempty intersections and directed unions. This is similar to the poset of 2.7(10). As is well known, the lattice Filt L is lattice isomorphic to the lattice of open subsets of the Stone space of the Boolean algebra L. □

In connection with 2.8(2) we note that the infinite intersection of ideals in a semilattice need not be an ideal (cf. 1.15 and figure).

EXERCISES

2.9. EXERCISE. Let X be a topological space and let $\Gamma\mathcal{O}(X) = \mathcal{O}(X) \cap \Gamma(X)$ be the sublattice of 2^X of all *closed-and-open sets* (sometimes: *clopen sets*). $\Gamma\mathcal{O}(X)$ is not complete in general, but it is always a Boolean algebra. For a compact totally disconnected space, $\Gamma\mathcal{O}(X)$ is complete iff the closure of every open set is open (such spaces are called *extremally disconnected*). (This complements Example 2.7(3).) □

2.10. EXERCISE (SEMICONTINUOUS FUNCTIONS). Let X be a topological space, and let $C(X, \mathbb{R}^*)$ be the set of continuous extended-real-valued functions. In general, under the point-wise ordering, $C(X, \mathbb{R}^*)$ is not complete, but it is a lattice with a top and bottom. For compact X, it is complete iff X is extremely disconnected.

Over an arbitrary space to have a complete lattice we must pass to a larger lattice. The *lower semicontinuous* functions $f \in LSC(X, \mathbb{R}^*)$ are characterized by the condition that the set $\{x \in X : r < f(x)\}$ is open in X for every $r \in \mathbb{R}^*$. (For *upper semicontinuous* functions we reverse the inequality.) $LSC(X, \mathbb{R}^*)$ is a complete lattice because it is closed under arbitrary point-wise sups. The lattices $LSC(X, \mathbb{R}^*)$ and $USC(X, \mathbb{R}^*)$ are anti-isomorphic and

$$C(X, \mathbb{R}^*) = LSC(X, \mathbb{R}^*) \cap USC(X, \mathbb{R}^*). \quad \square$$

In the next exercises, and many times elsewhere in the Compendium, we shall have occasion to discuss weaker forms of completeness. As these notions are important for applications and as the theory can generally include them easily, we provide a numbered definition; the theory which we are going to display will work under such weaker conditions.

2.11. DEFINITION. A poset is said to be *complete with respect to directed sets* (shorter: *up-complete*) iff every directed set has a sup.

A poset is called a *complete semilattice* iff every nonempty (!) subset has an inf and every directed subset has a sup. □

In order to compare the definitions of a complete lattice (2.1) and a complete semilattice (2.11) we suggest that the reader recall that a complete lattice is a poset with all conceivable completeness properties which a lattice may have and which are *symmetric* (i.e., remain invariant under passage to the opposite poset); whereas a complete semilattice has, coarsely speaking, the maximal completeness properties which a semilattice may have, short of becoming a lattice. Every *finite* semilattice of course is a complete semilattice. Every complete semilattice which is, in addition, unital is clearly a complete lattice (2.2(1)).

2.12. EXERCISE. Let S be a poset in which every nonempty subset has an inf. Then we have the following conclusions:

(i) Every $X \subseteq S$ with an upper bound has a sup.

(ii) Adjoin an identity by forming $S^1 = S \cup \{1\}$ with an element $1 \notin S$ and $x \leq 1$ for all $x \in S$. Then S^1 is a complete lattice. □

As a consequence, the adjunction of an identity to a complete semilattice will produce a complete lattice.

2.13. EXERCISE. Let S be the closed lower left triangle $\{(x,y) : 0 \leq x+y \leq 1\}$ in the square $[0,1]^2$.

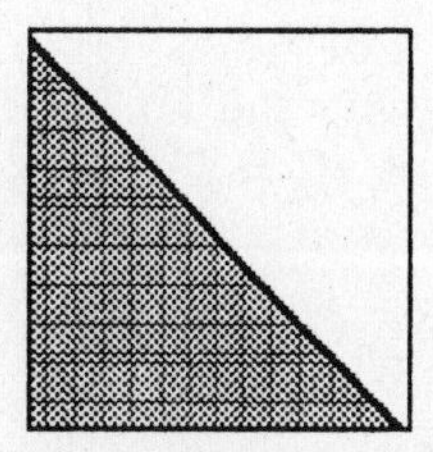

Then S is a complete semilattice but ***not*** a complete lattice. (Actually, the subsemilattice T of S consisting of the three corner points serves to illustrate this.) The interior of the triangle, $\{(x,y) : 0 \leq x+y < 1\}$, is a semilattice in which every subset has an inf, but it is ***not*** a complete semilattice.

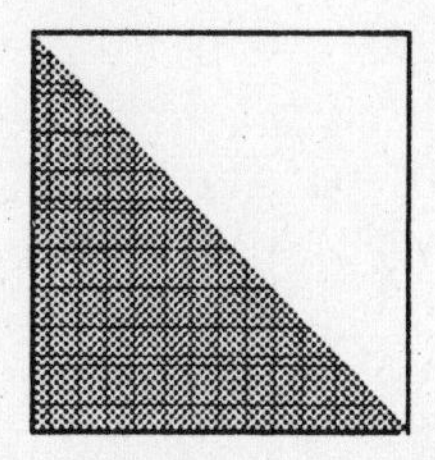

The half-open interval $]0,1]$ is an up-complete lattice, but it is ***not*** a complete semilattice. □

2.14. EXERCISE. (i) A semilattice is complete iff all nonempty filters have infs and all ideals have sups.

(ii) A poset is up-complete iff all ideals have sups. □

2.15. EXERCISE. (i) Every poset may be embedded into a complete lattice with the preservation of all existing infs.

(ii) Every lattice may be embedded into a complete lattice with the preservation of all finite lattice operations and all existing infs.

(iii) Every lattice may be embedded into a complete lattice with the preservation of all existing sups and infs.

(HINT: Parts (i) and (ii) are easily accomplished with the means available in Section 2. For (i) use the complete lattice of all lower sets and the embeding $x \mapsto \downarrow x$. For (ii) use the complete lattice Id L and the principal ideal embedding. Finally, (iii) is the so-called *MacNeille completion*, which is likewise constructed by using suitable ideals; we refer to the existing literature for details, e.g., Balbes and Dwinger [1974, p. 235].) □

2.16. EXERCISE. (i) For every semilattice S, the poset Id S is an up-complete semilattice.

(ii) If S is a semilattice in which every nonempty subset has an inf, then Id S is a complete semilattice. □

2.17. EXERCISE. In a Boolean algebra, is the lattice of finitely axiomatizable "theories" complete? up-complete? □

2.18. EXERCISE. Let G be a group and let H be any subgroup. Let L be the lattice of all subsets of G, that is, ($L = 2^G$). Let M be the collection of *double cosets* of H; that is, let

$$M = \{X \subseteq G : X = XH = HX\}.$$

Prove that M is a cBa, and discuss the closure properties of M within L with respect to sups and infs.

(HINT: Consider the map $X \mapsto HXH$.) □

2.19. EXERCISE. Let $\mathcal{F}$ be as in 2.7(10). Define $\mathcal{G} \subseteq \mathcal{F}$ to be the collection of all *one-one* partial functions. Is $\mathcal{G}$ a complete semilattice? □

NOTES

It would be inappropriate to attempt a history of the material contained in this introductory chapter; it belongs to the fundamentals of almost any kind of lattice theory and is therefore presented in most sources.

However, it may serve a useful purpose to give a guide to the existing textbook and monograph literature. We disclaim any ambition to be complete in this regard.

The classic sourcebook on lattice theory is, of course, the book by Garrett Birkhoff [1967] which has inspired many generations of lattice theoreticians. The latest edition is representative of the status of the theory in 1967. The date of the first edition in 1940 points up the truly classic character of this work.

The most modern source books on lattice theory are those of Grätzer [1978], of Balbes and Dwinger [1974] and of Crawley and Dilworth [1973]. As far as the topic of Boolean algebras is concerned, the book by Sikorski [1964] remains an effective source. The first edition dates back to 1957. 1955 is the date of appearance of the first edition of Hermes [1967], which experienced a second and revised edition in 1966. An introductory text to lattice theory was presented in 1953 by Dubreil-Jacotin, Lesieur and Croisot [1953].

Numerous easygoing textbooks for the student are available, too. Halmos' [1963] has become rather well known; other textbooks from the sixties are Gericke [1963], translated into English in 1966, and Abbot [1970].

For the most up-to-date bibliography and the most modern and comprehensive treatment of general lattice theory we recommend the book by Grätzer [1978].

3. GALOIS CONNECTIONS

We now introduce one of the most efficient tools in dealing with complete lattices; in this sense we continue the discussion of the previous section on complete lattices. One reason for this great efficiency is that the pairs of maps of the kind we are about to single out exist in great profusion. It is therefore very helpful to know in general what properties such maps have.

3.1. DEFINITION. Let S and T be two posets. We shall say that a pair (g,d) of functions $g : S \to T$ and $d : T \to S$ is *a Galois connection* or *an adjunction* between S and T provided that

(i) both g and d are monotone, and
(ii) the relations $g(s) \geq t$ and $s \geq d(t)$ are equivalent for all pairs of elements $(s,t) \in S \times T$.

In an adjunction (g,d), the function g is called the *upper adjoint* and d the *lower adjoint*. □

Notice that we have to keep the order straight. Then the upper adjoint is unambiguously determined by the "greater" side in the relation $g(s) \geq t$ of (ii) above (whence the letter g), whereas the lower adjoint is given by the lower or "downward" side in the relation $s \geq d(t)$ (whence the letter d).

Terminological difficulties may arise when we recognize that Galois connections are nothing but very special cases of pairs of *adjoint functors*. For this interpretation we need to construe S and T as *categories* with their respective elements as *objects*. The question is how to link the partial orders with *morphisms*. One is tempted to read an arrow $x \to y$ for $x,y \in S$ precisely when $x \geq y$, so that the arrow and the $\geq$-sign point in the *same* direction. This was done in Hofmann and Stralka [1976], and as a consequence upper adjoints were called *left* adjoints and lower adjoints *right* adjoints.

However, existing practice among category-theory oriented writers bears heavily upon us to choose the dual interpretation:

$$\begin{aligned} \operatorname{card}(\operatorname{Hom}(x,y)) &= 1, \qquad \text{if } x \leq y; \\ &= 0, \qquad \text{otherwise.} \end{aligned}$$

Thus $x \to y$ and $y \geq x$ are now equivalent statements. The "product" in a semilattice (that is, the inf) is then a product in the categorical sense. More generally, infs are limits, sups colimits. Order preserving maps are functors, and an adjunction (g,d) is a pair of adjoint functors with g being right adjoint and d left adjoint. The entire machinery of adjoint functors is now immediately available for Galois connections. (See, for example, MacLane [1971, Chapter IV].) But for the purposes of this work we wish to give a self-contained presentation and, therefore, we offer direct, elementary arguments for the essential facts. Moreover, we try to avoid the ambiguities involved in the use of "left" and "right" by using the to-be-hoped unambiguous words "upper" and "lower" instead.

3.2 THEOREM. *Let $g : S \to T$ and $d : T \to S$ be functions between posets. Then the following conditions are equivalent:*

(1) *(g,d) is a Galois connection.*
(2) *g is monotone and $d(t) = \min g^{-1}(\uparrow t)$ for all $t \in T$.*
(3) *d is monotone and $g(s) = \max d^{-1}(\downarrow s)$ for all $s \in S$.*

Consequently, in an adjunction one map uniquely determines the other.

Proof. (1) implies (2): Since $t \leq g(s)$ iff $d(t) \leq s$ by (1), we know that $d(t)$ is a lower bound of $g^{-1}(\uparrow t)$. But 3.1(ii) applied to $d(t) \leq d(t)$ gives us at once $t \leq g(d(t))$, that is, $d(t) \in g^{-1}(\uparrow t)$, whence (2).

(2) implies (1): Firstly, let $t \leq g(s)$. Then $s \in g^{-1}(\uparrow t)$, whence

$$s \geq \min g^{-1}(\uparrow t) = d(t).$$

Secondly, let $m = \min g^{-1}(\uparrow t)$, whence $m \in g^{-1}(\uparrow t)$, and thus $g(m) \geq t$. If now it holds that $s \geq d(t) = m$, then $g(s) \geq g(m) \geq t$, since g is monotone. The relation for d in (2) clearly makes d monotone, and thus the conditions of 3.1 are satisfied.

The proof of (1) iff (3) is analogous. (Or alternatively, we may observe that (g,d) is an adjunction between S and T iff (d,g) is an adjunction between T^{op} and S^{op}; thus, by duality, we can use what has already been proved.) □

3.3. THEOREM. *Any upper adjoint preserves infs, any lower adjoint, sups.*

Proof. Consider an adjunction (g,d) betweeen S and T. Let $\{s_j : j \in J\}$ be a family in S and let $s = \inf\{s_j : j \in J\}$.

Since g is order-preserving, we have $g(s) \leq g(s_j)$ for all $j \in J$. Now suppose that t is an arbitrary lower bound of $\{g(s_j) : j \in J\}$. Then for all $j \in J$ we have $g(s_j) \geq t$, which means $s_j \geq d(t)$ by 3.1(ii). Thus,

$$s = \inf\{s_j : j \in J\} \geq d(t),$$

whence $g(s) \geq t$. This shows that indeed $g(s) = \inf\{g(s_j) : j \in J\}$.

The proof that d preserves sups is dual. □

This result is very handy in establishing that certain functions preserve arbitrary infs or sups. In fact, in the presence of completeness, as we shall now see, the existence of a lower adjoint is *necessary* for the preservation of arbitrary infs.

3.4. THEOREM. *Let S be a complete lattice, and let T be a poset. Then every function $g : S \to T$ preserving all infs has a lower adjoint $d : T \to S$ given by either of the two formulae:*

(1) $d(t) = \inf g^{-1}(\uparrow t)$;
(2) $d(t) = \min g^{-1}(\uparrow t)$.

Proof. We define $d : \mathrm{T} \to \mathrm{S}$ by formula (1): this is possible since S is complete. Clearly, d is monotone. If $t \leq g(s)$, then $s \in g^{-1}(\uparrow t)$, and thus

$$d(t) = \inf g^{-1}(\uparrow t) \leq s.$$

Conversely, if $d(t) \leq s$, then $g(\inf\ g^{-1}(\uparrow t)) \leq g(s)$, since g is monotone (preserving infs); but, since g *preserves* infs, we also have

$$g(\inf g^{-1}(\uparrow t)) = \inf g(g^{-1}(\uparrow t)) \geq \inf \uparrow t = t.$$

This shows that (g,d) is an adjunction. We have also shown that $g(d(t)) \geq t$; that is, $d(t) \in g^{-1}(\uparrow t)$, which implies formula (2) in view of (1). □

3.5. **COROLLARY.** (i) *Let* $g : \mathrm{S} \to \mathrm{T}$ *be a function between posets of which* S *is a complete lattice. Then* g *preserves infs iff* g *is monotone and has a lower adjoint.*

(ii) *Let* $d : \mathrm{T} \to \mathrm{S}$ *be a function between posets of which* T *is a complete lattice. Then* d *preserves sups iff* d *is monotone and has a upper adjoint.*

Proof. This is clear from 3.3 and 3.4 and its dual. □

One can describe adjunctions in still other ways. We recall that a function $p : \mathrm{L} \to \mathrm{L}$ is *idempotent* iff $p^2 = p$.

3.6. **THEOREM.** *For every pair of order-preserving functions between posets,* $g : \mathrm{S} \to \mathrm{T}$ *and* $d : \mathrm{T} \to \mathrm{S}$, *the following conditions are equivalent:*

(1) *(g,d) is an adjunction;*
(2) *$dg \leq 1_\mathrm{S}$ and $1_\mathrm{T} \leq gd$.*

Moreover, these conditions imply

(3) *$d = dgd$ and $g = gdg$;*
(4) *gd and dg are idempotent.*

Proof. (1) implies (2): For all $s \in \mathrm{S}$ one has $g(s) \leq g(s)$, hence $d(g(s)) \leq s$ by (1); and for all $t \in \mathrm{T}$ one has $d(t) \geq d(t)$, hence $g(d(t)) \geq t$ by (1).

(2) implies (1): Let $t \leq g(s)$; then $d(t) \leq d(g(s))$, because d is monotone. By (2), $d(g(s)) \leq s$; whence, $d(t) \leq s$. Similarly $s \geq d(t)$ implies $g(s) \geq g(d(t)) \geq t$.

(2) implies (3): $dg \leq 1_\mathrm{S}$ implies $dgd \leq d$, since d is monotone; and $1_\mathrm{T} \leq gd$ implies $d \leq dgd$. Thus, $d = dgd$. The rest is similar.

(3) implies (4): Trivial. □

In an adjunction, injective and surjective maps are paired off as follows:

3.7. **PROPOSITION.** *For an adjunction (g,d) between posets* S *and* T, *the following conditions are equivalent:*

(1) *g is surjective.*
(2) *$d(t) = \min g^{-1}(t)$ for all $t \in \mathrm{T}$.*
(3) *$gd = 1_\mathrm{T}$.*

(4) *d is injective.*

Likewise, the following statements are equivalent:

(1*) *g is injective.*
(2*) $g(s) = \max d^{-1}(s)$ *for all* $s \in S$.
(3*) $dg = 1_S$.
(4*) *d is surjective.*

Proof. (1) implies (2): Now $d(t) = \min g^{-1}(\uparrow t)$ by 3.2. If g is surjective, then $g(g^{-1}(\uparrow t)) = \uparrow t$; and, since g is monotone,

$$g(d(t)) = \min g(g^{-1}(\uparrow t)) = \min \uparrow t = t.$$

Thus, $d(t) \in g^{-1}(t)$; whence $\min g^{-1}(t) = d(t)$.

(2) implies (3): From (2) we have $d(t) \in g^{-1}(t)$, i.e., $g(d(t)) = t$ for all $t \in T$.

(3) implies (4): By (3), d is a coretraction, hence, it is injective.

(4) implies (1): By 3.6 we have $d = dgd$, and if d is injective, we have $1_S = gd$. Thus, g is a retraction and hence surjective.

The equivalence of (1*)–(4*) is proved dually. □

We indicated in earlier examples how closure and kernel operators function in applications. Now we have a systematic framework for such maps:

3.8. DEFINITION. Let L be a poset.

(i) A *projection* is an idempotent, monotone self map $p : L \to L$.
(ii) A *closure operator* is a projection c on L with $1_L \leq c$.
(iii) A *kernel operator* is a protection k on L with $k \leq 1_L$.

Warning: This terminology deviates from that used in Scott [1976]. □

As to the nomenclature of (ii) and (iii), we remind the reader of Example 2.7(3): If X is a topological space, then $A \mapsto \mathrm{cl}\, A : 2^X \to 2^X$ is a closure operator and the map $A \mapsto \mathrm{int}\, A : 2^X \to 2^X$ is a kernel operator of 2^X. The image of the former is $\Gamma(X)$ and that of the latter $\mathcal{O}(X)$. We note that the map $U \mapsto \mathrm{int\, cl}\, U : \mathcal{O}(X) \to \mathcal{O}(X)$ is a closure operator with image $\mathcal{O}_{reg}(X)$.

3.9. NOTATION. For any function f: A→B, we denote the *corestriction* to the image as f°: $A \to f(A)$ and then the *inclusion* of the image into B accordingly as $f_\circ : f(A) \to B$. Thus, each f has the decomposition $f = f_\circ f^\circ$. If B = A, then $f^\circ f_\circ$ is the restriction *and* corestriction $f|f(A) : f(A) \to f(A)$. □

3.10. PROPOSITION. *Let* L *be a poset and* f: L→L *an order-preserving self-map of* L. *Then we have the following groups of equivalent statements:*

(1) *f is a projection operator*

(2) *f° is a retraction of* L *onto* f(L) *with* $f_\circ : f(\mathrm{L}) \to \mathrm{L}$ *as coretraction* (*that is,* $f^\circ f_\circ = 1_{f(\mathrm{L})}$)

(3) *There is a poset* T *and a monotone surjection* $q : \mathrm{L} \to \mathrm{T}$ *and a monotone injection* $i : \mathrm{T} \to \mathrm{L}$ *such that* $f = iq$ *and* $1_\mathrm{T} = qi$

(1_1) *f is a closure operator*

(2_1) *$(f_\circ, f^\circ)$ is an adjunction between* f(L) *and* L

(3_1) *There is an adjunction* (g,d) *between some* S *and* L *where* $f = gd$

(1_2) *f is a kernel operator*

(2_2) *$(f^\circ, f_\circ)$ is an adjunction between* L *and* f(L)

(3_2) *There is an adjunction* (g,d) *between* L *and some* T *where* $f = dg$

Proof. [We prove the equivalence of (1_2), (2_2), (3_2) only.]

(1_2) implies (2_2): If f is a projection, then we have $f^\circ f_\circ = 1_{f(\mathrm{L})}$ and $f_\circ f^\circ = f$; if in addition, f is a kernel operator, then $f \leq 1_\mathrm{L}$ and (2_2) follows by 3.6.

(2_2) implies (3_2): Trivial.

(3_2) implies (1_2): By 3.6(4), the map $f = dg$ is a projection. By 3.6(2) we have that $f = dg \leq 1_\mathrm{L}$, whence (1_2). □

We have in fact said that adjunctions on one hand and kernel and closure operators on the other are tightly linked: indeed, the corestriction to the image of every closure (resp., kernel) operator *is* the lower (resp., upper) adjoint of an adjunction. Conversely, whenever (g,d) is an adjunction, then gd is a closure operator and dg is a kernel operator.

Let us now note, however, that a mere projection is the "union" of a closure and a kernel operator:

3.11. LEMMA. *Let p be a projection on a poset* L. *We set*

$$\mathrm{L}_\mathrm{c} = \{x \in \mathrm{L} : x \leq p(x)\} \quad \textit{and} \quad \mathrm{L}_\mathrm{k} = \{x \in \mathrm{L} : p(x) \leq x\}.$$

Then we have the following conclusions:

(i) *If $p_\mathrm{c} : \mathrm{L}_\mathrm{c} \to \mathrm{L}_\mathrm{c}$ and $p_\mathrm{k} : \mathrm{L}_\mathrm{k} \to \mathrm{L}_\mathrm{k}$ are the two restrictions of p, then p_c is a closure operator and p_k is a kernel operator with*

$$\mathrm{im}\, p_\mathrm{c} = \mathrm{im}\, p_\mathrm{k} = \mathrm{im}\, p = \mathrm{L}_\mathrm{c} \cap \mathrm{L}_\mathrm{k}.$$

(ii) *L_c is closed under arbitrary sups and L_k under arbitrary infs.*

(iii) *If p preserves (filtered) infs, then L_c and* im p *are closed under (filtered) infs. Analogously, if p preserves (directed) sups, then L_k and* im p *are closed under (directed) sups.*

Proof. (i) Straightforward.

(ii) Let $X \subseteq L_c$ be such that sup X exists in L. Since $X \subseteq L_c$ and since p is monotone, $x \leq p(x) \leq p(\sup X)$ for all elements $x \in X$; therefore, we find that $\sup X \leq p(\sup X)$ and, consequently, $\sup X \in L_c$.

(iii) Now let X be a (filtered) subset of L_c for which inf X exists in L. If p preserves (filtered) infs, then inf $p(X)$ exists in L and

$$p(\inf X) = \inf p(X) \geq \inf X;$$

whence, $\inf X \in L_c$. Since im $p = L_c \cap L_k$ and since L_k is closed under arbitrary infs by (ii), then im p is also closed under (filtered) infs. □

The closure properties of im p may also be derived from 2.4 and 2.5. Notice that L_c and L_k are complete lattices as soon as L is a complete lattice by (ii) and 2.4. The second portion of (iii) will play a role when we discuss continuous lattices.

We remark next that the presence of projections makes certain preservation properties automatic:

3.12. PROPOSITION. (i) *The image of a closure operator is closed under the formation of infs, and that of a kernel operator is closed under the formation of sups (to the extent they exist).*

(ii) *The image of a complete lattice under a projection is a complete lattice.*

(iii) *The corestriction* $c^\circ : L \to c(L)$ *of a closure operator preserves arbitrary sups; hence,* $\sup_{c(L)} X = c\,(\sup_L X)$ *for* $X \subseteq c(L)$.

(iv) *The corestriction* $k^\circ : L \to k(L)$ *of a kernel operator preserves arbitrary infs; hence,* $\inf_{k(L)} X = k\,(\inf_L X)$ *for* $X \subseteq k(X)$.

Proof. (i) follows from 3.3 and 3.10(2_1), resp., 3.10(2_2).

(ii) By 3.11(ii) we know that L_c is a complete lattice if L is. From 3.11(i) and 3.12(i) it follows that $p(L) = pc(L_c)$ is complete.

(iii) and (iv) are consequences of 3.3 and 3.10. □

It will be useful to think about closure operators in alternate ways. One well-known way is to associate with a closure operator a "closure system". The specifics are as follows: Let L be a complete lattice, and let $\mathbb{C}(L)$ be the set of all subsets $S \subseteq L$ which are closed under arbitrary infs. (Compare 2.4.) We consider $\mathbb{C}(L)$ as a poset with respect to $\subseteq$. An element $S \in \mathbb{C}(L)$ will be called a *closure system* in this context.

3.13. PROPOSITION. *The function which assigns to a closure operator c on a complete lattice* L *the set* $c(L)$ *is an order isomorphism from the set of closure operators (under the pointwise order) onto* $\mathbb{C}(L)^{op}$. *Its inverse function* $S \mapsto c_S$ *associates with a closure system* $S \in \mathbb{C}(L)$ *the upper adjoint of the inclusion* $S \to L$ *followed by the inclusion* $S \to L$ *itself.*

Remark. We recall that the upper adjoint of the inclusion $S \to L$ is given by the formula $c_S{}^{\circ}(x) = \min(\uparrow x \cap S) = \inf(\uparrow x \cap S)$, and that $c_S{}^{\circ}(x) = c_S(x)$ for all elements $x \in L$.

Proof. The function $c \mapsto c(L)$ from the set of closure operators on L into $\mathbb{C}(L)$ is well defined by 3.12(i). It is readily verified that $c_S(L) = S$; conversely, given a closure operator c, then by 3.10(2_1) we know c° is the lower adjoint of the inclusion $c(L) \to L$, as is indeed the corestriction of $c_{c(L)}$; by the uniqueness of adjoints we have $c = c_{c(L)}$. Thus, the maps $c \mapsto c(L)$ and $S \mapsto c_S$ are inverses of each other. From the formula $c_S(x) = \inf(\uparrow x \cap S)$ it is clear that the function $S \mapsto c_S$ reverses order. □

3.14. COROLLARY. *The correspondence $c \mapsto c(L)$ between closure operators and closure systems on* L *maps the set of closure operators preserving directed sups bijectively onto the set of those closure systems which are closed (not only under arbitrary infs but also) under directed sups.*

Proof. If c preserves directed sups, then $c(L)$ is closed under directed sups by 3.11(iii). Conversely, suppose that $c(L)$ is closed under directed sups in L, and let D be a directed set in L. Then $\sup_L c(D) \leq c(\sup_L D)$ because c is monotone. But then, $c(\sup_L D) \leq c(\sup_L c(D))$, because c is a closure operator. Finally, we remark that $c(\sup_L c(D)) = \sup_L c(D)$ because $c(D)$ is a directed set in $c(L)$ whose sup, by hypothesis, is in $c(L)$. □

We conclude this section with some examples which will be of considerable importance in later chapters. In the first place, we return to Example 2.8.

3.15. PROPOSITION. (i) *For* L *complete, the map $I \mapsto \sup I : \mathrm{Id}\, L \to L$ is lower adjoint of the principal ideal map $x \mapsto \downarrow x : L \to \mathrm{Id}\, L$; in particular, it preserves sups.*

(ii) *The map $I \mapsto \downarrow \sup I : \mathrm{Id}\, L \to \mathrm{Id}\, L$ is a closure operator whose image is isomorphic to* L.

Proof. If $x \in L$ and $I \in \mathrm{Id}\, L$, then $I \subseteq \downarrow x$ iff x is an upper bound of I iff $\sup I \leq x$. This proves the adjointness; the rest follows from 3.3 and 3.10. □

Secondly, Galois connections also provide an access to a class of lattices which plays an important role in logic and also in the later developments of our theory.

3.16. LEMMA. *In a semilattice* S *the following two conditions are equivalent:*

(1) *For all $x \in S$, the function $s \mapsto xs : S \to S$ has an upper adjoint.*

(2) $\max\{s \in S : xs \leq t\}$ *exists for all $x, t \in S$.*

These conditions imply

(3) *For any family* $\{x_j : j \in J\}$ *with a sup and any* $x \in S$ *we have*

$$x \wedge \bigvee\{x_j : j \in J\} = \bigvee\{x \wedge x_j : j \in J\}.$$

If S *is a lattice, then* (3) *implies the distributive law* (D) of 2.6. *If* S *is a complete lattice, then* (1)–(3) *are equivalent and equivalent to* S *being a* cHa.

Remark. In (3) we can put $\leq$ for $=$.

Proof. The equivalence of (1) and (2) follows from 3.2. (3) follows by 3.3, and, trivially, (3) implies (D). If S is a complete lattice, then (3) implies (1) by 3.5(ii) and of course is just (ID) of 2.6 □

The point of the next definition is that completeness is ***not*** required.

3.17. DEFINTION. A *Heyting algebra* is a lattice H satisfying the equivalent conditions (1), (2) of 3.16.

The upper adjoint of the function $x \mapsto a \wedge x : H \to H$ is written

$$y \mapsto (a \Rightarrow y) : H \to H.$$

Thus, the conditions $x \geq a \wedge y$ and $(a \Rightarrow x) \geq y$ are equivalent in H. The binary operation $(a,b) \mapsto (a \Rightarrow b) : H \to H$ is called *implication.* Note that a Heyting algebra always has a unit, because $1 = a \Rightarrow a$.

If H has a zero, define $\neg a$ by $a \Rightarrow 0$ (that is, $\neg a = \max\{x \in H : a \wedge x = 0\}$). This unary operation is called *negation.* Notice that a Heyting algebra with zero satisfies $1 = \neg 0$ and $0 = \neg 1$. □

An example of a Heyting algebra without a zero is the half-open interval $]0,1]$, where $a \Rightarrow b = 1$ when $a \leq b$ but $= b$ otherwise.

EXERCISES

We continue in the next few exercises with the discussion of Heyting algebras and their relationship to Boolean algebras.

3.18. EXERCISE. Let H be a Heyting algebra with zero. Prove the following:

(i) $(\neg,\neg)$ is an adjunction between H^{op} and H. In other words, $\neg a \geq b$ iff $\neg b \geq a$ for all $a,b \in H$.

(ii) $\neg a \geq b$ iff $a \wedge b = 0$ for all $a,b \in H$.

(iii) $\neg\neg : H \to H$ is a closure operator, and $\neg\neg\neg = \neg$.

(iv) $\neg\neg$ preserves finite infs. □

For the following we recall from 2.6(C) that in a lattice L with 0 and 1 an element y is a *complement* of x iff $x \vee y = 1$ and $x \wedge y = 0$.

3.19. EXERCISE. Let L be a lattice with 0 and 1. Then the following conditions are equivalent:

(1) L is distributive and every element has a complement;

(2) L is a Heyting algebra in which negation is an involution (i.e., L satisfies $\neg\neg x = x$ for all x).

Moreover, if these conditions are satisfied, then $\neg x$ is the complement of x.

(HINT: The implication (2) implies (1) follows from the distributivity of a Heyting algebra and the fact that $x \wedge \neg x = 0$ implies $x \vee \neg x = 1$ whenever $\neg$ is an order reversing involution: hence, $\neg x$ is a complement. For the remaining implication we first observe that (1) trivially implies

(3) For every element x there is an x^* such that for all y

$$(y \vee x^*) \wedge x \leq y \text{ and } y \leq (y \wedge x) \vee x^* .$$

Next we observe that (3) implies (2): For, given x,y,z, if $x \leq y^* \vee z$, then $x \wedge y \leq (z \vee y^*) \wedge y \leq z$ by (3); conversely, if $x \wedge y \leq z$, then $x \leq (x \wedge y) \vee y^* \leq z \vee y^*$, again by (3). Thus, L is a Heyting algebra with $(y \Rightarrow z) = y^* \vee z$. Moreover, we find $\neg x = (x \Rightarrow 0) = x^* \vee 0 = x^*$.

Note that the proof in fact shows the equivalence of (1), (2) and (3).) □

These observations enable us to formally introduce the following definition (see also the remarks after 2.6).

3.20. DEFINITION. A *Boolean algebra* (or a *Boolean lattice*) is a lattice with 0 and 1 which satisfies the equivalent conditions of 3.19. □

Though we have not formulated numbered exercises, the reader should verify at this point that, in view of Lemma 3.16, a Heyting algebra that is complete as a lattice is a cHa; also, by a related argument, that a cBa is a cHa. Besides these obvious connections, it is useful to note that with every cHa there is canonically attached a cBa; the formalism of closure operators which we discussed in this section comes in handily for this purpose.

3.21. EXERCISE. Let H be a cHa and $c : \mathrm{H} \rightarrow \mathrm{H}$ a closure operator which preserves finite infs. Then $c(\mathrm{H})$ is also a cHa. If $c(\mathrm{H}) \subseteq \neg\mathrm{H}$ and $c(0) = 0$, then $c(\mathrm{H})$ is a cBa. In particular, $\neg\mathrm{H}$ is a cBa.

(HINT: By 3.12(ii), $c(\mathrm{H})$ is a complete lattice; and, since c° preserves finite infs by hypothesis and 3.12(i), and arbitrary sups by 3.12(iii), then equation 3.16(3) holds in $c(\mathrm{H})$. Whence, $c(\mathrm{H})$ is a Heyting algebra. Now suppose that $c(0) = 0$. If $a \in c(\mathrm{H})$ and $x \in \mathrm{H}$, then $a \wedge x \leq 0$ implies

$$a \wedge c(x) = c(a) \wedge c(x) = c(a \wedge x) \leq c(0) = 0.$$

Thus, $\max\,\{x \in \mathrm{H} : a \wedge x \leq 0\} \in c(\mathrm{H})$, and so $\neg_{c(\mathrm{H})} a = \neg a$. Hence, if we have $a = \neg b$ for some b, then

$$\neg_{c(H)}\neg_{c(H)}\, a = \neg\neg a = \neg\neg\neg b = \neg b = a,$$

by3.18(iii). Thus, if $c(0) = 0$ and $c(H) \subseteq \neg H$, then $c(H)$ is a Boolean algebra by 3.19(2). By 3.18(iii), we know that $c = \neg\neg$ is a closure operator with image $\neg H$, and $\neg\neg 0 = \neg 1 = 0$. Hence, the preceding applies to show that $\neg H$ is a cBa.) □

This allows us to produce some interesting complete Boolean algebras (as we have already remarked):

3.22. EXERCISE. (i) If H is a cHa and if L is a sublattice which is in fact closed under arbitrary sups, then L is itself a cHa. (HINT: If H satisfies 3.16(3), then so does L under the given hypotheses.)

(ii) For any set X the lattice 2^X is a cBa; hence, any sublattice L of 2^X which is closed under arbitrary unions is a cHa.

(iii) Let X be any topological space. Then $\mathcal{O}(X)$ is a cHa (cf. 2.7(3)). Moreover, $\neg\mathcal{O}(X) = \mathcal{O}_{reg}(X)$ is a cBa. □

The following examples of cHa's will be of interest in our later discussions.

3.23. EXERCISE. Let S be a semilattice equipped with a topology such that all translations $x \mapsto ax : S \to S$ are continuous. Let $L \subseteq \Gamma(S)$ be the lattice of all *closed* lower sets. Then L^{op} and L are cHa's. Dually, if $M \subseteq \mathcal{O}(S)$ is the lattice of all *open* upper sets, then both M and M^{op} are cHa's.

(HINT: Since M is closed under arbitrary unions and finite intersections in 2^S, equation 3.16(3) holds in M. M is complete, thus M and therefore $L^{op} \cong M$ are cHa's. In order to show that L is a cHa, let $\{A_j : j \in J\}$ be a family of closed lower sets; we have to show only that

$$A \cap (\bigcup\{A_j : j \in J\})^- \subseteq (\bigcup\{(A \cap A_j) : j \in J\})^-,$$

since the other containment is clear. Let $s \in A \cap (\bigcup\{A_j : j \in J\})^-$. Then

$$\begin{aligned} s &\in s(\bigcup\{A_j : j \in J\})^- \subseteq (s\bigcup\{A_j : j \in J\})^- \\ &= (\bigcup\{sA_j : j \in J\})^- \subseteq (\bigcup\{(A \cap A_j) : j \in J\})^- \end{aligned}$$

by the continuity of the translation by s and since the A_j are lower sets.) □

3.24. DEFINITION. A map $f : L \to M$ between Heyting algebras is called a *homomorphism* of Heyting algebras iff it preserves arbitrary sups and finite infs. A subset L of a Heyting algebra M is a *subalgebra* iff the inclusion $L \to M$ is a homomorphism (i.e., iff L is a sublattice which is closed under arbitrary sups). □

3.25. EXERCISE. The class of cHa's is closed under the formation of arbitrary direct products, subalgebras, and homomorphic images. □

We continue with some general remarks on adjunctions.

3.26. EXERCISE. Let S be a poset in which every *nonempty* subset has an inf and let T be a poset. Suppose further that $g : S \to T$ preserves all *existing* infs and also satisfies $T = \downarrow g(S)$. Then g has a lower adjoint given by the formula $d(t) = \inf g^{-1}(\uparrow t)$.

(HINT: Check the proof of 3.4 in the present situation.) □

3.27. EXERCISE. Let L be a lattice and diag : $L \to L \times L$ the diagonal map. Then diag is upper adjoint to the map $\vee : L \times L \to L$ and lower adjoint to the map $\wedge : L \times L \to L$. □

3.28. EXERCISE. Let $\{L_i\}_{i \in I}$ be a family of complete lattices, and let $L = \mathsf{X}_{i \in I} L_i$. Let $\pi_i : L \to L_i$ be the projection on the i th factor of L. Further define $\varepsilon_i : L_i \to L$ by

$$\pi_j \varepsilon_i(x) = x \quad \text{if } i = j,$$
$$= 1_j \quad \text{if } i \neq j,$$

and define $\delta_i : L_i \to L$ by

$$\pi_j \delta_i(x) = x \quad \text{if } i = j,$$
$$= 0_j \quad \text{if } i \neq j.$$

Then (π_i , δ_i) is a Galois adjunction between L and L_i ; while (ε_i , π_i) is a Galois adjunction between L_i and L. □

The next exercise concerns the adjointness property of the important map

$$I \mapsto \sup I : \operatorname{Id} L \to L$$

under weaker hypotheses than those given in 3.13:

3.29. EXERCISE. The conclusions of Proposition 3.15 persist in the case that L is an up-complete poset (rather than a complete lattice). □

3.30. EXERCISE. In the circumstances of 3.2, conditions (1), (2), and (3) are also equivalent to the following:

(2′) g is monotone and $g^{-1}(\uparrow t) = \uparrow g(t)$ for all $t \in T$;

(3′) d is monotone and $d^{-1}(\downarrow s) = \downarrow g(s)$ for all $s \in S$.

NOTES

A part of the literature on Galois connections deals with pairs (g,d) of (antitone) maps $g : S \to T^{op}$ and $d : T^{op} \to S$ such that the relations $g(s) \leq t$ and $d(t) \leq s$ are equivalent. (This is the same as saying that $s \leq dg(s)$ and $t \leq gd(t)$ hold for all $s \in S$ and all $t \in T$). It is this set-up which generalizes the formalism of classical Galois theory in which the order reversing correspondence is established between the lattice of fields F between two fields K and E, $K \subseteq F \subseteq E$, and the lattice of subgroups of the Galois group of $(E : K)$. In the antitone form, Galois connections were studied by Ore [1944]. Since that time, the general idea of Galois connections has become a pervasive theme in lattice theory literature, and it cannot be our objective to trace its precise history. A recent book on such matters is T.S. Blyth and M.F. Janowitz [1972.] Frequently cited contributions are C.J. Everett [1944], G. Pickert [1952], G. Aumann [1955], G.N. Raney [1960], J.C. Derderian [1967], J. Schmidt [1973], Shmuely [1974] and Bandelt [1979].

F.W. Lawvere noticed early on that Galois connections are quite special cases of the omnipresent situation of a pair of adjoint functors. This is pointed out in MacLane [1971], p. 93 ff. The consideration of Heyting algebras in this context is outlined in S. Eilenberg and G.M. Kelly [1966]; see in particular p. 555 ff. The authors credit F.W. Lawvere with this approach. There is a great variety of names under which Heyting algebras appear in the literature: Brouwerian logic, Brouwerian lattice, pseudo-boolean lattice and relatively pseudomplemented distributive lattice. For a complete Heyting algebra (cHa) the following names are also used: frame, local lattice, locale. We will have quite a bit to say on certain classes of Heyting algebras in Chapter V; in fact, our discussion will be exemplary for the connections between complete Heyting algebras and topological spaces.

The topic of closure and kernel operators is a lattice-theoretical classic. The systematic consideration of projections as the common generalization of both of these in 3.8–3.12 is due to Scott, and we will pursue this discussion for continuous lattices in Chapter I (see I-2.14, I-4.16). The systematic use of Galois connections in the study of continuous lattices advocated in Hofmann and Stralka [1976] particularly emphasized the importance of the sup-map on the ideal lattice (3.15) which will bear fruit in Chapter I, Section 2.

4. MEET-CONTINUOUS LATTICES

The inf operation $(x,y) \mapsto xy : L \times L \to L$ in a lattice preserves infs (to the extent they exist); in particular, all translations $s \mapsto xs : L \to L$ preserve infs. If L is complete, we have seen that it is precisely the complete Heyting algebras in which this translation preserves *all* sups. Frequently this is too much to ask, since many of the examples we have listed are not even distributive. However, it occurs rather often that $s \mapsto xs$ preserves *directed* sups. The class of lattices in which this is the case deserves a special designation:

4.1. DEFINITION. A lattice L is called *meet-continuous* if it is complete and satisfies:

$$\text{(MC)} \qquad x \sup D = \sup xD,$$

for all $x \in L$ and all ***directed*** sets $D \subseteq L$ (where we write xy in place of $x \wedge y$ as usual). We will say that L is *join-continuous* iff L^{op} is meet-continuous. □

Note that in (MC) the relation $\leq$ could replace $=$. (The same will be the case in 4.2(7) and (8) below.) In the literature, meet-continuous lattices are occasionally called "continuous lattices"; but we reserve this designation for those more special lattices which will be our principal topic. There are various equivalent ways of looking at meet-continuous lattices:

4.2. THEOREM. *In a complete lattice* L *the following conditions are equivalent:*

(1) *The sup-map for ideals* $I \mapsto \sup I : \mathrm{Id}\, L \to L$ *is a lattice homomorphism (preserving all sups: cf.* 3.15(i));
(2) *For two ideals* I_1, I_2 *we have* $(\sup I_1)(\sup I_2) = \sup I_1 I_2$;
(3) *For two directed sets* D_1, D_2 *we have* $(\sup D_1)(\sup D_2) = \sup D_1 D_2$;
(4) L *is meet-continuous;*
(5) *For each directed set* D *and each* $x \leq \sup D$ *we have* $x \leq \sup xD$ (*hence,* $x = \sup xD$);
(6) *The inf operation* $(x,y) \mapsto xy : L \times L \to L$ *preserves directed sups;*
(7) *For each* $x \in L$ *and each directed net* $(x_j)_{j \in J}$ *we have*
$$x \wedge \bigvee_{j \in J} x_j = \bigvee_{j \in J} (x \wedge x_j);$$
(8) *For each* $x \in L$ *and any family* $(x_j)_{j \in J}$ *we have*
$$x \wedge \bigvee_{j \in J} x_j = \bigvee_{A \in \mathrm{fin}\, J} (x \wedge \bigvee_{j \in A} x_j),$$
where fin J *is the set of all finite subsets of* J.

Proof. (1) iff (2): Use the definition of the sup-map for ideals and the fact that $I_1 I_2 = I_1 \cap I_2$ for two lower sets in a semilattice.

(2) iff (3): Notice $\downarrow(\downarrow D_1)(\downarrow D_2) = \downarrow(D_1D_2)$ and use 1.5 to calculate

$$\sup D_1D_2 = \sup \downarrow(D_1D_2) = \sup \downarrow(\downarrow D_1)(\downarrow D_2) = \sup (\downarrow D_1)(\downarrow D_2).$$

Remark 1.4 then establishes the desired equivalence.

Thus (1), (2), (3) are equivalent. Clearly (4) and (7) are equivalent, and the equivalence of (7) and (8) is easy by Remark 1.5.

The implications (6) implies (3) implies (4) implies (MC) implies (5) are trivial. The whole proof will be complete if we show:

(5) implies (6): Let $D \subseteq L \times L$ be directed and set $D_n = \pi_n D$, for $n = 1,2$. Then $D \subseteq D_1 \times D_2$. If, on the other hand, $(d,e) \in D_1 \times D_2$, then there are elements $x,y \in L$ with (d,y), $(x,e) \in D$. Since D is directed, we find some (d^*,e^*) majorizing (d,y) and (x,e); thus, $(d,e) \leq (d^*,e^*)$. We have thus proved that $D_1 \times D_2 \subseteq \downarrow D$.

If $m : L \times L \to L$ is the multiplication $m(x,y) = xy$, then

$$m(D) \subseteq m(D_1 \times D_2) = D_1D_2 \subseteq m(\downarrow D) \subseteq \downarrow m(D).$$

Thus $\sup m(D) \leq \sup D_1D_2 \leq \sup \downarrow m(D) = \sup m(D)$, by 1.5. If $d_n = \sup D_n$, $n=1,2$, then $(d_1,d_2) = \sup D$. It suffices therefore to prove $d_1d_2 = \sup D_1D_2$.

For every $x \in D_1$ we have $xd_2 \leq \sup D_2$, hence from (5) we know

$$xd_2 = \sup xd_2D_2 = \sup xD_2.$$

Since $d_1d_2 \leq \sup D_1$, once more by (5) we obtain

$$d_1d_2 = \sup d_1d_2D_1 = \sup D_1d_2.$$

But then

$$\sup D_1d_2 = \sup \{xd_2 : x \in D_1\} = \sup_{x \in D_1} \sup xD_2 =$$
$$= \sup (\bigcup_{x \in D_1} xD_2) = \sup D_1D_2.$$

This proves the claim. □

We point out that condition (8) is purely equational. If one imagines that directed sups are "limits" of sorts (a contention we will amply justify in Chapter II), then condition (6) is indeed a continuity assumption. This justifies the name "meet continuity". Condition (7), however, is a distributivity relation which readily compares with the distributivity relation 3.16(3) in Heyting algebras. In fact we have:

4.3. REMARK. *Let* L *be a lattice, then the following conditions are equivalent:*

(1) L *is a cHa;*

(2) L *is meet-continuous and distributive.*

Proof. (1) implies (2) is clear from 3.16.

(2) implies (1): By (MC), the function $s \mapsto xs : L \to L$ preserves directed sups; by 2.6(D) it preserves finite sups. Hence, it preserves arbitrary sups (see 1.10). Thus, 3.16(3) holds and (1) follows. □

While complete Heyting algebras are one source of meet-continuous lattices, compact topological semilattices are another. We will develop this subject at considerably greater length in Chapter VI. But it helps now to take note at least of the examples implied by the following.

4.4. PROPOSITION. *Let S be a lattice with a Hausdorff topology such that:*

(i) *Every directed net has a sup to which it converges, and S has a zero;*

(ii) *The translations* $s \mapsto xs : S \to S$ *are continuous for all* $x \in S$.

Then S is meet-continuous. If, moreover, S is compact, then condition (ii) *already implies* (i).

Proof. Let $x \in S$ and suppose that $(x_j)_{j \in J}$ is directed. Then $(xx_j)_{j \in J}$ is directed, and so $\sup_J x_j = \lim_J x_j$ and $\sup_J xx_j = \lim_J xx_j$. From (ii) we know that $\lim xx_j = x \lim x_j$, and, since limits are unique for a Hausdorff topology, we deduce $\sup_J xx_j = x \sup_J x_j$. Every directed subset has a sup by (i); hence, S is a complete lattice by the dual of 2.2(ii). Thus S is a meet-continuous lattice.

Assume now that S is compact Hausdorff and satisfies (ii); we have to verify (i). Let $(x_j)_{j \in J}$ be a directed net. Since the topology is compact, this net has at least one cluster point c. Let $i \in J$. Then eventually $x_i \leq x_j$, that is, $x_i = x_i x_j$. But, $\uparrow x_i = \{x \in S : x_i x = x_i\}$ is closed as translation by x_i is continuous, and since the net is eventually in $\uparrow x_i$, it follows that c is also. Hence $x_i c = x_i$ for each i, so c is an upper bound for the net. Moreover, if $b \in S$ is any upper bound of the net, then $bx_i = x_i$ for each i, so that $x_i \in Sb$ for each i. Again, since translation by b is continuous, Sb is closed in S, and so $c \in Sb$ also holds. Thus $c \leq b$ for each upper bound b of the net, whence $c = \sup x_i$. Dually, each filtered net converges to its inf. In particular, $0 = \inf S$ exists. □

Let us now, by contrast, look at a few simple complete lattices which *fail* to be meet-continuous.

4.5. COUNTEREXAMPLES. (1) Let L be the subset of the square $[0,1]^2$ consisting of its interior $]0,1[^2$ and the points $(0,0) = \bot$ and $(1,1) = \top$. Then L is a complete, distributive lattice which is isomorphic to its opposite. But L is ***not*** meet-continuous.

(HINT: Consider $D = \{1/3\} \times]0,1[$ with $\sup D = \top$; if $x = (2/3, 1/2)$, then we have $x \sup D = x$, but $\sup xD = (1/3, 1/2) \neq x$.)

(2) Let L be the following subset of the square $[0,1]^2$:

$$L = (\{1-1/n : n=1,2,3,\ldots\} \times \{0\}) \cup \{(0,1), (1,1)\}.$$

This lattice is complete but ***not*** meet-continuous. Because of its obvious compact topology, this lattice is a very useful counterexample.

(HINT: Consider $D = \{1-1/n : n=1,2,...\}\times\{0\}$ and $x = (0,1)$.) □

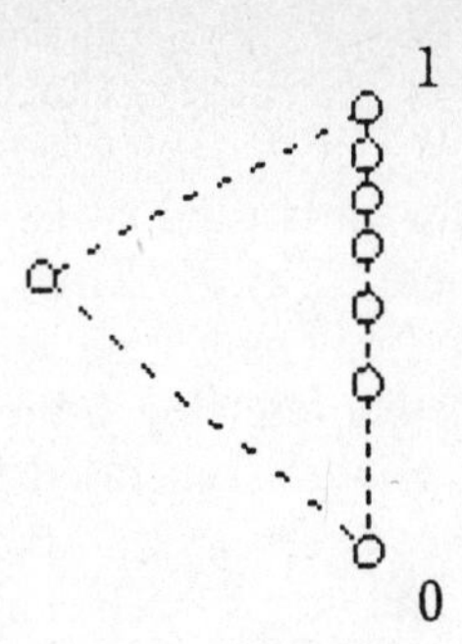

EXERCISES

We consider meet continuity under circumstances which are more general than that of complete lattices.

4.6. DEFINITION. A semilattice S is called *meet-continuous* iff it is up-complete (see 2.11) and satisfies (MC) of 3.1. □

4.7. EXERCISE. Theorem 4.2 holds for up-complete semilattices. □

4.8. EXERCISE. (i) If S is any semilattice, then Id S is a meet-continuous semilattice. (HINT: Utilize Exercise 2.16.)

(ii) If L is any lattice, then Id L is a meet-continuous lattice. □

4.9. EXERCISE. Let S be an up-complete semilattice (see 2.11). Then the following statements are equivalent (cf. G. Bruns [1967]):

(1) S is meet-continuous;

(2) $x\leq \sup C$ always implies $x\leq \sup xC$ for a **chain** $C\subseteq S$. □

4.10. EXERCISE. The class of all meet-continuous lattices is closed under the formation of the following operations:

(a) arbitrary products,

(b) subsets closed under finite infs and directed sups,

(c) sublattices which are complete with respect to the induced order and which are closed under directed sups,

(d) surjective images by functions preserving arbitrary infs and directed sups,

(e) images of projections (3.8(i)) preserving directed sups.

(HINT: Use 4.2(7) and model the proof after the one that is given for I-2.5 below and in the case of 3.11(iii).) □

4.11. Exercise. (i) Let L be a lattice. Then $\mathrm{Id}_0\,L$ is a meet-continuous lattice. If $f: L \to S$ is a function into a meet-continuous lattice preserving finite infs, then the function $F: \mathrm{Id}\,L \to S$ given by $F(I) = \sup f(I)$ preserves finite infs and directed sups. If f is a lattice morphism, then F preserves arbitrary sups.

(ii) Let L be the opposite of the lattice of finite subsets of a set X. (This is the free semilattice generated by X). Then $\mathrm{Id}_0\,L$ contains a copy of X and is in fact the free meet-continuous lattice over X in the category of meet-continuous lattices and maps preserving finite infs and directed sups.

(iii) Let L be the free lattice generated by a set X. Then $\mathrm{Id}_0\,L$ contains a copy of X and is the free meet-continuous lattice over X in the category of meet-continuous lattices and maps preserving finite infs and arbitrary sups. (See Isbell [1975b]) □

4.12. Exercise. Let L be a meet-continuous lattice and H the lattice of all equivalence relations on L whose graph in $L \times L$ is closed under finite infs and arbitrary (!) sups. Then H is a cHa. (See Isbell, loc. cit., p. 44). □

4.13. Exercise. Let L be an up-complete semilattice, and let

$$L' = \{I \subseteq L : \emptyset \neq I = I^+ = \downarrow I\},$$

where $I^+ = \{\sup D : D \subseteq I \text{ is directed}\}$. Then we have:

(i) L' is a lattice;

(ii) If L' is meet continuous, then so is L;

(iii) If L satisfies $I^{++} = I^+$ for each lower set $I \subseteq L$, then L' is meet continuous if L is.

(Hint: (i): Straightforward. (ii): Suppose that $D \subseteq L$ is directed where $\sup D = x$, and let $y \in L$. Then $\{\downarrow d : d \in D\} \subseteq L'$ is directed, and clearly $\downarrow d \subseteq \downarrow x$ for each $d \in D$. Moreover, if $\downarrow d \subseteq J$ for each $d \in D$, where $J \in L'$, then $J = J^+$ implies that $\sup D = x \in J$, and so $\downarrow x \subseteq J$. This shows $\downarrow x = \sup_{d \in D} \downarrow d$.

Now, L' meet continuous implies $\bigvee_d (\downarrow y \wedge \downarrow d) = \downarrow y \wedge \downarrow x$, and therefore we have $\bigvee_d \downarrow (y \wedge d) = \bigvee_d (\downarrow y \wedge \downarrow d) = \downarrow y \wedge \downarrow x$. Hence, if $z \in L$ with $y \wedge d \leq z$ for each $d \in D$, then $\downarrow (y \wedge d) \subseteq \downarrow z$ for each $d \in D$, and so $y \wedge x \in \downarrow z$ by the above. Thus $y \wedge x \leq z$. This shows that $y \wedge x = \sup_d y \wedge d$, and so L is meet continuous.

(iii): Suppose that L is meet continuous and satisfies the additional hypothesis above. Let $\{J_i\} \subseteq L'$ be directed with $\sup J_i = J$, and let $K \in L'$. Clearly $\sup(K \cap J_i) \subseteq K \cap J$ for each i, and so we only need show that the reverse inclusion holds. Suppose that $x \in K \cap J$. Then $x \in J = \sup J_i$. Moreover, the additional hypothesis under (iii) implies that $J = (\bigcup_i J_i)^+$, since $\bigcup_i J_i$ is a lower set in L. Thus, there is some directed set $D \subseteq \bigcup_i J_i$ with $x = \sup D$. Now, L is meet continuous, and so $x = x \wedge \sup D = \sup_d (x \wedge d)$. Moreover, $K = \downarrow K$ implies that $x \wedge d \in K$ for each $d \in D$ since $x \in K$. Since $J_i = \downarrow J_i$ also holds for each i, we conclude that $\{x \wedge d : d \in D\} \subseteq \bigcup_i (K \cap J_i)$, and so $x = \sup_d x \wedge d \in (\bigcup_i (K \cap J_i))^+$. Thus $K \cap J \subseteq \sup_i (K \cap J_i)$, and so these are equal.) □

NOTES

In the literature meet-continuous lattices (see Birkhoff [1967]) are sometimes called upper continuous lattices (Grätzer [1978]) or "nach oben stetige Verbände" (Hermes [1967]); lattices which are meet- and join-continuous have been called continuous (see, e.g., Hermes, *op. cit.*); this notation is in conflict with what we will call continuous lattices in this book; our nomenclature seems now widely accepted.

The role played by meet-continuous lattices in the literature seems to be somewhat implicit: they are rarely considered as a class by themselves. Usually it is observed that trivially all cHa's are meet-continuous and that all algebraic lattices (which we will consider in I-4 at some length) are meet-continuous. A coherent body of deep information does not appear to exist on the class of meet-continuous lattices per se. Some information is provided by Isbell [1975b] (see Exercises 4.11 and 4.12), but much of Isbell's paper is concerned with continuous lattices in our sense. The choice of morphisms for a category of meet-continuous lattices is not entirely clear. The definition would suggest that morphisms preserve finite infs and directed sups. In the case of the category of complete Heyting algebras one chooses morphisms the way we did in 3.24; Isbell considers this type of map for meet-continuous lattices, and this makes his category of meet-continuous lattices contain the category of cHa's as the full subcategory determined by the distributive objects (see 4.3). The characterization of meet-continuous lattices through the fact that the sup-map Id $L \to L$ is a lattice morphism is from Hofmann and Stralka [1976]. That meet continuity emerges in the context of compact topological semilattices (4.4) is well-known in topological algebra.

CHAPTER I

Lattice Theory of Continuous Lattices

Here we enter into the discussion of our principal topic. Continuous lattices, as the authors have learned in recent years, exhibit a variety of different aspects, some are lattice theoretical, some are topological, some belong to topological algebra and some to category theory—and indeed there are others. We shall contemplate these aspects one at a time, and this chapter is devoted entirely to the lattice theory surrounding our topic.

Evidently we have first to define continuous lattices, and, as we shall see from hindsight, there are numerous equivalent conditions characterizing them. We choose the one which is probably the simplest, but it does involve the consideration of an auxiliary transitive relation, definable in every complete lattice, by which one can say that an element x is "way below" an element y. We will write this as $x \ll y$. We devote Section 1 to the introduction of the way-below relation and of continuous lattices. We demonstrate that the occurrence of this particular additional ordering is not accidental and explain its predominant role in the theory.

In Section 2 we show that continuous lattices have a characterization in terms of (infinitary) equations. This gives us the important information that the class of continuous lattices, as an equational class, is closed under the formation of products, subalgebras, and homomorphic images—provided we recognize from the equations *which* maps ought to be considered as homomorphisms.

In Section 3 we explain why in a continuous lattice there are always sufficiently many meet-irreducible elements in the sense that every element is the infinum of the irreducibles dominating it. In Chapter V we will bring this result to fruition when we discuss the spectral theory of these lattices.

In Section 4 we show that the familiar concept of an algebraic lattice is subsumed under the more general one of a continuous lattice, and we review some of the known aspects of algebraic lattices in this light.

1. THE "WAY-BELOW" RELATION

It often happens that we encounter relations between elements of a given lattice which are stronger than the simple less-than-or-equal-to relation of the partial ordering. In a linearly ordered chain, for example, we usually have need to single out the strict less-than relation. In the non-linear case, however, this seldom proves to be a very interesting relation. Consider in this regard the lattice $\mathcal{O}(X)$ of open sets of a topological space X. To say $U \subseteq V$ but $U \neq V$ does not say very much, since the sets could differ at only one point. To say that U really is *inside* V we could say that the closure $U^- \subseteq V$. This means that U avoids the boundary of V even by limits, and in the case of compact Hausdorff spaces this is a well-known and useful relation (even though for clopen sets it is reflexive and does not imply $U \neq V$). If, on the other hand, the space is only locally compact, the relation is not as strong as it looks. In order to say that U is ***way inside*** V we could require that $U^- \subseteq V$ and U^- is ***compact***. This means that U avoids the boundary of V even in a compactification of the space. This relation, moreover, has a purely lattice-theoretical definition, since we can define it in $\mathcal{O}(X)$ as meaning that every open covering of V has a finite subcollection covering U. (At least this works in the locally compact Hausdorff case.) What we are now going to study is the abstract generalization of this relation on complete lattices and on continuous lattices, where the notion is non-trivial in an interesting way.

1.1. DEFINITION. Let L be a complete lattice. We say that *x is way below y*, in symbols $x \ll y$, iff for directed subsets $D \subseteq L$ the relation $y \leq \sup D$ always implies the existence of a $d \in D$ with $x \leq d$. An element satisfying $x \ll x$ is said to be *isolated from below* or *compact*. □

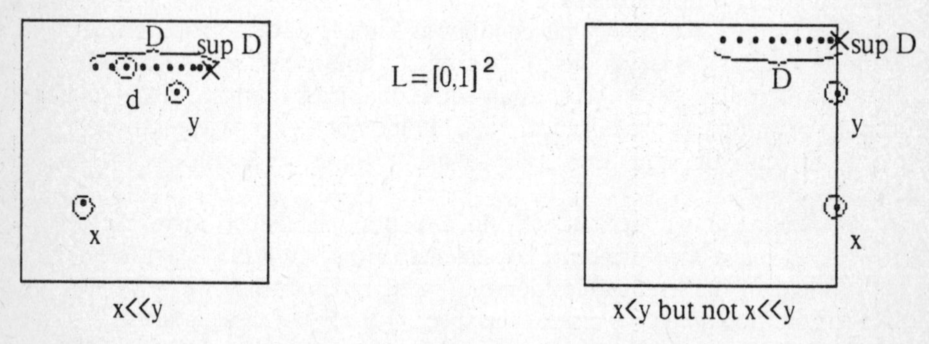

x<<y   x<y but not x<<y

Some authors prefer the term "relatively compact" to "way below", since in $\mathcal{O}(X)$ it is natural to read $U \ll V$ as "*U is relatively compact in V*". However, important as the topological example is, it is only one out of many examples of interesting lattices; we therefore choose to emphasize the lattice-theoretical view.

The following properties of the relation $\ll$ are immediate:

1.2. PROPOSITION. *In a complete lattice* L *one has the following statements for all* $u,x,y,z \in$ L:

(i) $x \ll y$ *implies* $x \leq y$;

(ii) $u \leq x \ll y \leq z$ *implies* $u \ll z$;

(iii) $x \ll z$ *and* $y \ll z$ *together imply* $x \vee y \ll z$;

(iv) $0 \ll x$. □

Clearly, $\ll$ is transitive and antisymmetric from (i) and (ii). In analogy with O-1.3 we write:

$$\mathop{\downarrow\!\!\downarrow} x = \{u \in \mathrm{L} : u \ll x\}, \quad \mathop{\uparrow\!\!\uparrow} x = \{v \in \mathrm{L} : x \ll v\},$$

and so on. We can then combine the four statements of 1.2 into the following single one:

For all x in a complete lattice, the set $\mathop{\downarrow\!\!\downarrow} x$ *is an ideal contained in* $\downarrow x$ *which depends monotonically on x.*

This is clear from 1.2 and the definition of an ideal (O-1.3).

1.3. EXAMPLES. (1) As a first example let L be a *complete chain.* Then $x<y$ obviously implies $x \ll y$. Conversely, if $x \ll y$, then either $x<y$ or $x=0$ or else $x=y$ is isolated from below—which in this case means simply that we have $\sup(\downarrow x \setminus \{x\}) < x$, so that x is the upper endpoint of a jump in the ordering. (Thus, if L is the ordinary unit interval [0,1], we have $x \ll y$ iff either $x<y$ or $x=y=0$.)

(2) The way-below relation generally behaves as a type of strict less-than relation, but, as we have just seen in the case of chains, the behavior is a little more subtle than that. If L is a complete chain, and we consider the partially ordered *direct power* L^I of L in the pointwise ordering, then in the complete lattice L^I we find $x \ll y$ iff $x_i \ll y_i$ for all $i \in I$ **and** $x_i = 0$ **for all but a finite number of indices** *i*. When I is infinite, this circumstance obviously justifies the "way" in "way below". (Perhaps "well below" would have been less colloquial, but we wanted to make the notion more memorable.) The reader can easily explain to himself the significance of the special case when L is just the two-element lattice and we can regard L^I as the powerset lattice.

(3) As a first bad example, consider the case of a *complete and atomless Boolean algebra.* Recall that "atomless" means that there are *no* minimal nonzero elements; thus, by the laws of Boolean complements, every nonzero element can be split as the join of two disjoint, nonzero parts. By employing completeness this splitting can be continued indefinitely to show that, in fact, every nonzero element is the sup of an *denumerable* family of pairwise disjoint, nonzero elements, each of which must necessarily be strictly smaller than the originally given element. By arranging these elements in a sequence

and taking the joins of the initial segments, we find that the original element is the sup of a *directed* family of strictly smaller elements.

Now suppose $x \ll y$. If $x \neq 0$, then the preceding construction can be carried out in the interval $[w,y]$ where w is the relative complement of x in y (one verifies directly that all nontrivial intervals are again atomless Boolean algebras). But no element of $[w,y]$ is above x except y. Thus, the way-below relation trivializes to: $x \ll y$ iff $x = 0$.

(4) For somewhat pathological examples, think what it means for $x \ll x$ to hold for *all* $x \in L$. Clearly, every *finite lattice* has this property. More generally it is necessary and sufficient that there be no strictly increasing infinite chains in the partial ordering, because the sup of such a chain cannot be isolated from below. This is just the *ascending chain condition* for L, and it is equivalent to saying that every nonempty subset contains a maximal element (and hence every directed set has a maximum). Note that if L satisfies this condition, there is no reason for L^{op} to do so: the definition of the way-below relation is, therefore, not at all symmetric with respect to the partial ordering.

(5) We should also recall at this point the examples of lattices from universal algebra, for instance those of O-2.7(4). The ring case will be sufficient for illustration: the lattice Id $\mathcal{A}$ of *two-sided ideals* of the ring $\mathcal{A}$ is complete. If $I,J \in \mathrm{Id}\,\mathcal{A}$, then, because each ideal is the directed union of the finitely generated ideals it contains, $I \ll J$ holds iff $I \subseteq F \subseteq J$ for some ***finitely generated*** $F \in \mathrm{Id}\,\mathcal{A}$. We note, too, that $F \ll F$ holds iff F is finitely generated. This and related examples will be studied in full detail in Section 4 on algebraic lattices. □

Topological spaces provide other good examples—in certain cases. That is to say, in certain cases it is easy to identify the way-below relation in topological terms. In our formulations we adopt the Bourbaki convention of calling a space (or subset) with the Heine-Borel property *quasicompact* and reserve the adjective *compact* for the Hausdorff setting.

1.4. PROPOSITION. *Let X be a topological space and let* $L = O(X)$.

(i) *If* $U,V \in L$ *and if there is a quasicompact subset* $Q \subseteq X$ *with* $U \subseteq Q \subseteq V$, *then* $U \ll V$.

(ii) *Suppose now that X is locally quasicompact (that is, every point has a basis of quasicompact neighborhoods). Then* $U \ll V$ *in L implies the existence of a quasicompact set Q with* $U \subseteq Q \subseteq V$.

Proof. (i): Indeed, any directed open cover of V is a directed open cover of Q, and, since Q is quasicompact, one of the covering sets contains Q, hence U. Thus, $U \ll V$ by 1.1.

(ii): Since X is locally quasicompact, each point $v \in V$ has a quasicompact neighborhood $Q_v \subseteq V$ with interior W_v containing v. Then

$$V = \bigcup\{W_v : v \in V\},$$

and the collection of finite unions $W_{v_1}\cup...\cup W_{v_n}$ is a directed cover of V. Hence, since $U\ll V$, there are elements $v_1,...,v_n$ such that

$$U\subseteq W_{v_1}\cup...\cup W_{v_n}\subseteq Q_{v_1}\cup...\cup Q_{v_n}\subseteq V.$$

The set $Q = Q_{v_1}\cup...\cup Q_{v_n}$ is the required quasicompact set. □

One notes immediately that in Hausdorff spaces the relation $U\subseteq Q\subseteq V$ for some quasicompact set is equivalent to saying that $U^-\subseteq V$ and U^- is compact.

The above examples and arguments suggest some alternatives to the definition of the way-below relation.

1.5. REMARK. (i) *In a complete lattice* L, *the following conditions are equivalent*:

(1) $x\ll y$;

(2) $y\leq\sup X$ *and* $X\subseteq$L *always implies the existence of a finite subset* $A\subseteq X$ *with* $x\leq\sup A$;

(3) $y\leq\sup I$, *where* I *is an ideal of* L, *always implies* $x\in I$.

(ii) *If, moreover,* L *is meet-continuous, we need only consider the case where* $y = \sup I$ *in* (3) *above.*

Proof. (i): (1) implies (2): If $x\ll y$ and X is given with $y\leq\sup X$, we let D be the set of all sup A with finite $A\subseteq X$. Then D is directed and $\sup D = \sup X$ by O-1.5; thus, by 1.1 there is a $d\in D$ with $x\leq d$; but d is of the form $d = \sup A$ for some finite $A\subseteq X$.

(2) implies (3): If $X=I$ is an ideal then $x\leq\sup A$ implies $x\in I$.

(3) implies (1): If $y\leq\sup D$, let $I = \downarrow D$.

For (ii) we have only to remark that $y\leq\sup I$ is equivalent to $y = \sup yI$ in a meet-continuous lattice. □

In an arbitrary complete lattice, as we have seen, we have no guarantee that the relation $x\ll y$ is satisfied for any pairs (x,y) other than those with $x=0$. Very roughly speaking, continuous lattices are those complete lattices for which the relation $x\ll y$ is "frequent". More precisely:

1.6. DEFINITION. A lattice L is called a *continuous lattice* if L is complete and satisfies the *axiom of approximation*:

(A) $\quad x = \sup\{u\in\mathrm{L} : u\ll x\}$

for all $x\in$L. □

We may write this condition above equivalently as $x = \sup \twoheaddownarrow x$ or as:

(A$_1$) Whenever $x \not\leq y$, then there is a $u \ll x$ with $u \not\leq y$.

In words 1.6 means that every element can be sufficiently well approximated by elements way below it. Indeed, the way-below relation completely determines the partial ordering, because in a continuous lattice it is the case that

$$x \leq y \text{ iff } \Downarrow x \subseteq \Downarrow y.$$

1.7. EXAMPLES. In view of the foregoing discussion in 1.3 and 1.4, we may assert that the following are continuous lattices:

(1) complete chains;

(2) the direct powers (products) of complete chains;

(3) complete lattices satisfying the ascending chain condition;

(4) the ideal lattice of a ring;

(5) the open-set lattice of a locally quasicompact space. □

Notice that for locally quasicompact X the lattice $\mathcal{O}(X)$ is in fact a continuous Heyting algebra by O-3.22. In the fourth section of this chapter we will see that $\mathcal{O}_{reg}(X)$ is *almost never* a continuous lattice, since Boolean algebras rarely are (cf. the atomless example of 1.3(3)). This shows that complete Heyting algebras need not be continuous lattices. In Chapter V we shall prove that every continuous cHa (distributive continuous lattice) is of the form $\mathcal{O}(X)$ for some locally quasicompact space X, even though not every cHa is a topology. We will also exhibit there some very bad spaces X for which $\mathcal{O}(X)$ is a continuous lattice, while every quasicompact subset of X has empty interior.

These examples cover the most immediate and obvious classes of continuous lattices. At the end of this present section we present some further types occurring "in nature", where the proof of continuity is not so quick. In the next section we will find construction methods allowing us to obtain a multitude of continuous lattices by using given ones as building blocks. Before turning to these concerns, however, we pause to tie up the lattice theory with the topology in an important case.

1.8. PROPOSITION. *If the space* X *is regular and* $\mathcal{O}(X)$ *is a continuous lattice, then* X *is locally quasicompact and, hence, locally compact if* T_0.

Proof. If $\mathcal{O}(X)$ is a continuous lattice and X regular, consider $x \in X$ and any open neighborhood V of x. Since $\mathcal{O}(X)$ is a continuous lattice, there is a $U \in \mathcal{O}(X)$ such that $x \in U \ll V$ by definition. Since X is regular, there is an open neighborhood W of x with $W^- \subseteq U$. Now let $\mathcal{W} = \{W_j : j \in J\}$ be an open cover of W^-. Then $\mathcal{W} \cup \{X \setminus W^-\}$ is an open cover of V. Since $U \ll V$, a finite subcover thereof covers U. But then a finite subcover of $\mathcal{W}$ covers W^-. Hence, W^- is quasicompact. □

To finish up our discussion of definitional matters, we take a closer look at the way-below relation and detect how it fits into a more general framework. We begin by reformulating as a definition something we already know for the way-below relation (cf. 1.2).

1.9. DEFINITION. We say that a binary relation $\prec$ on a complete lattice is an *auxiliary relation*, or *auxiliary order*, if it satisfies the following conditions:

(i) $x \prec y$ implies $x \leq y$;

(ii) $u \leq x \prec y \leq z$ implies $u \prec z$;

(iii) $x \prec z$ and $y \prec z$ together imply $x \vee y \prec z$;

(iv) $0 \prec x$.

The set of all auxiliary relations on L will be denoted Aux(L). □

Clearly, every auxiliary relation is transitive by (i) and (ii), and the way-below relation is an auxiliary relation by definition. The set Aux(L) is a poset relative to the containment of graphs on $L \times L$. The largest element is the relation $\leq$ itself, the smallest element is the relation $\bigcirc$ for which $x \bigcirc y$ iff $x = 0$. Aux(L) is obviously a complete lattice, because it is closed under arbitrary intersections of (graphs of) relations.

In order to gain better insight into the lattice Aux(L) we try to find an isomorphic copy.

1.10. PROPOSITION. *Let* L *be a complete lattice. Let* M *be the set of all monotone functions* $s : L \to \mathrm{Id}\, L$ *satisfying* $s(x) \subseteq \downarrow x$ *for all* $x \in L$*—considered as a poset relative to the pointwise ordering of monotone maps. Then the assignment*

$$\prec \mapsto s_{\prec} = (x \mapsto \{y : y \prec x\})$$

is a well-defined isomorphism from Aux(L) *onto* M, *whose inverse associates to each function* $s \in M$ *the relation* $\prec_s$ *given by:*

$$x \prec_s y \text{ iff } x \in s(y).$$

Proof. Let $\prec$ be an auxiliary relation. Then $s_{\prec}(x)$ is an ideal contained in $\downarrow x$ by 1.9(i)–(iv). If $x \leq y$, then $s_{\prec}(x) \subseteq s_{\prec}(y)$ by 1.9(ii). Thus $s_{\prec}$ is in M, and the assignment $\prec \mapsto s_{\prec}$ is clearly order preserving.

Conversely, if $s \in M$, then $s(x) \subseteq \downarrow x$ implies that $\prec_s$ satisfies 1.9(i). The relation $u \leq x \prec_s y \leq z$ implies $u \leq x$ and $x \in s(y) \subseteq s(z)$, since s is monotone. Because $s(z)$ is a lower set, $u \in s(z)$; whence, $u \prec_s z$. Thus 1.9(ii) is satisfied. Condition 1.9(iii) is immediate for $\prec_s$ from the fact that the ideal $s(z)$ is a sup-semilattice. Finally 1.9(iv) is clear from the fact that 0 is contained in any nonempty ideal. Thus, the assignment $s \mapsto \prec_s : M \to \mathrm{Aux}(L)$ is a well-defined function, and it is obviously order preserving also.

It remains to confirm that the two assignments are inverses of each other, but this is easy. □

After this proposition we know that an auxiliary relation is essentially the same thing as assigning in a monotone fashion an ideal bounded by x to each element x of L. The largest element in M is the function $x \mapsto \downarrow x$; the smallest is the constant function $x \mapsto \{0\}$. It is easy to see directly that M is a complete lattice.

Now we can raise the question how we might locate the auxiliary relation $\ll$ within Aux(L). From Remark 1.5(i) we know that

$$\Downarrow x = \bigcap\{I \in \mathrm{Id}\, \mathrm{L} : x \leq \sup I\}.$$

This does not yet express the function $x \mapsto \Downarrow x$ as an inf (in M) of a recognizable collection of other functions in M. In order to approach *this* goal, consider for an arbitrary $I \in \mathrm{Id}\,\mathrm{L}$ the function $m_I : \mathrm{L} \to \mathrm{Id}\ \mathrm{L}$ given by

$$\begin{aligned} m_I(x) &= xI = \downarrow x \cap I, && \text{if } x \leq \sup I, \\ &= \downarrow x, && \text{otherwise.} \end{aligned}$$

Then $m_I(x)$ is an ideal which is contained in $\downarrow x$, and $x \mapsto m_I(x)$ is monotone; in other words, $m_I \in \mathrm{M}$. Now we calculate inf $\{m_I : I \in \mathrm{Id}\ \mathrm{L}\}$in M:

$$\begin{aligned} (\inf_{I \in \mathrm{Id\,L}} m_I)(x) &= \bigcap_{I \in \mathrm{Id\,L}} m_I(x) \\ &= \bigcap_{x \leq \sup I} m_I(x) \cap \bigcap_{x \not\leq \sup I} m_I(x) \\ &= \bigcap_{x \leq \sup I} (\downarrow x \cap I) \cap \bigcap_{x \not\leq \sup I} \downarrow x \\ &= \bigcap\{I \in \mathrm{Id\,L} : x \leq \sup I\} \\ &= \Downarrow x. \end{aligned}$$

Definition 1.6 of a continuous lattice motivates us next to formulate the following definition:

1.11. DEFINITION. An auxiliary relation $\prec$ on a complete lattice L (and the function $s_\prec : \mathrm{L} \to \mathrm{Id}\ \mathrm{L}$ associated with it) is called *approximating* iff we have $x = \sup s_\prec(x) = \sup\{u \in \mathrm{L} : u \prec x\}$ for all $x \in \mathrm{L}$. The set of all such relations is denoted by App(L). □

The relation $\leq$ is trivially approximating, and, in a continuous lattice L, the relation $\ll$ is approximating by 1.6. Loosely speaking, the approximating auxiliary relations are those auxiliary relations which are "close" to $\leq$. One does not expect a rich supply of information for auxiliary relations which are not approximating, but they do occur.

1.12. LEMMA. *In a meet-continuous lattice* L, *all relations belonging to the functions* m_I *for* $I \in \mathrm{Id}$ L *are approximating. This holds also for continuous lattices, because they are all meet continuous.*

Proof. Let $x \in L$. If $x \leq \sup I$, then $\sup m_I(x) = \sup xI = x \sup I = x$ by O-4.2.5. If $x \not\leq \sup I$, then $\sup m_I(x) = \sup \downarrow x = x$.

Now suppose L is continuous. We use O-4.2(5). Assume that $x \leq \sup D$, where D is directed. To show $x \leq \sup xD$, it suffices to show $⇊x \subseteq \downarrow \sup xD$. But if $y \ll x$, then not only is $y \leq x$ but $y \leq z$ for some $z \in D$. As $y \leq xz \in xD$, it follows that $y \leq \sup xD$. □

We will remark in the next section that meet continuity is a corollary of a more general discussion (cf. I-2.2), but a direct proof in the present context is very elementary. Now we can have a lattice-theoretical description of $\ll$ within Aux(L)—at least for meet-continuous lattices:

1.13. PROPOSITION. *In a meet-continuous lattice, the way-below relation is the inf (that is, the intersection) of all approximating auxiliary relations.*

Proof. Define $s_0 \in M$ by the equation:

$$s_0(x) = \bigcap\{s_{\prec}(x) : \prec \text{ is in App(L)}\}.$$

Then obviously

$$s_0(x) \supseteq \bigcap\{I : x \leq \sup I\} = ⇊x.$$

On the other hand we have

$$\begin{aligned} ⇊x &= \bigcap\{m_I(x) : I \in \text{Id L}\} \\ &\supseteq \bigcap\{s_{\prec}(x) : \prec \text{ is in App(L)}\} \\ &= s_0(x) \end{aligned}$$

by Lemma 1.12. □

Notice that this does **not** say that $\ll$ is itself an approximating relation, because meet-continuous lattices need not be continuous lattices. More precisely, we have:

1.14. PROPOSITION. *Let* L *be a complete lattice. Then the following conditions are equivalent:*

(1) L *is a continuous lattice;*

(2) $\ll$ *is the smallest approximating auxiliary relation on* L;

(3) L *is meet-continuous, and there is a smallest approximating auxiliary relation on* L.

Proof. (1) implies (3): This is trivial in view of 1.12 and 1.13.

(1) implies (2): If s_0 is as in the proof of 1.13, we always have $⇊x \subseteq s_0(x)$. But, if $\ll$ is itself approximating, clearly equality holds.

(2) implies (3): Use 1.12 again. □

We turn the discussion now to the single most important property of the relation $\ll$ in continuous lattices.

1.15. DEFINITION. We say that an auxiliary relation $\prec$ on a complete lattice L satisfies the *strong interpolation property*, provided that the following condition is satisfied for all $x,z \in L$:

(SI) $\quad x \prec z$ and $x \neq z$ together imply $(\exists y)\,(x \prec y \prec z$ and $x \neq y)$.

We say that $\prec$ satisfies simply the *interpolation property* iff the following weaker condition holds for all $x,z \in L$:

(INT) $\quad x \prec z$ implies $(\exists y)\; x \prec y \prec z$. □

One may look at the interpolation property as a sort of order-density property. Notice that $x \prec z$ and $x = z$ (i.e., $x \prec x$) will trivially imply the existence of a y with $x \prec y \prec z$; namely, $y = x$. Clearly, then, (SI) implies (INT); if $\prec$ is approximating, then both conditions are equivalent. In the theory of continuous lattices much depends on the fact that $\ll$ satisfies the interpolation property. We are going to prove this now. The reason for making a distinction between the two interpolation properties is that $\ll$ is generally a very irreflexive relation and that the stronger one will be needed in Chapter IV.

1.16. LEMMA. *Any approximating auxiliary relation $\prec$ in a complete lattice* L *satisfies the following condition for all $x,z \in L$:*

$$x \prec z \text{ and } x \neq z \text{ together imply } (\exists y)\,(x \leq y \prec z \text{ and } x \neq y).$$

Proof. Since $z = \sup\{u : u \prec z\}$, there is a $u \prec z$ with $u \not\leq x$. If we now set $y = x \vee u$, then $y \prec z$ by 1.9(iii); and $x \neq y$, since $u \not\leq x$. □

1.17. LEMMA. *Any approximating auxiliary relation $\prec$ in a complete lattice* L *satisfies the following condition for all $x,z \in L$:*

$$x \ll z \text{ and } x \neq z \text{ together imply } (\exists y)\,(x \prec y \prec z \text{ and } x \neq y).$$

Proof. We define $I = \{u \in L : (\exists y)\; u \prec y \prec z\}$. We claim first that I is an ideal. By 1.9(ii), I is a lower set. If u_n, $n = 1,2$, are in I, then there are elements y_n with $u_n \prec y_n \prec z$, $n = 1,2$, and then $u_1 \vee u_2 \prec y_1 \vee y_2 \prec z$ by 1.9(iii). Therefore $u_1 \vee u_2 \in I$.

Next we claim that $\sup I = z$. Set $\sup I = z^*$ and assume that in fact $z^* \neq z$. Since $\prec$ is approximating (and thus $z = \sup\{y \in L : y \prec z\}$), there is a $y \prec z$ with $y \not\leq z^*$. Since $y = \sup\{u \in L : u \prec y\}$, there is a $u \prec y$ with $u \not\leq z^*$. But then $u \prec y \prec z$ implies $u \in I$ and thus $u \leq \sup I = z^*$, a contradiction. Thus $z^* = z$, as was claimed.

Finally we recall that $x \ll z$ and $z = \sup I$. By Remark 1.5(ii) this implies $x \in I$. Thus there is an element y^* with $x \prec y^* \prec z$. Then, by 1.16, there is an element y^{**} with $x \leq y^{**} \prec z$. Set $y = y^* \vee y^{**}$. Then $x \neq y$ and $x \prec y$, by 1.9(ii), and $y \prec z$ by 1.9(iii). □

In view of these lemmas, the following result is immediate.

1.18. THEOREM. *In a continuous lattice the way-below relation satisfies the strong interpolation property.* □

We remark that for many purposes (INT) suffices. We also note that in the formulation of the property we could not strengthen (SI) to include $y \neq z$, because the two-element chain offers a trivial counterexample.

1.19. COROLLARY. *In a continuous lattice* L *the following conditions are equivalent*:

(1) $x \ll y$.

(2) *For each directed set D of* L *the relation* $y \leq \sup D$ *implies the existence of* $d \in D$ *with* $x \ll d$.

Proof. By 1.2(i), clearly (2) implies (1). Now suppose (1) and $y \leq \sup D$ for a directed set $D \subseteq L$. By Theorem 1.18 there is an x^* such that $x \ll x^* \ll y$. Then we find a $d \in D$ with $x^* \leq d$ by 1.1. Hence, $x \ll d$ by 1.2(ii). □

Let us now consider three examples of types of continuous lattices which can be fairly said "to occur in (mathematical) nature." In connection with Section 4 of this chapter, where we introduce algebraic lattices, the wide classes of lattices described here will in particular furnish examples of continuous lattices which are not generally algebraic.

1.20. EXAMPLE. (CLOSED IDEALS OF C*-ALGEBRAS). Recall the concept of a C*-algebra, which is of central importance in functional analysis in the context of operators on Hilbert space and of operator norm closed involutive algebras of operators. Abstractly a C*-algebra is a complex Banach algebra $\mathcal{A}$ with an involution $a \mapsto a^*$ satisfying $||a^*a|| = ||a||^2$. We record the following fact (cf. Laursen-Sinclair [1975, esp. p. 168]):

FACT. *Each C*-algebra* $\mathfrak{B}$ *contains a unique smallest dense two-sided ideal* $\mathfrak{B}_0$, *called the* ***Pedersen ideal*** *of* $\mathfrak{B}$. □

If $\mathfrak{B}$ has an identity, then $\mathfrak{B}_0 = \mathfrak{B}$; if $\mathfrak{B}$ is the algebra $\mathbf{C}_0(X)$ of all continuous complex-valued funcions on a locally compact and noncompact space vanashing at infinity, then $\mathfrak{B}_0$ is the ideal K(X) of all continuous functions of compact support; it was for the purposes of generalizing integration theory to the noncommutative situation that Pedersen found and investigated the ideal $\mathfrak{B}_0$. If $\mathfrak{B}$ is the algebra $\mathbf{LC}(\mathcal{H})$ of compact operators on the Hilbert space $\mathcal{H}$, then $\mathfrak{B}_0$ is the ideal of all finite-rank operators.

For any subset X in the C*-algebra $\mathcal{A}$ we let $\langle X \rangle$ denote the closed (!) two-sided ideal generated by X in $\mathcal{A}$. Recall that $\mathrm{Id}^-\mathcal{A}$ denotes the lattice of closed two-sided ideals (O-2.7(7)). Each $I \in \mathrm{Id}^-\mathcal{A}$ is in itself a C*-algebra. We now have the following proposition on the complete lattice $L = \mathrm{Id}^{-}\mathcal{A}$.

1.20.1. PROPOSITION. *For $I,J \in \mathrm{Id}^{-}\mathcal{A}$ the following statements are equivalent:*

(1) $I \ll J$ *(in* L).

(2) *There is an element* $a \in J_0$ *with* $0 \leq a$ *such that* $I \subseteq \langle a \rangle$

(3) *There is a finite subset* $F \subseteq J_0$ *with* $I \subseteq \langle F \rangle$.

Proof. (2) implies (3) is trivial. For the proof of (3) implies (1) we formulate a lemma:

LEMMA. *Let* $P \in \mathrm{Id}^{-}\mathcal{A}$ *and* $Q \in \mathrm{Id}\,\mathcal{A}$ *with* $P \subseteq Q^{-}$. *Then* $(P \cap Q)^{-} = P$.

Proof. Let $0 \leq x \in P$; then $x = \lim x_n$ with $x_n \in Q$; thus $xx_m \in P \cap Q$; hence $x^2 = \lim xx_m \in (P \cap Q)^2$ and thus $x \in (P \cap Q)^{-}$ by the functional calculus for C*-algebras. □

Now we prove (3) implies (1): Let D be any directed subset of L with $J \subseteq \sup D = (\bigcup D)^{-}$. By the Lemma, $J \cap \bigcup D$ is a dense two-sided ideal of J; hence, it contains J_0 by the Fact quoted above. If $F \subseteq J_0$ is as in (3), we may therefore conclude that there is some member $K \in D$ with $F \subseteq K$, and thus $I \subseteq \langle F \rangle \subseteq K$. This proves (1).

For a proof of (1) implies (2), we take $0 \leq x,y \in J_0$, and so $0 \leq x+y \in J_0$; the observation $0 \leq x \leq x+y$ allows us to conclude $x \in \langle x+y \rangle$, since closed two sided ideals are "hereditary". Thus the collection $\{\langle x \rangle : 0 \leq x \in J_0\}$ is directed in L, and its union contains J_0; whence, its sup dominates J by the Fact quoted above. Condition (1) now implies that we find some $a \in J_0$ with $0 \leq a$ and $I \subseteq \langle a \rangle$. □

We notice that the crux in defining the way-below relation in a lattice of closed ideals (or congruences, subgroups, etc.) in topological rings (algebras, groups, etc.) is the fact that the formation of the sup of a directed collection involves an eventual closure of the union, and this creates all the complication. Once again one notices that C*-algebras constitute a class of topological algebras with particularly desirable properties. We now arrive at the the main conclusion that is the point of this discussion.

1.20.2. PROPOSITION. *The lattice* $L = \mathrm{Id}^{-}\mathcal{A}$ *of closed two-sided ideals in the C*-algebra* $\mathcal{A}$ *is a continuous lattice.*

Proof. For $I \in L$ we have $I = (\bigcup\{\langle F \rangle : F \text{ finite in } I_0\})^{-}$ by the Fact quoted above. Then apply Proposition 1.20.1 above. □

Other proofs of this result exist, but this one relates most directly to the definitions. We remark that in fact $\mathrm{Id}^{-}\mathcal{A}$ is a Heyting algebra, hence indeed a continuous Heyting algebra by the preceding proposition.

1.21. EXAMPLE. (LOWER SEMICONTINUOUS FUNCTIONS). Let LSC(X) = LSC(X,$\mathbb{R}^*$) denote the complete lattice of all lower semicontinuous functions on a topological space X with values in the extended real line $\mathbb{R}^*$ (see O-2.10). For any function $f: X \to \mathbb{R}^*$ we set $G_f = \{(x,r) : r < f(x)\}$. Then f is lower semicontinuous iff G_f is open in $X \times \mathbb{R}^*$. We use the notation $x \ll y$ in $\mathbb{R}^*$, a continuous lattice itself, which of course means that $x < y$ or $x = -\infty$.

1.21.1. PROPOSITION. *Suppose that* X *is a compact space. Then for functions* $f,g \in$ LSC(X) *The following statements are equivalent:*

(1) $f \ll g$ *in* LSC(X);

(2) *There is an open cover* $\{U_j : j \in J\}$ *of* X *and a family* $\{y_j : j \in J\}$ *of elements in* $\mathbb{R}^*$ *where* $f(x) \leq y_j \ll g(x)$ *for all* $j \in J$ *and* $x \in U_j$;

(3) *For each element* $x \in X$ *there is an open neighborhood* U *in* X *and an element* $y \in \mathbb{R}^*$ *where* $f(x_1) \leq y \ll g(x_1)$ *for all* $x_1 \in U$;

(4) $G_f^- \subseteq G_g$ *in* $X \times \mathbb{R}^*$;

(5) *There is a* **continuous** *function* $h \in C(X,\mathbb{R}^*)$ *where for all* $x \in X$ *we have* $f(x) \leq h(x) \ll g(x)$.

Remark. The implications (1) implies (2) and (2) iff (3) maintain their validity for regular X.

Proof. (2) iff (3): Clear.

(1) implies (3): Let $\mathfrak{F}(g)$ be the set of all functions $s\chi_U$ such that

(i): χ_U is the characteristic function of an open set U, and

(ii): $s \ll g(x)$ for all $x \in U^-$.

Then $\mathfrak{F}(g)$ is a subset of LSC(X) with $g = \sup \mathfrak{F}(g)$, since X is regular and g is lower semicontinuous. Because $f \ll g$ we find functions

$$s_1\chi_{U_1},\ldots,s_n\chi_{U_n} \in \mathfrak{F}(g) \text{ with } f \leq \sup s_j\chi_{U_j}.$$

Let $x \in X$ be arbitrary, and set $I(x) = \{j : x \in U_j^-\}$ Then $i \in I(x)$ implies $s_i \ll g(x)$. Set $s(x) = \max\{s_j : j \in I(x)\}$, then $s(x) \ll g(x)$. The set

$$V(x) = X \setminus \bigcup\{U_j^- : j \notin I(x)\}$$

is an open neighborhood of x. Since g is lower semicontinuous, there is an open neighborhood $U(x)$ of x in $V(x)$ such that $u \in U(x)$ implies $s(x) \ll g(u)$. But $u \in U(x)$ entails that $u \notin U_j$ for $j \notin I(x)$; whence,

$$f(u) \leq \sup\{s_j\chi_{U_j}(u) : j = 1,\ldots,n\} = \sup\{s_j\chi_{U_j} : j \in I(x)\} \leq s(x).$$

(3) implies (1): Let h_j be a directed net in LSC(X) with $g \leq \sup h_j$. For each $x \in X$ we find an index $j = j(x)$ and an element $s(x) \in \mathbb{R}$ where $s(x) \ll h_j(x)$, and $f(y) \leq s(x)$ for all y in an open neighborhood $U(x)$ of x. Since h_j is lower semicontinuous, there is an open neighborhood $V(x) \subseteq U(x)$ with $s(x) \ll h_j(v)$ for all $v \in V(x)$. By the compactness of X, we find finitely many $x_1,\ldots,x_n$ with

$X = V(x_1)\cup\ldots\cup V(x_n)$. Let k be an index with $j(x_1),\ldots,j(x_n)\leq k$. Then for each $x\in X$ there is an i with $f(x)\leq s(x_i)\leq h_k(x)$; whence, $f\leq h_k$.

(4) iff (3): Direct verification.

(5) implies (3): Straightforward enough.

(2) implies (5): Let $\{V_i : i\in I\}$ be a finite cover such that for each $i\in I$ we have a $j(i)\in J$ with $V_i^-\leq U_{j(i)}$. We write $z_i = y_{j(i)}$. Let $\{p_i : i\in I\}$ be a partition of unity subordinate to $\{V_i : i\in I\}$. Define $h = \Sigma z_i\, p_i$. For $x\in X$ note

$$h(x) = \Sigma\{z_i p_i(x) \mid x\in V_i\} \ll \Sigma\{g(x)\, p_i(x) \mid x\in V_i\} = g(x).$$

Similarly $f(x)\leq h(x)$. □

If X is **N** with the discrete topology and f is the constant function with value 1/2 and g that with the value 1, then (3) is evidently satisfied. But (1) fails: Consider the directed family of characteristic functions of finite subsets of **N**; its sup is g, but no member dominates f. Thus the compactness of X is indispensable to conclude (3) implies (1).

1.21.2. PROPOSITION. *If* X *is a compact space, then* LSC(X) *is a continuous lattice.*

Proof. Consider that $f = \sup\{r\chi_U : r< f(u) \text{ for all } u\in U.\}$ □

The description we have given for $f\ll g$ in LSC(X) and the proofs we used were those of an analyst. Proposition 1.21.2 will be vastly generalized in Chapter II (see II-4.6); the methods will be more lattice theoretical and topological and will not be based on a frontal attack via the way-below relation. (See also Exercises 2.16 and 2.17 below.) Indeed, the examples of 1.20 and 1.21 illustrate that in many circumstances the way-below relation is difficult to describe explicitly, and this points to the need of these other tools in order to deal effectively with continuous lattices.

1.22. EXAMPLE. (CONVEX COMPACT SUBSETS OF A COMPACT CONVEX SET). Let K be a compact convex subset of a locally convex topological vector space. Denote by Con(K) the lattice of all closed convex subsets of K (including the empty set). Recall that $\mathrm{Con}(K)^{op}$ is the lattice with the reverse inclusion.

1.22.1. PROPOSITION. *The lattice* $\mathrm{Con}(K)^{op}$ *is a continuous lattice, in which we have* $A\ll B$ *iff* $B\subseteq \mathrm{int}(A)$ *with the interior being taken in the relative topology of* K.

Proof. Since Con(K) is closed under arbitrary intersections, it—and hence $\mathrm{Con}(K)^{op}$—is a complete lattice. If $B\subseteq\mathrm{int}(A)$, then if the intersection of a descending family of closed convex sets is contained in B, it follows by compactness that one of them is a subset of A. Thus $A\ll B$ in $\mathrm{Con}(K)^{op}$. The local convexity of the vector space implies that B is the intersection of its compact convex neighborhoods in K. Hence $\mathrm{Con}(K)^{op}$ is continuous.

Finally assume that $A \ll B$ in Con(K)^{op}. Since we have just seen that B is the intersection of its compact convex neighborhoods in K, and since this family of neighborhoods is descending, we conclude from the definition of $\ll$ that one of them must be a subset of A. Therefore, A is a compact convex neighborhood of B, and this completes the proof. □

We remark that this example illustrates how we sometimes encounter lattices L where L^{op} instead of L is the continuous lattice even though L is more naturally given. However, there is no point of taking the dual definition, since there are just as natural examples that conform to the convention we have adopted.

EXERCISES

We begin by mentioning a few general results on the interpolation property.

1.23. EXERCISE. Let L be a complete lattice and $\prec$ an auxiliary relation. Let $s_\prec$: L→Id L be the function in M corresponding to $\prec$ according to 1.10.

(i) The following statements are equivalent:

(1) $\prec$ satisfies (SI);

(2) No ideal of the form $s_\prec(x)$ has a maximal element with respect to $\prec$ unless $x \prec x$.

(ii) Also the following statements are equivalent:

(3) $\prec$ satisfies (INT);

(4) Each ideal of the form $s_\prec(x)$ is directed with respect to $\prec$.

(iii) If $\prec$ is approximating, then (INT) implies (SI). □

1.24. EXERCISE. Let L be a complete lattice and $\prec$ an auxiliary relation on L. Define a binary relation $\prec\bullet$ as follows:

$x \prec\bullet y$ iff there is a $\prec$-chain C such that $x,y \in C$, $x \prec y$, and $\prec$ restricted to C satisfies (SI).

Then we have:

(i) $\prec\bullet$ is an auxiliary relation which satisfies (SI);

(ii) The given relation $\prec$ satisfies (SI) iff $\prec \; = \; \prec\bullet$;

(iii) Moreover, $\prec\bullet$ is the largest auxiliary relation contained in $\prec$ satisfying (SI). □

1.25. EXERCISE. Let L be a complete lattice and $\prec$ an auxiliary relation on L. Define a binary relation $\prec^{sup}$ as follows:

$x \prec^{sup} y$ iff there exist systems of elements $\{x_i : i \in I\}$ and $\{y_i : i \in I\}$ such that $x_i \prec y_i$ for all $i \in I$ and $x = \sup x_i$ and $y = \sup y_i$.

We can call $\prec^{\sup}$ the *sup-closure* of $\prec$. Then we have:

(i) If L is meet-continuous, then $\prec^{\sup}$ is an auxiliary relation;
(ii) $\prec$ is approximating iff $\prec^{\sup} = \leq$. □

In the following exercises we wish to demonstrate (a) that the concept of continuity applies not only to complete lattices but also to up-complete posets and semilattices, and (b) that in fact there are good reasons to consider such generalizations. For an up-complete poset (cf. O-2.11) we can take over, word for word, the definition of $\ll$ from 1.1. Conditions (i) and (ii) of 1.2 are equally valid in this more general setting (as well as (iii) and (iv), if applicable). Also, conditions (1) and (3) of 1.5(i) remain equivalent so that

$$\Downarrow x = \bigcap\{I \in \mathrm{Id}\, \mathrm{L} : x \leq \sup I\}.$$

1.26. DEFINITION. A poset L is called *continuous* iff it is up-complete and the following two conditions are satisfied:

(i) For all $x \in \mathrm{L}$, the set $\Downarrow x$ is directed;
(ii) $x = \sup \Downarrow x = \sup \{u \in \mathrm{L} : u \ll x\}$ for all $x \in \mathrm{L}$.

A semilattice is called *continuous* iff it is continuous as a poset. It is called *complete-continuous* if it is, in addition, a complete semilattice (see O-2.11). □

Remark. Each complete semilattice automatically satisfies condition (i) above, because the sets $\Downarrow x$ are closed under finite sups; hence, it is complete-continuous iff it satisfies condition (ii) above.

1.27. EXERCISE. In a continuous poset L the way-below relation $\ll$ satisfies the strong interpolation property and the conclusions of 1.19. □

1.28. EXERCISE. (i) Every closed interval of a continuous poset is in itself a continuous poset; it is a continuous lattice if the poset is a complete-continuous semilattice.

(ii) If L is a continuous poset and $\mathrm{M} \subseteq \mathrm{L}$ satisfies

$$\mathrm{M} = \uparrow \mathrm{M} = \Uparrow \mathrm{M} = \bigcup\{\Uparrow u : u \in \mathrm{M}\}$$

then M is also a continuous poset. □

1.29. EXERCISE. (i) Let S be a complete semilattice and let $\mathrm{S}^1 = \mathrm{S} \cup \{1\}$ be the complete lattice obtained from S by adjoining an identity (whether S has one or not—see O-2.12). Then the following statements are equivalent:

(1) S is a continuous semilattice (and thus a complete-continuous semilattice).

(2) S^1 is a continuous lattice.

(ii) Conversely, if L is a continuous lattice with $1 \ll 1$ (that is, 1 is isolated from below), then $S = \mathrm{L} \setminus \{1\}$ is a complete-continuous semilattice.

(iii) More generally, if L is a continuous lattice and $X = \uparrow X$ is an upper set where $L \backslash X$ is closed under directed sups, then $L \backslash X$ is a complete-continuous semilattice. □

These observations show that, from the standpoint of the general theory, complete-continuous semilattices are very close to continuous lattices. Indeed if a complete-continuous semilattice fails to be a continuous lattice, then it only fails by lacking a top. The complete-continuous semilattices, therefore, do not deserve too much separate attention, since they can easily be fitted into the general theory of continuous lattices by adjoining an identity. Or at least this is true when only *one* semilattice is being considered at a time, and then it does not matter whether the "ideal" elements like 1 are "really there" or considered fictions (as with $\pm\infty$ in the reals). When *several* semilattices are being combined (in, say, a direct product), then the inclusion of ideal elements makes a big difference to the outcome by their entering into combination with other elements (recall, for example, LSC(X), where the values $\pm\infty$ can be utilized in a function quite often). The exact way extra elements can occur sometimes subtly affects the properties of a construction; one example is contained in Exercise 1.30. There are in addition important applications in which the adjunction of an identity is simply *unnatural*, namely, Example O-2.7(10). We review this matter in the light of the definitions of this section by discussing a more general situation in 1.31 below.

1.30. EXERCISE. Let L and M be two posets. Define the following five kinds of "disjoint" sums:

(1) $L+_1M =$ The disjoint union of L and M (with the obvious partial ordering);

(2) $L+_2M = L+_1M$ with the 0-elements identified (if they have them);

(3) $L+_3M = (L+_2M)^1$;

(4) $L+_4M = L+_2M$ with the 1-elements identified;

(5) $L+_5M = (L+_1M)_0^1$ (that is, with *new* 0- and 1-elements adjoined).

Under the various assumptions that L and M are continuous lattices, complete-continuous semilattices, or continuous posets, investigate whether $L+_iM$ has any of these properties. (Some pictures will help!) In particular, show that if $L = M = [0,1]$, then $L+_4M$ is **not** a continuous lattice; whereas, if L and M are continuous lattices, then so are $L+_3M$ and $L+_5M$. □

1.31. EXERCISE. Let X be a topological space and Y a set. Let L be the set of partial functions with values in Y defined on an *open* subset $\operatorname{dom} f \subseteq X$. Consider on L the partial order given by

$$f \leq g \text{ iff } \operatorname{dom} f \subseteq \operatorname{dom} g \text{ and } f = g \mid \operatorname{dom} f.$$

(i) The nowhere-defined function $\varnothing : \varnothing \to Y$ is the bottom of L; every nonempty subset has an inf and every directed subset has a sup.

(ii) If $\mathcal{O}(X)$ is a continuous lattice, then L is a complete-continuous semilattice. (Recall that $\mathcal{O}(X)$ is a continuous lattice if X is locally quasicompact by 1.7.)

(iii) The semilattice L has no top if Y has more than one element. □

Continuous semilattices are substantially more general than continuous lattices. The following example indicates one of the main motivations for considering continuous semilattices from time to time, rather than continuous lattices.

1.32. EXERCISE. Let X be a set which is at the same time a topological space and a semilattice in such a fashion that the following two conditions are satisfied:

(F) If U is an open filter and $u \in U$, then there is an open filter V with $u \in V$ and $\min V \in U$;

(P) All principal filters $\uparrow x$ are quasicompact.

(i) Then the poset $\mathcal{O}\mathrm{Filt}_0$ X of open filters (together with the empty set) is a continuous semilattice with respect to $\subseteq$.

(ii) If X has a top, then $\mathcal{O}\mathrm{Filt}$ X $= (\mathcal{O}\mathrm{Filt}_0\ X)\backslash\{\emptyset\}$ is a continuous semilattice, too.

(HINT: Note $V \ll U$ in $\mathcal{O}\mathrm{Filt}$ X iff $\min V \in U$, using (F) and (P); then verify the conditions of 1.26.) □

The example of the poset of open filters of a semilattice with a topology is a very natural object to study. It exhibits, however, some characteristic deficiencies: Since the interior of a filter may not have a unique maximal open subfilter (see the example following 3.2 below!), $\mathcal{O}\mathrm{Filt}_0$ X is not complete in general.

NOTES

Continuous lattices were introduced by Dana Scott, who discovered the idea as a generalization of algebraic lattices in the fall of 1969. He presented the first coherent picture at the Dalhousie Category Theory Conference in 1971; this presentation appears in Scott [1972a] and is the first source on continuous lattices in the accessible literature. In an expository paper Scott [1973] details his motivation for the invention of continuous lattices; there he repeats his original definition and says: "Such lattices I call continuous lattices, and their mathematical theory is highly satisfactory." What he meant was that everything seemed to fall neatly in place, and considering the extensive mathematical development of the theory of continuous lattices since 1974, this claim is a modest understatement.

Whether the choice of nomenclature was an entirely wise one will remain contested in some quarters. The name is reminiscent of that of von Neumann's continuous geometries (a certain type of lattice), which—strictly speaking—have nothing to do with continuous lattices in our sense. Nevertheless, the passage from the more discrete (more precisely: zero-dimensional) algebraic lattices to the continuous lattices has a certain analogy to the passage from a discrete range of dimensions to a continuous dimension function. Furthermore, Scott had in mind the circumstance that continuous functions (in what we call in Chapter II the Scott topology) on a continuous lattice are well behaved and exist in profusion; in particular, the lattice operations are continuous. There is considerable sense to calling a lattice "continuous" just when its lattice operations are continuous, but actually the known classes of such lattices are very wide (e.g. meet-continuous lattices). Continuous lattices in the sense of this Compendium have the advantages of being restricted enough to have a good theory, general enough to capture important examples, and natural enough that we can argue that the class ought to be singled out for many different reasons.

It is noteworthy that the concept of continuous lattice was rediscovered independently by other authors working quite independently in other areas. In 1973-74, Karl Hofmann and Albert Stralka studied the algebraic (i.e., lattice-theoretical) foundations of a class of compact topological semilattices known to workers in the field of compact semigroups as *Lawson semilattices*; they found that continuous lattices and *compact* Lawson semilattices were one and the same thing, although at the time they were not aware of Scott's article and, thus, did not phrase their results in this language. Their paper appeared as Hofmann-Stralka [1976], and the not too succinct title was soon contracted for everyday use to ATLAS (using the initial letters of "Algebraic Theory of Lawson Semilattices"). Reference should also be made to the extensive work of Ju. L. Ershov (see the Bibliography), part of which was independent of Scott's work and part of which answered many questions Scott left open.

In 1975 Alan Day identified the monadic algebras associated with the so-called *filter monad* (in the same sense as compact spaces emerge as monadic algebras of the ultrafilter monad) and found them to be continuous lattices (cf. Day [1975]). Independently, Oswald Wyler also studied these algebras in 1975 and described and discussed them thoroughly; only towards the end of 1976, due to a hint by John Isbell, was it discovered that indeed Wyler's algebras were precisely the continuous lattices.

The definition given here, strictly speaking, is not the principal one given by Scott, nor does it correspond explicitly to either the characterization given in ATLAS or that given by Day and Wyler. The first published version of this definition is to be found in the note Lea [1976/77] (though see the remark in the next papagraph). The point is that the way-below relation of 1.1 is not Scott's auxiliary relation $\prec$ derived from a topology (which we will consider extensively in Chapter II). It is immediately seen from Scott's definition that $x \prec y$ implies $x \ll y$ in any complete lattice. Scott *defines* a continuous lattice to be a complete lattice in which his relation $\prec$ is approximating. By 1.14 this means that *on continuous lattices*, Scott's relation and the way-below relation agree. On complete lattices they are different in general (see II-1.24).

The way-below relation had been implicitly introduced by Hofmann and Stralka by saying that "*x is relatively compact under y*" iff $x \ll y$ (loc. cit. p. 27) and they also introduced the notation $\ll$ (loc. cit. p. 42). Isbell used the terminology "*x is compact in y*" in his paper on meet-continuous lattices,

Isbell [1975b]; his identification of Scott's relation with the way-below relation on meet-continuous lattices is not convincing, and in fact to our knowledge it is not known in general on which class of complete lattices the two agree. That is one good reason for our employing the more understandable relation here in the definition. Scott, however, had originally defined continuous lattices in this more lattice-theoretic way and refers to the characterization in his paper (loc. cit. p.110). In writing up his paper he chose the other definition in order to emphasize the topological simplicity of the notion, and he did not feel the need to consider whether the two definitions of the way-below relation agreed in a wider context.

Propositions 1.13 and 1.14 are new. The strong interpolation property has been recognized as useful in the last years, although 1.18 and a complete proof have not been published. But J. Isbell had recognized the interpolation property in his paper (loc. cit.).

The axioms of auxiliary orders with the interpolation property were formulated by M. B. Smyth (cf. Scott [SCS-4] and Smyth [1978]) for sup-semilattices; in later exercises (III-4.19 ff.) we will describe his motivation. The equivalence between auxiliary relations and functions $L \to \mathrm{Id}\, L$ discussed in 1.10 ff. was anticipated by Smyth (loc. cit.) and by Gierz, Hofmann, Keimel and Mislove [SCS-12]; in this report the details of Exercise 1.21 are introduced and elaborated.

The literature contains several forerunners of the way-below relation in the context of the representation theory of lattices; see in particular Raney [1953, notably p. 520]; Papert [1959, notably p. 174 ff.]; also Bruns [1961/62, notably Part II, p. 4 ff.].

The fact that locally quasicompact spaces X give rise to a continuous lattice $\mathcal{O}(X)$ was known to Day and Kelly [1970, Proposition 5].

The observation that the lattice L of closed two-sided ideals in a C*-algebra is a continuous lattice and its proof via Pedersen's ideals is from Hofmann [SCS-31]. There are at least two alternative proofs, one requiring the entire spectral theory of C*-algebras and the results of Chapter V, another requiring an observation due to J.M.G. Fell [1961, 1962] to the effect that L is a compact subsemilattice of a product of unit-interval semilattices together with the equivalence of continuous lattices with compact Lawson semilattices (see Chapter VI).

Example 1.21 of classical lower semicontinuous functions and its discussion is from Hofmann [SCS-17].

The concept of a continuous semilattice (1.26) is from Lawson [SCS-31], which theory of duality we will explore further in later exercises. Continuous posets are also discussed in Markowski [1976], R.-E. Hoffmann [1979a], and Wilson [1978].

2. THE EQUATIONAL CHARACTERIZATION

The characterization theorem for meet-continuous lattices (O-4.2) contains several descriptions which are of an equational character, notably 4.2(7) and (8). Here we characterize continuous lattices in a similar, although more technical vein. The significance of the equational description of continuous lattices is that it enables us to discern clearly what kind of homomorphisms should be considered between continuous lattices and to find many operations and constructions under which the class of continuous lattices is closed. We begin the discussion by considering the way-below relation from a new perspective.

2.1. THEOREM. *Let* L *be a complete lattice. Then the following conditions are equivalent*:

(1) L *is a continuous lattice*;

(2) *For each* $x \in \mathrm{L}$, *the set* $\Downarrow x$ *is the smallest ideal* I *with* $x \leq \sup I$;

(3) *For each* $x \in \mathrm{L}$ *there is a smallest ideal* I *with* $x \leq \sup I$;

(4) *The sup map* $r = (I \mapsto \sup I) : \mathrm{Id\ L} \to \mathrm{L}$ *has a lower adjoint*;

(5) *The sup map* $r : \mathrm{Id\ L} \to \mathrm{L}$ *preserves arbitrary* (!) *infs and sups.*

Proof. For $x \in \mathrm{L}$ set $J(x) = \{I \in \mathrm{Id\ L} : x \leq \sup I\}$. Recall that by 1.5(i) we have $\Downarrow x = \bigcap J(x)$. Now (1) holds iff $\Downarrow x \in J(x)$ by Definition 1.6; this is precisely (2). (2) trivially implies (3); but if $J(x)$ has a smallest element, then that element is necessarily $\Downarrow x$, whence (3) implies (1). Thus (1), (2) and (3) are equivalent.

(3) iff (4): By O-3.2, the map r has a lower adjoint iff $\min r^{-1}(\uparrow x)$ exists for all x. But $\min r^{-1}(\uparrow x)$ is precisely the smallest element of $J(x)$.

(4) implies (5): The sup-map preserves sups by O-3.15 and infs by O-3.3.

(5) implies (4): Clear by O-3.4. □

2.2. COROLLARY. *Every continuous lattice is meet continuous.*

Proof. Trivially, condition 2.1(5) implies condition O-4.2(1). □

We are now ready for the equational description of continuous lattices. The type of equation we will use is a form of the infinite distributive law which we shall call the *directed distributive law.* Later in 2.4 we will discuss in more precise terms the *complete distributive law.* It is well known that the lattice of all subsets of a set is completely distributive (with respect to arbitrary unions and intersections). Note that in 2.1(5) above we have shown that every continuous lattice L is the image of a lattice of sets—namely, Id L—by a map preserving infs and sups. But Id L is closed under set-theoretical intersection and directed union; hence, any equations these operations satisfy on set-theoretical grounds will transfer to L. What we now exhibit is a basis for these equations.

2.3. THEOREM. *For a complete lattice* L, *the following conditions are equivalent*:

(1) L *is continuous.*

(2) *Let* $\{x_{j,k} : j \in J,\ k \in K(j)\}$ *be a family of elements in* L *such that* $\{x_{j,k} : k \in K(j)\}$ *is directed for all* $j \in J$. Let M *be the set of all functions* $f : J \to \bigcup_{j \in J} K(j)$ *with* $f(j) \in K(j)$ *for all* $j \in J$. *Then the following identity holds:*

$$\text{(DD)} \qquad \bigwedge_{j\in J} \bigvee_{k\in K(j)} x_{j,k} = \bigvee_{f\in M} \bigwedge_{j\in J} x_{j,f(j)}$$

(3) *Let* $\{x_{j,k} : (j,k) \in (J \times K)\}$ *be any family in* L. *Let* N *be the set of all functions* $f : J \to \mathrm{fin}\ K$ *into the finite subsets of* K. *Then the following identity holds:*

$$\text{(DD*)} \qquad \bigwedge_{j\in J} \bigvee_{k\in K} x_{j,k} = \bigvee_{f\in N} \bigwedge_{j\in J} \bigvee_{k\in f(j)} x_{j,k}$$

Remark. Note that all the sups in (DD) are directed sups. In the sequel for easy recognition of such sups we shall write $\bigvee^{\mathrm{up}}$ for any directed sup we wish to call attention to. Strictly speaking, (DD) ***is not*** an equation because its validity requires the hypothesis that certain sets are directed. The point of formulating (DD*) is that it, on the other hand, ***is*** a pure lattice equation in (infinite) infs and sups. Note, too, that we could write $\leq$ in place of $=$ in (DD) and (DD*) since the reverse inequality always holds in any complete lattice.

Proof. (1) implies (2): For convenience let *lhs* denote the left-hand side of (DD) and *rhs* the right-hand side. It is obvious that in any complete lattice $lhs \geq rhs$. Assuming that L is continuous, all we have to do to prove the reverse inequality is to show that whenever $t \ll lhs$, then $t \leq rhs$.

Suppose then that $t \ll lhs$; we conclude that $t \ll \bigvee^{\mathrm{up}}_{k\in K(j)} x_{j,k}$ for all $j \in J$. By the definition of $\ll$, we can therefore choose a $g(j) \in K(j)$ with $t \leq x_{j,g(j)}$ for all $j \in J$. But then we see that $t \leq \bigwedge_{j\in J} x_{j,g(j)}$, and so $t \leq rhs$ must follow.

(2) implies (1): We are going to establish the approximation axiom (A) of the original Definition 1.6: namely, $x = \sup\{u \in \mathrm{L} : u \ll x\}$. Indeed let $x \in \mathrm{L}$ be a given element, and let J be the set of all up-directed subsets j of L with $\sup j \geq x$. For each $j \in J$ let $K(j) = j$, in other words j is indexing itself. Further, consider the family of elements $x_{j,k} = k$ for $j \in J$ and $k \in K(j)$. The hypothesis of (2) is thus satisfied.

Suppose $f \in M$ and let $t = \bigwedge_{j\in J} x_{j,f(j)} = \bigwedge_{j\in J} f(j)$. Then we claim that $t \ll x$. Indeed if D is a directed set with $x \leq \sup D$, then $t \leq f(D) \in D$—because $D \in J$ and t is defined to make this so.

Looking now at (DD), we see that $x = lhs$, because $\{x\} \in J$. But the equation $x = rhs$ implies (A) in view of what we checked in the last paragraph. Hence, L is continuous.

(2) iff (3): We note first that condition (2) is equivalent to the following variant:

(2′) For any family $\{x_{j,k} : (j,k)\in J\times K\}$ in L such that $\{x_{j,k} : k\in K\}$ is directed for all $j\in J$, the following identity holds:

$$(\mathrm{DD}_0) \qquad \bigwedge_{j\in J}\bigvee_{k\in K} x_{j,k} = \bigvee_{f\in K^J}\bigwedge_{j\in J} x_{j,f(j)}$$

Indeed (DD) obviously implies (DD_0); conversely suppose that (DD_0) is satisfied and that $\{x_{j,k} : j\in J,\ k\in K(j)\}$ is given as in (2). Then define the set $K = \bigcup_{j\in J} K(j)$ and for $(j,k)\in J\times K$ define

$$\begin{aligned} y_{j,k} &= x_{j,k}, && \text{if } k\in K(j);\\ &= 0, && \text{otherwise.}\end{aligned}$$

Then $\bigvee_{k\in K(j)} x_{j,k} = \bigvee_{k\in K} y_{j,k}$ for all $j\in J$, and

$$\bigvee_{f\in M}\bigwedge_{j\in J} x_{j,f(j)} = \bigvee_{g\in K^J}\bigwedge_{j\in J} y_{j,g(j)},$$

and thus the desired equation (DD) for the $x_{j,k}$ follows from (DD_0) for the $y_{j,k}$.

The equivalence of (DD_0) and (DD*) is easily seen via O-1.5, and we leave the details to the reader. □

We remark that an alternate proof that (2) implies (1) in 2.3 can be given utilizing the characterization of continuous lattices of 2.1(3). For this purpose let J be the set if all ideals j of L with $x\leq\sup j$. Then (DD) shows that the ideal I that is the intersection of all the ideals in J also satisfies $x\leq\sup I$. This is just what we need to apply 2.1(3).

Before we move on to applications, we observe how the equations in Theorem 2.3 relate to traditional lattice-theoretical concepts. In O-4.3 we noted that meet-continuity generalized the property of being a complete Heyting algebra (that is, of satisfying a general distributive law O-3.16(3)). In a similar vein we observe that 2.3(2) generalizes the most restrictive type of general distributivity, which we put on record here:

2.4. DEFINITION. A lattice is called *completely distributive* iff it is complete and for any family $\{x_{j,k} : j\in J,\ k\in K(j)\}$ in L the identity

$$(\mathrm{CD}) \qquad \bigwedge_{j\in J}\bigvee_{k\in K(j)} x_{j,k} = \bigvee_{f\in M}\bigwedge_{j\in J} x_{j,f(j)}$$

holds, where M is the set of functions defined on J with values $f(j)\in K(j)$. As usual, we could write $\leq$ in place of $=$. □

2.5. COROLLARY. *Every completely distributive lattice is continuous.* □

Completely distributive lattices are fairly special; various characterizations are known which do not properly belong to our topic. However, later we will give a characterization theorem for complete distributivity in continuous lattices.

Now we exploit the fundamental Theorem 2.3. Our first observation is that the algebraic operations occurring in the equational characterization of continuous lattices through equation (DD) are *arbitrary* infs and *directed* sups. If we want to define homomorphisms between continuous lattices at this point, it is then clear that these are the algebraic infinitary operations which should be preserved by such homomorphisms. Therefore we make the following definition.

2.6. DEFINITION. If S and T are continuous lattices, then a function $g : S \to T$ is called a *morphism of continuous lattices*, or, briefly, a *homomorphism*, if it preserves arbitrary infs and directed sups. A continuous lattice T is called a *homomorphic image* of a continuous lattice S iff there is a surjective homomorphism $g : S \to T$. A subset S of a continuous lattice T is called a *subalgebra* (*of the continuous lattice* T) iff the inclusion map $S \to T$ is a homomorphism (that is, iff S is closed in T under the formation of arbitrary infs and directed sups.) □

In due time, notably in Chapter IV, we will use more systematically the language of category theory; the present terminology suffices for the purposes at hand, and we are able to formulate the most important consequences of 2.3.

2.7. THEOREM. *The class of continuous lattices is closed under the formation of arbitrary products, subalgebras and homomorphic images. Specifically we have the following conclusions*:

(i) *If* $\{L_j : j \in J\}$ *is a family of continuous lattices, then the cartesian product* $\times_{j \in J} L_j$ *is a continuous lattice (relative to the componentwise partial order)*;

(ii) *If* L *is a continuous lattice and* S *a subalgebra of* L, *then* S *is a continuous lattice in the induced order*;

(iii) *If* L *is a continuous lattice and* S *a poset and if* $g : L \to S$ *is a surjective homomorphism, then* S *is a continuous lattice.*

Proof. (i) Since all operations in the cartesian product are componentwise, then any equation which holds in each factor holds in the product. Since the product of complete lattices is complete, 2.3 proves the claim.

(ii): Firstly, S is a complete lattice, since it is closed in L under arbitrary infs (O-2.4). If $x_{j,k} \in S$ is a family satisfying the conditions of 2.3(2), then both sides of the equation (DD) in 2.3(2) are contained in S (since S is closed under infs and directed sups). Since the equation holds in L, it then holds in S, and thus S is a continuous lattice by 2.3.

(iii): Let $X \subseteq S$ and set $Y = g^{-1}(X)$. Since g is surjective, $X = g(Y)$. As L is a complete lattice, $y = \inf Y$ exists in L. Since g preserves infs, then

$$g(y) = g(\inf Y) = \inf g(Y) = \inf X.$$

Hence, S is a complete lattice by O-2.2. Now let $x_{j,k}$ be a family in S satisfying the conditions of 2.3(2). Let d be the lower adjoint of g, which exists by O-3.5.

If we set $y_{j,k} = d(x_{j,k})$, then the family $y_{j,k}$ satisfies the hypotheses of 2.3(2) in L, since d is order preserving. Thus the $y_{j,k}$ satisfy equation (DD) of 2.3. Now we apply g to both sides of this equation and obtain equation (DD) for the family $x_{j,k}$ (since g commutes with all $\bigwedge$ and all $\bigvee^{\mathrm{up}}$ and satisfies $gd = 1$ by O-3.7). It follows that S is a continuous lattice by 2.3. □

2.8. PROPOSITION. *Let $g : \mathrm{L} \to \mathrm{T}$ be a homomorphism between continuous lattices. Then the kernel $R = (g\times g)^{-1}(\Delta)$, where Δ = diagonal in $\mathrm{T}\times\mathrm{T}$, is a subalgebra of $\mathrm{L}\times\mathrm{L}$.*

Proof. Let $X \subseteq R$ and $(x,y) = \inf X$. Then $(g(x),g(y)) = (g\times g)(\inf X) = \inf (g\times g)(X)$, since g preserves infs. But $(g\times g)(X) \subseteq \Delta$, whence we see that $(g(x),g(y)) \in \Delta$, since the diagonal is closed under infs. Thus $g(x) = g(y)$, and so $(x,y) \in R$. The proof that R is closed under directed sups is completely analogous. □

We will see to what extent the converse holds in 2.11 below.

Some very important and frequently used consequences of these results pertain to projections and closure and kernel operators (O-3.8). The remainder of the section is devoted to them.

We turn to closure operators first. We recall from O-3.13 and the definition preceding it that a closure system in a complete lattice L is a subset which is closed under arbitrary infs. In O-3.14 it was established that there is a bijective correspondence between the closure operators of L which preserve directed sups and the closure systems which are closed under directed sups. If L is a continuous lattice, then these closure systems are precisely the subalgebras by 2.6. This gives immediately the following classification of closure operators preserving directed sups by subalgebras.

2.9. PROPOSITION. *Let L be a continuous lattice. Then the assignment $c \mapsto c(\mathrm{L})$, which associates with a closure operator $c : \mathrm{L} \to \mathrm{L}$ of L its image, induces a bijection from the set of all closure operators of L which preserve directed sups onto the set of subalgebras of L.* □

The next step is to consider kernel operators.

2.10. LEMMA. *If $k : \mathrm{L} \to \mathrm{L}$ is a kernel operator on a continuous lattice and if k preserves directed sups, then $k(\mathrm{L})$ is a continuous lattice.*

Proof. Apply 2.7(iii) to the corestriction $k^\circ : \mathrm{L} \to k(\mathrm{L})$, recalling O-3.12(i) and (iv). □

We now can give a more complete picture of congruences on continuous lattices and how they relate to kernel operators (while, as we recall from 2.9, closure operators relate to subalgebras).

2.11. THEOREM. *Let L be a continuous lattice and $R \subseteq \mathrm{L}\times\mathrm{L}$ an equivalence relation. Let L/R be the quotient set. Then the following conditions are* *equivalent:*

(1) *R is a subalgebra of* $L \times L$.

(2) L/R *is a continuous lattice in such a way that the quotient map* $g : L \to L/R$ *is a homomorphism.*

(3) *There is a kernel operator k on* L *preserving sups of directed sets such that* $R = (k \times k)^{-1}(\Delta)$ *where Δ is the diagonal of* $L \times L$.

Proof. (1) implies (3): Let $R(x) = \{y \in L : xRy\}$. Then $\{x\} \times R(x) = (\{x\} \times L) \cap R$, and, by (1), $R(x)$ is closed under infs of nonempty sets. Define the map $k : L \to L$ by $k(x) = \min R(x)$. Then $k(x) \in R(x)$ and $k(x) \leq x$.

Suppose next that $x \leq y$, then $xy = x$. Since R is a semilattice congruence, $R(x)R(y) \subseteq R(xy)$; whence

$$k(x) = k(xy) \leq k(x)k(y) \leq k(y).$$

This shows that k is monotone. Since $R(k(x)) = R(x)$ we have $k^2 = k$. Thus k is a kernel operator.

Let D be directed in L and set $d = \sup D$. Let $d^* = \sup k(D) \leq k(d)$. We claim that $k(d) \leq d^*$. Indeed for all $x \in D$ we have $(x, k(x)) \in R$ and so the set $\{(x, k(x)) : x \in D\}$ is directed. Since R is closed with respect to sups of directed sets we conclude $(d, d^*) = \sup_{x \in D}(x, k(x)) \in R$. Thus $d^* \in R(d)$, whence $k(d) = \min R(d) \leq d^*$ as was claimed. Thus k preserves sups of directed sets.

Finally, if $(x,y) \in R$, then $k(x) = k(y)$, and vice versa, whence we see that $(k \times k)^{-1}(\Delta) = R$.

(3) implies (2): The corestriction $k^\circ : L \to k(L)$ factors canonically through the quotient map $g : L \to L/R$ with a bijection $f : L/R \to k(L)$. This means that $k^\circ = fg$ with $f(R(x)) = k(x)$. By 2.10, $k(L)$ is a continuous lattice, and by O-3.12(iv), k° preserves infs. Since k preserves directed sups and $k(L)$ is sup-closed in L by O-3.12(i), then k° preserves sups of directed sets. If we transport the lattice structure of $k(L)$ to L/R via f^{-1}, then L/R is a continuous lattice such that g preserves all infs and sups of directed sets.

(2) implies (1): Immediate from 2.8. □

2.12. DEFINITION. If L is a continuous lattice, then a subset R of $L \times L$ which satisfies the conditions of 2.11 is called a *congruence* (*of continuous lattices*), and L/R is called a (*continuous lattice*) *quotient.* □

Evidently, the homomorphic images of a continuous lattice and its quotients are practically the same thing on account of the canonical factorization theorem for homomorphisms.

The following is now a parallel to 2.9:

2.13. COROLLARY. *Let* L *be a continuous lattice. Then* $k \mapsto (k \times k)^{-1}(\Delta)$, *where Δ is the diagonal of* $L \times L$, *associates with a kernel operator its kernel congruence and induces a bijection from the set of all kernel operators of* L *which preserve directed sups onto the set of* ***all*** *congruences of* L. □

Lastly we consider projections, the common generalization of closure and kernel operators. The next theorem is of remarkable generality and will turn out to be very useful.

2.14. THEOREM. *If* L *is a continuous lattice and* $p : L \to L$ *a projection operator* (O-3.8) *preserving sups of directed sets, then* $p(L)$ *is a continuous lattice (relative to the induced order).*

Proof. By O-3.11, the set $L_k = \{x \in L : p(x) \leq x\}$ is closed under infs and directed sups, hence is a continuous lattice by 2.7(ii). By O-3.11(i), the restriction and corestriction $p_k : L_k \to L_k$ of p is a kernel operator with im p_k = im $p = p(L)$. But p_k preserves directed sups, since p does and L_k is closed under directed sups; hence 2.10 shows that im p_k is a continuous lattice. □

It is useful to recall from O-3.12(i), that $p(L)$ = im p_k is sup-closed in L_k, hence is closed in L under sups of directed sets. There is an analog to 2.14 which is (as is usual in our context) not exactly symmetric to it:

2.15. PROPOSITION. *If* L *is a continuous lattice and* $p : L \to L$ *is a projection operator preserving infs, then* $p(L)$ *is a continuous lattice. Moreover* $p(L)$ *is closed in* L *under infs.*

Proof. By O-3.11(ii) and (iii) L_c is closed under sups and infs. Thus L_c is a continuous lattice by 2.7(ii). The map $p_c : L_c \to L_c$ is a closure operator which preserves infs, since p preserves infs and L_c is inf closed.

The corestriction $p_c^\circ : L_c \to \text{im } p_c$ then preserves infs since im p_c is inf closed in L_c by O-3.12(i), and it preserves sups by O-3.12(iii). Thus 2.7(iii) applies to show that im p_c is continuous lattice. But im $p_c = p(L)$ by O-3.11(i). By O-3.12(i), $p(L)$ = im p_c is closed in L_c under infs, and L_c is closed in L under infs, thus $p(L)$ is closed in L under infs. □

EXERCISES

In the following exercises we construct some further examples of continuous lattices by utilizing kernel operators. We will, however, reprove these results in Chapter II with different methods.

2.16. EXERCISE. (i) Let S be any poset and T a continuous lattice. Let $(S \to T)$ denote the poset of all order-preserving maps $f : S \to T$ with the pointwise order. Then $(S \to T)$ is a continuous lattice.

(ii) Let S be a continuous poset. For $f \in (S \to T)$ define $k(f) : S \to T$ by

$$k(f)(s) = \sup f(\Downarrow s).$$

Then $k(f) \in (S \to T)$ and $k : (S \to T) \to (S \to T)$ is a kernel operator. Further, k preserves directed sups. In particular, im k is a continuous lattice (2.10).

(iii) Let S be a continuous poset. $[S \to T]$ denotes the poset of all those maps $f \in (S \to T)$ with $f = k(f)$. Each such f preserves directed sups.

(HINT: (i): Note that $(S \to T)$ is a subalgebra of T^S, which is a continuous lattice by 2.7(1).

(ii): Use the interpolation property of $\ll$ for the proof of $k^2 = k$.

(iii): Let s_j be a monotone net with $\sup_j s_j = s$ in S. Clearly we have then $\sup_j f(s_j) \leq f(s)$. The converse requires a frequently used trick: Take an arbitrary element

$$x \ll f(s) = k(f)(s) = \sup f(\downarrow s)$$

and show $x \leq \sup_j f(s_j)$; since x is arbitrary, $f(s) \leq \sup_j f(s_j)$ will follow. To accomplish the claim, use 1.1 to find $s^* \ll s$ with $x \leq f(s^*)$. Since $s^* \ll \sup_j s_j$, there is an i with $s^* \ll s_i$ (1.27). Thus

$$x \leq f(s^*) \leq k(f)(s_i) = f(s_i) \leq \sup_j f(s_j).) \quad \square$$

We summarize the main outcome of this discussion in the following exercise and indicate some applications:

2.17. EXERCISE. (i) If S is a continuous poset and T a continuous lattice, then the poset $[S \to T]$ of *all* functions $S \to T$ preserving directed sups is a continuous lattice. (Note that it is closed in T^S under arbitrary sups but not under infs in general!)

(ii) Since $\mathbb{R}$ is a continuous semilattice and $\mathbb{I} = [0,1]$ a continuous lattice, then $[\mathbb{R} \to \mathbb{I}]$ is a continuous lattice. This is the space of all monotone, lower semi-continuous functions $\mathbb{R} \to \mathbb{I}$. Those functions $F \in [\mathbb{R} \to \mathbb{I}]$ which are such that $F(x) \to 0$ as $x \to -\infty$ and $F(x) \to 1$ as $x \to +\infty$ are precisely the distribution functions of real random variables. They form a sublattice $P(\mathbb{R})$ which is closed under arbitrary sups of sets with upper bounds and under arbitrary infs of sets with lower bounds. But there are no elements $F, G \in P(\mathbb{R})$ with $F \ll G$. Thus $P(\mathbb{R})$ fails totally to be a continuous poset.

(iii) By contrast, however, the set $P(\mathbb{I}) \subseteq [\mathbb{I} \to \mathbb{I}]$ consisting of all $F : \mathbb{I} \to \mathbb{I}$ with $F(1) = 1$ is the set of distribution functions of probability measures on the unit interval, and it is closed under arbitrary infs and sups; hence, it is a continuous lattice. $\square$

One could call $P(\mathbb{I})$ *the random unit interval.* Notice, however, that the partial order on the corresponding probability measures is the opposite of the partial order induced from $[\mathbb{I} \to \mathbb{I}]$; indeed if X and Y are random variables on $\mathbb{I}$ with distribution functions F_X and F_Y, then $F_X \leq F_Y$ means that X is likely to take larger values than Y: we should write $Y \leq X$.

We comment next on continuous semilattices in the light of the topic of the present section.

2.18. EXERCISE. Let L be an up-complete semilattice (O-2.11). Then the following conditions are equivalent:

(1) L is a continuous semilattice.

(2) For each $x \in L$, the set $\downarrow x$ is the smallest ideal I with $x \leq \sup I$.

(3) For each $x \in L$ there is a smallest ideal I with $x \leq \sup I$.

(4) The sup map $r = (I \mapsto \sup I) : \mathrm{Id}\, L \to L$ has a lower adjoint.

These conditions imply (5) below, and if L is a complete semilattice, then (1)–(5) are equivalent:

(5) The sup map $r : \mathrm{Id}\, L \to L$ preserves existing infs of nonempty sets and all existing sups.

(HINT: Adjust the proof of 2.1. For (5) implies (4) use O-3.26.) □

2.19. EXERCISE. Every continuous semilattice is meet-continuous.

(HINT: See O-4.6 and use 2.18 as a substitute for 2.1 in 2.2. Or argue directly.) □

2.20. EXERCISE. Let S be a complete semilattice (O-2.11). Then S is a complete-continuous semilattice (1.26) iff condition (2) of 2.3 is satisfied. □

2.21. EXERCISE. Theorems 2.7, 2.11, 2.14 and Propositions 2.9 and 2.13 persist for complete-continuous semilattices. □

Despite these successes, there remain some questions:

PROBLEM. Is there an equational characterization of continuous semilattices?

PROBLEM. Do any of the conclusions of Theorems 2.7, 2.11 and 2.14 and of Propositions 2.9 and 2.13 persist for continuous semilattices?

The following exercise discusses some material which is related to the context of large distributive laws and auxiliary relations.

2.22. EXERCISE. Let L be a complete lattice. Let $\mathcal{M}$ be a set of subsets of L satisfying the following conditions:

(A_1) If $\mathcal{A} \subseteq \mathcal{M}$ and $\{\sup X : X \in \mathcal{A}\} \in \mathcal{M}$, then $\bigcup \mathcal{A} \in \mathcal{M}$;

(A_2) If $\mathcal{A} \subseteq \mathcal{M}$, then $\{\inf f(\mathcal{A}) : f \in \mathrm{Sel}(\mathcal{A})\} \in \mathcal{M}$,

where $\mathrm{Sel}(\mathcal{A}) = \{f : f \text{ is a selection function } \mathcal{A} \to \bigcup \mathcal{A}\}$.

G. Bruns [1962] calls such a set *distributively closed.* One calls a lattice *$\mathcal{M}$-distributive* if for any set $\mathcal{A} \subseteq \mathcal{M}$

$$\inf\{\sup X : X \in \mathcal{A}\} = \sup\{\inf f(\mathcal{A}) : f \in \mathrm{Sel}(\mathcal{A})\}.$$

(If $\mathcal{M} = 2^L$, then we retrieve complete distributivity. If $\mathcal{M}$ is the smallest distributively closed subset containing all finite sets, then one obtains the (F)-distributivity of S. Papert [1959]. The set $\mathfrak{D}$ of all directed subsets of L is *not* distributively closed; however, the distributive law (DD) in 2.3 would be called $\mathfrak{D}$-distributivity in our present context.)

Define two relations on L as follows (the first follows S. Papert, loc. cit., p. 174, the second G. Bruns, loc. cit., p. 4):

$x \swarrow y$ iff for all $X \in \mathcal{M}$ with $y = \sup X$ one has $x \leq a$ for some $a \in X$;

$x \dashv y$ iff for all $X \in \mathcal{M}$ with $y \leq \sup X$ one has $x \leq a$ for some $a \in X$.

(i) Then the relations $\swarrow$ and $\dashv$ satisfy 1.9(i), (ii), and (iv) and (INT) (but not in general 1.9(iii)).

(ii) Both relations are approximating in any $\mathcal{M}$-distributive lattice.

(iii) For $x \in \mathrm{L}$ there is a set $X(x) \in \mathcal{M}$ such that $\downarrow X(x) = \{y : y \prec x\}$ with $\prec$ equal to $\swarrow$ or $\dashv$, respectively. □

We have seen in Corollary 2.2 that every continuous lattice is meet continuous; the converse is incorrect. Indeed not even a complete Boolean algebra need be a continuous lattice. It is therefore useful to have sufficient conditions which will ensure that meet continuity implies continuity.

2.23. EXERCISE. A meet-continuous lattice L is continuous if it satisfies at least one of the following two conditions:

(a) L does not contain a free semilattice with infinitely many generators.

(b) L does not contain an infinite antichain.

(HINT: (a): This is established by proving the following theorem:

THEOREM. *Let* L *be a meet-continuous lattice and let* x *and* y *be elements such that* $\downarrow x \subseteq \downarrow y$ *but* $x \notin \downarrow y$. *Then* $(\downarrow x \backslash \downarrow y) \cup \{1\}$ *contains a free semilattice with infinitely many generators.*

Proof. Suppose that F is a free semilattice such that $F \backslash \{1\}$ is contained in $\downarrow x \backslash \downarrow y$ and that X is the generating set in $F \backslash \{1\}$. Suppose that X is finite. Then $z = \inf F$ is not in $\downarrow x$. Hence, since L is meet-continuous, there is a directed set D with $\sup D = x$ but $D \cap \uparrow z = \emptyset$. For each $f \in F$ we know that $\sup fD = f$. Since F is finite, we find an element $b \in D$ such that $b \leq d \in D$ implies $fd \neq gd$ for all $f \neq g$ in F. Since $\sup zD = z$ and $z \not\leq y$, there is some $p \in D$ with $b \leq p$ such that $pz \not\leq y$. Then $F \cup \{p\}$ is a free set, and the semilattice F' generated by this set is contained in $\downarrow x \backslash \downarrow y$. By induction we obtain a countably generated infinite free semilattice contained in $\downarrow x \backslash \downarrow y$.

(b): The generating set of a free semilattice is an antichain.) □

NOTES

The lead results in Theorems 2.1 and 2.3 again have two different sources. Theorem 2.1 is, in essence, the principal result of Hofmann and Stralka [1976] from which most of the other results in that paper are derived, notably the one we will encounter in Chapter VI. The characterization of continuous lattices through equations as expressed in Theorem 2.3 is due to Alan Day [1975]. His proof is different from ours. He obtained the equational characterization in the course of identifying the class of continuous lattices and the class of homomorphisms of continuous lattices (2.6) as a category which is equivalent to the category of algebras of the filter monad over sets (and the open filter monad over T_0-spaces). Of course one may interpret this set-up as identifying free continuous lattices over a set, as was pointed out by D. Scott [SCS-15]. A. Day explored the issue further in [SCS-18]. Independently of these developments, O. Wyler also identified continuous lattices as algebras of filter monads sometime in 1975-76.

The distributive law (DD) of 2.3 is of a type considered in a systematic fashion by G. Bruns [1961] and [1962]. However, the case of continuous lattices is not subsumed in Bruns' work, and before him S. Papert used certain (almost) auxiliary relations which are approximating and satisfy the interpolation property (see 2.22). In the case of completely distributive lattices, the relation $\dashv$ of 2.22 was introduced by G.N. Raney [1953]. He showed that it was approximating iff the lattice was completely distributive, and he observed that it satisfied the interpolation property in this case. With these tools he showed that a complete lattice is completely distributive iff it can be embedded into a product of complete chains under preservation of arbitrary sups and infs. (See 3.19 and IV-2.29 below.)

That the closure properties of the class of continuous lattices which are expressed in Theorem 2.7 would be important from the viewpoint of universal algebra was remarked in Scott [SCS-15]. The fact that quotients of continuous lattices are continuous is probably the hardest of the closure properties; the stability under formation of products and subalgebras could be derived relatively easily directly from our Definition 1.7; this is not the case with the quotients. This had also been the harder part of the theory of compact semilattices with small sub-semilattices, Lawson [1969], which we know today is equivalent to the theory of of continuous lattices. (See 3.15 ff. and IV-2.29 below.)

Scott had emphasized all along the significance of projections (retracts) on continuous lattices, notably those which preserve directed sups. Theorem 2.14 is from [SCS-15]. The useful statements 2.9 and 2.15 concerning closure operators are published here for the first time.

The result in Exercise 2.17 is a mild generalization of a principal result of Scott [1972a, p. 112, Theorem 3.3]. A systematic treatment will follow in Chapter II. The random unit interval was discussed by Hofmann and Liukkonen in [SCS-16].

The fact that all continuous lattices are meet-continuous (2.2) was in principle known in Scott [1972a, p. 106, Prop. 2.7] (see also the direct proof in 1.12), as Isbell points out [1975b, p. 46].

The results in Exercise 2.23 are due to Lawson [SCS-23]; however, the sufficient condition (b) was independently rediscovered by Heiko Bauer with an independent (and more complicated) proof in [SCS-45].

3. IRREDUCIBLE ELEMENTS

In a semilattice an element is irreducible if it is not the meet of two larger elements. These elements play an important role in lattice theory, notably for distributive lattices, where they are exactly the prime elements; they are at the basis of all of the spectral theory and of the representation theorems of distributive lattices. Irreducible elements exist abundantly in all finite lattices, and it is one of the important features of continuous lattices that this property persists.

We first introduce some necessary machinery to prepare the way for the development of the theory of irreducible elements. This early material will be better motivated in Chapter II when topologies are introduced.

3.1. DEFINITION. Let L be a complete lattice. An upper set $U = \uparrow U$ will be called *open* iff for each $u \in U$ there is an element $v \in U$ with $v \ll u$ (equivalently: $\Downarrow u \cap U \neq \emptyset$). □

3.2. REMARK. *An upper set U in a complete lattice is open iff*

$$U = \bigcup\{\Uparrow v : v \in U\}. \ \square$$

In particular, a filter is open iff it is open in the sense of 3.1. Using the interpolation property (1.18) we see immediately that in a continuous lattice, all sets $\Uparrow x$ are open. One must be careful to note that even in a continuous lattice the sets $\Uparrow x$ are not generally filters:

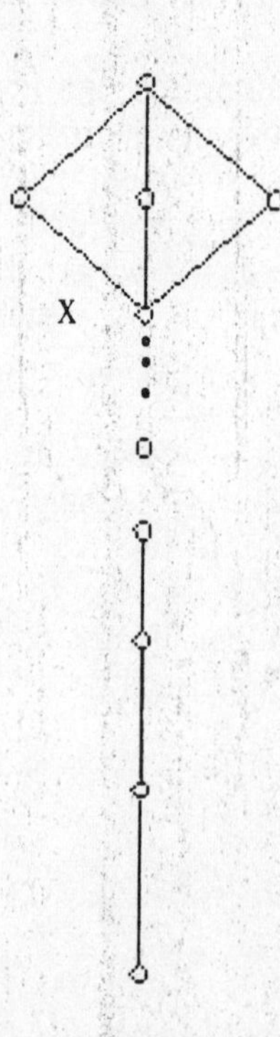

Nevertheless, there are still enough open filters:

3.3. PROPOSITION. *Let* $x \ll y$ *in a continuous lattice. Then there is an open filter* U *with* $y \in U \subseteq \Uparrow x$.

Proof. By the interpolation property (1.18) we construct inductively a decreasing sequence of elements y_n with

$$x \ll \ldots \ll y_n \ll y_{n-1} \ll \ldots \ll y_1 = y.$$

Set $U = \bigcup\{\uparrow y_n : n = 1,2, \ldots\}$. As an ascending union of filters, U is a filter, and it clearly contains $y = y_1$. Since each $\uparrow y_n$ is in $\Uparrow x$ by 1.2(ii) we have $U \subseteq \Uparrow x$. Finally let $u \in U$. Then $u \in \uparrow y_n$ for some n; whence $y_{n+1} \ll y_n \leq u$, and so $\Downarrow u \cap U \neq \emptyset$. □

The significance of the openness of an upper set U is that its complement must have maximal elements:

3.4. LEMMA. *Let* U *be an open upper set in a complete lattice. Then for any* $x \in L \setminus U$ *there is an* $m \in L \setminus U$ *with* $x \leq m$ *and* m *maximal in* $L \setminus U$.

Proof. By the Hausdorff Maximality Principle, there exists a maximal chain $C \subseteq L \setminus U$ containing x. Let $m = \sup C$. If $m \in U$, then by 3.1 there is a $u \in U$ with $u \ll m$. But then by 1.1, there is a $d \in C$ with $u \leq d$ and, since U is an upper set, $d \in U$ would follow contradicting $C \subseteq L \setminus U$. Thus $m \in L \setminus U$. Since C is a maximal chain, not only is $m \in C$ but m is maximal in $L \setminus U$. □

3.5. DEFINITION. An element p in a lattice (or semilattice) is called (*meet*) *irreducible* iff the relation $p = xy$ always implies $x = p$ or $y = p$. The set of all irreducible elements is written IRR L. □

Note that the elements p for which $\uparrow p \setminus \{p\}$ is a filter are precisely the irreducible elements. *Join irreducibles* are defined dually.

3.6. PROPOSITION. *Let* L *be a lattice and* $p \in L$. *If* p *is maximal in* $L \setminus F$, *for some filter* F, *then* p *is irreducible.*

Proof. Since p is maximal in $L \setminus F$, we see that $\uparrow p \setminus \{p\} = \uparrow p \cap F$. Since this is a filter, we conclude that p is irreducible. □

The next theorem is important because it guarantees an abundance of irreducibles in any continuous lattice.

3.7. THEOREM. *Suppose that* x *and* y *are elements of a continuous lattice with* $y \not\leq x$. *Then there is an irreducible element* p *with* $x \leq p$ *and* $y \not\leq p$.

Proof. By 1.6, there is an element $u \ll y$ with $u \not\leq x$. By Proposition 3.3, there is an open filter U with $y \in U$ and $U \subseteq \uparrow u$. Then $x \notin U$ and, by 3.4 and 3.6, there is an irreducible element p with $x \leq p \notin U$. Since $y \in U$, we then have $y \not\leq p$. □

This result may be rephrased in another convenient fashion (3.10).

3.8. DEFINITION. A subset X of a lattice L is said to be *order generating* iff $x = \inf (\uparrow x \cap X)$ for all $x \in L$. □

Note that it is also true that $\inf (\uparrow x \cap X) = \inf (\uparrow x \cap (X \setminus \{1\}))$.

3.9. PROPOSITION. *For a subset X of a complete lattice L, the following statements are equivalent:*

(1) *X is order generating;*

(2) *Every element of L can be written as an inf of a subset of X;*

(3) *L is the smallest subset containing X closed under arbitrary infs;*

(4) *Whenever $y \not\leq x$, then there is a $p \in X$ with $x \leq p$ but $y \not\leq p$.*

Proof. (1) implies (2): Immediate from the definition.

(2) iff (3): A standard lattice-theoretical argument.

(2) implies (4): Let $y \not\leq x$. By assumption $x = \inf P$ for some $P \subseteq X$. But then $y \not\leq p$ for some $p \in P$, and the conclusion follows.

(4) implies (1): Let $y = \inf (\uparrow x \cap X)$ for a given $x \in L$. If $y \neq x$, then $y \not\leq x$; and so we can take p as in (4). But $p \in (\uparrow x \cap X)$, and therefore $y \leq p$. This contradiction proves that $x = y$; hence, X is order generating. □

3.10. COROLLARY. *In a continuous lattice L, the set* IRR $L \setminus \{1\}$ *of nonidentity irreducibles is order generating.* □

At this point we specialize our discussion mainly to ***distributive*** lattices. In this context we introduce a type of element, called prime; in arbitrary lattices every prime element is irreducible, but the converse fails without distributivity.

3.11. DEFINITION. An element p in a lattice L is called *prime* iff the relation $xy \leq p$ always implies $x \leq p$ or $y \leq p$. The set of prime elements is denoted PRIME L. An element is *co-prime* iff it is a prime of L^{op}. □

Notice that we allow 1 to be prime—not all authors agree. For the record we note some general facts about primality.

3.12. REMARK. *Let $p \neq 1$ in a lattice L. Then the following statements are equivalent:*

(1) *p is prime;*

(2) *the function $f\colon L \to 2$ (where $2 = \{0,1\}$ is the two-element lattice) defined by $f(x) = 0$ iff $x \leq p$ is a lattice homomorphism;*

(3) *$L \setminus \downarrow p$ is a filter;*

If L is distributive, then the above are equivalent to:

(4) *p is irreducible;*

If L *is Boolean, then the above are equivalent to*:

(5) *p is a co-atom (that is, a maximal nonunit element)*;

If L *is continuous and distributive, then the above are equivalent to*:

(6) *p is maximal in the complement of an open filter.*

Proof. We leave the equivalence of (1)–(5) and the implication (6) implies (4) as exercises (cf. 3.6).

(4) implies (6): Assume now that L is continuous and distributive and set $U = L\setminus\downarrow p$. Then $U = \{x\in L \mid p\vee x\in\uparrow p\setminus\{p\}\}$. Since p is irreducible by (4) and $s \mapsto p\vee s$ preserves finite infs by distributivity, U is a filter. If $u\in U$, then $u\not\leq p$, and thus by continuity of L, there is an $x\ll u$ with $x\not\leq p$ (see 1.6(A_1)). Then $x\in U$ and thus U is open. Clearly p is maximal in $\downarrow p = L\setminus U$. □

3.13. LEMMA. *If* PRIME L *is order generating in a complete lattice* L, *then* L *is distributive and meet-continuous—hence, a cHa.*

Proof. By 3.12(2) and 3.9 the lattice homomorphisms $f : L\to 2$ determined by primes separate the points of L. If H is the set of all of these homomorphisms, then the function $x \mapsto (f(x))_{f\in H} : L\to 2^H$ is an injective lattice homomorphism. But 2^H is distributive, hence so is L. By the special choice of H, this mapping also preserves arbitrary sups. So L is meet-continuous since 2^H is. □

Because $L^{op} \cong L$ for any Boolean algebra, we find nontrivial primes in a Boolean algebra iff we find atoms. The Boolean algebra $\mathcal{O}_{reg}([0,1])$ is a complete Boolean algebra without primes. Hence, the converse of 3.13 fails even in very special complete distributive lattices. It is fundamental for the theory of continuous distributive lattices that this converse holds.

3.14. THEOREM. *Let* L *be a continuous lattice. Then the following conditions are equivalent*:

(1) L *is distributive*;

(2) L *is a Heyting algebra*;

(3) PRIME L *is order generating.*

Proof. (1) implies (3): By 3.10 and 3.12(4).
(3) implies (2): By 3.13.
(2) implies (1): By definition. □

The sublattice $[0,1[^2\cup\{(1,1)\}$ of the square $[0,1]^2$ in the product order is a complete distributive lattice in which (1,1) is the only prime (cf. O-4.5(1))—thus, 3.14 may very well fail for a complete distributive lattice. In fact, it may fail for a complete Heyting algebra and even for a complete Boolean algebra. (Boolean algebras which are continuous lattices are characterized in the next section.) All of this shows that continuity is quite essential in 3.14.

If L is a chain, then L = IRR L, and this property is evidently characteristic for chains. If L = [0,1], let X be the set of rationals, Y the set of irrationals. Then each of the disjoint sets X and Y is order generating and neither is minimal relative to this property. In general, the property of IRR L in 3.8 does not characterize the set, neither is it a minimal set which order generates L. We will see later that *algebraic* lattices have a minimal order generating set (4.23). We will also show that the closure of IRR L relative to a suitable topology is the smallest *closed* order generating set for this topology.

We saw in 2.5 that every completely distributive lattice is continuous. Now we can give a sharper description of completely distributive lattices in terms of continuous lattices and primes and co-primes (primes of the dual).

3.15. Theorem. *Let* L *be a complete lattice. Then the following statements are equivalent*:

(1) L *is continuous and every element is the sup of co-primes*;

(2) L *is completely distributive*;

(3) L *is distributive and both* L *and* L^{op} *are continuous lattices.*

Proof. (2) implies (3): One knows that L is completely distributive iff L^{op} is completely distributive. Then 2.5 yields the desired implication.

(3) implies (1): By 3.14 applied to L^{op}.

(1) implies (2): We remark first that, in a lattice satisfying (1), every element is the sup of co-primes way below it; because *all* the elements way below it are sups of co-primes. Thus, to verify the equation (CD) of 2.4, it is sufficient to show that every *co-prime* way below the left-hand side (*lhs*) is less than or equal to the right-hand side (*rhs*).

Let $x_{j,k}$ be given as in 2.4 and suppose that p is a co-prime element with $p \ll lhs$. Then $p \ll \bigvee_{k \in K(j)} x_{j,k}$ for all $j \in J$. By 1.5 we find, for each $j \in J$, a finite set $F \subseteq K(j)$ with $p \leq \bigvee_{k \in F} x_{j,k}$. But, since p is co-prime, there is in fact some element $k \in F$ with $p \leq x_{j,k}$; we denote this k by $f(j)$. By these choices we have found a function $f \in M$ such that $p \leq \bigwedge_{j \in J} x_{j,f(j)}$. This proves that $p \leq rhs$, and the proof is complete. □

It is interesting to remark that from G.N. Raney's classical theory of completely distributive lattices one knows that the equivalent conditions of 3.15 are also equivalent to embedding L into a direct product of complete chains under preservation of arbitrary sups and infs. (See e.g., Balbes and Dwinger [1974, p. 248]. We will present this result as Exercise IV-2.29 and 2.30.)

We also know that for any lattice L the complete lattice Id L of ideals is distributive iff L is distributive. In view of the importance for continuous lattices of the sup map on ideals, $I \mapsto \sup I$: Id L→L, (see 2.1 and compare (4.16), we need to view Id L in our present context. First a simple remark:

3.16. REMARK. *For an ideal I of a lattice* L *the following statements are equivalent*:

(1) $I \in \mathrm{PRIME}(\mathrm{Id}\, L)$;

(2) $L \backslash I$ *is a filter or is empty*;

(3) *If $xy \in I$, then $x \in I$ or $y \in I$ for all $x, y \in L$.*

Proof. (2) iff (3): Trivial.

(1) implies (3): Let I be a prime element of Id L and suppose $xy \in I$. Then $\downarrow x \cap \downarrow y = \downarrow xy \subseteq I$. Since I is a prime element in Id L, $\downarrow x \subseteq I$ or $\downarrow y \subseteq I$, which implies $x \in I$ or $y \in I$.

(3) implies (1): Let J and K be ideals of L such that $J \not\subseteq I$ and $K \not\subseteq I$. Then we find elements $x \in J \backslash I$ and $y \in K \backslash I$. By (3) we know $xy \notin I$, and it follows that $xy \in (J \cap K) \backslash I$. Hence, I is a prime element of Id L. □

3.17. DEFINITION. A *prime ideal* in a lattice is an ideal satisfying the equivalent conditions of 3.16. *Prime filters* are defined dually. □

It will be useful at this point to recall a basic mathematical concept. The well-known notion of a *filter of sets* is just what we mean by a filter in a powerset lattice. From any standard reference the reader may extract a proof of the following:

3.18. REMARK. *Let $\mathfrak{F}$ be a filter on a set X. The following statements are equivalent*:

(1) *$\mathfrak{F}$ is a prime filter in 2^X*;

(2) *For any subset of X, either it or its complement belongs to $\mathfrak{F}$*;

(3) *$\mathfrak{F}$ is a maximal proper filter in 2^X.* □

Recall, too, that filters satisfying the equivalent conditions of 3.18 are usually called *ultrafilters*. We record a standard Lemma:

3.19. LEMMA. *Let* L *be a distributive lattice, I an ideal and F a filter in* L *with $I \cap F = \emptyset$. Then there is a prime ideal P in* L *with $I \subseteq P$ and $P \cap F = \emptyset$.*

Proof. By Zorn's Lemma we find a maximal ideal P containing I and missing F. We claim that P is a prime ideal. To prove this we let $x, y \notin P$. The ideal generated by P and x as well as the ideal generated by P and y both meet F by the maximality of P. Thus there are some elements $u, v \in P$ with $u \vee x$, $v \vee y \in F$. Let $w = u \vee v$. Then $w \in P$, since P is an ideal; and we also have $w \vee x$, $w \vee y \in F$, since F is an upper set. From the fact that L is distributive and F is a filter we conclude that $w \vee (x \wedge y) = (w \vee x) \wedge (w \vee y) \in F$. Since $w \in P$, we cannot have $x \wedge y \in P$, because otherwise we would have $w \vee (x \wedge y) \in P \cap F = \emptyset$; but $x \wedge y \notin P$ is what we had to show. □

Note that if $\mathfrak{F}$ is a filter on X and $\mathfrak{I}$ is an ideal of 2^X disjoint from $\mathfrak{F}$, then by applying 3.19 to $(2^X)^{\mathrm{op}}$, we conclude that there is an ultrafilter containing the given filter $\mathfrak{F}$ but missing the ideal $\mathfrak{I}$. In the special case where $\mathfrak{I} = \{\emptyset\}$, we obtain the well-known fact that *every* proper filter may be extended to an ultrafilter.

Ultrafilters are frequently useful tools in the theory of continuous lattices. We illustrate this with the following two propositions, which also show how compactness theorems in topology often have more general formulations in terms of the way-below relation.

3.20. PROPOSITION. *Let U and V be open subsets in a topological space* X, *with $U \subseteq V$. The following statements are equivalent*:

(1) $U \ll V$;

(2) *Every proper filter containing U has a cluster point in V*;

(3) *Every ultrafilter containing U has a convergence point in V.*

Proof. (1) implies (2): Let $\mathfrak{F}$ be a proper filter with $U \in \mathfrak{F}$. If no member of V is a cluster point, then for every element $x \in V$, we can find an open set W_x containing x and a set $F_x \in \mathfrak{F}$ such that $W_x \cap F_x = \emptyset$. By hypothesis, because the W_x cover V, there are finitely many of them covering U. The finite intersection of the corresponding F_x has an empty intersection with U, which is a contradiction since this finite intersection is in $\mathfrak{F}$.

(2) implies (3): Immediate—since by 3.18(2) cluster points are points of convergence.

(3) implies (1): Let $\mathfrak{U}$ be an open cover of V, and assume that U has no finite subcover. Then the family of sets $U \setminus W$ with $W \in \mathfrak{U}$ generates a proper filter. Extend this to an ultrafilter $\mathfrak{F}$; it is the case that $U \in \mathfrak{F}$. By assumption, let $p \in V$ be a point of convergence. Now for some $W \in \mathfrak{U}$ we have $p \in W$; but $\mathfrak{F}$ converges to p, so $W \in \mathfrak{F}$. It then follows that both $U \cap W$ and $U \setminus W$ belong to $\mathfrak{F}$, a contradiction. □

The following result is a mild generalization of the classical Alexander's Lemma. The reader should recall the difference between a base and a subbase for a topology.

3.21. PROPOSITION. *Let $\mathfrak{B}$ be a collection of open subsets forming a* ***subbase*** *for the topology of a space* X, *and let U and V be open sets with $U \subseteq V$. Then a necessary and sufficient condition for $U \ll V$ is that every cover of V by members of $\mathfrak{B}$ has a finite subcover of U.*

Proof. The necessity is clear. For the sufficiency we use 3.20(3). Let $\mathfrak{F}$ be an ultrafilter with $U \in \mathfrak{F}$. Suppose no element of V is a convergence point of $\mathfrak{F}$. Then, if $x \in V$, there is a ***basic*** open set W_x containing x but not in $\mathfrak{F}$. Since W_x is a finite intersection of elements of $\mathfrak{B}$, and since $\mathfrak{F}$ is a filter, we can assume W_x to be ***subbasic***; that is, $W_x \in \mathfrak{B}$. Because $\mathfrak{F}$ is an ultrafilter, it follows that $U\setminus W_x \in \mathfrak{F}$. Because the W_x cover V, it follows by assumption that finitely many cover U. This means that a finite intersection of the $U\setminus W_x$ is empty, which is impossible because $\mathfrak{F}$ is a proper filter. □

The next proposition is essentially an abstract version of 3.21.

3.22. PROPOSITION. *Let x and y be elements in a complete distributive lattice. Then $x \ll y$ if and only if for every prime ideal P with $y \leq \sup P$ we have $x \in P$.*

Remark. Compare this statement with 1.5 in order to note that prime ideals suffice here to test the relation $x \ll y$.

Proof. "Only if" is clear from 1.5. In order to see that the new condition is sufficient, let I be an arbitrary ideal with $y \leq \sup I$; we must show that $x \in I$. Suppose not. Then we set $F = \uparrow x$ and apply 3.19 to find a prime ideal P with $I \subseteq P$ and $x \notin P$. But $I \subseteq P$ implies $y \leq \sup I \leq \sup P$. Hence, by our hypothesis, $x \in P$, and this is a contradiction which proves the claim. □

Each prime element p of L gives rise to a prime ideal $\downarrow p$ (as is immediate from 3.16(3)); but, conversely, if P is a prime ideal, then sup P ***need not*** be a prime element. If we look at the example following 3.2, then $x = \sup \Downarrow x$ is not prime, but $\Downarrow x$ is a prime ideal. This motivates the formulation of the following definition:

3.23. DEFINITION. An element p of a complete lattice is called *pseudoprime* if $p = \sup P$ for some prime ideal P. The set of pseudoprimes is called ΨPRIME L. □

By the preceding remarks PRIME L $\subseteq$ ΨPRIME L, and the containment is proper in general. In view of 2.1(2) an element p in a continuous lattice is pseudoprime iff there is a prime ideal P with $\Downarrow p \subseteq P \subseteq \downarrow p$. It is clear that for any complete lattice L the sup map *on ideals* maps PRIME (Id L) onto ΨPRIME L. In continuous lattices we have the following characterization of pseudoprimes:

3.24. PROPOSITION. *Let* L *be a distributive continuous lattice and $p \in$*L*. Then the following conditions are equivalent:*

(1) *p is pseudoprime;*

(2) *In any finite collection $x_1,\ldots,x_n \in$*L *with $x_1 \cdots x_n \ll p$ there is one of the elements with $x_j \leq p$;*

(3) *The filter generated by* L$\setminus\downarrow p$ *does not meet $\Downarrow p$.*

Remark. The implications (1) implies (2) and (2) iff (3) do not require distributivity.

Proof. Condition (2) says that no finite product of elements from $L\setminus\downarrow p$ is ever way below p. Therefore (2) and (3) are always equivalent.

(1) implies (2): Let p be pseudoprime and suppose that $x_1 \ldots x_n \ll p$. Let I be a prime ideal with $\sup I = p$. By 2.1(2) we have $\Downarrow p \subseteq I$, hence $x_1 \ldots x_n \in I$. Since I is prime, there is one x_j with $x_j \in I \subseteq \downarrow p$.

(3) implies (1): Suppose now that L is distributive. Let F be the filter generated by $L\setminus\downarrow p$; by (3) we have $\Downarrow p \cap F = \emptyset$. By Lemma 3.19, there is a prime ideal P with $\Downarrow p \subseteq P$ and $P \cap F = \emptyset$. Since $L\setminus\downarrow p \subseteq F$, we have $P \subseteq L\setminus F \subseteq \downarrow p$, whence $p = \sup \Downarrow p \leq \sup P \leq \sup \downarrow p = p$. Thus $p = \sup P$ (where we used continuity of L via 1.6), and so p is pseudo-prime. □

We draw the reader's attention to the fact that condition (2) is a "weak" analogue of the definition of a prime, which may be formulated by saying that p is prime if in any collection $x_1, \ldots, x_n \in L$ with $x_1 \ldots x_n \leq p$ there is one of these elements with $x_j \leq p$. The hard implication, involving the Axiom of Choice, is (3) implies (1).

At a later point we will give yet another characterization of pseudoprimes in a continuous lattice, but topology will be needed for that result; it will in effect say that pseudo-primes are exactly those elements which can be approximated by primes in a suitable sense. (See V-2.)

It is a natural question to ask for circumstances under which every pseudoprime is in fact prime. In order to establish a sufficient condition we record the following Lemma:

3.25. LEMMA. *In a lattice* L *the following conditions are all equivalent for any auxiliary relation* $\prec$ *(see* 1.9):

(1) *For all* $a,x,y \in L$, *the relations* $a \prec x$ *and* $a \prec y$ *imply* $a \prec xy$;

(2) *For all* $x \in L$, *the set* $\{y \in L : x \prec y\}$ *is a filter*;

(3) *For all* $a,b,x,y \in L$ *the relations* $a \prec x$ *and* $b \prec y$ *imply* $ab \prec xy$;

(4) *The graph of* $\prec$ *is a subsemilattice of* $L \times L$;

(5) *The function* $x \mapsto s_\prec x : L \to \mathrm{Id}\, L$ *is a semilattice morphism, where* $s_\prec x = \{y \in L : y \prec x\}$.

Proof. The connections (1) iff (2), (3) implies (1), and (3) iff (4) are trivial. If $a \prec x$ and $b \prec y$, then $ab \prec x$ and $ab \prec y$. If (1) holds then this implies $ab \prec xy$. Thus (1) implies (3). For $x,y \in L$ one has $s_\prec xy \subseteq s_\prec x \cap s_\prec y$. Thus (5) means that for all $a,x,y \in L$ with $a \in s_\prec x \cap s_\prec y$ one has $a \in s_\prec xy$; but this is the same as (1). □

3.26. DEFINITION. We will say that an auxiliary relation $\prec$ on L is *multiplicative* iff it satisfies the equivalent conditions of 3.25.

3.27. PROPOSITION. *Let* L *be a continuous lattice. If* $\ll$ *is multiplicative, then the following conditions are equivalent for an element* $p \in L$:

(1) *p is pseudoprime;*

(2) *If $ab \ll p$, then $a \leq p$ or $b \leq p$ for all $a,b \in$ L;*

(3) *p is prime.*

Conversely, if L *is, in addition, distributive, then* PRIME L $=$ ΨPRIME L *implies that $\ll$ is multiplicative.*

Proof. Firstly assume that $\ll$ is multiplicative, then (3) implies (1) is clear. By 3.24 and Remark we have (1) implies (2).

(2) implies (3): By way of contradiction suppose that p is not prime. Then there are elements $x,y \not\leq p$ with $xy \leq p$. By 1.6 we find elements $a,b \not\leq p$ with $a \ll x$ and $b \ll y$. Since $\ll$ is multiplicative, we conclude $ab \ll xy \not\leq p$; that is, $ab \ll p$ by 1.2(ii). But this contradicts (2).

Secondly, assume that L is distributive and $\ll$ is not multiplicative. We wish to show PRIME L $\neq$ ΨPRIME L. There are elements a,x,y with $a \ll x$ and $a \ll y$ but not $a \ll xy$. The ideal $\downarrow xy$ and the filter $\uparrow a$ are disjoint; hence, by 3.19, there is a prime ideal P containing $\downarrow xy$ missing a. Then it follows that $p = \sup P \in$ ΨPRIME L. But $xy = \sup \downarrow xy$ (1.6) $\leq \sup P = p$. Consider the representative case $x \leq p = \sup P$. Then $a \ll x$ implies $a \in P$ by 1.2, which is impossible. Thus, $x \not\leq p$ and $y \not\leq p$; whence, $p \notin$ PRIME L. □

EXERCISES

The first exercise is a variant of Proposition 3.6.

3.28. EXERCISE. Let L be a modular lattice and $p \in$ L. The following conditions are equivalent:

(1) p is irreducible;

(2) p is maximal in the complement of every filter maximal with respect to missing p;

(3) p is maximal in the complement of some filter maximal with respect to missing p. □

PROBLEM. Is every irreducible element in a continuous lattice maximal in the complement of some open filter? □

3.29. EXERCISE. A semilattice S is defined to be *distributive* if $ab \leq x$ always implies the existence of c,d with $a \leq c$, $b \leq d$, and $x = cd$.

(i) Show that if S is actually a lattice, then S is distributive as a semilattice iff it is distributive as a lattice;

(ii) Show that if S is a sup-semilattice, then S is distributive iff Id S, the lattice of ideals, is distributive.

(iii) The condition of distributivity is equivalent to $\uparrow(uv) = (\uparrow u)(\uparrow v)$ for all $u,v \in$ S. □

The concept of open upper sets (3.1) carries over to the case of continuous posets as the following exercises show.

3.30. DEFINITION. An upper set U in an up-complete poset is called *open* iff for all $u \in U$ we have $\downarrow u \cap U \neq \emptyset$. □

3.31. EXERCISE. Let $x \ll y$ in a continuous poset. Then there is an open filter U with $y \in U \subseteq \uparrow x$.

(HINT: Check the proof of 3.3, using 1.27.) □

3.32. EXERCISE. Let S be an up-complete poset. Let ($\mathcal{O}$Filt S,$\subseteq$) denote the poset of all nonempty *open* filters (in the sense of 3.30).

(i) If U is an open upper set, then

$$U = \bigcup\{V: V \in \mathcal{O}\text{Filt S and } V \subseteq U\}.$$

(ii) In $\mathcal{O}$Filt S one has $U \ll V$ if there is a $v \in V$ with $U \subseteq \uparrow v$ (iff $U \subseteq \uparrow v$). Moreover, this condition is also necessary provided $\ll$ satisfies the interpolation property.

(iii) If S is an up-complete poset, then $\mathcal{O}$Filt S is a continuous poset—provided $\ll$ satisfies the interpolation property.

(iv) If S is a continuous poset, then so is $\mathcal{O}$Filt S.

(v) Note that the above hold if S is a continuous semilattice.

(HINT: (i): Let $u \in U$ and construct inductively a sequence $....u_2 \ll u_1 \ll u_0$ with $u_0 = u$ and $u_n \in U$. Then $V = \bigcup\{\uparrow u_n : n = 0, 1, 2, ...\}$ is an open filter contained in U and containing u.

(ii): The first part is straightforward. For the second, use the interpolation property to construct, for each open upper set V and each $v \in V$, a $w \in V$ and a sequence of elements $w ... \ll v_2 \ll v_1 \ll v_0 = v$; this yields an open filter V_v contained in V and containing v such that V_v has a lower bound in V. The collection of all V_v is directed and its union is V. If $U \ll V$, then $U \subseteq V_v$ for some v.

(iii): Use (ii) to verify definitions.

(iv): This follows from (iii) and 1.27.) □

Notice that we use the same notation $\mathcal{O}$Filt S which we used in 1.32 for a concept of open filters in the sense of a suitable topology. In Chapter II we will see that both concepts are compatible. We should in any case compare 3.32 and 1.32. It follows from 3.32 that for the open filters considered in 3.32, condition 1.32 (F) is satisfied if S is a continuous semilattice.

It is noteworthy that $\mathcal{O}$Filt L is in general not a continuous lattice even if we start from a continuous lattice L. On the other hand, the importance of the concept $\mathcal{O}$Filt S introduced here will become apparent later when we show that any continuous poset can be represented as $\mathcal{O}$Filt S for some continuous poset S.

3.33. EXERCISE. Let S be an up-complete poset and U an open upper set. If we have $s \in S \setminus U$, then $\uparrow s \setminus U$ has maximal elements.

(HINT: Verify the proof of 3.4.) □

Next we note that all the concepts of irreducible, prime and pseudoprime elements make perfectly good sense in any up-complete semilattice; in fact, irreducibility and primeness may be considered in any semilattice. (See definitions 3.5, 3.11, 3.23).

3.34. EXERCISE. Theorem 3.7 holds in a continuous semilattice. Hence,

$$s = \inf(\uparrow s \cap \mathrm{IRR}\ S) \text{ for all } s \in S$$

holds in any continuous semilattice S.

(HINT: Verify the proof of 3.7 and 3.10.) □

3.35. EXERCISE. In a distributive continuous semilattice (cf. 3.29), any element is the inf of primes.

(HINT: Use 3.34 and the fact that irreducibles in distributive semilattices are primes.) □

Remark. The following example shows that the converse of 3.35 fails:

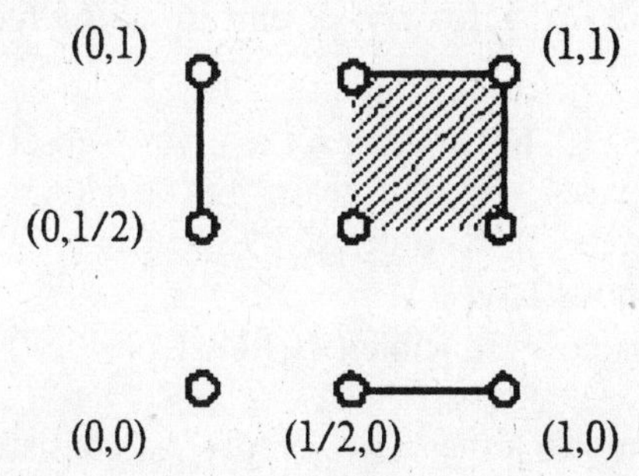

Indeed $(1/2, 1/2) \in \uparrow((1,0) \wedge (0,1)) \setminus (\uparrow(1,0) \cap \uparrow(0,1))$; see 3.29(iii).

PROBLEM. Is $\mathcal{O}$Filt S distributive for a distributive continuous semilattice?

Let us identify the primes in a few specific examples which we discussed earlier.

3.36. EXERCISE. Let $L = \mathcal{O}(X)$ for a tonological space, and let $U \in \mathcal{O}(X)$ and $A = X \setminus U$. Then the following statements are equivalent:

(1) $U \in \mathrm{PRIME}\ \mathcal{O}(X)$.
(2) A is a closed set which is not the union of two proper closed subsets. □

A closed set $A \subseteq X$ is called *irreducible* iff it satisfies the condition in 3.36(2). Any set of the form $\{x\}^-$ is irreducible. A space is called *sober* if every nonempty irreducible set has a unique dense point. (We will return to these concepts in a more systematic way in Chapter II and in Chapter V.) Clearly every Hausdorff space is sober. An infinite set in which the finite sets and the whole space are the only closed sets is T_1 but not sober, because the whole space is irreducible. From what has been said it is clear that a space is sober iff the function

$$x \mapsto X\backslash\{x\}^- : X \to (\text{PRIME } \mathcal{O}(X))\backslash\{X\}$$

is bijective.

3.37. EXERCISE. Let $L = \text{Id}^-\mathcal{A}$ be the lattice of closed two-sided ideals of a C*-algebra. Then $I \in \text{PRIME } L$ iff I is a closed prime ideal in the ring-theoretical sense. □

Note that in a C*-algebra, for two closed two sided ideals one has $IJ = I \cap J$. Every primitive ideal (i.e., the kernel of an irreducible representation) is prime and thus is an element of PRIME L. If Prim $\mathcal{A}$ denotes the set of all primitive ideals of $\mathcal{A}$, we have Prim $\mathcal{A} \subseteq \text{PRIME } L$. For separable C*-algebras equality is known to hold (and we will give an independent proof V-5.27). It is unknown whether this is true for nonseparable algebras. One knows from the theory of C*-algebras that Prim $\mathcal{A}$ is an order generating set for L (in the sense of 3.8).

3.38. EXERCISE. Let $\{L_j : j \in J\}$ be a family of lattices. Then an element $(x_j)_{j \in J} \neq 1$ is irreducible (resp., prime) in the product $\mathsf{X}_{j \in J} L_j$ iff there is an index $k \in J$ such that $x_j = 1$ for $j \neq k$ and x_k is irreducible (resp., prime) in the lattice L_k. □

3.39. EXERCISE. Let K be a compact convex subset of a locally convex topological vector space over the reals. Let Con(K) be the lattice of all compact convex subsets of K ordered by inclusion; by 1.22 we know that $\text{Con(K)}^{\text{op}}$ is a continuous lattice.

(i) $A \in \text{Con(K)}$ is co-irreducible ($A \in \text{IRR Con(K)}^{\text{op}}$) iff A has at most one point;

(ii) A is coprime iff either A is empty or consists of a single element which is an extreme point of K.

(HINT: (i): Obviously a singleton is co-irreducible. Conversely, suppose $A \in \text{Con(K)}$ has more than one point. Let f be a continuous linear functional into the reals which is nonconstant on A. Find a real number r such that the inverse images under f of the intervals $]-\infty, r]$ and $[r, +\infty[$ have nonempty intersections with A. This will give a decomposition of A that will show it is not co-irreducible in Con(K).

(ii): If A is the singleton of an extreme point, then it is easy to check that, when it is contained in the convex hull of the union of two sets in Con(K), it must be contained in one of them. Conversely, if A is co-prime, it is co-irreducible and so must consist of at most one point. If it does have a point and is not extreme, then it lies between two other points. This will show that A is not co-prime.) □

3.40. EXERCISE. Let L be a lattice and $m : L \times L \to L$ be the meet function given by $m(x,y) = xy$. Let $G = \{(x,y) \in L \times L : x \leq y\}$, the graph of the relation $\leq$, and let G^{-1} be the graph of $\geq$. Then

(i) $m((L \times L)\backslash(G \cup G^{-1})) = L\backslash\text{IRR } L$;

(ii) $m^{-1}(\text{IRR } L) \subseteq G \cup G^{-1}$.

(HINT: Clearly (ii) follows from (i). If $(x,y) \notin G \cup G^{-1}$, then xy is different from x and from y, and thus $xy \notin \mathrm{IRR}\ \mathrm{L}$. If $z \notin \mathrm{IRR}\ \mathrm{L}$, then there are elements $z < x$ and $z < y$ with $z = xy$. Then $(x,y) \notin G$ and $(x,y) \notin G^{-1}$.) □

The following exercise concerns pseudoprimes in a continuous lattice. We know that a lattice must be distributive if it is order generated by primes. How nearly is a continuous lattice distributive if it is order generated by pseudoprimes? 3.41 gives some sort of answer.

We say that $p \in \mathrm{L}$ is a *weak prime* iff condition 3.24(2) is satisfied. (Thus if L is distributive, then p is a weak prime iff it is a pseudoprime by 3.24.) The set of weak primes is denoted WPRIME L (cf. V-3.1 and 3.4).

3.41. EXERCISE. Let L be a continuous lattice. Then the following statements are equivalent:

(1) IRR L$\subseteq$WPRIME L;

(2) WPRIME L is order generating;

(3) For all finite sequences of elements $a_1,\ldots,a_n,x \in \mathrm{L}$ the relation $a_1 \ldots a_n \ll x$ implies $(a_1 \vee x)\ldots(a_n \vee x) = x$;

(4) For all finite sequences of elements $a_1,\ldots,a_n,x \in \mathrm{L}$ the relation $a_1 \ldots a_n \ll x$ implies the existence of elements $b_k \geq a_k$, $k=1,\ldots,n$, such that $b_1 \ldots b_n = x$;

Moreover these conditions imply:

(WH) For all $X \subseteq \mathrm{L}$ and $x \in \mathrm{L}$ we have $x \sup X \doteq \sup xX$,

where we write $a \doteq b$ iff $\Uparrow a = \Uparrow b$.

(HINT: See Hofmann and Lawson [1976/77, p. 337].) □

Notice that, according to our definitions

$$\mathrm{PRIME\ L} \subseteq \Psi\mathrm{PRIME\ L} \subseteq \mathrm{WPRIME\ L}.$$

In Chapter V we will return to these concepts.

In the following exercise we begin to discuss the connection between completely distributive lattices (see 2.4, 2.5, 3.15) and continuous posets; only in Chapter V will we be able to complete this discussion (V-3.5 ff.).

3.42. EXERCISE. Let L be a complete lattice. Let P = COPRIME L$\setminus\{0\}$ be the set of nonzero coprimes with the induced partial order. Then:

(i) P is an up-complete poset.

(ii) If L is a completely distributive lattice, then for each $x \in \mathrm{P}$

$$x = \sup_{\mathrm{L}} (\Downarrow_{\mathrm{L}} x \cap P) = \sup_{\mathrm{P}} \Downarrow_{\mathrm{P}} x.$$

(HINT: (i): An element $p \in \mathrm{L}$ is coprime iff $\mathrm{L} \setminus \uparrow p$ is an ideal. If $(p_j)_{j \in J}$ is a monotone net of coprimes (see O-1.2), then for $p = \sup p_j$ the set

$$\mathrm{L} \setminus \uparrow p = \bigcup_{j \in J} \mathrm{L} \setminus \uparrow p_j$$

is an ideal since the union is ascending.

(ii): Since $\Downarrow_L x \cap P \subseteq \Downarrow_P x$, we need only show the first equality. Let $x \not\leq y$. Then there is a $u \ll x$ with $u \not\leq y$. Now $u = \sup(\downarrow u \cap P)$ by 3.15; but $\downarrow u \cap P \subseteq \Downarrow_L x \cap P$, and so $\sup(\Downarrow_L x \cap P) \not\leq y$.) □

If we compare this with Definition 1.26, then we see that only 1.26(i) is lacking for P to be recognized as a continuous poset. We will show this condition in the exercises to Section 1 of Chapter V.

3.43. EXERCISE. (GENERALIZED BAIRE CATEGORY THEOREM). This exercise presents in a sequence of results an application of the theory of irreducible elements to derive a very general Baire category theorem. However, we need some of the early results of this section in a sharpened form, and we state them in the first four propositions.

Proposition 3.3 was one of the keys to the principal results. It says in effect that, in a continuous lattice, for any open upper set V of the form $\Uparrow x$ and for any element $v \in V$, there is an open filter $U \subseteq V$ with $v \in U$. A considerably stronger result is true—but it is noteworthy that countability enters here.

3.43.1. PROPOSITION. *Let* L *be a continuous lattice and V an open upper set. If F is a countably generated filter in* L *with $VF \subseteq V$, then for each $v \in V$ there is an open filter $U \subseteq V$ containing both v and F.*

Remark. With $F = \{1\}$ we obtain a version of 3.3.

(HINT: We define $G = \{x \in L : Vx \subseteq V\}$. Then G is a filter containing F. Suppose $a_1, a_2, \ldots$ is a sequence of generators of F, which without loss of generality we may suppose decreasing. Let $v \in V$ be given—we must find an open filter U so that $va_n \in U$ for all n. Inductively select a sequence $b_n \in V$ as follows: Let $b_0 = v$. Since $a_1 \in G$, we have $va_1 \in V$; and since V is open, we find a $b_1 \ll va_1$. Suppose that $b_0, \ldots, b_n$ have been selected. Since $a_{n+1} \in G$ we have $b_n a_{n+1} \in V$; and thus, since V is open, we find a $b_{n+1} \ll b_n a_{n+1}$. Since

$$\ldots b_n \ll b_{n-1} \ll \ldots \ll b_1 \ll v,$$

then U, the filter generated by the b_n, is an open filter containing v. Because of $b_n \leq a_n$, all a_n are contained in U.)

3.43.2. COROLLARY. *Let* L *be a continuous lattice and V an open upper set. If N is any countable subset of* L *with $VN \subseteq V$, then for each $v \in V$ there is an open filter U with $vN \subseteq U$.*

(HINT: The set $G = \{x \in L : Vx \subseteq V\}$ is a filter. As $N \subseteq G$, the countably generated filter F generated by N is contained in G. Hence 3.43.1 applies to F and yields the assertion.)

The basic result 3.10 may be formulated as follows: If V is the upper set $L \setminus \downarrow s$, then for every $y \in V$ and every $x \notin V$ there is a $p \in \mathrm{IRR}\ L$ such that $x \leq p$ and $y \not\leq p$. This we now sharpen:

3.43.3. PROPOSITION. *Let* L *be a continuous lattice and* V *an open upper set. If* N *is a countable set with* $VN \subseteq V$, *then for every* $y \in V$ *and every* $x \notin V$ *there is an irreducible element* p *such that* $x \leq p$ *and* $p \notin \uparrow yN$ (*that is,* $yn \not\leq p$ *for all* $n \in N$).

(HINT: Let U be the open filter constructed in 3.43.2. Then by 3.4 there is a maximal element p in $\uparrow x \setminus U$, and by 3.6 p is irreducible.)

3.43.4. COROLLARY. *Let* L *be a continuous lattice,* $x \in$ L, *and* $N \subseteq$ L *a countable set such that* $y \not\leq x$ *and* $n \in N$ *always imply* $yn \not\leq x$. *Then for any* y *with* $y \not\leq x$ *there is an irreducible* p *with* $x \leq p$ *and* $yn \not\leq p$ *for all* $n \in N$.

(HINT: Apply 3.43.3 with $V =$ L$\setminus\downarrow x$.)

Now consider the lattice of open subsets $\mathcal{O}$(X) of a topological space X. An open set $U \in \mathcal{O}$(X) is *dense* in X iff for all $V \in \mathcal{O}$(X) with $V \neq \emptyset$ we have $U \cap V \neq \emptyset$. Accordingly we are motivated to make the following definition:

3.43.5. DEFINITION. An element u in a complete lattice L is called *dense* if $v \neq 0$ implies $uv \neq 0$ for all $v \in$ L (equivalently $0 \notin u$(L$\setminus\{0\}$)).

Let us momentarily assume that X is a Hausdorff space. Then by 3.36 there is a bijection between the points $x \in X$ and the nontrivial primes of $\mathcal{O}(X)$ given by $x \mapsto X \setminus \{x\}$. For an element $U \in \mathcal{O}(X)$ we have $x \in U$ iff $U \not\subseteq X \setminus \{x\}$. This motivates the following definition:

3.43.6. DEFINITION. In a complete lattice L we define a binary relation $\mathcal{E}$ between IRR L$\setminus\{1\}$ and L by setting $p \mathcal{E} u$ iff $u \not\leq p$.

At this point we return to Corollary 3.43.4 which we specialize to the case $x = 0$. The hypothesis on N then says that all members of N are dense. This leads to:

3.43.7. THEOREM. (BAIRE CATEGORY THEOREM FOR CONTINUOUS LATTICES). *Let* L *be any continuous lattice and* D *a countable collection of dense elements. Then for any nonzero element* u *there is a point* $p \in$ IRR L$\setminus\{1\}$ *such that* $p \mathcal{E} u \wedge v$ *for all* $v \in D$.

As a consequence we have the following result:

3.43.8. THEOREM. (BAIRE CATEGORY THEOREM FOR LOCALLY QUASICOMPACT SPACES). *Let* X *be any locally quasicompact space and* $\mathfrak{D}$ *a countable collection of dense open sets. Then for any nonempty open set* U *there is an irreducible set* A *such that* $A \cap U \cap V \neq \emptyset$ *for all* $V \in \mathfrak{D}$.

(HINT: Recall that $\mathcal{O}$(X) is a continuous lattice by 1.7(5).)

Notice that if $A = \{x\}^-$, then the conclusion of 3.43.8 implies that $x \in A \cap U \cap W$ and that $U \cap W \neq \emptyset$, where W is the intersection of the sets in $\mathfrak{D}$. We record next a definition from general topology:

3.43.9. DEFINITION. A topological space X is called a *Baire space* iff the intersection of any countable collection of dense open subsets is dense (or, equivalently, iff the union of a countable collection of nowhere dense closed sets is nowhere dense).

As an immediate consequence of the above we have a generalization of Baire's well known classical theorem on locally compact Hausdorff spaces:

3.43.10. COROLLARY. *Every locally quasicompact sober space is a Baire space.*

The following examples are instructive; both are first countable:

Let X = **N** with upper sets in the usual ordering open. Then X is locally quasicompact T_0, but it is not a Baire space.

Let X be the set of all ordinals less than the first uncountable ordinal with upper sets open. Then X is a locally quasicompact T_0 Baire space which is not sober. □

NOTES

The study of irreducibles and primes in continuous lattices was begun in Hofmann and Lawson [1976/77] and [1979]. Substantial contributions were made by Gierz and Keimel [1977]; this paper as well as the second one mentioned above will contribute to Chapter V.

The proof of Proposition 3.3 is due to Lawson (folklore tradition in SCS). The results in Exercise 3.43 are due to Hofmann [SCS-43]. The result that a locally quasicompact space is a Baire space can also be excavated from Isbell [1975a, p. 334, Section 4.2]. Isbell asks for a "pointless" generalization of Baire category; our Baire category theorem for continuous lattices and for locally quasicompact spaces may be considered such a "pointless" theory. The characterization of completely distributive lattices in terms of continuous lattices in Theorem 3.15 is a result of Kamara [1978]. The theory of pseudoprimes was first developed in the second of the two papers by Hofmann and Lawson. Proposition 3.27, however, in slightly different language is due to Keimel and Mislove [SCS-19].

The current chapter deals exclusively with the purely lattice-theoretical aspects of the theory of continuous lattices. Many of the finer results on the spectra of continuous lattices require a better understanding of various topologies on a continuous lattice and will be treated in Chapter V.

4. ALGEBRAIC LATTICES

In universal algebra, algebraic lattices have become familiar objects as lattices of congruences and lattices of subalgebras of an algebra. As a consequence, they have been extensively studied and it cannot be our purpose here to survey this classical field. However, algebraic lattices are continuous, and they fit perfectly into the general theory. It is this fit which is the object of our present discussion.

We have noticed that the auxiliary relations need not be reflexive, and that in fact the way-below relation rarely is. Nevertheless there are elements x in a complete lattice such as $x = 0$ which satisfy $x \ll x$. It is those elements which now come into focus. Recall that in Definition 1.1 they were called *isolated* (from below) or *compact*.

4.1. DEFINITION. The *subset of all compact elements* is denoted K(L). □

Clearly $x \leq k \leq y$ with $k \in \mathrm{K(L)}$ implies $x \ll y$ by 1.2(ii).

4.2. REMARK. *For an element x in a complete lattice* L *the following statements are equivalent*:

(1) $\uparrow x$ *is an open filter* (*in the sense of* 3.1);

(2) x *is compact.*

Proof. (2) implies (1): If $u \in \uparrow x$ and $x \ll x$, then $x \ll u$ by 1.2(ii).

(1) implies (2): From (1) there is an $y \in \uparrow x$ with $y \ll x$. But $x \leq y \ll x$ implies $x \ll x$ by 1.2(ii). □

4.3. REMARK. *In a complete lattice*, K(L) *is a sup-subsemilattice with smallest element* 0.

Proof. If $x \ll x$ and $y \ll y$, then $x \vee y \ll x \vee y$ by 1.2(iii), and $0 \in \mathrm{K(L)}$ by 1.2(iv). □

If L is the unit interval, then K(L) $= \{0\}$. If L is the standard Cantor chain in the unit interval, then K(L) consists exactly of those elements which are isolated from below in the topological sense. In this example they are so abundant that every element is approximated from below by them. The general idea of this kind of abundance is formalized in the following definition:

4.4. DEFINITION. A lattice L is called *algebraic* iff it is complete and the following axiom:

(K) $\quad x = \sup(\downarrow x \cap \mathrm{K(L)})$

holds for all $x \in \mathrm{L}$. □

4.5. PROPOSITION. *Let* L *be a lattice. Then the following statements are equivalent:*

(1) L *is algebraic.*

(2) L *is continuous and for all* $x \ll y$ *there is a* $k \in K(L)$ *with* $x \leq k \leq y$.

In particular, every algebraic lattice is continuous.

Proof. (1) implies (2): Under the hypothesis of (1), L is complete. Since $\downarrow x \cap K(L) \subseteq \Downarrow x$, by 1.2(ii) we have $x = \sup(\downarrow x \cap K(L)) \leq \sup \Downarrow x \leq x$. Thus L is continuous. Now let $x \ll y$. Since K(L) is a sup-semilattice of L (cf. 4.3), $\downarrow y \cap K(L)$ is directed and has sup y by 4.4. Thus, there is a $k \in \downarrow y \cap K(L)$ with $x \leq k$ by 1.1. Then $x \leq k \leq y$ and $k \in K(L)$.

(2) implies (1): Again L is complete, and by (2) the set $\downarrow y \cap K(L)$ is cofinal in $\Downarrow y$. Hence $\sup(\downarrow y \cap K(L)) = \sup \Downarrow y = y$ by (2) and 1.6. □

A frequently encountered subclass of algebraic lattices is introduced in the next definition.

4.6. DEFINITION. A lattice L is called *arithmetic* iff it is algebraic and K(L) is a sublattice of L (that is, closed under finite infs). □

4.7. PROPOSITION. *Let* L *be an algebraic lattice. Then the following statements are equivalent.*

(1) L *is arithmetic.*

(2) K(L) *is a lattice.*

(3) $\ll$ *is multiplicative* (3.26).

Proof. (1) implies (2): Trivial.

(2) implies (1): Let $a, b \in K(L)$, $c = a \wedge_{K(L)} b$. Then $c \leq ab$ $(= a \wedge_L b)$. But if $X = \downarrow(ab) \cap K(L)$, then $c = \sup_{K(L)} X$, and $ab = \sup_L X$, since L is algebraic. Thus $ab \leq c$, whence $a \wedge_{K(L)} b = ab$.

(1) implies (3): Let $a \ll x$ and $a \ll y$. Then there are $c, k \in K(L)$ with $a \leq c \leq x$ and $a \leq k \leq y$ by 4.5. Thus $a \leq ck \leq xy$, and since $ck \in K(L)$ by (1), we have $a \ll xy$ by 4.5.

(3) implies (1): If $a, b \in K(L)$, then $a \ll a$, $b \ll b$, hence $ab \ll ab$ by (3), Thus $ab \in K(L)$. □

4.8. COROLLARY. *Every pseudo-prime in an arithmetic lattice is prime. Conversely, if in an algebraic distributive lattice* L *we have* ΨPRIME L = PRIME L, *then* L *is arithmetic.*

Proof. 4.7 and 3.27. □

We will look at some characteristic examples presently. But first we recall that if a closure operator (O-3.8) c on a continuous lattice L preserves directed sups, then c (L) is a continuous lattice (and in fact a subalgebra of L; see 2.9).

We have a parallel for the algebraic case:

4.9. PROPOSITION. *If* L *is an algebraic lattice and* $c : \mathrm{L} \to \mathrm{L}$ *a closure operator preserving sups of directed sets, then*

(i) $c(\mathrm{L})$ *is an algebraic lattice (relative to the induced order);*
(ii) $c(\mathrm{K}(\mathrm{L})) = \mathrm{K}(c(\mathrm{L}))$.

Proof. (i): By 2.9, $c(\mathrm{L})$ is continuous. We claim that $c(\mathrm{K}(\mathrm{L})) \subseteq \mathrm{K}(c(\mathrm{L}))$. If so, then since c preserves directed sups, we have

$$\begin{aligned} c(x) &= c(\sup_{\mathrm{L}} (\downarrow x \cap \mathrm{K}(\mathrm{L}))) \\ &= \sup_{c(\mathrm{L})} c(\downarrow x \cap \mathrm{K}(\mathrm{L})) \leq \sup_{c(\mathrm{L})} (\downarrow c(x) \cap \mathrm{K}(c(\mathrm{L}))) \leq c(x), \end{aligned}$$

and this will show that $c(\mathrm{L})$ is algebraic.

Now let $k \in \mathrm{K}(\mathrm{L})$. We show $c(k) \in \mathrm{K}(c(\mathrm{L}))$. Let D be directed in $c(\mathrm{L})$ with $c(k) \leq \sup_{c(\mathrm{L})} D$. Since c is a closure operator we have $k \leq c(k)$ and since c preserves directed sups, we have $c(\sup_{\mathrm{L}} D) = \sup_{\mathrm{L}} D$. But by O-3.12(iii) we have $c(\sup_{\mathrm{L}} D) = \sup_{c(\mathrm{L})} D$. Thus $k \leq \sup_{\mathrm{L}} D$, and so $k \leq d$ for some $d \in D$, since k is compact. Thus $c(k) \leq c(d) = d$, and this proves the claim.

(ii): We have seen $c(\mathrm{K}(\mathrm{L})) \subseteq \mathrm{K}(c(\mathrm{L}))$. It remains to show that the converse containment holds. For this purpose let $a \in \mathrm{K}(c(\mathrm{L}))$; in particular $c(a) = a$. Since L is algebraic, $a = \sup_{\mathrm{L}}(\downarrow a \cap \mathrm{K}(\mathrm{L}))$ by 4.4. In view of O-3.12(iii) we calculate

$$a = c(a) = \sup_{c(\mathrm{L})} c(\downarrow a \cap \mathrm{K}(\mathrm{L})) \leq \sup_{c(\mathrm{L})} (\downarrow a \cap c(\mathrm{K}(\mathrm{L}))) \leq a.$$

Thus $a = \sup_{c(\mathrm{L})}(\downarrow a \cap c(\mathrm{K}(\mathrm{L})))$ with a directed set $\downarrow a \cap c(\mathrm{K}(\mathrm{L}))$ in $c(\mathrm{L})$. But since a is compact in $c(\mathrm{L})$ this implies the existence of a $k \in \mathrm{K}(\mathrm{L})$ where we have $a = c(k)$ (by 4.1 and 1.1). Thus $a \in c(\mathrm{K}(\mathrm{L}))$. □

4.10. COROLLARY. *Let* L *be an algebraic lattice. Then the assignment* $c \mapsto c(\mathrm{L})$, *which associates with a closure operator* $c : \mathrm{L} \to \mathrm{L}$ *its image, induces a bijection from the set of all closure operators of* L *preserving directed sups onto the set of subalgebras of* L. *Moreover, all of the subalgebras of* L *are algebraic.*

Proof. This is immediate from 4.9 and 2.9. □

4.11. EXAMPLES. (1) For any set X the lattice 2^X is algebraic, and $F \in \mathrm{K}(2^X)$ iff F is finite.

(2) If L is a subalgebra of 2^X (that is, L is a subset which is closed under arbitrary intersections and directed unions), then

$$E \in \mathrm{K}(\mathrm{L}) \text{ iff } E = \bigcap\{Y \in \mathrm{L} : F \subseteq Y\} \text{ for some finite } F \in 2^X.$$

(3) As a special case of this last example, consider 2^{2^X}. There are many well-known subalgebras: Filt 2^X and Id 2^X, to name two. In the case of filters,

$$P \in \mathrm{K}(\mathrm{Filt}\,(2^X)) \text{ iff } P \text{ is a principal filter.}$$

It is important to observe that this example generalizes considerably. The details follow in the next two results. □

4.12. PROPOSITION. (i) *Let* S *be a sup-semilattice with minimal element* 0. *Then* L = Id S *is an algebraic lattice whose compact elements are the principal ideals*;

(ii) *The principal ideal map is an isomorphism* $x \mapsto \downarrow x : S \to K(L)$;

(iii) *If* S *is a lattice, then* L *is arithmetic.*

(iv) *Conversely, if* L *be an algebraic lattice, then* S = K(L) *is a sup-semilattice with minimal element* 0;

(v) *The map* $x \mapsto \downarrow x \cap S : L \to \mathrm{Id}\ S$ *is an isomorphism*;

(vi) *The ideal* $\downarrow x \cap S$ *is principal iff* $x \in S$ *and* $\sup(\downarrow x \cap S) = x$.

Proof. Most of the facts recorded here have already been established. We add a few hints. (i): By invoking O-2.8(3) we note that L is a complete lattice. The set L is closed in 2^S under intersections and directed unions. Hence L is algebraic by 4.10. Now 4.11(2) shows that K(L) consists of finitely generated ideals, and since ideals are sup-semilattices, these are precisely the principal ideals. The proofs of (ii) and (iii) are clear.

(iv): In 4.3 it was shown that S = K(L) is a sup-semilattice with minimum 0.

(v): To prove that $x \mapsto \downarrow x \cap S : L \to \mathrm{Id}\ S$ is bijective, we claim that $\sup : \mathrm{Id}\ S \to L$ is the inverse of this map. Since the latter is clearly surjective, it suffices to show that $\downarrow(\sup I) \cap S = I$ for each $I \in \mathrm{Id}\ S$, and since $\supseteq$ is clear, we must show $\subseteq$. Let $k \in \downarrow(\sup I) \cap S$; that is, $k \ll k \leq \sup I$. Thus $k \ll \sup I$ by 1.2(ii). Hence by 4.1 we have an $i \in I$ with $k \leq i$. But since I is an ideal in S, we have $k \in I$. □

In particular, we have the following corollary:

4.13. COROLLARY. *Every algebraic lattice* L *allows an injection* $g : L \to 2^{K(L)}$ *preserving infs and directed sups such that* $(pr_k\ g)^{-1}(1) = \uparrow k$ *for* $k \in K(L)$.

Proof. By 4.12 $L \cong \mathrm{Id}(K(L))$, and $\mathrm{Id}(K(L)) \subseteq 2^{K(L)}$ is closed under intersections and under sups of directed sets. If the injection g is interpreted in terms of characteristic functions, then $g(x)(k) = 1$ iff $k \in \downarrow x \cap K(L)$ iff $k \leq x$, and this is the assertion. □

It is now natural to investigate the closure properties of the class of algebraic lattices within the class of continuous ones, following the lines of 2.6 and 2.7.

4.14. PROPOSITION. (i) *If* $\{L_j : j \in J\}$ *is a family of algebraic lattices, then the cartesian product* $\mathsf{X}_{j\in J} L_j$ *is an algebraic lattice*;

(ii) *If* L *is an algebraic lattice, then all subalgebras are algebraic lattices.*

Proof. (i): An element $(x_j)_{j\in J}$ of the product is compact iff $x_j \in K(L_j)$ for all $j \in J$ and $x_j = 0$ for all but a finite number if indices. Since every factor is algebraic, every element of L is the sup of such elements.

(ii): Immediate from 4.10. □

It is noteworthy that the class of algebraic lattices is not closed under the formation of homomorphic images (see 2.6). The ordinary Cantor set C in the unit interval $\mathbf{I} = [0,1]$ is an algebraic lattice. The Cantor function $g : C \to \mathbf{I}$ which maps C continuously, surjectively and in a monotone fashion onto the unit interval illustrates this phenomenon. Of course, all homomorphic images of an algebraic lattice are continuous by 2.7(iii), and in 4.16 below we will see that all continuous lattices are so obtained. From this viewpoint it is correct to say that the class of continuous lattices is the smallest class closed under the formation of products, subalgebras, and homomorphic images and which contains all algebraic lattices (or even just the two-element lattice as a generator).

If $d : \mathbf{I} \to C$ is the lower adjoint of the Cantor function g just mentioned, then $k = dg : C \to C$ is a kernel operator preserving sups whose image is not algebraic. Thus, a sharp analog of Theorem 2.14 for algebraic lattices is not available. Proposition 4.9 provides a substitute. We utilize this observation further in giving the next characterization theorems for algebraic and continuous lattices.

4.15. THEOREM. *Let* L *be a lattice. Then the following statements are equivalent:*

(1) L *is algebraic;*

(2) *For some set* X, *the lattice* L *is isomorphic to a subset of* 2^X *which is closed under arbitrary infs and directed sups;*

(3) L *is isomorphic to the image of some closure operator* $c : 2^X \to 2^X$ *which preserves directed sups.*

Proof. (1) implies (2): 4.13.
(2) implies (3): 4.11(1) and 4.14(ii).
(3) implies (1): 4.9. □

The following shows how continuous lattices may be derived from algebraic ones.

4.16. THEOREM. *Let* L *be a lattice. Then the following statements are equivalent:*

(1) L *is continuous.*

(2) *There is an arithmetic lattice* A *and a surjective map* $g : A \to$ L *preserving (arbitrary) infs and directed sups.*

(3) *There is an algebraic lattice* A *and a surjective map* $g : A \to$ L *preserving infs and directed sups.*

(4) *There is a set* X *and a projection operator* $p : 2^X \to 2^X$ *preserving directed sups such that* L $\cong$ im p.

Remark. We could rephrase (4) in words as: L is (isomorphic to) a retract of some 2^X under some projection operator preserving directed sups.

Proof. (1) implies (2): Take $A = \operatorname{Id} L$ and let $g(I) = \sup I$. Then 4.12 and 2.1, (1) implies (5), prove the assertion.

(2) implies (3): Trivial.

(3) implies (4): Let $d : L \to A$ be the lower adjoint of g, then $k = dg$ is a kernel operator preserving directed sups. By 4.15 there is a closure operator c on 2^X preserving directed sups such that $A \simeq \operatorname{im} c$. Define $p : 2^X \to 2^X$ by $p = c_\circ k c^\circ$ (see O-3.9). By O-3.12(iii), c° preserves arbitrary sups, and by O-3.11(iii), $c_\circ$ preserves directed sups. Hence p preserves directed sups, and $\operatorname{im} p \simeq g c_\circ(X) = g(A) = L$.

(4) implies (1): 2.14. □

In proving 4.16 we obtained a given continuous lattice L as a quotient of its ideal lattice $A = \operatorname{Id} L$, which is arithmetic. In this construction A depends rather heavily on L, but there is in fact a choice of an arithmetic lattice that depends only on the *cardinality* of L. As we have already proved that the class of continuous lattices is equationally characterizable in Section 2, we could guess at which lattice this is: the free continuous lattice on card(L) generators. Indeed, it follows from quite general theorems that such a lattice exists. Instead of invoking the general theory, however, we can construct free lattices for this class directly and see at once why they are arithmetic.

4.17. THEOREM. *Let **m** be a cardinal number and X be a set of cardinality **m**. The free continuous lattice with **m** generators is isomorphic to* Filt 2^X *and the set of generators $\{\mathfrak{F}(x) : x \in X\}$ consists of the fixed ultrafilters.*

Proof. Recall that $\mathfrak{F}(x) = \{Y \subseteq X : x \in Y\}$. Clearly $\{\mathfrak{F}(x) : x \in X\}$ is in a one-one correspondence with X, so this subset of the filter lattice has the right cardinality. If F is any filter, then

$$F = \sup\{\inf\{\mathfrak{F}(x) : x \in Y\} : Y \in F\},$$

because the set in the inside is just $\uparrow Y$ in 2^X. As the sup (which in 2^X is just a union) is directed, this shows that F belongs to the subalgebra generated by the $\mathfrak{F}(x)$. Therefore, any map on Filt 2^X which preservs infs and directed sups is uniquely determined by its action on the $\mathfrak{F}(x)$. This remark does not prove, however, that sufficiently many of these maps exist.

To prove that the filter lattice is free, we must show that an *arbitrary* map

$$f : \{\mathfrak{F}(x) : x \in X\} \to L$$

into a continuous lattice L can be extended to a map

$$f^* : \operatorname{Filt} 2^X \to L$$

preserving arbitrary infs and directed sups. (By what we remarked, this extension is unique.) The definition of f^* can be given on a filter F as follows:

$$f^*(F) = \sup\{\inf\{f(\mathfrak{F}(x)) : x \in Y\} : Y \in F\}$$

On the right-hand side the sups (directed!) and infs are to be calculated in L. (Note this is a generalization of the previous formula where the map was the identity function.) f^* is well defined and it obviously preserves directed sups, because in the filter lattice directed sups *are* unions. We must prove that f^* preserves infs; it will be better to calculate backwards.

$$\begin{aligned}\inf\{f^*(F_i) : i \in I\} &= \bigwedge_{i\in I} \bigvee_{Y\in F_i} \bigwedge_{x\in Y} f(\mathcal{F}(x)) \\ &= \bigvee_{Z\in P} \bigwedge_{i\in I} \bigwedge_{x\in Z_i} f(\mathcal{F}(x)) \\ &= \bigvee_{Z\in P} \bigwedge \{f(\mathcal{F}(x)) : x \in \bigcup_{i\in I} Z_i\}\end{aligned}$$

Here P is the cartesian product of the F_i for $i\in I$, and we have applied the distributive law 2.3 (DD) to the lattice L. Now note that for $Z\in P$ we have

$$(\bigcup_{i\in I} Z_i) \in (\bigcap_{i\in I} F_i),$$

and that every element of the intersection of the filters comes up in this way. Thus, the right-hand side of the above equation reduces to $f^*(\bigcap_{i\in I} F_i)$ as desired. □

It follows at once from what we have done that if A = Filt 2^{L}, then L is the quotient of the arithmetic lattice A by the map which sends $\mathcal{F}(x)$ to x for $x\in$L.

This is a good time to characterize continuous Boolean algebras:

4.18. THEOREM. *Let* L *be a Boolean algebra. Then the following statements are equivalent:*

(1) L $\cong 2^X$ *for some set* X;

(2) L *is arithmetic*;

(3) L *is algebraic*;

(4) L *is continuous*;

(5) L *and* L^{op} *are continuous*;

(6) L *is completely distributive*;

(7) *Every element in* L *is the sup of atoms and* L *is complete.*

Proof. (1) implies (2) implies (3) implies (4) is trivial, and since in a Boolean algebra $x \mapsto \neg x : \mathrm{L}\to\mathrm{L}^{\mathrm{op}}$ is an isomorphism, (4) implies (5) is clear. Since a Boolean algebra is distributive, and since coprimes are precisely atoms, the equivalences of (5), (6), and (7) follow from 3.15.

(7) implies (1): Let X be the set of atoms. Define two functions:

$$f = (A \mapsto \sup A) : 2^X \to \mathrm{L};$$
$$g = (x \mapsto {\downarrow}x \cap X) : \mathrm{L} \to 2^X.$$

Then $fg = 1_L$ by (7). In order to show that f if an isomorphism it is sufficient to understand that f is injective. For this it suffices to observe that for $A \subseteq X$, $a \in X \cap A$, one has $\sup(\{a\} \cup A) = a \vee \sup A > \sup A$; and this may be deduced from the fact that in a Boolean algebra the function

$$x \mapsto (x \wedge a, x \wedge (\neg a)) : \mathrm{L} \to [0,a] \times [0,\neg a]$$

is an isomorphism. □

This shows that continuous Boolean algebras are trivial. Continuous Heyting algebras are much less trivial; they can nevertheless be completely characterized, as is shown in Chapter V.

The issue of irreducibility and order generation which we discussed in Section 3 for continuous lattices can be rendered even more precise for algebraic lattices. The key is the fact that algebraic lattices contain an ample supply of special irreducibles which we introduce in the next definition.

4.19. DEFINITION. Let L be a lattice. An element $p \in \mathrm{L}$ is called *completely irreducible* iff $\min(\uparrow p \setminus \{p\})$ exists and is different from p (that is, $\uparrow p = \{p\} \cup \uparrow p^+$ with a (unique) element $p^+ > p$.) The set of all completely irreducible elements of L will be written Irr L. □

Clearly by definition $1 \notin \mathrm{Irr\ L} \subseteq \mathrm{IRR\ L}$.

4.20. REMARK. *If X is an order generating subset of a lattice* L, *then* $\mathrm{Irr\ L} \subseteq X$.

Proof. Let $p \in \mathrm{Irr\ L}$. Then $p = \inf(\uparrow p \cap X)$ since X is order generating. Then $p \in \uparrow p \cap X \subseteq X$ by 4.19. □

We now establish a characterization of complete irreducibility under suitable conditions:

4.21. PROPOSITION. *Let* L *be a complete lattice and* $p \in \mathrm{L}$. *Consider the following statements:*

(1) *There is a* $k \in K(\mathrm{L})$ *such that* p *is maximal in* $\mathrm{L} \setminus \uparrow k$.

(2) p *is completely irreducible.*

Then (1) *implies* (2) *and if* L *is join-continuous and distributive, both conditions are equivalent.*

Proof. (1) implies (2): We note that

$$\min(\uparrow p \setminus \{p\}) = \min(\uparrow p \cap \uparrow k) = p \vee k$$

exists and that $p \vee k \neq p$, since $p \notin \uparrow k$. This proves (2).

(2) implies (1): Let $U = \mathrm{L} \setminus \downarrow p$; then U is an open filter. Set $k = \inf U$. Since L is join continuous (O-4.1), then

$$p \vee k = p \vee \inf U = \inf(p \vee U) \geq \min(\uparrow p \setminus \{p\}),$$

since $p\vee U \subseteq \uparrow p \setminus \{p\}$. Thus $k \in U$, and so $\uparrow k = U$. This shows $k \in K(L)$ by 4.2. □

We will see presently (4.24) that (1) and (2) are equivalent in algebraic lattices, too.

If L is a complete chain, then Irr L $=$ K(L^{op})$\setminus\{1\}$. Thus, for the unit interval, Irr L is empty. The important fact for algebraic lattices is that there are enough complete irreducibles:

4.22. THEOREM. *Suppose that x and y are elements of an algebraic lattice with $y \not\leq x$. Then there is a completely irreducible element p with $x \leq p$ and $y \not\leq p$.*

Proof. The proof is analogous to that of 3.9. By 4.4, there is an element $k \in K(L)$ with $k \leq y$ and with $k \not\leq x$. By 4.2 and 3.4 there is a maximal element p in $\uparrow x \setminus \uparrow k$. By 4.21, p is completely irreducible. □

4.23. THEOREM. *In any algebraic lattice,* Irr L *is the unique smallest order generating set. In particular,* $s = \inf(\uparrow s \cap \mathrm{Irr}\ L)$ *for all* $s \in L$.

Proof. By 4.22 and 3.9, Irr L is order generating. By 4.20, Irr L is the unique smallest order generating set. □

Recall that in a continuous lattice in general there is no smallest order generating set, as the example of the unit interval demonstrates. As the example of the Cantor lattice C shows, we have in general Irr L $\neq$ IRR L (since Irr C $=$ K(C^{op}) $\neq$ C $=$ IRR L). In Chapter V, Section 2 we will learn more about generating sets in continuous lattices.

4.24. COROLLARY. *The conditions* (1) *and* (2) *of* 4.21 *are equivalent if* L *is algebraic.*

Proof. (2) implies (1): Let $p \in$ Irr L. By the proof of 4.22,

$$p = \inf\{x \in \uparrow p: \text{ there is a } k \in K(L) \text{ with } x \text{ is maximal in } L \setminus \uparrow k\}.$$

Since p is completely irreducible, p is maximal in $L \setminus \uparrow k$ for some $k \in K(L)$. □

EXERCISES

We know from 1.7(5) that for a locally quasicompact topological space X the lattice $\mathcal{O}$(X) is continuous, and that in the case of Hausdorff spaces by 1.8 that X is locally compact iff $\mathcal{O}$(X) is a continuous lattice. Let us look at these facts in the light of the algebraic lattices considered in the present section.

4.25. EXERCISE. Let X be a topological space.

(i) An open set is a compact element of the lattice $\mathcal{O}$(X) iff it is quasicompact.

(ii) The lattice $\mathcal{O}$(X) is algebraic iff the space X has a basis of quasicompact open sets.

(iii) The lattice $\mathcal{O}(X)$ is arithmetic iff the space X has a basis of quasi-compact open sets which is closed under finite intersections.

(iv) If X is Hausdorff, then $\mathcal{O}(X)$ is algebraic iff $\mathcal{O}(X)$ is arithmetic iff X is totally disconnected and locally compact. Moreover, $K(\mathcal{O}(X))$ is a complete lattice iff X is compact extremally disconnected.

(HINT: We note that there is a small point in the proof of the characterization of arithmetic topologies which needs to be observed. One may wish first to note the following **Lemma**: Let L be a distributive algebraic lattice (i.e., an algebraic Heyting algebra). If $B \subseteq K(L)$ is such that $BB \subseteq B$ and $x = \sup(\downarrow x \cap B)$ for all $x \in L$, then L is arithmetic.) □

We remark that the terminology of calling the elements x in a lattice with $x \ll x$ *compact* is motivated by the example of $\mathcal{O}(X)$. Example 4.11 would suggest to call these elements *finite*, and indeed this terminology has also been utilized.

4.26. EXERCISE. Let G be a locally compact group with identity component G_0 and suppose that G/G_0 is compact. Let L be the lattice of compact normal subgroups. Then L^{op} is algebraic.

(HINT: This requires considerable insight into the structure of locally compact groups. First, L is a complete lattice; in particular, it has a maximal element. Second, one shows that N is compact in L^{op} iff G/N is a Lie group. Third, one applies the fact that every locally compact group H with H/H_0 compact (e.g., H = G/N) is a projective limit of Lie groups. This is tantamount to saying that every compact normal subgroup N is the intersection of compact normal subgroups M for which G/M is a Lie group.) □

4.27. EXERCISE. A *gap* in a totally ordered set is a pair of elements $u < v$ with nothing strictly in between. Show that a totally ordered and complete set L is an algebraic lattice iff whenever $x < y$ there is a gap where $x \leq u < v \leq y$. (In this case there is no distinction between algebraic and arithmetic.) □

4.28. EXERCISE. Define an *algebraic poset* (resp., *semilattice*) L to be an up-complete poset (resp., semilattice) in which the following two conditions are satisfied for all $x \in L$:

(K_1) $\downarrow x \cap K(L)$ is directed;

(K_2) $x = \sup(\downarrow x \cap K(L))$.

We say that the poset (resp., semilattice) is *arithmetic* if also K(L) is an inf-semilattice.

Show that every algebraic poset (semilattice) is continuous. □

4.29 EXERCISE. Let L be an algebraic semilattice (poset). Then $\mathcal{O}$Filt (L) is an algebraic semilattice (poset). (See 3.32). □

Notice that even if we start with an algebraic lattice L, then $\mathcal{O}$Filt (L) in general is just an algebraic semilattice. We will see later that every algebraic lattice can be represented as $\mathcal{O}$Filt (S) for some algebraic semilattice S.

4.30. EXERCISE. Let L be a poset. Then Id L, the poset of all nonempty ideals (directed lower sets, see O-1.3) is an algebraic poset. If L is a semilattice, then Id L is an arithmetic semilattice. □

From 2.18 we know that the sup-map r : Id L→L preserves all infs of nonempty subsets (and all existing sups), if L is a continuous semilattice. In this sense, each continuous semilattice is a quotient of an arithmetic one.

4.33. EXERCISE. State and prove an analog of Proposition 4.15 for posets and algebraic posets. □

4.34. EXERCISE. Let L be a poset. If P is an up-complete poset and f: L→P is an order preserving function, then there exists a unique F : Id L→P such that F preserves directed sups and $F(\downarrow x) = f(x)$ for each $x \in$L. (See Markowsky and Rosen [1976]). □

NOTES

This section links the framework of continuous lattices which we discussed in Sections 1, 2 and 3 with the classical theory of algebraic lattices. These appear now as a special case of continuous lattices. Algebraic lattices were invented in the forties by G. Birkhoff and O. Frink [1948] and L. Nachbin [1949], who independently and in their own way conceived of the idea of compact elements in a lattice. In the thirty years of their history, algebraic lattices have become a part of the textbook literature of lattice theory and universal algebra, notably because of their applications to the theory of congruence lattices and lattices of subalgebras in universal algebras. The close relationship between algebraic lattices and the topological algebra of compact semilattices and their character theory was emphasized in Hofmann-Mislove-Stralka [1974]. These matters will be touched upon in Chapters III and IV; in the meantime, 4.15 gives a flavor of this theory.

The examples of algebraic lattices given in 4.11 and 4.12 are more or less standard. The fact that the class of algebraic lattices is not closed under the formation of quotients (4.14 ff.) is the source of complications which were recognized in topological algebra by A.D. Wallace and R.J. Koch. The classification of those algebraic lattices, all of whose quotients are likewise algebraic was accomplished by Hofmann-Mislove-Stralka [1973] and by Hofmann-Mislove [1977]. The facts about closure operators on algebraic lattices (4.9, 4.10, 4.15) are classical. Closure operators preserving directed sups have been called *inductive* or *algebraic* in the literature (see also Scott [1976, notably pp. 549-553]). The representation theorem 4.16 of continuous lattices is a combination of results by Scott [1972] and Hofmann-Stralka [1976]. The results in 4.21 and 4.22 are classics due to R.P. Dilworth and P. Crawley [1960]. The concept of an algebraic poset (Exercise 4.28) was formulated by R.-E. Hoffmann [1979a].

We conclude the chapter by depicting the hierarchy of some of the classes of complete lattices which we discussed.

THE HIERARCHY OF LATTICES ACCORDING TO THEIR DISTRIBUTIVITY AND CONTINUITY PROPERTIES

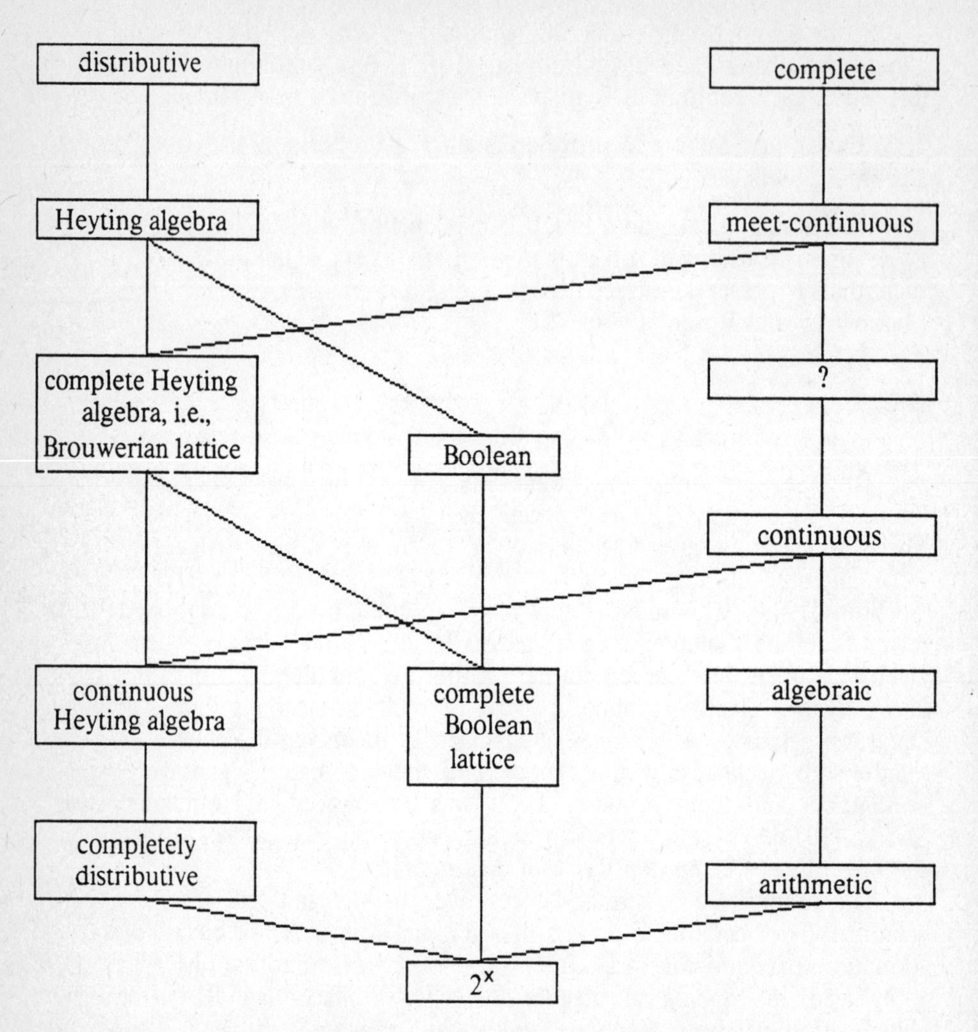

In the place of the question mark the hierarchy will be refined in Chapters V and VI. The most noteworthy block in this subhierarchy will be that of compact semilattices.

CHAPTER II

Topology of Continuous Lattices: The Scott Topology

In Chapter I we encountered the rich lattice-theoretic structure of continuous lattices. Perhaps even more typical for these lattices is their wealth of topological structure. The aim of the present chapter is to introduce topology into the study—a program to be continued in Chapter III.

Section 1 begins with a discussion of the Scott topology and its connection with the convergence given in lattice-theoretic terms by lower limits, or lim infs. This leads to a characterization theorem for continuous lattices in terms of properties of its lattice of Scott-open sets (1.14), which is a type of theorem that will be a recurrent theme (see Chapter VII).

In Section 2 we determine that the functions continuous for the Scott topology are those preserving directed sups. We can thus express one and the same property of a function between complete lattices either in topological or in algebraic (i.e., lattice-theoretical) terms. The space $[S \to T]$ of all Scott-continuous functions between *continuous* lattices is itself a continuous lattice, and the category of continuous lattices proves to be *cartesian closed.*

At this point we know that every continuous lattice is a topological space in the Scott topology; it is T_0, quasicompact, locally quasicompact, and sober. But exactly which T_0-spaces arise in this fashion? Section 3 presents the answer: they are precisely the *injective* ones.

In Section 4 we consider spaces of continuous functions $[X,\Sigma L]$ from a space X into a nonsingleton complete lattice L equipped with the Scott topology. The function space $[X,\Sigma L]$ is contained in the complete lattice L^X, hence it carries a partial order. Relative to this partial order $[X,\Sigma L]$ is a continuous lattice if and only if both $\mathcal{O}(X)$, the lattice of open sets of X, and L are continuous lattices. Spaces X for which $\mathcal{O}(X)$ is a continuous lattice have been given various names in the literature: quasi locally compact (as opposed to "locally quasicompact"), semi-locally bounded, core-compact, *CL*-spaces. In this section we take a new tack: we refuse to name them at all—even though they appear in several significant places (4.7, 4.10, 4.11).

1. THE SCOTT TOPOLOGY

The definition of the Scott topology on a complete lattice will characterize rather than exhibit open sets; in general topology this type of definition is common in associating open sets with a class of nets given as convergent. Since we wish to make a strong case for this parallel and illustrate at the same time the relation of the Scott topology with the classical idea of semi-continuity, we take some time at first to dwell on the concept of lower semi-continuous functions.

Consider an extended real-valued function $f: X \to \mathbb{R}^*$ on, say, a metric space X. It is *lower semicontinuous* (cf. also O-2.10 and I-1.21) if and only if it satisfies any of the following equivalent conditions:

(1) For each real number t, the set $f^{-1}(]t,\infty])$ is open in X;

(2) For any sequence x_n converging to x in X, the cluster points c of the sequence $f(x_n)$ satisfy $f(x) \leq c$;

(3) For any sequence x_n converging to x in X, $f(x) \leq \underline{\lim}_n f(x_n)$, where $\underline{\lim}_n f(x_n) = \sup_n \inf_{m \geq n} f(x_m)$.

In the above, sequences are adequate because X is metric; in more abstract settings nets would be required. Note that the range $\mathbb{R}^*$ is a complete (and, of course, continuous) lattice. In order to treat the concepts emerging in the conditions (1), (2) and (3) in a systematic fashion, we describe on an arbitrary complete lattice that structure of convergence (with its associated topology) which pertains precisely to the idea of lower semicontinuity. Evidently, the lower limit (often referred to as lim inf or $\underline{\lim}$) is a vital ingredient. We make it the subject of our first definition.

1.1. DEFINITION. Let L be a complete lattice. For any net $(x_j)_{j \in J}$ on L write:

$$\underline{\lim}_j x_j = \sup_j \inf_{i \geq j} x_i,$$

and we call this element the *lower limit* (or the *lim inf*) of the net.

For the time being denote by $\mathcal{S}$ the class of all those pairs $((x_j)_{j \in J}, x)$ consisting of a net on L and an element $x \in L$ with $x \leq \underline{\lim}\, x_j$; for each such pair we say that x *is* ***an*** *$\mathcal{S}$-limit* of $(x_j)_{j \in J}$ and write briefly $x \equiv_{\mathcal{S}} \lim x_j$. □

We remark that for any (eventually) constant net x_j with value x we have $x = \underline{\lim}\, x_j$, and that more generally for any net with $x = \underline{\lim}\, x_j$, if eventually $x_j \leq y$, then $x \leq y$ (the same holds with $\leq$ replaced by $\geq$). In the case of monotone nets (cf. O-1.2), $\underline{\lim}$ is just sup. Keep in mind that $\mathcal{S}$-limits, by this definition, are far from being unique (the $\underline{\lim}$ is only the *largest*).

We must recall next the general relation between *convergence* and *topology*. If, on any set L one is given an arbitrary class $\mathcal{L}$ of pairs $((x_j)_{j \in J}, x)$ consisting of a net and an element of L, then associated with $\mathcal{L}$ is a family of sets

$$\mathcal{O}(\mathcal{L}) = \{U \subseteq L: \text{whenever } ((x_j)_{j \in J}, x) \in \mathcal{L} \text{ and } x \in U, \text{ then eventually } x_j \in U\}.$$

Clearly both $\emptyset$ and L belong to $\mathcal{O}(\mathcal{L})$, which is closed under the formation of arbitrary unions and finite intersections; that is to say, $\mathcal{O}(\mathcal{L})$ is a topology.

By the very definition we know that, for any $((x_j)_{j\in J},x)\in\mathcal{L}$, the element x is a limit of the net x_j relative to the topology $\mathcal{O}(\mathcal{L})$. Since, however, $\emptyset$ and L may very well be the only elements of $\mathcal{O}(\mathcal{L})$, we are obviously not saying very much; specific information on $\mathcal{L}$ must become available before one can hope to get a close link between $\mathcal{L}$ and $\mathcal{O}(\mathcal{L})$. (A canonical reference for the relation between convergence and topology in this framework is Kelley [1955, Chapt. II].) Fortunately, in our present situation, we do have specific information about our class $\mathcal{S}$. We begin exploiting it by characterizing the sets $U\in\mathcal{O}(\mathcal{S})$.

1.2. LEMMA. *Let* L *be a complete lattice and* $U\subseteq$L. *Then* $U\in\mathcal{O}(\mathcal{S})$ *iff the following two conditions are satisfied:*

(i) $U = \uparrow U$;
(ii) $\sup D \in U$ *implies* $D\cap U \neq \emptyset$ *for all directed sets* $D\subseteq$L.

In (ii) *directed sets may be replaced by ideals.*

Proof. First, suppose $U\in\mathcal{O}(\mathcal{S})$. To prove (i), assume $u\in U$ and $u\leq x$. Then $u\leq x = \underline{\lim}\ x$ with the constant net (x) with value x. Since we have that $u\in U\in\mathcal{O}(\mathcal{S})$, we conclude from the definition of $\mathcal{O}(\mathcal{S})$ that the net (x) must be eventually in U. This means $x\in U$.

In order to prove (ii), let D be a directed set in L with $\sup D\in U$. Consider the net $(x_d)_{d\in D}$ with $x_d = d$. Now $\inf_{c\geq d} x_c = d$, and thus $\underline{\lim}\ x_d = \sup D \in U\in\mathcal{O}(\mathcal{S})$. Since $((x_d)_{d\in D}, \sup D) \in \mathcal{S}$, we conclude that $d = x_d$ is eventually in U; whence $D\cap U \neq \emptyset$.

Second, suppose that (i) and (ii) are satisfied. We take $((x_j)_{j\in J},x)$ with $x\in U$, and we must show that x_j is eventually in U. By the definition of $\mathcal{S}$, we have $x\leq\sup_j \inf_{i\geq j} x_j$. We set $D = \{\inf_{i\geq j} x_i : j\in J\}$; this gives a directed set in L. We have $x\leq\sup D$, whence $\sup D\in U$ by (i). Thus, by (ii), we find an element $d\in D\cap U$. By the definition of D we note $d = \inf_{i\geq j} x_i$ for some j. Since $d\leq x_i$ for all $i\geq j$, we conclude from (1) that $x_i\in U$ for all $i\geq j$. This is what we wanted to show.

The equivalence of (ii) with ideals in place of directed sets is immediate in the presence of condition (i). □

From our previous remarks we know that the sets U satisfying the conditions in 1.2 form a topology, and it is simple enough to verify this directly. The point of our discussion was to show that this is a naturally arising topology, because lim inf convergence is natural in any complete lattice. This topology will thus be officially named.

1.3. DEFINITION. A subset U of a complete lattice L is called *Scott open* iff it satisfies the conditions of 1.2. The complement of a Scott-open set is called *Scott closed*. The collection of all Scott-open subsets of L will be called the *Scott topology* of L and will be denoted σ(L).

We say that a subset X of a complete lattice L *has the property* (S) provided that the following condition is satisfied:

(S) If $\sup D \in X$ for any directed set D, then there is a $y \in D$ such that $x \in X$ for all $x \geq y$ with $x \in D$. □

1.4. REMARK. *In any complete lattice* L *we have the following conclusions:*

(i) *A set is Scott-closed iff it is a lower set closed under directed sups;*
(ii) $\downarrow x = \{x\}^-$ *(closure with respect to* $\sigma(\mathrm{L})$*) for all* $x \in \mathrm{L}$;
(iii) $\sigma(\mathrm{L})$ *is a* T_0*-topology;*
(iv) *Every upper set is the intersection of its Scott-open neighborhoods;*
(v) *A set is Scott-open iff it is an upper set satisfying* (S);
(vi) *Every lower set has property* (S);
(vii) *The collection of all subsets having property* (S) *is a topology.*

Proof. (i): $A \subseteq \mathrm{L}$ is a lower set iff $\mathrm{L} \setminus A$ is an upper set, and $\mathrm{L} \setminus A$ satisfies 1.2(ii) iff A is closed under directed sups.

(ii): $\downarrow x$ is the smallest lower set containing x, and it happens to be closed under directed sups.

(iii): If $\{x\}^- = \{y\}^-$, then $\downarrow x = \downarrow y$ by (ii); thus $x = y$.

(iv): Every upper set B is the intersection of the sets $\mathrm{L} \setminus \downarrow x$ where $x \in \mathrm{L} \setminus B$. These sets are open in view of (ii).

(v) and (vi): Immediate.

(vii): The intersection of two sets satisfying (S) will again satisfy (S), and any union of sets satisfying (S) will satisfy (S). Since $\varnothing$ and L clearly satisfy (S), the assertion follows. □

The definition of Scott-open sets of a complete lattice provides a characterization but not a procedure for building Scott-open sets in general—except for the rather meager information of 1.4(ii) above. It is therefore important that we familiarize ourselves with some examples.

1.5. EXAMPLES. Let L be a complete lattice.

(1) If L is finite, then the Scott-open sets are just the upper sets.

(2) If L is a chain, then the sets $]x,1] = \uparrow x \setminus \{x\} = \mathrm{L} \setminus \downarrow x$ are Scott open for any $x \in \mathrm{L}$; and together with L these are all the Scott-open sets.

(3) For the chain $2 = \{0,1\}$, we have $\sigma(2) = \{\varnothing, \{1\}, \{0,1\}\}$. The space 2 with this topology is well-known under the name of the *Sierpinski space.*

(4) If $\mathrm{L} = 2^X$, the power-set space, we have $\sigma(\mathrm{L})$ equal to the well-known collection of families of sets called *families of finite character.* These are families that contain a set iff they contain some finite subset.

(5) If L is a continuous lattice, all sets $\twoheaduparrow x$ for $x \in \mathrm{L}$ are Scott-open. This is exactly the content of I-1.19. We show below in 1.10 that these sets are a basis for $\sigma(\mathrm{L})$, and that $\twoheaduparrow x$ is in fact the interior of $\uparrow x$ with respect to this

topology. (Note that it suffices for the openness of all $\Uparrow x$ that $\ll$ satisfy the interpolation property (I-1.15).)

(6) If $L = [0,1]^2$, the square with the componentwise order, a subset U is Scott open iff it is an upper set and is open in the ordinary topology induced by the plane. Here is a typical picture:

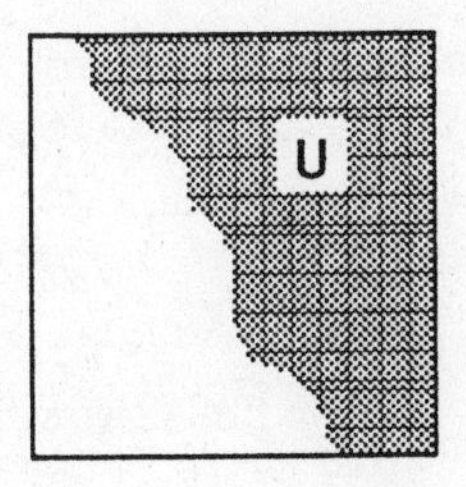

We leave this characterization as an exercise, since later on we will have enough theory to make this an easy consequence. We can see at this point, however, that every Scott-open set of $[0,1]^2$ is the union of open upper rectangles. Note that these rectangles are the intersection of two sets of the form $L \setminus \downarrow x$. □

Notice 1.5(5) indicates at this early stage that a *continuous* lattice has a good supply of relatively small Scott-open sets (small, given the restriction that they all have to be upper sets). Observe also, that, in general, the Scott topology on a complete lattice is neither the coarsest nor the finest of all of the T_0-topologies for which $\downarrow x = \{x\}^-$.

In order to complete the story we must return to the discussion of the concept of convergence and investigate whether the Scott topology (which we derived from a convergence concept) is in fact adequate to describe in topological terms lim inf convergence. The topologists tell us how to deal with this type of question in general (see Kelley, loc. cit.). Namely, we have:

FACT. Given a class $\mathcal{L}$ we always have

$$((x_j)_{j\in J}, x) \in \mathcal{L} \text{ iff } x = \lim x_j \text{ with respect to } \mathcal{O}(\mathcal{L})$$

precisely when the following axioms are satisfied:

(CONSTANTS) For every constant net one has $((x)_{j\in J}, x) \in \mathcal{L}$.

(SUBNETS) If $(y_i)_{i\in I}$ is a subnet of $(x_j)_{j\in J}$ and $((x_j)_{j\in J}, x) \in \mathcal{L}$, then $((y_i)_{i\in I}, x) \in \mathcal{L}$.

(DIVERGENCE) If $((x_j)_{j\in J}, x)$ is ***not*** in $\mathcal{L}$, then $(x_j)_{j\in J}$ has a subnet $(y_i)_{i\in I}$ no subnet $(z_k)_{k\in K}$ of which ever has $((z_k)_{k\in K}, x) \in \mathcal{L}$.

(ITERATED LIMITS) If $((x_i)_{i\in I}, x)\in\mathcal{L}$, and if $((x_{i,j})_{j\in J(i)}, x_i)\in\mathcal{L}$ for all $i\in I$, then $((x_{i,f(i)})_{(i,f)\in I\times M}, x)\in\mathcal{L}$, where $M = \mathsf{X}_{i\in I}J(i)$ is a product of directed sets. □

We now discuss the conditions under which our notion of lim inf convergence is topological using this general criterion on an arbitrary convergence class.

1.6. PROPOSITION. (i) *For any complete lattice* L, *the class* $\mathcal{S}$ *of* 1.1 *always satisfies the axioms* (CONSTANTS) *and* (SUBNETS).

(ii) *If* L *is a continuous lattice, then it also satisfies the axioms* (DIVERGENCE) *and* (ITERATED LIMITS). *Consequently, for a continuous lattice, we have*

$$x \equiv_{\mathcal{S}} \lim x_j \text{ iff } x = \lim x_j \text{ with respect to } \sigma(\mathrm{L}).$$

Remark. Remember that limits are not unique with respect to $\sigma(\mathrm{L})$; thus, convergence in this topology means $x \le \underline{\lim}\, x_j$.

Proof. (i): (CONSTANTS): This is trivial as we have observed before.

(SUBNETS): Suppose that $x\le \sup_j \inf_{i\ge j} x_i$ in L and that $f\colon I\to J$ gives a subnet $y_i = x_{f(i)}$ of x_i. For any $j\in J$, there is an $i_j\in I$ such that $i_j\le k$ implies $j\le f(k)$, and thus

$$\inf_{i\ge j} x_i \le \inf_{i^*\ge i_j} y_{i^*}.$$

Forming the sup over j gives on the left side $\underline{\lim}\, x_j$, which dominates x by assumption. The sup on the right side is certainly majorized by $\sup_i \inf_{i^*\ge i} y_{i^*} = \underline{\lim}\, y_i$, whence $x\le\underline{\lim}\, y_i$, which is what we had to show.

(ii): We assume for the remainder that L is continuous.

(DIVERGENCE): Let $(x_j)_{j\in J}$ be a net such that ***not*** $x\le\underline{\lim}\, x_j$. We must produce a subnet $(y_i)_{i\in I}$, such that for any of its subnets $(z_k)_{k\in K}$ we also have ***not*** $x\le\underline{\lim}\, z_k$.

Since L is continuous, we find an element $u\ll x$ with $u\not\le\underline{\lim}\, x_j$ (I-1.7). From the last relation and $\inf_{j^*\ge j} x_{j^*}\le\underline{\lim}\, x_{j^*}$ for all $j\in J$, we can now deduce $\inf_{j^*\ge j} x_j \not\ge u$ for all j. Hence, for each $j\in J$, we find an $f(j)\ge j$ such that $x_{f(j)}\not\ge u$. Set $y_j = x_{f(j)}$, and observe that $(y_j)_{j\in J}$ is a subnet of $(x_j)_{j\in J}$.

Now let $(z_k)_{k\in K}$ be any subnet of $(y_j)_{j\in J}$. If $k\in K$, then for some $k^*\ge k$ we have $z_{k^*} = x_{f(j)}$, with a suitable $j\in J$, and thus

$$\inf_{k^*\ge k} z_{k^*}\not\ge u.$$

If the directed net $d_k = \inf_{k^*\ge k} z_{k^*}$ had its sup dominating x, then for some $k\in K$ we would have $u\le d_k$ by I-1.1; but, as we have just remarked, this is impossible. Hence $\underline{\lim}\, z_k = \sup_k d_k\not\ge x$, which is what we had to show.

(ITERATED LIMITS): Suppose that we have nets $(x_{i,j})_{j\in J(i)}$, $(x_i)_{i\in I}$ such that $x\le\underline{\lim}\, x_i$ and $x_i\le\underline{\lim}_j\, x_{i,j}$ for all $i\in I$. If we abbreviate $M = \mathsf{X}_{i\in I} J(i)$, then we have to show

$$x \leq \underline{\lim}_{(i,f)\in I\times M}\, x_{i,f(i)}.$$

For this purpose, we take an arbitrary $u \ll x$; we need to show that it is $\leq$ the lim inf. Since we have by assumption convergence of the given nets with respect to $\mathcal{O}(\mathcal{S}) = \sigma(\mathrm{L})$, then, since topological spaces satisfy **(ITERATED LIMITS)**, it follows that

$$x = \lim_{(i,f)\in I\times M} x_{i,f(i)} \text{ with respect to } \sigma(\mathrm{L}).$$

But $\uparrow u$ is a Scott-open neighborhood of x by 1.5(5), and thus $u \ll x_{i,f(i)}$ eventually; in particular $u \leq x_{i,f(i)}$, eventually. Hence $u \leq \underline{\lim}\, x_{i,f(i)}$, as was to be shown.

The remainder is then clear from Kelley, loc. cit. □

It is conceivable at this point that the assumption of the continuity of L in proving the **(DIVERGENCE)** and **(ITERATED LIMITS)** axioms was purely a matter of convenience for the proof. That this is not so (at least as far as the second is concerned) is shown in the following lemma:

1.7. LEMMA. *If, in a complete lattice* L, *the class* $\mathcal{S}$ *satisfies* **(ITERATED LIMITS)**, *then* L *is continuous.*

Proof. We use the axiom to verify the validity of equation (DD) of I-2.3. Thus, let $\{x_{j,k} : j\in J,\ k\in K(j)\}$ be a family of elements in L such that $(x_{j,k})_{k\in K(j)}$ is directed for each $j\in J$. We set $M = \mathsf{X}_{j\in J} K(j)$, and we set $x_j = \sup_k x_{j,k}$ and $x = \inf_j x_j$.

If we consider on the set J the quasiorder $j \leq k$ for *all* $(j,k)\in J\times J$, then $(x_j)_{j\in J}$ becomes a net in such a fashion that

$$\inf_{k\geq j} x_k = \inf_{k\in J} x_k = x,$$

whence $x = \underline{\lim}\, x_j$. Now we apply **(ITERATED LIMITS)** and conclude

$$x \leq \underline{\lim}\, x_{j,f(j)} = \bigvee_{(j,f)\in J\times M} \bigwedge_{(j^*,f^*)\geq(j,f)} x_{j^*,f^*(j^*)}.$$

Since $(x_{j,k})_{k\in K(j)}$ is directed, we have $x_{j^*,f^*(j^*)} \geq x_{j^*,f(j^*)}$ whenever $(j^*,f^*) \geq (j,f)$. Thus the element

$$s_{j,f} = \bigwedge_{(j^*,f^*)\geq(j,f)} x_{j^*,f^*(j^*)}$$

equals the element

$$\bigwedge_{j^*\geq j} x_{j^*,f(j^*)} = \bigwedge_{i\in J} x_{i,f(i)}$$

by virtue of the trivial ordering of J. Hence, $s_{j,f}$ does not depend on j, and we may abbreviate it as s_f. We conclude that

$$x \leq \bigvee_{(j,f)\in J\times M} s_{j,f} = \bigvee_{f\in M} s_f.$$

Recalling the definition of x and s_f, we obtain

$$\bigwedge_{j\in J}\bigvee_{k\in K(j)} x_{j,k} \leq \bigvee_{f\in M}\bigwedge_{j\in J} x_{j,f(j)},$$

and this is sufficient for (DD). □

What we now have in effect proved is the following characterization of continuous lattices:

1.8. THEOREM. *For a complete lattice* L *the following statements are equivalent:*

(1) *lim-inf convergence is topological convergence for the Scott topology; that is, for all* $x \in L$ *and all nets* (x_j) *on* L

$$x \leq \underline{\lim}\, x_j \text{ iff } x = \lim x_j \text{ with respect to } \sigma(L);$$

(2) L *is a continuous lattice.* □

Having recognized lim inf convergence as the essential ingredient in the study of lower semicontinuity, we can say after Theorem 1.8 that among complete lattices it is precisely the *continuous* ones which allow the study of lower semicontinuity completely in topological terms. Nevertheless, the Scott topology in itself remains a highly useful tool in treating arbitrary complete lattices.

Back in Section 3 of Chapter I we introduced as a tool a definition of openness; we pause for a moment now to tie up the loose ends and to show that the terminology was appropriate.

1.9. REMARK. *In a complete lattice, an upper set open in the sense of* I-3.1 *is open with respect to the Scott topology* $\sigma(L)$.

Proof. Let U be an upper set open in the sense of I-3.1. We need only verify 1.2(ii). Suppose $\sup D \in U$, where D is directed. Then there is a $v \ll \sup D$ with $v \in U$. It follows that $v \leq u$ for some $u \in D$. Thus, since U is upper, we have $u \in U$ as desired. □

1.10. PROPOSITION. *Let* L *be a continuous lattice.*

(i) *Each point* $x \in L$ *has a* $\sigma(L)$*-neighborhood basis consisting of the sets* $\twoheaduparrow u$ *with* $u \ll x$.

(ii) *The open upper sets in the sense of* I-3.1 *are exactly the open sets of the Scott topology.*

(iii) *With respect to* $\sigma(L)$, *we have* $\operatorname{int} \uparrow x = \twoheaduparrow x$.

(iv) *With respect to* $\sigma(L)$, *we have for any subset* $X \subseteq L$

$$\operatorname{int} X = \bigcup\{\twoheaduparrow u : \twoheaduparrow u \subseteq X\}.$$

Proof. (i): As we remarked in 1.5(5), the sets $\twoheaduparrow u$ are open neighborhoods in $\sigma(L)$. Conversely, if U is a Scott-open neighborhood of x, then, since

$$\sup \twoheaddownarrow x = x \in U,$$

there is a $u \in \twoheaddownarrow x \cap U$ by 1.2(ii). Since U is an upper set by 1.2(i), we conclude $x \in \twoheaduparrow u \subseteq U$.

(ii): Half of the statement is 1.9. Conversely, if U is Scott open, then by (i) we have

$$U = \bigcup\{\Uparrow u : u \in U\};$$

whence, it is open in the sense of I-3.1.

(iii): If $y \in \operatorname{int} \uparrow x$, then by (i) there is a $u \in \uparrow x$ with $u \ll y$. But then $y \in \Uparrow x$. Obviously $\Uparrow x \subseteq \operatorname{int} \uparrow x$.

(iv): This follows directly from (i). □

We recall at this point that every topology is a lattice, and indeed a complete Heyting algebra (remember O-3.22!). It is therefore meaningful to search for prime and co-prime elements in $\sigma(L)$ (see I-3.11, -3.15). To formulate one of our conditions it is necessary to speak of the continuity of an operation (the main topic of the next section). We say that the sup operation is *jointly continuous with respect to the Scott topology* provided that the mapping

$$(x,y) \mapsto x \vee y : (L, \sigma(L)) \times (L, \sigma(L)) \to (L, \sigma(L))$$

is continuous in the product topology.

1.11. PROPOSITION. *Let* L *be a complete lattice and* $U \subseteq L$.

(i) *U is a co-prime in $\sigma(L)$ iff $U \in \sigma(L)$ and U is a filter;*

(ii) *U is a prime in $\sigma(L)$ and $U \neq L$ if $U = L \setminus \downarrow u$ for some $u \in L$;*

(iii) *This last condition is also necessary, provided that the sup operation is jointly continuous;*

(iv) *This is the case if* L *is a continuous lattice.*

Proof. (i): Firstly suppose that $U \in \sigma(L)$ is a filter and that U is not a co-prime in $\sigma(L)$. Thus there are $V, W \in \sigma(L)$ such that $U \subseteq V \cup W$ and elements $v \in U \setminus V$ and $w \in U \setminus W$. Since V and W are upper sets we have $vw \notin V \cup W$. But $vw \in U$ since U is a filter. This is a contradiction.

Secondly, suppose that U is a co-prime in $\sigma(L)$. To show that U is a filter, note first that it is an upper set. Suppose $v, w \in U$. Then $U \not\subseteq L \setminus \downarrow v$ and $U \not\subseteq L \setminus \downarrow w$. By 1.4(ii), the sets $L \setminus \downarrow v$ and $L \setminus \downarrow w$ are Scott open. Thus, since U is co-prime,

$$U \not\subseteq ((L \setminus \downarrow v) \cup (L \setminus \downarrow w)) = L \setminus (\downarrow v \cap \downarrow w) = L \setminus \downarrow vw.$$

Thus $U \cap \downarrow vw \neq \emptyset$. Since U is an upper set, this means $vw \in U$.

(ii): Recall that $L \setminus \downarrow u = L \setminus \{u\}^-$ for all u by 1.4(ii). But the complements of singleton closures are prime in any topology.

(iii): Assume that the sup-operation is jointly continuous relative to the Scott topology, and let $U \in \mathrm{PRIME}\ \sigma(L)$, $U \neq L$. Set $A = L \setminus U$ and $u = \sup A$. In order to show that $U = L \setminus \downarrow u$ it suffices to verify $u \in A$. Since A is closed under directed sups by 1.4(i), all we have to verify is that A is directed.

By way of contradiction assume that it is not. Then there are elements $a, b \in A$ with $a \vee b \in U$. By the continuity of the sup-operation we would find Scott-open neighborhoods V and W of a and b, respectively, such that $V \vee W \subseteq U$. But since V and W are upper sets, we have $V \vee W = V \cap W$. Since

U is prime, the relation $V\cap W\subseteq U$ implies $V\subseteq U$ or $W\subseteq U$. This would entail $a\in U$ or $b\in U$, which would contradict $a,b\in A = L\setminus U$.

(iv): Finally suppose that L is continuous. In order to show the continuity of the sup-operation at (a,b) we pick some $u\ll a\vee b$. By I-1.6 we have

$$a\vee b = (\sup \Downarrow a)\vee(\sup \Downarrow b) = \sup (\Downarrow a\vee\Downarrow b).$$

Since $\Downarrow a\vee\Downarrow b$ is directed (I-1.2(iii)), we find some $x\ll a$ and $y\ll b$ with $u\ll x\vee y$ (by I-1.19). But then $\Uparrow x$ and $\Uparrow y$ are Scott-open neighborhoods of a and b, respectively, such that

$$\Uparrow x\vee\Uparrow y\subseteq\uparrow(x\vee y)\subseteq\Uparrow u.$$

Since the $\Uparrow u$ with $u\ll a\vee b$ form a basis of σ(L)-neighborhoods of $a\vee b$ by Proposition 1.10(i), the desired continuity is established. □

We can immediately rephrase 1.11 in topological terminology, if we recall the concept of a *sober space* (see I-3.36 and the discussion). Remember that a non-empty subset A of a topological space X is called *irreducible* iff it is closed and not the union of two proper closed subsets (that is, the complementary set $X\setminus A \in$ PRIME $\mathcal{O}(X)$; see I-3.11). A space X is called *sober* iff every irreducible set A has a unique dense point (that is, $A = \{a\}^-$ with a unique $a\in A$). Clearly all singleton closures are irreducible. (Notice that a sober space is automatically T_0 since $\{x\}^- = \{y\}^-$ always implies $x = y$.) We now have the following corollaries of 1.11 with a slight sharpening:

1.12. COROLLARY. *If* L *is a complete lattice such that the sup-operation is jointly Scott-continuous, then* (L,σ(L)) *is a sober space.*

Proof. Immediate from 1.11 and the definitions. □

1.13. COROLLARY. *If* L *is a continuous lattice, then* (L,σ(L)) *is a quasicompact and locally quasicompact sober space. In particular,* (L,σ(L)) *is a Baire space.*

Proof. We have to show that a point $x\in$L has a basis of quasicompact neighborhoods. By 1.10 the sets $\Uparrow y$ with $y\ll x$ form a basis for the neighborhoods of the point. But as we know, if $x\in U\in\sigma$(L), then actually we have a $y\in U$ with $y\ll x$; hence, $\uparrow y\subseteq U$, and so the sets $\uparrow y$ can be used as neighborhoods. Since $\uparrow y$ (and hence, in particular L $= \uparrow 0$) is trivially quasicompact with respect to any topology whose open sets are upper sets, the assertion is proved. That (L,σ(L)) is a Baire space follows from I-3.43.10. □

We wish to warn the reader about a subtlety concerning the joint continuity of the sup-operation above. We cannot be satisfied by saying that the sup-operation is a continuous function (L$\times$L, σ(L$\times$L))$\rightarrow$(L,σ(L)); this continuity is weaker, since in general we have a proper containment of topologies: σ(L$\times$L)$\supset\sigma$(L)$\times\sigma$(L). We will return to this question at greater length in Section 4 below (see 4.11 ff).

We know enough about the Scott topology now to use it for yet another characterization theorem for continuous lattices.

1.14. THEOREM. *For any complete lattice, the following conditions are equivalent:*

(1) L *is continuous*;

(2) *If* $U \in \sigma(\mathrm{L})$, *then* $U = \bigcup\{\Uparrow x : x \in U\}$;

(3) *Each point has a neighborhood basis (in the Scott topology) of Scott-open filters, and* $\sigma(\mathrm{L})$ *is a continuous lattice*;

(4) *For each point* $x \in \mathrm{L}$ *we have* $x = \sup\{\inf U : x \in U \in \sigma(\mathrm{L})\}$;

(5) $\sigma(\mathrm{L})$ *has enough co-primes and is a continuous lattice*;

(6) $\sigma(\mathrm{L})$ *is completely distributive*;

(7) *Both* $\sigma(\mathrm{L})$ *and* $\sigma(\mathrm{L})^{\mathrm{op}}$ *are continuous.*

Proof. (1) implies (2): Use 1.10.

(2) implies (1): Let $x \in \mathrm{L}$ and set $y = \sup \Downarrow x \leq x$. If $y < x$, then $\mathrm{L}\backslash\downarrow y$ is a Scott-open neighborhood of x; hence by (2) it contains an open neighborhood $\Uparrow z$ of x with $z \in \mathrm{L}\backslash\downarrow y$. But then $z \ll x$, and thus $z \leq \sup \Downarrow x = y$, a contradiction.

(1) implies (3): By (1) iff (2), x has arbitrarily small neighborhoods of the form $\Uparrow y$ with $y \ll x$. By 1.10 and I-3.3, we know then that x has arbitrarily small Scott-open neighborhoods which are filters. In order to prove the continuity of $\sigma(\mathrm{L})$, we let U be Scott open. For any $x \in U$ we find a $y \in U$ with $y \ll x$ by (2). Then $x \in \Uparrow y \in \sigma(\mathrm{L})$, and we claim that $\Uparrow y \ll U$: Indeed, if D is a directed family of Scott-open sets covering U, then one of its members must contain y, hence $\uparrow y$, since Scott-open sets are upper sets, and thus it contains $\Uparrow y$. We have shown $U = \bigcup\{V : V \ll U\}$.

(3) implies (4): For $x \in \mathrm{L}$ set

$$y = \sup\{\inf U : x \in U \in \sigma(\mathrm{L})\} \leq x.$$

If $y < x$, then $\mathrm{L}\backslash\downarrow y$ is a Scott-open neighborhood of x. Let V be a Scott-open neighborhood of x with $V \ll \mathrm{L}\backslash\downarrow y$, which exists since $\sigma(\mathrm{L})$ is continuous by (3). Now use (3) to find a Scott-open filter neighborhood U of x within V. By the definition of y we have $\inf U \leq y$. Then

$$\mathrm{L}\backslash\downarrow y \subseteq \mathrm{L}\backslash\downarrow \inf U = \mathrm{L}\backslash\bigcap\{\downarrow u : u \in U\} = \bigcup\{\mathrm{L}\backslash\downarrow u : u \in U\}.$$

Since U is a filter, the $L\backslash\downarrow u$ for $u \in U$ form a directed family of Scott-open sets. Since $V \ll \mathrm{L}\backslash\downarrow y$, there must be a $u \in U$ such that $V \subseteq \mathrm{L}\backslash\downarrow u$ and so $u \notin V$. This is a contradiction to $U \subseteq V$. Thus $x = y$.

(4) implies (1): Clear since for every Scott-open neighborhood U of x one has $\inf U \ll x$.

(3) iff (5): Clear from 1.11.

(5) iff (6) iff (7): Consequence of I-3.15. □

Note that condition (2) is stronger than saying that $\sigma(\mathrm{L})$ has a basis of sets $\Uparrow x$. A parallel result to 1.14 for algebraic lattices reads as follows:

1.15. COROLLARY. *For any complete lattice* L *the following conditions are equivalent:*

(1) L *is algebraic.*

(2) *The Scott topology has a basis of sets* $\uparrow k$ *where* $k \in K(L)$.

(3) $\sigma(L)$ *is algebraic and has enough co-primes.*

Proof. (1) implies (2): Let U be a Scott-open neighborhood of x. We recall that $x = \sup(\downarrow x \cap K(L))$ and $\downarrow x \cap K(L)$ is directed. Hence by 1.2(ii), we find a $k \in \downarrow x \cap K(L) \cap U$. Then $\uparrow k = \Uparrow k$ is a Scott-open neighborhood of k, hence of x, with $\uparrow k \subseteq U$.

(2) implies (3): If $k \in K(L)$, then $\uparrow k \in \sigma(L)$ since $\Uparrow k = \uparrow k$. Now $\uparrow k$ is a quasicompact set (if $\uparrow k$ is covered by Scott-open sets, then one of them must contain k, hence $\uparrow k$ by 1.2(i)). Thus $\sigma(L)$ is algebraic by I-4.4. Since all $\uparrow k$ are filters, hence co-primes by 1.11, we are done.

(3) implies (1): Since $\sigma(L)$ is algebraic, the Scott topology has a basis of quasicompact sets U. Since there are enough co-primes, U is a union of open filters by 1.11, and thus, by quasicompactness, $U = U_1 \cup \ldots \cup U_n$ with open filters U_k. It is no loss of generality to assume that none of the U_k is contained in the union of the others. Then $V = U_1 \setminus (U_2 \cup \ldots \cup U_n)$ is quasicompact and filtered. We claim that V has a smallest element u_1.

For if not, then $V \subseteq \bigcup\{L \setminus \downarrow v : v \in V\}$; and by quasicompactness and the fact that the $L \setminus \downarrow v$ form a directed family, there would be a $v \in V$ with $V \subseteq L \setminus \downarrow v$, notably $v \notin V$, which is impossible. Since $U_2 \cup \ldots \cup U_n$ is an upper set, it cannot contain $\inf U_1$. Hence $u_1 = \min U_1$. Since U_1 is an upper set, $U_1 = \uparrow u_1$. Since U_1 is Scott open, $u_1 \in K(L)$.

We have shown that $\sigma(L)$ has a basis of sets $\uparrow k$ where $k \in K(L)$. Now let $x \in L$, set $y = \sup(\downarrow x \cap K(L)) \leq x$. If $y < x$, then the Scott-open neighborhood $L \setminus \downarrow y$ of x would contain a basic neighborhood $\uparrow k$ of x. Then there would be a $k \in K(L)$ with $k \leq x$ and $k \not\leq y$ which contradicts the definition of y. □

EXERCISES

1.16. EXERCISE. (i) Definition 1.1 of the lim-inf applies reasonably within any complete semilattice (O-2.11).

(ii) Definition 1.3 of the Scott topology $\sigma(L)$ applies within any poset L and reasonably so in any up-complete poset (O-2.11).

(iii) The relation $\mathcal{O}(\mathcal{S}) = \sigma(L)$ holds in any complete semilattice. (See 1.2.)

(iv) The observations of 1.4 and 1.5(1) hold in any up-complete poset, and 1.5(5) holds in a continuous poset. □

DEFINITION. Let L be an up-complete poset. A topology on L is said to be *order consistent* iff

(i) $\{x\}^- = \downarrow x$ for all $x \in L$.

(ii) If $x = \sup_j x_j$ for a directed net, then $x = \lim_j x_j$.

In (ii) we could say equivalently: if $x = \sup I$ for an ideal I, then $x = \lim I$. We call the *upper topology* on L the topology generated by the collection of sets $L \setminus \downarrow x$, and we denote it by $\upsilon(L)$. (Cf. O-4.11 and III-3.21 ff.) □

1.17. EXERCISE. (i) Both the Scott and the upper topologies are order consistent, and for any order consistent topology τ on L, we have

$$\upsilon(L) \subseteq \tau \subseteq \sigma(L).$$

In other words, the upper topology is the coarsest and the Scott topology is the finest of all order consistent topologies.

(ii) If L is a complete semilattice and τ an order consistent topology on L, then $\downarrow(\underline{\lim}_j x_j)$ is contained in the set of all τ-cluster points of the net x_j.

(iii) If $(x_j)_{j \in J}$ is a directed net, then the set of all limit points of this net is precisely $\downarrow \sup x_j$.

(iv) If, in addition, the translations $x \mapsto a \wedge x : L \to L$ are τ-continuous for all $a \in L$ (in this case we say that L is a *semitopological semilattice* (see also VI-1.11)), then L is meet continuous. □

In III-3.23 we will describe those complete lattices for which $\upsilon(L) = \sigma(L)$.

1.18. EXERCISE. (i) Show 1.6 and 1.7 for complete-continuous semilattices.

(ii) Formulate and prove the analogue of Theorem 1.8 for complete semilattices.

(iii) Observe that complete-continuous semilattices are precisely the Scott-closed subsemilattices of continuous lattices. □

1.19. EXERCISE. In a continuous poset, a filter is open in the sense of I-3.30 iff it is open with respect to the Scott topology $\sigma(L)$ (cf. 1.9, 1.10). □

This links Exercise I-1.32 with Exercise I-3.30, since evidently condition I-1.32 (P) is satisfied for the Scott topology. In continuous semilattices, one may use the expression "open filter" without much ambiguity. Moreover, in an up-complete poset L, the open filters are still the co-primes of $\sigma(L)$. (See 1.11(ii).) The following will, among other things, yield a different proof of the fact that a continuous lattice is a sober space relative to its Scott topology. (See 1.12.)

1.20. EXERCISE. (i) If L is a continuous poset, then $(L, \sigma(L))$ is a locally quasicompact sober space.

(ii) Moreover, $U \in \sigma(L)$ where $U \neq L$ is prime in $\sigma(L)$ iff $U = L \setminus \downarrow u$ for some $u \in L$ (cf. 1.11).

(HINT: (i): Local quasicompactness is shown as in 1.13. Sobriety will follow from (ii).

(ii): Now let $U \in \sigma(L)$, $U \neq L$. If $U = L \setminus \downarrow u$, then U is prime in $\sigma(L)$, since $\downarrow u = \{u\}^-$ with respect to $\sigma(L)$. Conversely, assume that U is prime in $\sigma(L)$ and set $A = L \setminus U$; we have to show that A has a largest element u;

since A is a lower set, this will show $A = \downarrow u$ as desired. Now let

$$A^* = \{a \in A\colon \text{there is a } b \in A \text{ with } a \ll b\} = \bigcup\{\twoheaddownarrow a : a \in A\} = \twoheaddownarrow A.$$

We claim that A^* is directed: Let $a,b \in A^*$; we must show $\twoheaduparrow a \cap \twoheaduparrow b \cap A \neq \emptyset$. If not, then $\twoheaduparrow a \cap \twoheaduparrow b \subseteq U$ for some $a,b \in A^*$; but $\twoheaduparrow a, \twoheaduparrow b \in \sigma(\mathrm{L})$ by 1.16(iv). Since U is prime, we conclude $\twoheaduparrow a \subseteq U$ or $\twoheaduparrow b \subseteq U$; but $\twoheaduparrow a$ contains a $c \in A = \mathrm{L} \backslash U$ which is impossible; similarly $\twoheaduparrow b \subseteq U$ is impossible. As A^* is a directed set, $u = \sup A^*$ exists. Since A is Scott closed, $u \in A$. Now let $a \in A$. Then $a = \sup \twoheaddownarrow a$ since L is a continuous poset; but $\twoheaddownarrow a \subseteq A^*$ implies

$$a = \sup \twoheaddownarrow a \leq \sup A^* = u.$$

Thus $u = \max A$ as was desired.) □

1.21. EXERCISE. Let L be a continuous poset. Then each $x \in \mathrm{L}$ has a $\sigma(\mathrm{L})$-neighborhood basis of the form $\twoheaduparrow u$, where $u \ll x$, and a $\sigma(\mathrm{L})$-neighborhood basis of open filters.

(HINT: For the first assertion, verify the proof of 1.10(i); for the second, use the first, I-3.31 and 1.19 above.) □

It is noteworthy, that Theorem 1.14 generalizes to continuous semilattices:

1.22. EXERCISE. Let L be an up-complete poset, and consider the conditions:

(2′) If $U \in \sigma(\mathrm{L})$, then $U = \bigcup\{\twoheaduparrow x : x \in U\}$ and all sets $\twoheaddownarrow x$ are directed.

(4′) For each point $x \in \mathrm{L}$ the set $\twoheaddownarrow x$ is directed and $x = \sup\{y : y$ is a lower bound of some U with $x \in U \in \sigma(\mathrm{L})\}$.

Then the conditions (1), (3), (5), (6) and (7) of 1.14 and (2′) and (4′) are equivalent. □

PROBLEM. Can one characterize those complete lattices on which the Scott topology has the property that each point has a neighborhood basis of open filters, or, alternately, those complete lattices L for which $\{\twoheaduparrow x : x \in \mathrm{L}\}$ is a basis of $\sigma(\mathrm{L})$? □

1.23. EXERCISE. Corollary 1.15 generalizes to up-complete posets. □

1.24. EXERCISE. In a complete lattice L define $x \prec y$ iff $y \in \mathrm{int}_\sigma \uparrow x$ where $\mathrm{int}_\sigma X$ denotes the $\sigma(\mathrm{L})$-interior of a set. Then

(i) $\prec$ is an auxiliary relation;

(ii) $x \prec y$ implies $x \ll y$;

(iii) We have the equivalence of $x \prec y$ and $x \ll y$ for all $x,y \in \mathrm{L}$ iff $\twoheaduparrow x$ is Scott open for all $x \in \mathrm{L}$. (This is the case if $\ll$ satisfies the interpolation property (cf. 1.5(5)).)

(iv) The relation $\prec$ is approximating (see I-1.11) iff $\ll$ is approximating (that is, iff L is a continuous lattice).

(HINT: For the proof of (iv) use the equivalence of (1) and (4) in Theorem 1.14.) □

PROBLEM. For which complete lattices do we have $x \prec y$ iff $x \ll y$?

1.25. EXERCISE. The following example shows that $x \ll y$ need not imply $x \prec y$ in a complete lattice: Let $L = \{0\} \cup (\mathbf{N} \times [0,1]) \cup \{1\}$ with 0 as zero, 1 as one and $\uparrow(n,r) = (\{n\} \times [r,1]) \cup \{(p,1) : p \geq n\} \cup \{1\}$. Then $(1,0) \ll 1$ but not $(1,0) \prec 1$. In fact $\mathrm{int}_\sigma \uparrow(1,0) = \emptyset$. □

NOTES

The topology introduced on a complete lattice L (or, more generally, on any up-complete poset L) in this section was first concisely formulated for the lattice $L = \mathcal{O}(X)$ of open sets of a topological space by B.J. Day and G.M. Kelly [1970, see p. 51]. But it is clearly the merit of D. Scott [1972a] to have defined this topology in all generality and to have demonstrated its usefulness in his article on "Continuous Lattices". The name *Scott topology* was first used by Isbell [1975b, p. 41] and [1975a, p. 317], and the name has been used in the Seminar on Continuity in Semilattices (SCS) for several years. Our detailed discussion of the relation between the Scott topology, lim-inf convergence and lower semicontinuity is an elaboration of a motive proffered by Scott in his 1972 article (see p. 104), by J.D. Lawson [1973] and by K.H. Hofmann and A. Stralka [1976, see p. 16].

The results on the sobriety of the Scott topology discussed in 1.11 ff. are new as is most of the Characterization Theorem 1.14. For further results on sobriety and order, see Isbell [1975a] and Hoffmann [1979a,b]. The equivalence of (1) and (4) in 1.14, however, was (practically) used by Scott to define a continuous lattice in Scott [1972a]. Thus it is really Theorem 1.14 which establishes the equivalence of our definition of continuous lattices in I-1.6 with Scott's original definition. Scott used the auxiliary relation $\prec$ of 1.24 for his definition, and the precise statement of the equivalence of Scott's definition of a continuous lattice with Definition I-1.6 is given in 1.24. The example in 1.25 is due to C.E. Clark [SCS-21]. Theorem 1.15 is new.

2. SCOTT-CONTINUOUS FUNCTIONS

The next task is to characterize those functions between complete lattices which are continuous with respect to the Scott topology. Our original motivations must now bear fruit: after all the Scott topology was introduced to describe the classical concept of lower semicontinuous functions.

2.1. PROPOSITION. *For a function f from a complete lattice* S *into a complete lattice* T, *the following conditions are equivalent:*

(1) *f is continuous with respect to the Scott topologies, that is, $f^{-1}(U)\in\sigma(S)$ for all $U\in\sigma(T)$;*

(2) *$f(\underline{\lim}\, x_j)\leq\underline{\lim}\, f(x_j)$, for any net x_j on* S;

(3) *$f(\sup D) = \sup f(D)$, for all directed subsets D of* S.

If S *and* T *are continuous lattices, then the above are equivalent to each of the following two conditions:*

(4) *$y\ll f(x)$ iff for some $w\ll x$ one has $y\ll f(w)$, for all $x\in S$ and $y\in T$;*

(5) *$f(x) = \sup\{f(w) : w\ll x\}$, for all $x\in S$.*

If S *and* T *are algebraic, then the following conditions are also equivalent to the preceding ones:*

(6) *$k\leq f(x)$ iff for some $j\leq x$ with $j\in K(S)$ one has $k\leq f(j)$, for all $x\in S$ and $k\in K(T)$;*

(7) *$f(x) = \sup\{f(j) : j\leq x$ and $j\in K(S)\}$, for all $x\in S$.*

Moreover, in arbitrary complete lattices, Scott-continuous functions are always monotone.

Proof. (1) implies (2): First we show that (1) implies that f is monotone: If $x\leq y$ in S, but ***not*** $f(x)\leq f(y)$, then the Scott-open set $V = T\setminus\downarrow f(y)$ contains $f(x)$. Thus $U = f^{-1}(V)$ is a Scott-open neighborhood of x by (1). But then $y\in\uparrow U = U$, and so $f(y)\in T\setminus\downarrow f(y)$, a contradiction.

Next suppose that $x = \underline{\lim}\, x_j$; we claim $f(x)\leq\underline{\lim}\, f(x_j)$. Set $t = \underline{\lim}\, f(x_j)$ and assume ***not*** $f(x)\leq t$. The Scott-open set $T\setminus\downarrow t$ contains $f(x)$; thus, the inverse image $U = f^{-1}(T\setminus\downarrow t)$ is a Scott-open neighborhood of $x = \underline{\lim}\, x_j$ in S by (1). It follows that $\inf_{i\geq j} x_i\in U$ for all sufficiently large j. As f is monotone, by what we showed above, we see that

$$f(\inf_{i\geq j} x_i)\leq\inf_{i\geq j} f(x_i)\leq t.$$

Whence,

$$\inf_{i\geq j} x_i \in f^{-1}(\downarrow t) = S\setminus U \text{ for all } j,$$

and this is a contradiction.

(2) implies (3): First we remark that (2) implies that f is monotone. (HINT: if $a \leq b$ in S, define $x_1 = a$, $x_2 = b$, $x_3 = a$, $x_4 = b$, etc. Then $\underline{\lim}\, x_j = a$ and $f(a) = f(\underline{\lim}\, x_j) \leq \underline{\lim}\, f(x_j) = f(a) \wedge f(b)$.)

Now let D be a directed subset of S. Then $\sup f(D) \leq f(\sup D)$, since f is monotone. If we set $x_d = d$, $d \in D$, then

$$\underline{\lim}\, x_d = \sup_d \inf_{c \geq d} c = \sup D;$$

and hence,

$$f(\sup D) = f(\underline{\lim}\, x_d) \leq \underline{\lim}\, f(x_d) = \sup_d \inf_{c \geq d} f(c) \leq \sup_d f(d) = \sup f(D).$$

(3) implies (1): Again we have to check that f is monotone. (HINT: If $a \leq b$ in S, then the set $\{a,b\}$ is directed. It follows that $f(b) = f(a) \vee f(b)$.)

Let A be a Scott-closed subset of T. In order to show that $f^{-1}(A)$ is Scott closed in S we take a directed subset D of $f^{-1}(A)$. Then by (3)

$$f(\sup D) = \sup f(D).$$

But $\sup f(D) \in A$ by 1.4(i), since A is Scott closed and $f(D)$ is directed owing to the monotonicity of f. Hence $\sup D \in f^{-1}(A)$, and thus $f^{-1}(A)$ is Scott-closed by 1.4(i).

From now on we assume that S and T are continuous lattices.

(3) implies (5): Clear, since $x = \sup \Downarrow x$ by I-1.6.

(5) implies (4): From (5) we can conclude that f is monotone. (HINT: If $x \leq y$, then $\Downarrow x \subseteq \Downarrow y$.)

Now let $y \ll f(x) = \sup f(\Downarrow x)$; since f is monotone, $f(\Downarrow x)$ is directed. Thus, by I-1.19, there is an $w \ll x$ with $y \ll f(w)$. Conversely, if $y \ll f(w)$ for some $w \ll x$, then $y \ll f(x)$ by monotonicity of f and I-1.2(ii) for $\ll$.

(4) implies (1): Let $U \in \sigma(\mathrm{T})$ and $x \in f^{-1}(U)$. By 1.9 there is a $u \in U$ with $u \ll f(x)$. By (4) we find a $w \ll x$ such that $u \ll f(w)$; we will have finished the proof if we show that $f(\Uparrow w) \subseteq U$.

Now, take a z with $w \ll z$. For every $y \ll f(w)$ we have $y \ll f(z)$ by (4); and consequently $f(w) = \sup \Downarrow f(w) \leq f(z)$. But $u \leq f(w)$ by I-1.2(i) and $u \in U$; hence, $f(z) \in U$ by 1.2(i).

Now let S and T be algebraic. Note that in an algebraic lattice we have $x \ll y$ iff there is a compact element k with $x \leq k \leq y$. Thus the equivalences (4) iff (6) and (5) iff (7) follow easily. □

2.2. DEFINITION. A function f : S→T between complete lattices is *Scott continuous* iff it satisfies the equivalent conditions 2.1(1)-(3). The category whose objects are complete lattices and whose morphisms are Scott-continuous maps will be denoted by ***UPS*** (preservation of ***UP***-directed ***S***ups).

The full subcategory of continuous lattices is called ***CONT***. The full subcategory of algebraic lattices is called ***ALG***. □

In Chapter IV we will talk about the category ***SUP*** of all complete lattices and sup-preserving maps. This will be a proper subcategory of ***UPS***. There the reader will also find other necessary categories of continuous and algebraic lattices that are not full subcategories of ***CONT***.

Let us pause to record next those examples of continuous functions which, implicitly, we have encountered before.

2.3. EXAMPLES. (1) Every map preserving arbitrary sups is Scott continuous. This remark applies in particular to each lower adjoint (Section O-3). Notably, the co-restriction of any closure operator to its image is Scott continuous (O-3.12(iii)).

(2) We had specific occurrences of maps preserving directed sups in O-3.11, O-3.14, O-4.2(6), I-2.6 through 2.14, I-2.16, I-2.17, I-4.9, I-4.10, I-4.15, I-4.16. (It might be a useful exercise for the reader to rephrase these propositions and theorems in terms of Scott continuity.)

(3) A function $f: X \to \mathbb{R}^*$ from a topological space into the extended set of reals is lower semicontinuous iff it is continuous with respect to the Scott topology on $\mathbb{R}^*$. With this remark we have closed the circle which we began with the motivating observations preceding 1.1. It is in this light that we prefer to view Examples like I-1.21.

(4) As we remarked in 2.1, every Scott-continuous function is monotone. The converse is obviously false; counterexamples $f: \mathbb{R}^* \to \mathbb{R}^*$ are trivial to construct. (This is so whenever the domain of the function is to contain proper limits: on a finite lattice monotonicity and continuity come to the same thing.) However, there is an interesting circumstance where "in effect" monotonicity implies continuity. Consider the question of defining continuous functions $f: S \to T$, where S is *algebraic* (and T is just assumed to be complete). In view of 2.1(7), the function f is completely determined by its restriction to the sup-semilattice K(S); on K(S), moreover, all we can say about f is that it is monotone. To see this suppose we are given any monotone f_0: K(S)→T. We then employ the formula of 2.1(7) as a *definition* of an extension to all of S:

$$f(x) = \sup\{f_0(j) : j \leq x \text{ and } j \in K(S)\}, \text{ for all } x \in S.$$

The reader can easily prove that the f so defined is continuous and agrees exactly with f_0 on K(S). □

Now we are in a position to build a general theory which subsumes the investigation of objects like LSC(X,$\mathbb{R}^*$). This will be accomplished in the remainder of this Chapter.

One of the most noteworthy features of the category ***UPS*** is that it is *cartesian closed* (we will say presently what this means explicitly). This is not only fundamental for the applications of continuous lattices to logic and computing, but it also provides evidence of the mathematical naturalness of the notion.

As a first step toward showing why the category is cartesian closed, we discuss hom sets. Let S and T be complete lattices. As is common, we denote by ***UPS***(S,T) the *set* of Scott-continuous functions from S into T, that is, the set of maps in the category ***UPS*** with the indicated domain and codomain. This set is also a poset under the pointwise partial ordering. For the moment we will not distinguish notationally between ***UPS***(S,T) as an object in ***SET***, the category of sets, and as an object in some other (concrete) category.

2.4. LEMMA. *Let* S *and* T *be complete lattices. Let* L *be the poset* ***UPS***(S,T) *of Scott-continuous functions from* S *into* T. *Then* L *is closed in* T^S *under the formation of arbitrary sups; consequently* L *is a complete lattice.*

Proof. Let $F \subseteq L$ and $f = \sup F$. (The sup is pointwise!) Now take an arbitrary directed set D in S. Then

$$\sup f(D) = \sup_{d \in D} \sup_{g \in F} g(d) = \sup_{g \in F} \sup g(D)$$

$$= \sup_{g \in F} g(\sup D) = f(\sup D). \ \square$$

The following will fix our notation for the complete lattice we have just constructed by putting it in the proper category.

2.5. DEFINITION. We denote by $[S \rightarrow T]$ the set ***UPS***(S,T) considered as a complete lattice. Furthermore if $f : S_1 \rightarrow S_2$ and $g : T_1 \rightarrow T_2$ are ***UPS***-maps, then we denote by $[f \rightarrow g] : [S_2 \rightarrow T_1] \rightarrow [S_1 \rightarrow T_2]$ the map $h \mapsto ghf$. $\square$

2.6. PROPOSITION. *The construction of 2.5 defines a functor*

$$[\cdot \rightarrow \cdot] : \boldsymbol{UPS}^{op} \times \boldsymbol{UPS} \rightarrow \boldsymbol{UPS}.$$

Proof. The map $[f \rightarrow g]$ is certainly well defined in 2.5 and carries functions in the first set to functions in the second set. It also obviously behaves properly under composition. We need only observe that $[f \rightarrow g]$ preserves directed sups to make sure the map resides in the right category. But if $h = \sup H$, then

$$[f \rightarrow g](h)(s) = gh(f(s)) = g(\sup_{k \in H} k(f(s))$$

$$= \sup_{k \in H} g(k(f(s))) \text{ (since } g \text{ preserves directed sups)}$$

$$= \sup_{k \in H} [f \rightarrow g](k)(s);$$

and since sups are calculated pointwise, the assertion follows. $\square$

We wish to show in addition the important result that this functor restricts to the categories ***CONT*** and ***ALG***. Before we give the argument, however, it is useful to identify the lattice $[S \rightarrow 2]$, where of course 2 is the two-element chain. The easy proof is left to the reader. (He should recall that in the Scott-topology of 2 the point $\{1\}$ is *open.*)

2.7. LEMMA. *For a complete lattice* S *the function* $f \mapsto f^{-1}(1) : [S \to 2] \to \sigma(S)$ *is an isomorphism of lattices.* □

2.8. THEOREM. *The functor* $[\cdot \to \cdot]$ *maps* ***CONT***$^{op} \times$ ***CONT*** *into* ***CONT*** *and* ***ALG***$^{op} \times$ ***ALG*** *into* ***ALG***. *In particular,* $[S \to T]$ *is a continuous* [*resp., algebraic*] *lattice if* S *and* T *are.*

Proof. Suppose that T is a continuous (resp., algebraic) lattice. Then T is the image of some 2^X under a Scott-continuous projection (resp., closure) operator by I-4.16 (resp., I-4.15). Every functor preserves idempotent morphisms; hence so does $[S \to \cdot]$. By 2.6 $[S \to \cdot]$ therefore preserves Scott-continuous projection operators. If $c^* \leq c$ in $[T^* \to T]$, then $[1_S \to c^*] \leq [1_S \to c]$ in $[[S \to T^*] \to [S \to T]]$ by 2.6; hence, $[S \to \cdot]$ also preserves Scott-continuous closure operators (recall O-3.8(ii)). Hence $[S \to T]$ is the image of $[S \to 2^X]$ under a Scott-continuous projection (resp., closure) operator. But $[S \to \cdot]$ preserves products, and so $[S \to 2^X] \simeq [S \to 2]^X \simeq \sigma(S)^X$ by Lemma 2.7. Since $\sigma(S)$ is continuous if S is continuous by 1.14 (resp., algebraic if S is algebraic by 1.15), and since products of continuous (resp., algebraic) lattices are continuous (resp., algebraic) by I-2.7(i) (resp., I-4(i)), then $[S \to 2^X]$ is continuous (resp., algebraic). Thus $[S \to T]$ is continuous by I-4.16 (resp., algebraic by I-4.15). □

The argument for 2.8 is typical of proofs that give the *answer* without very directly giving the *reason*. Another proof of the last statement of 2.8 was indicated in I-2.14 and I-2.15. We return to the question again in the exercises below; see 2.15 and 2.16.

Having thus provided a function-space construction for ***UPS***, ***CONT*** and ***ALG***, we now turn to a discussion of some properties of products appropriate to the proof that all these categories are cartesian closed.

2.9. LEMMA. *Let* R,S,T *be complete lattices and* $f : R \times S \to T$ *a function. Then the following conditions are equivalent:*

(1) *f is Scott continuous on the product lattice* $R \times S$;

(2) *f is Scott continuous in each variable separately; that is,*

(a) *For all* $s \in S$, *the function* $r \mapsto f(r,s) : R \to T$ *is Scott continuous;*
(b) *For all* $r \in R$, *the function* $s \mapsto f(r,s) : S \to T$ *is Scott continuous.*

Proof. Clearly (1) implies (2). Now assume (2), and let D be a directed subset of $R \times S$. For $d \in D$ we write $d = (d_1,d_2)$. It is easy to check that $\sup D = (a_1,a_2)$ with $a_k = \sup_{d \in D} d_k$, $k = 1,2$. Then

$$f(\sup D) = f(a_1,a_2) = f(\sup_{d \in D} (d_1,a_2))$$

$$= \sup_{d \in D} f(d_1,a_2) \text{ (by (2(a)))} = \sup_{d \in D} f(d_1, \sup_{c \in D} c_2)$$

$$= \sup_{d \in D} \sup_{c \in D} f(d_1,c_2) \text{ (by (2(b)))}.$$

Since D is directed, we find for $d,c \in D$ an $e \in D$ with $d,c \leq e$. Since f is monotone by (2), we conclude

$$f(\sup D) \leq \sup_{e \in D} f(e_1,e_2) = \sup f(D).$$

The inequality $\sup f(D) \leq f(\sup D)$ holds for any monotone map f. □

A category A with finite limits is called *cartesian closed* iff there is a functor $(Y,Z) \mapsto Z^Y : A^{op} \times A \to A$ such that there is a natural isomorphism $A(X \times Y, Z) \simeq A(X, Z^Y)$. The principal properties on the cartesian closure of the relevant categories are now collected in the following theorem. In each of the categories ***UPS***, ***CONT***, and ***ALG*** we remark that $(R \times S)$ provides the cartesian product functor and $[R \to S]$ the internal hom-functor. (More information on cartesian closed categories may be found in Mac Lane [1971, pp. 95 ff.].)

2.10. THEOREM. *Let* R,S,T *be complete lattices, and let*

$$(T^S)^R \underset{\psi}{\overset{\varphi}{\rightleftarrows}} T^{(R \times S)}$$

be the canonical pair of mutually inverse bijections given by

$$\varphi(g)(r,s) = g(r)(s) \text{ and } \psi(f)(r)(s) = f(r,s).$$

Then φ and ψ induce mutually inverse bijections that are in fact isomorphisms of complete lattices:

$$[R \to [S \to T]] \rightleftarrows [(R \times S) \to T].$$

*In particular, **UPS**, **CONT**, and **ALG** are cartesian closed categories.*

Proof. If $g \in UPS(R,[S \to T])$, then g preserves directed sups, that is, for each $s \in S$, the function $r \mapsto g(r)(s)$ preserves directed sups. Since $g(r) \in UPS(S,T)$ for all $r \in R$, then $g(r)$ preserves directed sups. Hence $\varphi(g)$ satisfies the hypotheses (2) of 2.9, whence $\varphi(g) \in UPS(R \times S, T)$.

On the other hand, let $f \in UPS(R \times S, T)$; then f preserves directed sups in each argument separately, whence $\psi(f)(r) \in UPS(S,T)$ for each $r \in R$. Since sups of functions are calculated pointwise, and $r \mapsto \psi(f)(r)(s) = f(r,s)$ preserves directed sups for each $s \in S$, then $\psi(f) \in UPS(R,[S \to T])$.

This proves that the restrictions of φ and ψ relate the desired lattices. But φ and ψ are clearly monotone, and so they are isomorphisms, since they are obviously inverse to one another. The reader may check that the isomorphism of functors obtained in this way is natural. □

It is convenient to record in conclusion a few obvious functors between the various categories which we have already encountered implicitly or explicitly and which we will often use in the following developments.

We know that a function $f: S \to T$ between complete lattices belongs to ***UPS***(S,T) iff it is Scott-continuous (see 2.1). The assignment, which associates with a complete lattice L the topological space $(L, \sigma(L))$ defines in an obvious way the functor Σ from ***UPS*** into the category ***TOP*** of T_0-topological spaces. The restrictions of this functor to ***CONT*** and ***ALG*** are very interesting, and in the next section we shall describe the subcatories of ***TOP*** thereby obtained.

If X is a T_0-space, then its topology $\mathcal{O}(X)$ is a complete Heyting algebra (O-3.22). In view of the identity O-3.16(3), which singles out cHa's in the class of all algebras, it is reasonable to consider the *category* ***HEYT*** of complete Heyting algebras and functions preserving *arbitrary* sups and *finite* infs (cf. O-3.24.). If we are given a continuous map $f\colon X \to Y$, then the map $U \mapsto f^{-1}(U)$, which we shall call $\mathcal{O}(f) : \mathcal{O}(Y) \to \mathcal{O}(X)$, preserves arbitrary unions and finite intersections. Thus $\mathcal{O} : \boldsymbol{TOP} \to \boldsymbol{HEYT}$ is a well-defined contravariant functor.

If L is a complete lattice, then we have

$$\mathcal{O}(\Sigma(\mathrm{L})) = \mathcal{O}(\mathrm{L},\sigma(\mathrm{L})) = \sigma(\mathrm{L}).$$

Thus we also have a contravariant functor $\sigma : \boldsymbol{UPS} \to \boldsymbol{HEYT}$ (which on functions operates just like $\mathcal{O}$). By Theorem 1.14 the functor σ maps ***CONT*** contravariantly into the category ***CONT***$\cap$***HEYT*** of continuous Heyting algebras, and by 1.15, it maps ***ALG*** contravariantly into the category ***ALG***$\cap$***HEYT*** of algebraic Heyting algebras.

2.11. NOTATION. We record the following functors:

$\sigma : \boldsymbol{UPS}^{\mathrm{op}} \to \boldsymbol{HEYT}$,	$\sigma(\mathrm{L})$ = Scott topology,	$\sigma(f)(U) = f^{-1}(U)$;
$\Sigma : \boldsymbol{UPS} \to \boldsymbol{TOP}$,	$\Sigma(\mathrm{L}) = (\mathrm{L},\sigma(\mathrm{L}))$,	$\Sigma(f) = f$;
$\mathcal{O} : \boldsymbol{TOP}^{\mathrm{op}} \to \boldsymbol{HEYT}$,	$\mathcal{O}(X)$ = topology of X,	$\mathcal{O}(f)(U) = f^{-1}(U)$.

We note $\sigma = \mathcal{O}\Sigma$. □

EXERCISES

2.12. EXERCISE. Let L be a complete lattice and let I be a given directed set. Then each of the following three infinitary operations can be viewed as a mapping defined on the direct power:

$$\underline{\lim},\ \sup,\ \inf : \mathrm{L}^I \to \mathrm{L}.$$

All these functions are monotone, but which are Scott continuous? □

2.13. EXERCISE. Consider functions of several variables which for simplicity are defined on and take values in a fixed complete lattice S. Such a function is called *Scott continuous* iff it is Scott continuous in each of its variables ***separately***.

Let f be an n-place function, and let $g_1, \ldots, g_n$ each be m-place functions. Define the m-place function h by composition:

$$h(x_1, \ldots, x_m) = f(g_1(x_1, \ldots, x_m), \ldots, g_n(x_1, \ldots, x_m)).$$

(i) Using a suitable generalization of 2.9 (and possibly other lemmas), prove that h is Scott continuous if f and the g_i are.

(ii) Give a direct proof of the same result.

(HINT: For (ii) it is sufficient to consider the two special cases $f(x,x)$ and $f(g(x))$ by first making all variables distinct and then identifying them one occurrence at a time.) □

2.14. EXERCISE. For each of the categories ***UPS***, ***CONT***, and ***ALG***, we have:

(i) There is a natural isomorphism

$$[R \to (S \times T)] \cong [R \to S] \times [R \to T].$$

(ii) The evaluation map $(f,x) \mapsto f(x)$, which we can call:

$$\mathrm{eval} : [R \to S] \times R \to S,$$

is Scott continuous.

(iii) The composition map $(f,g) \mapsto f \circ g : [S \to T] \times [R \to S] \to [R \to T]$ is Scott continuous.

(iv) Analogous results hold for *all* cartesian closed categories. □

2.15. EXERCISE. Let L and M be algebraic lattices. Define maps

$$\mathrm{fun} : 2^{K(L) \times K(M)} \to [L \to M] \text{ and } \mathrm{graph} : [L \to M] \to 2^{K(L) \times K(M)}$$

by the formulas:

$$\mathrm{fun}(F)(x) = \sup\{j : (k,j) \in F \text{ and } k \leq x\},$$

$$\mathrm{graph}(f) = \{(k,j) : j \leq f(k)\}.$$

(i) These maps are Scott continuous (assuming only that we know that $[L \to M]$ is a complete lattice);

(ii) (fun, graph) is an adjunction with the first map surjective and the second injective (cf. O-3.7).

(iii) $[L \to M]$ is isomorphic to the range of a Scott-continuous closure operator on $2^{K(L) \times K(M)}$; hence, it is algebraic.

(iv) Part (iii) can be used to describe $K([L \to M])$ explicitly. □

2.16. EXERCISE. Let L and M be continuous lattices. For $x \in L$ and $y \in M$ define a function on L by the formula:

$$[x \Rightarrow y](z) = y, \text{ if } x \ll z, \text{ and } = 0_M \text{ otherwise.}$$

(i) This function is always Scott continuous;

(ii) If $f \in [L \to M]$ and $y \ll f(x)$, then $[x \Rightarrow y] \ll f$ in $[L \to M]$;

(iii) If $f \in [L \to M]$, then $f = \sup\{\, [x \Rightarrow y] : y \ll f(x)\}$;

(iv) Hence, $[L \to M]$ is a continuous lattice. □

2.17. EXERCISE. Carry out the suggestion mentioned in 2.3(2). □

2.18. EXERCISE. Proposition 2.1 holds for up-complete posets—except that part (2) must be left out because the required infs do not always exist. □

2.19. EXERCISE. The category of up-complete posets and Scott-continuous functions between them is cartesian closed. And so is the category of complete-continuous semilattices. □

PROBLEM. Is the category of continuous semilattices and Scott-continuous functions cartesian closed?

2.20. EXERCISE. Prove that the functor $\boldsymbol{UPS}^{op} \rightarrow \boldsymbol{HEYT}$ of the Scott topology (see 2.11 and preceding remarks) extends to a functor from the category of all (up-complete) posets into ***HEYT*** in such a fashion that the full subcategory of all continuous posets is mapped into the full subcategory of ***CONT*** of all completely distributive lattices. (cf. Exercise 1.22.) □

2.21. EXERCISE. Let L be a complete semilattice. Then the function

$$\text{fix} = (f \mapsto \min\{x \in L : f(x) = x\}) : [L \rightarrow L] \rightarrow L$$

preserves directed sups; hence, it is in $[[L \rightarrow L] \rightarrow L]$.

(HINT: Show that $\text{fix}(f) = \sup\{f^n(0) : n = 1,2,...\}$ and observe that directed sups commute. Note that the Tarski fixed-point theorem (O-2.3) cannot be used directly here, because L is not a lattice.) □

2.22. EXERCISE. (i) Let ***SEMI*** be the category of sup-semilattices with 0 and monotone maps. Then we can construe the construction of I-4.12 as a functor

$$\text{Id} : \boldsymbol{SEMI} \rightarrow \boldsymbol{ALG}$$

provided we define $\text{Id}(f)$ to be the map $I \mapsto \downarrow f(I)$.

(ii) This construction is ***not*** an equivalence of categories, because not every continuous map is obtained.

(iii) Expand ***SEMI*** to the category ***GRAPH*** using the idea of 2.15 by adding more maps. Specifically, for $R,S \in \boldsymbol{SEMI}$ define $F : R \rightarrow S$ to mean that F is a sub-sup-semilattice of $S \times R$ which is monotone in the sense that yFx and $y_1 \leq y$ always imply $y_1 Fx$. (Every $f : R \rightarrow S$ in ***SEMI*** is represented in ***GRAPH*** by the relation $\{(y,x) : y \leq f(x)\}$.) Composition GF for $G : S \rightarrow T$ is just the ordinary composition of relations. In this way ***GRAPH*** is a category and ***SEMI*** is a sub-category with the same collection of objects.

Now define $\text{Id} : \boldsymbol{GRAPH} \rightarrow \boldsymbol{ALG}$ where $\text{Id}(F)$ is the map

$$I \mapsto \{y : yFx \text{ for some } x \in I\}.$$

This construction provides an equivalence of categories. □

NOTES

The results in this section are largely based on Scott's ideas. (See Scott [1972a] and also Scott [1976] and the bibliography contained therein.)

3. INJECTIVE SPACES

In the previous sections we associated with a complete lattice a canonical topology, and we further pursue the relation between topological spaces and lattices in this section. Our goal here is to characterize *continuous* lattices via the Scott topology in purely topological terms. The question is: which topological spaces are of the form $\Sigma(L)$ for L a continuous lattice? We find a complete (and brief!) answer. Furthermore we show that $\Sigma(L)$ as a space completely determines L as a lattice.

In the remainder of this chapter we work entirely in the category ***TOP*** of T_0-spaces and continuous maps and will never consider topological spaces which do not at least satisfy the T_0-separation axiom.

We begin by recalling the idea of *relative injectives* in a category $\boldsymbol{A}$. One is given a class J of monomorphisms which is closed under the pre- and post-multiplication with isomorphisms. Then an object Z is called a *J-injective* iff for any map $j: X \to Y$ in J and every morphism $f: X \to Z$ in $\boldsymbol{A}$ there is a morphism $f^*: Y \to Z$ with $f = f^*j$, that is, the following diagram commutes.

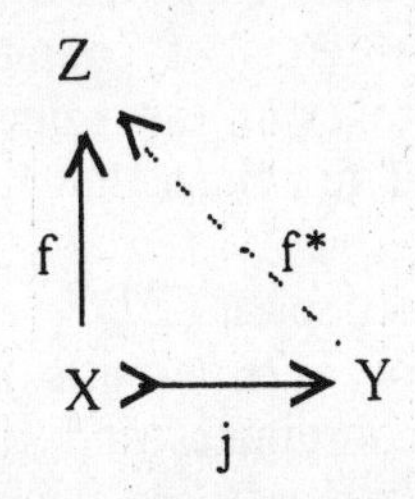

In the category ***TOP*** we wish to consider J-injectives for the class J of all *subspace embeddings* (that is, continuous maps whose co-restriction to their image is a homeomorphism). For future reference we restate the definition in this special case.

3.1. DEFINITION. A T_0-space Z is called *injective* iff every continuous map f: X→Z extends continuously to any space Y containing X as a subspace. □

It is useful to record that there are some purely arrow-theoretical facts about relative injectives (whose proof we leave as an exercise on the manipulation of injectives in any category).

3.2. LEMMA. (i) *Products of J-injectives are J-injectives;*

(ii) *Retracts of J-injectives are J-injectives;*

(iii) *If Z is a J-injective and $j: Z \to Y$ is a J-monomorphism, then Z is a retract of Y.* □

The immediate question is now whether in ***TOP*** we have any injectives. We give at first a rather modest answer, but it will be the key for all that follows.

3.3. LEMMA. *The Sierpinski space $\Sigma(2)$ is injective.*

Proof. Suppose that X is a subspace of Y and that $f: X \to \Sigma(2)$ is a continuous map. Then $U = f^{-1}(1)$ is open in X since $\{1\} \in \sigma(2)$. By the definition of the induced topology on X, there is an open set V on Y with $U = V \cap X$. Define $g: Y \to \Sigma(2)$ to be the characteristic function of V (that is, $g^{-1}(1) = V$); it is continuous, and clearly $g|X = f$. □

In order to see how far this will lead us we make the following remarks:

3.4. LEMMA. (i) *For every set M we have $\Sigma(2^M) = (\Sigma 2)^M$; that is, the Scott topology on 2^M and the product topology agree. Moreover, $\Sigma(2^M)$ is injective;*

(ii) *Every T_0-space X is embedded in some $(\Sigma 2)^M$;*

(iii) *Every injective T_0-space X is a retract of some $(\Sigma 2)^M$; that is, there is a continuous $f: (\Sigma 2)^M \to (\Sigma 2)^M$ with $f^2 = f$ and $\operatorname{im} f$ homeomorphic to X.*

Proof. (i): As 2^M is an algebraic lattice, we recall by 1.15 that $\Sigma(2^M)$ has as a basis for its topology the sets of the form $\uparrow k$ where k is compact in 2^M. But by I-4.11(i)—expressed in terms of characteristic functions rather than sets—k is just a function that takes on the value 1 only finitely often. The set $\uparrow k$, then, is exactly the class of functions that take the value 1 at least at the places that k does.

Turning now to the product space $(\Sigma 2)^M$, we remark that, because $\{1\}$ is the only nontrivial open set of $\Sigma(2)$, a basis for the open sets is given by putting $\{1\}$ on finitely many coordinates and $\{0,1\}$ on the remainder. But as we just noted the sets formed this way are the sets of the form $\uparrow k$. Thus, the two topologies have the same basis and must be the same.

The last assertion of (i) then follows from 3.3 and 3.2(i).

(ii): For a given space X we take $M = \mathcal{O}(X)$ and define $j: X \to 2^M$ by having $j(x)(U) = 1$ iff $x \in U$. Since X is T_0, it follows that j is injective.

Let W be a basic open set of 2^M. By our description in the proof of (i), W is determined by a finite number of coordinates $U_1, \ldots, U_n \in M$. We have

$$j(x) \in W \text{ iff } x \in U_1 \cap \ldots \cap U_n;$$

whence, j is continuous.

Let V be any open subset of X. Then it is easy to see that

$$j(V) = \{f \in \operatorname{im} j : f(V) = 1\}.$$

As this is the intersection of a basic open subset of 2^M with $\operatorname{im} j$, this shows that j is an embedding.

(iii): This is now a consequence of (ii), (i) and 3.2(iii). □

It is useful at this point to recall the various formal aspects of the concept of a retract. If $j: X \to Y$ and $e: Y \to X$ are morphisms in a category with $ej = 1_X$, then X is called a *retract* of Y. The map e is a *retraction*, the map j a *coretraction.*

If, in a given category, every morphism $f: A \to B$ may be decomposed into a composition $f = f_\circ f^\circ$ with an epimorphism f° and a monomorphism $f_\circ$, then any projection $f = f^2$ on an object Y gives rise to a retract where X = domain $f_\circ$ = codomain f°. Indeed $f_\circ f^\circ = f = f^2 = f_\circ f^\circ f_\circ f^\circ$ implies that $f^\circ f_\circ = 1_X$, since $f_\circ$ is monic and f° is epic. In such categories the retracts X of an object Y are, up to canonical isomorphism, in bijective correspondence with the projections on Y. We have made use of this situation in Section I-2 (in 2.9, 2.14, 2.15) and in 3.4(iii) above. We will use it further now. All the categories we consider have the required epic-monic factorization property. A direct proof of the next proposition is given in Exercise 3.4.

3.5. PROPOSITION. *If* L *is a continuous lattice, then* Σ(L) *is an injective space.*

Proof. By I-4.16(4), L is a retract in ***UPS*** of some 2^M. Since functors preserve retracts, ΣL is a retract of $\Sigma(2^M)$. By 3.4(i), $\Sigma(2^M)$ is injective. Hence, by 3.2(ii), ΣL is injective. □

So far we operated exclusively in terms of topology, using, where lattices arose, the canonical Scott topology. We now associate with each T_0-space a canonical (and well-known) poset structure.

If X is a T_0-space, then for two elements x and y in X the following relations are easily seen to be equivalent:

(1) $\{x\}^- \subseteq \{y\}^-$;

(2) $x \in \{y\}^-$;

(3) $x \in U$ implies $y \in U$, for all open sets U.

This relation is clearly reflexive and transitive, and, since X is a T_0-space, it is antisymmetric. Hence, we have a partial order. Furthermore if $f: X \to Y$ is a continuous map in ***TOP*** , then it is obvious from (3) that the relation is preserved; that is, f is a monotone map. We thus have a functor from ***TOP*** into the category ***POSET*** of posets and monotone maps.

3.6. DEFINITION. The partial order $\leq$ defined on a T_0-space X by

$$x \leq y \text{ iff } x \in \{y\}^-$$

is called the *specialization order.*

We denote by Ω : ***TOP*** → ***POSET*** the functor which associates with a space X the poset (X,$\leq$), where $\leq$ is the specialization order, and with a continuous function itself. □

If L is a complete lattice, then $\Omega\Sigma$L = L by 1.4(ii); that is to say, the Scott topology determines the partial ordering by means of a purely topological definition. We are now ready for a counterpart of 3.5.

3.7. PROPOSITION. *If* X *is an injective* T_0*-space, then* $\Omega(X)$ *is a continuous lattice.*

Proof. By 3.4(iii), there is a continuous function $f = f^2: (\Sigma 2)^M \to (\Sigma 2)^M$ such that we may identify the space X with im f. We apply the functor Ω and note $\Omega(\Sigma 2)^M = \Omega\Sigma(2^M) = 2^M$ by 3.4(i). We thus obtain a projection operator $f: 2^M \to 2^M$ which preserves directed sups by 2.1. But then im f is a continuous lattice in the *induced* partial order by I-4.16. However, the *specialization* order of a space induces on a subspace the specialization order of this subspace (indeed if A is a subspace of B and $P \subseteq A$, then the closure of P in A is $P^- \cap A$, where P^- is the closure of P in B). Thus, ΩX is a continuous lattice. □

If we apply to the diagram

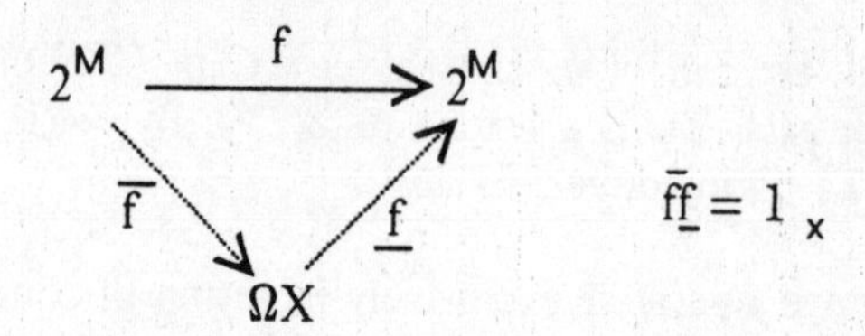

of ***UPS***-maps the functor Σ, we obtain, in view of 3.4(i), the commutative diagram

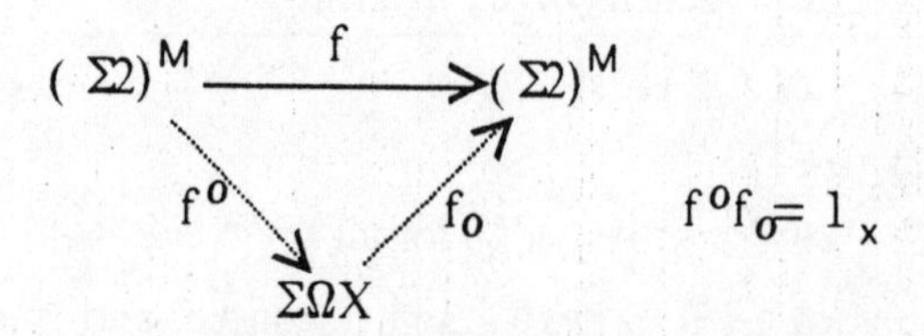

of continuous maps. Since $f_o : X \to (\Sigma 2)^M$ is an embedding, then the identity map $1_X : \Sigma\Omega X \to X$ is continuous; since the retraction $f^o : (\Sigma 2)^M \to X$ is a quotient map (as all retractions are), the identity map in the other direction $1_X : X \to \Sigma\Omega X$ is continuous. Hence $\Sigma\Omega X = X$.

Taking this remark into account, we may summarize the principal results of this section in the following theorem:

3.8. THEOREM. (i) *If* L *is a continuous lattice, then* $\Sigma L = (L, \sigma(L))$ *is an injective space and* $\Omega\Sigma X = L$.

(ii) If X *is an injective* T_0*-space, then* $\Omega X = (X, \leq)$ *is a continuous lattice* (*with respect to the specialization order*) *and* $\Sigma\Omega X = X$. □

There is, therefore, a canonical bijection between continuous lattices and injective topological T_0-spaces. In fact we have shown that ***INJ***, the full subcategory of ***TOP*** consisting of injective spaces and all continuous maps, is essentially the same category as *CONT*. This allows not only a purely topological description of continuous lattices (injective spaces under the specialization order), but also a complete answer to the question as to which spaces are of the form $\Sigma(L)$ with L continuous.. The results of the next section will serve as a first illustration of how useful this knowledge can be. (Note that by 3.8 and 1.13 all injective T_0-spaces are locally quasicompact and sober. The converse is clearly incorrect as the two point *discrete* space shows.)

For an arbitrary space X, very little can be said in general of the poset structure of ΩX. Directed nets need not have sups. Even if directed nets always have sups, they need not converge to their sups (as the unit interval in which all upper sets are open shows). The following definition is therefore natural:

3.9. DEFINITION. A T_0-space X is called a *monotone convergence space* iff every subset D directed relative to the specialization order (3.6) has a sup, and the relation sup $D \in U$ for any open set U of X implies $D \cap U \neq \emptyset$. □

Clearly, a space is a monotone convergence space iff every directed net in ΩX has a sup and converges to this sup; whence the name. By 1.2(ii), every space ΣL for a complete lattice L is a monotone convergence space. In particular, all injective spaces are monotone convergence spaces.

3.10. LEMMA. *A continuous function $f: X \to Y$ from a monotone convergence space X to any space Y preserves directed sups in the specialization orders (that is, Ωf preserves directed sups).*

Proof. Let D be a directed subset of ΩX. Since f is monotone, it follows that sup $f(D) \leq f(\sup D)$. Suppose sup $f(D) \neq f(\sup D)$ and set $U = Y \setminus \downarrow \sup f(D)$. Then U is an open neighborhood of $f(\sup D)$; that is, $f^{-1}(U)$ is an open neighborhood of sup D, by the continuity of f. Since X is a monotone convergence space, there is a $d \in D$ with $d \in f^{-1}(U)$; that is, $f(d) \in U$. Since f is monotone, we have $f(d) \leq \sup f(D)$; hence, sup $f(D) \in U = Y \setminus \downarrow \sup f(D)$, which is a contradiction. □

3.11. LEMMA. *Let X be a space and Y a monotone convergence space. Let $(f_j)_{j \in J}$ be a net of continuous functions $f_j : X \to Y$ such that $(f_j(x))_{j \in J}$ is a directed net of ΩY for each x. Let $f: X \to Y$ be the pointwise sup of the net f_j. Then f is continuous.*

Proof. Let U be an arbitrary open set of Y. Then $x \in f^{-1}(U)$ iff $f(x) \in U$, and since $f(x) = \sup f_j(x)$ and Y is a monotone convergence space, this is the case iff there is a $j \in J$ with $f_j(x) \in U$; that is, iff $x \in f_j^{-1}(U)$. Thus we have
$f^{-1}(U) = \bigcup_{j \in J} f_j^{-1}(U)$; and since all f_j are continuous, this is an open set. □

As an immediate corollary we obtain the following result:

3.12. PROPOSITION. (i) *Let* X *be a space and* Y *a monotone convergence space. Then the subset* ***TOP***(X,Y) *of* $(\Omega Y)^X$ *is closed under directed sups.*

(ii) *If* Z *is a monotone convergence space and* $f : Y \to Z$ *a continuous map, then the function* ***TOP***$(X,f) = (g \mapsto fg) :$ ***TOP***$(X,Y) \to$ ***TOP***(X,Z) *preserves directed sups relative to the pointwise orders.*

Proof. Part (i) follows immediately from 3.11. In order to prove part (ii) it suffices to point out that f preserves directed sups by Lemma 3.10 and that sups of functions are calculated pointwise. □

EXERCISES

3.13. EXERCISE. For a T_0-space X, show that the following conditions are equivalent:

(1) X is injective;

(2) X is a retract of every space of which it is a subspace;

(3) $X = \Sigma\Omega X$, $\Omega(X)$ is complete, and $\mathcal{O}(X)$ is completely distributive.

(HINT: Use 3.4(ii) and 3.2(i) for (1) iff (2). For (1) iff (3) recall 1.14.) □

We remark that 3.13(3) gives a characterization of injective spaces that is completely intrinsic, in the sense that relationships to other spaces are not involved in the statement of the property.

3.14. EXERCISE. Let L be a continuous lattice and let X and Y be T_0-spaces. If X is a subspace of Y and $f_0 : X \to \Sigma L$ is continuous, then we know that there is a continuous extension of f_0 to $f : Y \to \Sigma L$. Show that in particular the function

$$f(y) = \sup\{\inf f_0(X \cap U) : y \in U \in \mathcal{O}(Y)\}$$

is one such extension. How does it compare to others in the pointwise ordering? □

3.15. EXERCISE. QUESTION: Which spaces are of the form $\Sigma(L)$ for some *algebraic* lattice L? ANSWER: They are the injective spaces for which the supercompact open sets form a basis for the topology. (A subset is called *supercompact* if every covering by a family of open sets can be reduced to ***one*** set in that family. This is the same as being quasicompact and a co-prime in the lattice of open sets.)

(HINT: Recall 1.15.) □

3.16. EXERCISE. Let X be a sober space (see paragraph preceding 1.12). Then $\Omega X = (X, \leq)$ (see 3.6) is an up-complete poset (O-2.11). If D is directed in ΩX, then $x = \sup D$ is also $\lim D$ and $D^- = \{x\}^-$.

(HINT: Suppose D is directed in ΩX. Then D^- is irreducible. We must show that the nonempty open subsets of D^- are a filter basis: Let $U, V \in \mathcal{O}(X)$ with $U \cap D^- \neq \emptyset \neq V \cap D^-$. Then there are $d, c \in D$ with $c \in U$ and $d \in V$. Since D is directed there is $e \in D$ with $d, c \leq e$; whence $e \in U \cap V \cap D^-$. Since X is sober, $D^- = \{x\}^-$ for some x. Show that $x = \sup D = \lim D$.) □

3.17. EXERCISE. Each sober space is a monotone convergence space. □

3.18. EXERCISE. Characterize those sober spaces for which ΩX is a continuous poset.

(HINT: See R.-E. Hoffmann [1979] and also V-5.24 below.) □

3.19. EXERCISE. In the definition of injective space replace the subspace embeddings by ***dense*** subspace embeddings. As this is a smaller class of monomorphisms, there are more injectives. Show that the injectives in this sense are just the spaces of the form $\Sigma(L)$ where L is a complete-continuous semilattice.

(HINT: Use the characterization of I-1.29 that connects continuous semilattices with continuous lattices.) □

NOTES

The idea of characterizing continuous lattices exclusively in terms of T_0-spaces as injectives in the category of T_0-spaces and continuous functions was one of the basic results in Scott [1972a]; thus, Theorem 3.8 was a core result of that treatise.

The specialization order of 3.6 has been traditionally used in the spectral theory of rings. The idea of monotone convergence spaces (3.9) was used by O. Wyler in a seminar report [SCS-35]; he called them *d*-spaces.

4. FUNCTION SPACES

In Section 2 we introduced the poset $[S \to T]$ for two complete lattices as the set *UPS*(S,T) equipped with the pointwise order induced from T. By 2.1 we have *UPS*(S,T) = *TOP*(ΣS,ΣT), so this suggests that there is a topological description of the poset. Indeed, the partial order is the pointwise order relative to the specialization order of ΣT, since we know $\Omega\Sigma T = T$. This construction can therefore be generalized to arbitrary T_0-spaces.

4.1. DEFINITION. For two T_0-spaces X and Y let [X,Y] denote the *poset* defined on the set *TOP*(X,Y) by the pointwise order induced from ΩY. Let, in addition, $[f,h](g) = hgf$, thus defining a functor:

$$[\cdot,\cdot] : \mathbf{TOP}^{op} \times \mathbf{TOP} \to \mathbf{POSET}. \quad \square$$

We leave as an exercise the check that $[f,h]$ is actually a monotone function. Clearly $[S \to T] = [\Sigma S, \Sigma T]$ for complete lattices by 2.1.

4.2. LEMMA. (i) *If* Y *is a monotone convergence space, then* [X,Y] *is closed in* $(\Omega Y)^X$ *under the formation of directed sups;*

(ii) *If* $f : Y \to Z$ *is a continuous function between monotone convergence spaces and* [X,Y], [X,Z] *are complete lattices, then* $[X,f] : [X,Y] \to [X,Z]$ *is in* ***UPS***.

Proof. The first part is a reiteration of 3.12(i). As to the second, the function $[X,f]$ preserves directed sups by 3.12(ii) and so is in ***UPS***. $\square$

This allows the following application:

4.3. LEMMA. (i) *Let* X *be a space and* Y *a retract of the monotone convergence space* Z. *If* [X,Z] *is a complete lattice, then* [X,Y] *is a retract of* [X,Z] *in* ***UPS***.

(ii) *If, in addition,* [X,Z] *is a continuous lattice, then* [X,Y] *is a continuous lattice.*

Proof. Let $f : Z \to Z$ be a continuous map with $f^2 = f$ and $Y = f(Z)$. If we set $F = [X,f] : [X,Z] \to [X,Z]$, then $F = F^2$. The functions in im F are precisely those $g : X \to Z$ with $fg = g$, that is, with $g(X) \subseteq Y$; thus $F([X \to Z])$ is order isomorphic with [X,Y]. If [X,Z] is a complete lattice, then F is in ***UPS*** by 4.2(ii); hence, [X,Y] is a complete lattice by O-2.5. If [X,Z] is a continuous lattice, then [X,Y] is likewise continuous by I-2.14. $\square$

We now clarify when a poset [X,Y] is a continuous lattice.

4.4. PROPOSITION. *If* X *is a space, and if* Y *is a* **nonsingleton** *monotone convergence space such that* [X,Y] *is a continuous lattice, then* O(X) *is a continuous lattice.*

Remark. In later parts of this section (see 4.10) and notably in Chapter V we undertake a detailed study of spaces X for which $\mathcal{O}(X)$ is continuous. By I-1.7(5) and I-1.8 we know that this class contains all locally quasicompact spaces, and, among regular T_0-spaces, only the locally compact spaces are in this class. If L is a continuous lattice, then ΣL is in this class because $\mathcal{O}(\Sigma L) = \sigma(L)$ is continuous by 1.14.

Proof. Since [X,Y] is a lattice and Y is nonsingleton, ΩY cannot have the trivial order (for which $y \leq y^*$ iff $y = y^*$). Hence Y is not T_1. Thus, there are elements $y \leq y^*$ with $y \neq y^*$ in Y. Let U be any open neighborhood of y^* which does not contain y. Its characteristic function is a retraction $Y \to \Sigma 2$ with right inverse $i : \Sigma 2 \to Y$, where $i(1) = y^*$ and $i(0) = y$. (Note that for any open set $V \subseteq Y$ with $0 \in i^{-1}(V)$ we have $y \in V$; hence $y^* \in V$, and thus $1 \in i^{-1}(V)$!) Now Lemma 4.3 applies and shows that $[X, \Sigma 2]$ is a continuous lattice. But $\mathcal{O}(X)$ is isomorphic to $[X, \Sigma 2]$. □

4.5. PROPOSITION. *If* X *is a space and if* Y *is a* T_0*-space such that* [X,Y] *is a continuous lattice, then* ΩY *is a continuous lattice.*

Proof. We fix a point $b \in X$ and define a function $F : [X,Y] \to [X,Y]$ by $F(g)(x) = g(b)$ for all $x \in X$. Then $F^2 = F$ and $F \in$ ***UPS***. Hence im F is a continuous lattice by I-2.14. But im F is order isomorphic to ΩY. □

Let us observe that the continuity of the lattice ΩY ***does not*** allow us to conclude that Y is an injective space; by 3.8, it is indeed the case that $\Sigma\Omega Y$ is an injective space, but the given topology of Y may be either coarser or finer than the Scott topology of ΩY. If Y is a *monotone convergence space*, then its topology is coarser than or equal to the Scott topology.

4.6. PROPOSITION. *If* X *is a space such that* $\mathcal{O}(X)$ *is a continuous lattice and* Y *is an injective space, then* [X,Y] *is a continuous lattice.*

Proof. By 3.4(iii), Y is a retract of some $(\Sigma 2)^M$. If $\pi_m : (\Sigma 2)^M \to 2$ denotes the mth projection, then the function

$$f \mapsto (\pi_m f)_{m \in M} : [X,(\Sigma 2)^M] \to [X, \Sigma 2]^M$$

is an order isomorphism. But $[X, \Sigma 2]$ is isomorphic to $\mathcal{O}(X)$ and hence is a continuous lattice. Thus, by I-2.7(i), the lattice $[X, \Sigma 2]^M$ is continuous; whence, $[X,(\Sigma 2)^M]$ is continuous. Now it follows that [X,Y] is a continuous lattice by Lemma 4.3(ii). □

If S is a complete lattice such that $\sigma(S)$ is a continuous lattice (e.g., if S is continuous as in (1.14), and if T is a continuous lattice, we deduce that $[S \to T] = [\Sigma S, \Sigma T]$ is a continuous lattice. This reproves 2.8.

From our main propositions we can now extract the following theorem:

4.7. THEOREM. *Let* X *be a space and* L *a complete* ***nonsingleton*** *lattice. Then the following statements are equivalent:*

(1) $[X,\Sigma L]$ *is a continuous lattice*;

(2) *Both* $\mathcal{O}(X)$ *and* L *are continuous lattices.*

Proof. We set $Y = \Sigma L$; then Y is a monotone convergence space. If condition (1) holds, then the hypotheses of 4.4 and 4.5 are satisfied. Hence, $\mathcal{O}(X)$ and $\Omega Y = \Omega\Sigma L = L$ are continuous lattices; thus (2) follows. If (2) holds, then, by 3.8 and 4.6, condition (1) follows. □

The function space posets of the form $[X,\Sigma L]$ deserve more attention.

4.8. LEMMA. *Let* X *and* Y *be topological spaces and* $f: X \to \Sigma\mathcal{O}(Y)$ *be a function. If the set* $G_f = \{(x,y)\in X\times Y : y\in f(x)\}$ *is open in* $X\times Y$, *then f is continuous.*

Proof. Let $U\in\sigma(\mathcal{O}(Y))$ and suppose that $f(x)\in U$. Since G_f is open, for each $y\in f(x)$ there is an open neighborhood $A(y)$ of x in X and an open neighborhood $B(y)$ of y in Y such that $A(y)\times B(y)\subseteq G_f$. If J is the directed set of finite subsets of $f(x)$ and $F\in J$, we set $B(F) = \bigcup_{y\in F}B(y)$. Then $(B(F))_{F\in J}$ is a directed net in $\mathcal{O}(Y)$ with $\sup_F B(F) = f(x)\in U$. By 1.2(ii) we then find a finite set $F\subseteq f(x)$ such that $B(F)\in U$. Set $A(F) = \bigcap_{y\in F}A(y)$. Then $A(F)$ is an open neighborhood of x in X, and for each $a\in A(F)$ we have

$$\{a\}\times B(F)\subseteq A(F)\times B(F)\subseteq G_f;$$

hence, $B(F)\subseteq f(a)$. But also $B(F)\in U$, hence $f(a)\in U$. □

4.9. PROPOSITION. *For topological spaces* X *and* Y, *there is a natural monotone function* $\theta : \mathcal{O}(X\times Y)\to[X,\Sigma\mathcal{O}(Y)]$ *given by*

$$\theta(W)(x) = \{y\in Y : (x,y)\in W\}.$$

Proof. We have to show that $\theta(W) : X\to\Sigma\mathcal{O}(Y)$ is continuous. But

$$G_{\theta(W)} = \{(x,y)\in X\times Y : y\in\theta(W)(x)\} = W\in\mathcal{O}(X\times Y).$$

Hence, $G_{\theta(W)}$ is open, and Lemma 4.8 shows that $\theta(W)$ is continuous. It is evident that θ is monotone. □

In general, there is no reason to believe that the map θ is an isomorphism. We will now give necessary and sufficient conditions for this to be the case. We will make use of the canonical pair of mutually inverse bijections given by the formulas:

$$\psi(f)(x)(y) = f(x,y) \text{ and } \varphi(g)(x,y) = g(x)(y).$$

We have a diagram as in Theorem 2.10:

$$(L^Y)^X \underset{\psi}{\overset{\varphi}{\rightleftarrows}} L^{X\times Y}.$$

4.10. THEOREM. *Let* Y *be a* T_0*-space. Then the following statements are equivalent:*

(1) *For all spaces* X *and all continuous lattices* L, *the canonical pair* φ,ψ *of mutually inverse bijections induce by restriction bijections*

$$TOP(X,\Sigma[Y,\Sigma L]) \rightleftarrows TOP(X\times Y,\Sigma L);$$

(1′) *For all spaces* X *and all continuous lattices* L, *the canonical pair* φ,ψ *of mutually inverse bijections induce by restriction order isomorphisms*

$$[X,\Sigma[Y,\Sigma L]] \rightleftarrows [X\times Y,\Sigma L];$$

(2) *For all spaces* X, *the function* $\theta : \mathcal{O}(X\times Y)\rightarrow[X,\Sigma\mathcal{O}(Y)]$ *of 4.9 is an isomorphism*;

(3) *For all continuous* $f : X\rightarrow\Sigma\mathcal{O}(Y)$ *the set* G_f *of 4.8 is open in* $X\times Y$;

(4) *The set* $\{(U,y)\in\mathcal{O}(Y)\times Y : y\in U\}$ *is open in* $\Sigma\mathcal{O}(Y)\times Y$;

(5) *For each* $y\in U\in\mathcal{O}(Y)$ *there is a Scott-open neighborhood* $H\in\sigma(\mathcal{O}(Y))$ *containing* U *such that* $\cap H$ *is a neighborhood of* y *in* Y;

(6) $\mathcal{O}(Y)$ *is a continuous lattice.*

Proof. (1) iff (1′): Clear.

(1) iff (2): For a space Z we denote with $\alpha_Z : [Z,\Sigma 2]\rightarrow\mathcal{O}(Z)$ the isomorphism given by $\alpha_Z(f) = f^{-1}(1)$. A straight-forward calculation shows the following diagram to be commutative:

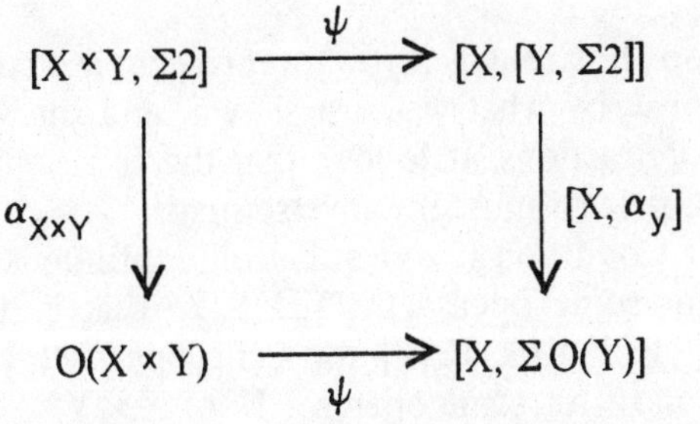

$$\begin{array}{ccc}
[X\times Y,\Sigma 2] & \xrightarrow{\ \psi\ } & [X,[Y,\Sigma 2]] \\
\downarrow{\scriptstyle \alpha_{X\times Y}} & & \downarrow{\scriptstyle [X,\alpha_Y]} \\
\mathcal{O}(X\times Y) & \xrightarrow[\ \psi\ ]{} & [X,\Sigma\mathcal{O}(Y)]
\end{array}$$

Condition (2) is therefore equivalent to the following:

(2′) For all spaces X

$$TOP(X\times Y,\Sigma 2) \xrightarrow{\psi} TOP(X,\Sigma[Y,\Sigma 2])$$

is a bijection with inverse φ.

Thus (1) clearly implies (2′). We now show the converse: First we prove (1) for the special case that $L = 2^M$. This follows from (2′) and the pair of commutative diagrams

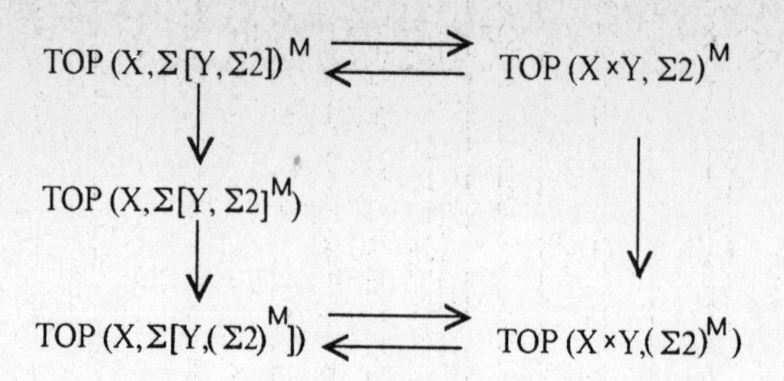

in which the vertical arrows are natural bijections, and the pairs of horizontal arrows at the top are mutually inverse maps by (2′).

Now suppose that L is an arbitrary continuous lattice. Then ΣL is injective by 3.8, and hence is a retract of $(\Sigma 2)^M$ for some M by 3.4(iii). Then $[Y,\Sigma L]$ is a retract of $[Y,(\Sigma 2)^M]$ in ***UPS*** by 4.3. Thus, $\Sigma[Y,\Sigma L]$ is retract in ***TOP*** of $\Sigma[Y,(\Sigma 2)^M]$, and there is a pair of commutative diagrams

$$\begin{array}{ccc} \mathrm{TOP}(X,\Sigma[Y,(\Sigma 2)^M]) & \rightleftarrows & \mathrm{TOP}(X\times Y,(\Sigma 2)^M) \\ \downarrow\uparrow & & \downarrow\uparrow \\ \mathrm{TOP}(X,\Sigma[Y,\Sigma L]) & \rightleftarrows & \mathrm{TOP}(X\times Y,\Sigma L) \end{array}$$

in which the top horizontal arrows represent mutually inverse functions induced by φ and ψ by what was just shown, and the vertical pairs represent retractions and coretractions. It follows that the functions φ and ψ induce the bottom horizontal pair of mutually inverse maps.

(2) iff (3): Condition (2) says that all continuous $f: X\to\Sigma\mathcal{O}(Y)$ are of the form $\theta(W)$ for some open set $W\subseteq X\times Y$; that is, they are given by the equation $f(x) = \{y\in Y : (x,y)\in W\}$ for some open set W of $X\times Y$. This says precisely that $G_f = W$ for some open set W of $X\times Y$.

(3) implies (4): Take $X = \Sigma\mathcal{O}(Y)$ and $f = 1_{\sigma\mathcal{O}(Y)}$, then we find that $G_f = \{(U,y) : y\in U\}$.

(4) implies (5): Let $Y\in U\in\mathcal{O}(Y)$; then by (4) there is an open neighborhood H of U in $\Sigma\mathcal{O}(Y)$ and an open neighborhood V of y in Y such that $(W,v)\in H\times V$ implies $v\in W$. Thus $V\subseteq\bigcap H$.

(5) implies (3): Let $f: X\to\Sigma\mathcal{O}(Y)$ be continuous. Take $(x,y)\in G_f$, that is, $y\in f(x)$. By (5) there is an open neighborhood H of $f(x)$ in $\Sigma\mathcal{O}(Y)$ and an open neighborhood V of y in Y such that $V\subseteq\bigcap H$. Since f is continuous, there is an open neighborhood U of x in X with $f(U)\subseteq H$. If now $(u,v)\in U\times V$, then $v\in V\subseteq\bigcap H\subseteq f(u)$ since $f(u)\in H$. Thus $U\times V\subseteq G_f$, and $U\times V$ is an open neighborhood of (x,y) in $X\times Y$.

(5) implies (6): We have to show that for each $U \in \mathcal{O}(Y)$ and each Y there is a $V \in \mathcal{O}(Y)$ with $y \in V \ll U$ in order to satisfy I-1.6 for L = $\mathcal{O}(Y)$. By (5) there is an open neighborhood H of U in $\Sigma\mathcal{O}(Y)$ and an open neighborhood V of y such that $V \subseteq \bigcap H$. Let D be a directed set in $\mathcal{O}(Y)$ with $U \subseteq \bigcup D$. Then $\bigcup D \in H$ by 1.2(i); thus by 1.2(ii) there is a $W \in D$ with $W \in H$, and then $V \subseteq \bigcap H \subseteq W$. This shows that $V \ll U$.

(6) implies (5): Let $y \in U \in \mathcal{O}(Y)$. Since $\mathcal{O}(Y)$ is a continuous lattice, there is a $V \in \mathcal{O}(Y)$ with $y \in V \ll U$. Then $H = \Uparrow V$ is an open neighborhood of U in $\Sigma\mathcal{O}(Y)$ by 1.5(5), and evidently $V \subseteq \bigcap \Uparrow V$. □

We apply this information in the proof of the following result (cf. Exercises 4.23-25 for an alternate, more direct proof):

4.11. THEOREM. *Let* L *be a complete lattice. Then the following statements are equivalent:*

(1) $\sigma(L)$ *is a continuous lattice.*

(2) *For every complete lattice* S *one has* $\sigma(S \times L) = \mathcal{O}(\Sigma S \times \Sigma L)$.

(3) *For every complete lattice* S *one has* $\Sigma(S \times L) = \Sigma S \times \Sigma L$.

Proof. (2) iff (3): Obvious.

(1) implies (2): We have $\sigma(S \times L) \cong [S \times L \to 2]$ (sending a Scott-open set of $S \times L$ to its characteristic function) $\cong [S \to [L \to 2]]$ (by 2.10) $\cong [S \to \sigma(L)]$ (since $[L \to 2] \cong \sigma(L)$) $= [S, \Sigma\mathcal{O}(\Sigma L)]$ (since $\sigma(L) = \mathcal{O}(\Sigma L)$, and in view of 2.1); and the isomorphism $\psi : \sigma(S \times L) \to [S, \Sigma\mathcal{O}(\Sigma L)]$ is given by $\psi(W)(S) = \{y \in L : (s,y) \in W\}$. But by Theorem 4.10 we have $\mathcal{O}(\Sigma S \times \Sigma L) \cong [S, \Sigma\mathcal{O}(\Sigma L)]$ under the same map. Hence $\mathcal{O}(\Sigma S \times \Sigma L) = \sigma(S \times L)$.

(2) implies (1): We apply condition (2) with $S = \sigma(L)$. Then the topology of $\Sigma\sigma(L) \times \Sigma L$ is the Scott topology of $\sigma(L) \times L$. Now we verify condition 4.10(4) with $Y = \Sigma L$ which will prove the continuity of $\mathcal{O}(Y) = \sigma(L)$. We must show that the set W of all $(U,y) \in \sigma(L) \times L$ with $y \in U$ is open in $\sigma(L) \times L$; but by the preceding this is tantamount to showing that this set is Scott open in $\sigma(L) \times L$. This is not hard to see: If the net (U_j, y_j) is directed with $U = \bigcup_j U_j$ and $y = \sup y_j$, and if $(U,y) \in W$, that is, $y \in U$, then $y_j \in U$ for some j, since U is Scott open, and thus $y_j \in U_k$, for some k. But then, if $m > j,k$ we obtain $y_m \in U_m$, and thus $(U_m, y_m) \in W$. □

We reformulate this in a slightly weaker fashion:

4.12. COROLLARY. *The functor* Σ : ***CONT*** $\to$ ***TOP*** *preserves finite products.*

Proof. This follows from 4.11, since $\sigma(L)$ is continuous whenever L is continuous by 1.14. □

There are other useful corollaries:

4.13. COROLLARY. *If* L *is a complete lattice such that* $\sigma(L)$ *is continuous, then the sup-operation* $\vee : \Sigma L \times \Sigma L \to \Sigma L$ *is continuous.*

Proof. By 4.11, $\Sigma L \times \Sigma L = \Sigma(L \times L)$. But we know $\vee : \Sigma(L\times L) \to \Sigma L$ is continuous, since $\vee : L \times L \to L$ preserves arbitrary sups. □

4.14. COROLLARY. *If* L *is a complete lattice such that* $\sigma(L)$ *is continuous, then* ΣL *is a sober space.*

Proof. 1.12 and 4.13. □

In 4.13 we found a sufficient condition for a complete lattice to be a topological sup-semilattice with respect to the Scott topology. What do we know about the inf-operation? By O-4.2(6) and 2.1 we know that a complete lattice L is meet-continuous (O-4.1) iff $\wedge : L\times L \to L$ is Scott-continuous; that is, $\wedge : \Sigma(L\times L) \to \Sigma L$ is continuous. Thus, if $\sigma(L)$ is a continuous lattice, then L is meet-continuous iff ΣL is a topological inf-semilattice iff (in view of 4.13) ΣL is a topological lattice. The question remains whether the meet continuity of L can be recognized from properties of $\sigma(L)$. The answer is yes:

4.15. PROPOSITION. *Let* L *be a complete lattice. Then the following conditions are equivalent:*

(1) L *is meet-continuous* (O-4.1);

(2) $\sigma(L)$ *is join-continuous* (O-4.1);

(3) $\sigma(L)^{op}$ *is a complete Heyting algebra.*

Remark. $\sigma(L)$ is always a complete Heyting algebra by O-3.22.

Proof. (1) implies (3): By O-4.1 and 2.1, all translations $x \to a\wedge x$: $\Sigma L \to \Sigma L$ are continuous. We apply O-3.23 with $S = \Sigma L$ and $M = \sigma(L) = \mathcal{O}(\Sigma L)$ and conclude (3).

(2) iff (3): By O-4.3.

(2) implies (1): By 1.4(i), $\sigma(L)^{op}$ is isomorphic to the lattice T of all lower sets which are closed under directed sups. Let $S \subseteq T$ be the subset of all principal ideals $\downarrow x$ for $x \in L$. Then S is closed in T under directed sups: indeed, if t_j is a directed net in L with $\sup t_j = t$, then $\downarrow t$ is an upper bound in T of the net $\downarrow t_j$. And if $A \in T$ is an upper bound on all $\downarrow t_j$, then $t_j \in A$ and thus $t \in A$, since A is closed under directed sups; hence, $\downarrow t \subseteq A$.

Since $\bigcap_{x\in X} \downarrow x = \downarrow \inf X$, then S is closed in T under arbitrary infs. Thus, if T satisfies the relation (MC) of O-4.1, then so does S. But L is isomorphic to S under the map $x \mapsto \downarrow x$. □

4.16. COROLLARY. *Let* L *be a complete lattice such that* $\sigma(L)$ *is a continuous lattice. Then the following statements are equivalent:*

(1) L *is meet-continuous;*

(2) ΣL *is a topological lattice;*

(3) $\sigma(L)$ *is join-continuous*;

(4) $\sigma(L)$ *is a continuous Heyting algebra and* $\sigma(L)^{op}$ *is a complete Heyting algebra.* □

In Chapter VII we will identify those L satisfying the conditions of 4.16 as precisely the underlying lattices of compact topological semilattices with identity.

Let us note in passing, that 4.15 allows us to express the following fact: If $\tau(L)$ denotes the lattice of all lower sets which are closed under directed sups, then for a complete lattice L, the following statements are equivalent:

(1) L is meet-continuous.

(2) For some $n = 1,2,3,...$, the lattice $\tau^n(L)$ is meet-continuous.

(3) For all $n = 1,2,3,...$, the lattice $\tau^n(L)$ is meet-continuous.

Here, of course, we set $\tau^{n+1}(L) = \tau(\tau^n(L))$, inductively.

The results in 4.15 and 4.16 should be compared with 1.14 above. These theorems express properties, such as the continuity, or meet continuity of L, exclusively in terms of properties of the Scott topology $\sigma(L)$. We will pursue this further in Chapter VII. But dwelling just a bit longer on complete lattices L with continuous Scott topology $\sigma(L)$ and on function spaces $[X,\Sigma L]$, we note a companion to 4.7.

4.17. PROPOSITION. *Let* X *be a topological space and* L *a complete lattice such that* $\sigma(L)$ *is a continuous lattice. Then*

(i) $[X,\Sigma L]$ *is a meet continuous lattice iff* L *is meet continuous*;

(ii) $[X,\Sigma L]$ *is a cHa if and only if* L *is a cHa.*

Proof. If L is meet continuous, then ΣL is a topological lattice by 4.16 and thus $[X,\Sigma L]$ is closed in L^X under finite infs and sups. Since ΣL is a monotone convergence space, $[X,\Sigma L]$ is closed under directed sups in L^X by 3.12. Hence $[X,\Sigma L]$ is closed in L^X under *arbitrary* sups (O-1.10), and thus it is a complete lattice. Since L^X as a product of meet-continuous lattices is meet continuous and since $[X,\Sigma L]$ is closed in L^X under finite infs and arbitrary (hence, in particular, directed) sups, then $[X,\Sigma L]$ is meet-continuous. If, in addition, L is distributive (see O-4.3), then so are L^X and the sublattice $[X,\Sigma L]$. Conversely if, $[X,\Sigma L]$ is meet continuous (and distributive), then so is the sublattice of all constant functions. This last is isomorphic to L. □

We conclude the section by identifying the irreducible elements of $[X,\Sigma L]$. The reader should recall I-3.5.

4.18. LEMMA. *Let* L *be a continuous lattice and* X *a space. For a function* $f \in [X,\Sigma L]$ *the following assertions are equivalent:*

(1) $f \in \mathrm{IRR}\,[X,\Sigma L]$;

(2) *There is a prime element U of $\mathcal{O}(X)$ and an irreducible element p of L such that $f = \chi_U \vee \mathrm{const}_p$, where χ_U is the characteristic function of the set U and const_p is the constant function on X with value p in L.*

Proof. (2) implies (1): Suppose $f = a \wedge b$. Let $A = a^{-1}(\downarrow p)$ and $B = b^{-1}(\downarrow p)$. Then A and B are closed, and if $p < 1$ is irreducible, $A \cup B = X \setminus U$. But as U is prime, that is, $X \setminus U$ is an irreducible closed set, $A = X \setminus U$ or $B = X \setminus U$ follows, and so $f = a$ or $f = b$. The case $p = 1$ yields $f = \mathrm{const}_1$.

(1) implies (2): We may assume that $f \neq \mathrm{const}_1$. Hence we find a $t < 1$ in $f(X)$. We take an arbitrary $s \ll t$ in L and set $U = f^{-1}(\Uparrow s)$. Since $\Uparrow s$ is open in L by 1.5(5) and f is continuous, U is open and the two functions $a = f \vee \chi_U$ and $b = f \vee \mathrm{const}_s$ are in $[X, \Sigma L]$.

If $x \in U$ then $a(x) = f(x) \vee 1 = 1$ and $b(x) = f(x) \vee s = f(x)$, since $s \ll f(x)$; hence $(a \wedge b)(x) = f(x)$. If, however, $x \notin U$, then $a(x) = f(x) \vee 0 = f(x)$, and $b(x) = f(x) \vee s$; that is,

$$(a \wedge b)(x) = f(x) \wedge (f(x) \vee s) = f(x).$$

Hence $a \wedge b = f$, and since f is irreducible by (1), we have $a = f$ or $b = f$.

In the first case, $f = a = f \vee \chi_U$, that is, $\chi_U \leq f$. Let x be such that we have $t = f(x)$. Then $s \ll f(x)$, that is, $x \in U$, and thus $1 \leq f(x) = t < 1$, a contradiction. Hence we must have $f = b = f \vee \mathrm{const}_s$. Thus $s \leq f(x)$ for all $x \in X$. Since we chose $s \ll t$ arbitrarily, and since L is continuous, we conclude $t \leq f(x)$ for all $x \in X$ (see I-1.6). This means that f takes at most one value $t < 1$. Thus we have shown that $f \neq 1$ implies $f = \chi_U \vee \mathrm{const}_p$ for a $p < 1$ and for the open set $U = f^{-1}(1)$.

We claim that $p < 1$ is irreducible: If $p = vw$, set $a = \chi_U \vee \mathrm{const}_v$ and set $b = \chi_U \vee \mathrm{const}_w$, and observe $a \wedge b = f$. By (1) we have $f = a$ or $f = b$; that is, $p = v$ or $p = w$. Now we claim that U is prime: Indeed if $U = V \cap W$, then set $a = \chi_V \vee \mathrm{const}_p$ and $b = \chi_W \vee \mathrm{const}_p$. If $x \in U$, then $(a \wedge b)(x) = 1 \wedge 1 = 1$; if $x \notin U$, then $x \notin V$, say, and then $(a \wedge b)(x) = p \wedge 1 = p$. Hence, we have shown that $a \wedge b = f$. The irreducibility of f implies either $f = a$ (that is, $U = V$) or $f = b$ (that is, $U = W$). The proof is complete. □

We recall that in a sober space (see remarks preceding 1.12) the prime elements of $\mathcal{O}(X)$ are precisely the sets X and $X \setminus \{x\}^-$ where $x \in X$. We will note that each T_0-space X can be naturally embedded into a sober space Y such that every continuous function $f : X \to S$ into a sober space extends uniquely to Y (see Chapter V). For a continuous lattice L, the space ΣL is sober by 1.12. Hence $[X, \Sigma L] = \{Y, \Sigma Y]$, and it is therefore no loss of generality if we now talk about sober spaces only in our present context.

4.19. PROPOSITION. *Let X be a sober space and L a continuous lattice. Then there is a bijection*

$$(x,p) \mapsto \chi_{X\setminus\{x\}} - \vee \text{const}_p : X \times ((\text{IRR } L)\setminus\{1\}) \to \text{IRR}[X,\Sigma L]\setminus\{1\}. \quad \square$$

This allows us to identify the irreducible elements <1 in $[X,\Sigma L]$ in a canonical way with the product $X \times (\text{IRR } L\setminus\{1\})$. From Theorem 4.7 we know that $[X,\Sigma L]$ is a continuous lattice iff $\mathcal{O}(X)$ is continuous. (If X is arbitrary, we can still assert meet continuity by 1.14 and 4.17). Thus Theorem I-3.10 applies to $[X,\Sigma L]$ under these circumstances. However, the explicit characterization of $\text{IRR}[X,\Sigma L]$ in 4.19 allows us to draw the conclusion of Theorem I-3.10 even without hypothesis on X, other than its sobriety. In fact, if for each $x \in X$ we have a set $P(x) \subseteq \text{IRR } L\setminus\{1\}$ with $s = \inf(\uparrow s \cap P(x))$ for all $s \in L$, then the image P under the canonical function in 4.19 of the set $\bigcup\{\{x\} \times P(x) : x \in X\}$ satisfies $f = \inf(\uparrow f \cap P)$ for all $f \in [X,\Sigma P]$. This allows the construction of rather bizarre order-generating sets, which we will use in Chapter V for the construction of pathological spaces X with $\mathcal{O}(X)$ continuous.

EXERCISES

The entire function-space theory of this section was developed without recourse to the relation $\ll$ and thus without referring directly to the original definition of a continuous lattice. In the following we elucidate the way-below relation on function spaces. In a special case we undertook an investigation of this kind in Exercise I-1.21.

4.20. EXERCISE. Let X be a space such that $\mathcal{O}(X)$ is a continuous lattice, and let L be a continuous lattice.

(i) If $f \ll g$ in $[X,\Sigma L]$, then f vanishes outside a set $U \in \mathcal{O}(X)$ with $U \ll X$.

Remark. Roughly speaking, this means that f has quasicompact support; in view of I-1.4(ii) this is indeed the case if X is locally quasicompact. Later we see that under our present conditions X has to be locally quasicompact as soon as it is a sober space (Chapter V).

(ii) If $s \in L$ we identify s with the constant function $X \to L$ with value s so that $s\chi_U \in [X,\Sigma L]$ for all $U \in \mathcal{O}(X)$. For each $f \in [X,\Sigma L]$ we have:

$$f = \sup\{s\chi_U : U \in \mathcal{O}(X) \text{ and there is a } V \in \mathcal{O}(X) \text{ with } U \ll V \text{ and } s \ll \inf f(V)\}.$$

(iii) If $s \ll t$ in L and $U \ll V$, then $s\chi_U \ll t\chi_U \ll t\chi_V$ in $[X,\Sigma L]$.

Remark. Note that (ii) and (iii) together yield another proof of the continuity of $[X,\Sigma L]$ which uses directly the definition of the continuity of a lattice. See also 2.16.

(iv) Let f be a finitely valued function in $[X,\Sigma L]$. Then there is a finite family $(U_j, s_j) \in \mathcal{O}(X) \times L$ such that $f = \sup s_j\, \chi_{U_j}$.

(v) Let $\text{Fin}[X,\Sigma L]$ be the sup-semilattice of finitely valued functions generated by all $s\chi_U$ with $U \ll X$. Then the following statements are equivalent for two functions $f,g \in [X,\Sigma L]$:

(1) $f \ll g$ in $[X, \Sigma L]$;

(2) There is an $h \in \mathrm{Fin}[X, \Sigma L]$ such that $f \leq h$ and that for each nonzero value s of h there is $t \in S$ and $U, V \in \mathcal{O}(X)$ such that $s \ll t$ and $f^{-1}(s) \subseteq U \ll V$ and $t \leq g(x)$ for $x \in V$;

(3) $f^{-1}(L \setminus \{0\}) \ll X$ and for each $x \in X$ there are open neighborhoods $U \ll V$ of x and elements $s \ll t$ in L such that $f(u) \leq s$ for $u \in U$ and $t \ll g(v)$ for $v \in V$;

(4) There is a finite sequence of open sets $U_i \ll V_i$ in $\mathcal{O}(X)$, $i=1,\ldots,n$, and a finite sequence of elements $s_i \ll t_i$ in L such that f vanishes outside $\bigcup_i U_i$ and for each $i=1,\ldots,n$ one has $f(u) \leq s_i$ for $u \in U_i$ and $t_i \ll g(v)$ for $v \in V_i$.

(HINTS: (i): Consider the directed family $(\chi_U)_{U \ll X}$ of characteristic functions and note $g \leq \sup \chi_U$.

(ii): If $f \in [X, \Sigma L]$ and $y \in X$, let $s \ll f(y)$ be arbitrary. By the interpolation property I-1.5, find t with $s \ll t \ll f(y)$. The set $f^{-1}(\uparrow t)$ is open, hence we find $U \in \mathcal{O}(X)$ with $y \in U \ll f^{-1}(\uparrow t)$, using the continuity of $\mathcal{O}(X)$. With the continuity of L we arrive at the assertion.

(iii): If $(f_j)_{j \in J}$ is a directed family with $t\chi_U \leq \sup f_j$ in $[X, \Sigma L]$, then $U_j = f_j^{-1}(\uparrow s)$ is a directed family with $V \leq \sup U_j$ in $\mathcal{O}(X)$. Hence find $j \in J$ with $U \leq U_j$, then $s\chi_U \leq f_j$.

(iv): Without loss of generality we can assume $f \neq 0$. Let F be the finite set $f(x) \cup \{0\}$ and let m be a maximal element and $a \ll m$ in F. Let $U = f^{-1}(m)$, then U is open. (Indeed let V be an open filter in L with $V \cap F = \{m\}$, then $U = f^{-1}(V)$). Define g by $g(x) = f(x)$ for $x \notin U$ and $g(x) = a$ for $x \in U$. Then $g \in [X, \Sigma L]$, and the number of nonzero values taken by g is less than the corresponding one for f. By induction, g is the sup of finitely many fuinctions of the form $s\chi_V$. But $f = g \vee m\chi_U$.

(v): Use the preceding information.) □

Notice that in part (v) above we have a characterization of the way-below relation in $[X, \Sigma L]$ entirely in terms of the way-below relations in X and L.

4.21. EXERCISE. Let X be a quasicompact space such that $\mathcal{O}(X)$ is a continuous lattice, and let S be a continuous poset in which every bounded finite set has a sup (this is, e.g., satisfied if S is a sup-semilattice). Then $[X, \Sigma S]$ is a continuous poset.

(HINT: Verify those portions of the proof of 4.20 which carry over to the present case and make the necessary modifications.)

PROBLEM. Investigate systematically under which circumstances $[X, \Sigma L]$ is a continuous poset or semilattice.

Notice that in this present section we have presented a general theory of function spaces of lower semicontinuous functions. If X is a (locally) compact space, then by 2.3(3) we have $\mathrm{LSC}(X, \mathbb{R}^*) \simeq [X, \Sigma \mathbf{I}]$, where $\mathbf{I} = [0,1]$, the unit interval with its natural order. In this sense, the present theory supersedes earlier discussions such as Exercises I-1.21 (but notice the variation of the theme "$f \ll g$" in I-1.21 and 4.20 above!), I-2.16, I-2.17.

PROBLEM. To which extent do 4.11 and 4.12 extend to continuous posets and semilattices?

4.22. EXERCISE. Proposition 4.15 holds for up-complete semilattices.

The following three exercises give an alternate direct route to an analog of Theorem 4.11 for posets.

4.23. EXERCISE. Let P and Q be (up-complete) posets. If the lattice of Scott-open sets on P, $\sigma(\mathrm{P})$, is a continuous lattice, then $(\mathrm{P}\times\mathrm{Q},\sigma(\mathrm{P}\times\mathrm{Q})) = (\mathrm{P},\sigma(\mathrm{P}))\times(\mathrm{Q},\sigma(\mathrm{Q}))$ (that is, the Scott topology on $\mathrm{P}\times\mathrm{Q}$ is the product of the Scott topologies).

(HINT: The product topology is always contained in $\sigma(\mathrm{P}\times\mathrm{Q})$. Conversely, suppose W is Scott open in $\mathrm{P}\times\mathrm{Q}$. Let $(p,q)\in W$. Then

$$\mathrm{M} = \{x\in \mathrm{P} : (x,q)\in W\}$$

is a Scott-open set in P containing p. Let $\mathrm{N}\in\sigma(\mathrm{P})$ such that $p\in\mathrm{N}\ll\mathrm{M}$.

Now let $V = \{y\in\mathrm{Q} : U\times\{y\}\subseteq W$ for some $U\in\sigma(\mathrm{P})$ with $\mathrm{N}\ll U\}$. Note $q\in V$. If V is shown to be Scott open, then $(p,q)\in\mathrm{N}\times V\subseteq W$, and hence we see that W is open in the product topology, completing the proof.

It is easily seen $V = \uparrow V$. Let D be a directed set in Q with $b = \sup D\in V$. Then $U\times\{b\}\subseteq W$ for some open U with $\mathrm{N}\ll U$. For each $d\in D$, let $U_d = \{x\in\mathrm{P}: (x,d)\in W\}$. Then each U_d is Scott open, and $U\subseteq\bigcup\{U_d : d\in D\}$ (since $(x,b) = \sup\{(x,d) : d\in D\}$ and $(x,b)\in W$ for each $x\in U$). Hence $\mathrm{N}\ll U_d$ for some $d\in D$ (by I-1.18). Thus $d\in V$ and hence V is Scott open.) □

4.24. EXERCISE. Let P be a(n) (up-complete) poset and let $\mathrm{L} = \sigma(\mathrm{P})$ be the lattice of Scott-open sets on P.

(i) Show that $G = \{(x,U)\in\mathrm{P}\times\mathrm{L} : x\in U\}$ is Scott open in the poset $\mathrm{P}\times\mathrm{L}$.

(ii) Show that G is open in the product topology (that is, in $(\mathrm{P},\sigma(\mathrm{P}))\times(\mathrm{L},\sigma(\mathrm{L}))$) iff $\sigma(\mathrm{P}) = \mathrm{L}$ is a continuous lattice.

(HINT: Part (i) is straightforward. One implication of (ii) follows from (i) and Exercise 4.23. The other implication follows by combining together the proofs for (4) implies (5) and (5) implies (6) in 4.11.) □

4.25. EXERCISE. Using 4.23 and 4.24, devise an analog to Theorem 4.11 for posets. □

4.26. EXERCISE. If L is a complete lattice such that $\sigma(\mathrm{L})$ is continuous (such as is the case if L is continuous by 1.14), then $\Sigma[\mathrm{L}\to\mathrm{L}]$ is a topological monoid relative to composition $(f,g)\mapsto f\circ g$.

(HINT: 2.14 (iii) and 4.11.). □

NOTES

The train of thought leading to the main result 4.7 and this result itself are due to John Isbell [1975a,b]. (The former of the two sources is the one to consult according to Isbell's own recommendation.)

In 4.10 we have a first characterization theorem for spaces Y to have a continuous lattice as topology $\mathcal{O}(Y)$; later we will see others. This result is new, although certain equivalences were known: (4) iff (5) is in the paper of B.J. Day and G.M. Kelly [1970]; in fact this paper as well as Isbell's second paper above contain additional information concerning the context of 4.10. Various names were used in the literature for spaces Y for which $\mathcal{O}(Y)$ is a continuous lattice: "semi-locally bounded" in Isbell [1975b], "quasi-locally compact" in A.S. Ward, [1968], "*CL*-spaces" in Hofmann [1977] and "core compact" in Hofmann and Lawson [1978].

The results in 4.11 through 4.14 are new. For continuous lattices one finds them in Scott [1972a, see p. 107, 2.9]. Results such as 4.11 are vital if one wishes to determine the (joint) continuity of finitary operations (as in the example of 4.13). Also 4.15 and 4.16 are new; these results will be applied in Chapter VII. They belong to the type of statement in which lattice theoretical properties of a complete lattice L are characterized in terms of lattice theoretical properties of the Scott topology $\sigma(L)$. An earlier example of this is 1.14. The identification of irreducible (or prime) elements of a function space in 4.18 and 4.19 is new. We will refer back to this result in Chapter V. The results in the exercises through 4.22 are new.

CHAPTER III

Topology of Continuous Lattices: The Lawson Topology

The first topologies defined on a lattice directly from the lattice ordering (that is, Birkhoff's order topology and Frink's interval topology) involved "symmetrical" definitions—the topologies assigned to L and to L^{op} were identical. The guiding example was always the unit interval of real numbers in its natural order, which is of course a highly symmetrical lattice. The initial interest was in such questions as which lattices became compact and/or Hausdorff in these topologies. The Scott topology stands in strong contrast to such an approach. Indeed it is a "one-way" topology, since, for example, all the open sets are always upper sets; thus, for nontrivial lattices, the T_0-separation axiom is the strongest it satisfies. Nevertheless, we saw in Chapter II that the Scott topology provides many links between continuous lattices and general topology in such classical areas as the theory of semicontinuous functions and in the study of lattices of closed (compact, convex) sets (ideals) in many familiar structures.

In this chapter we introduce a new topology, called the Lawson topology, which is crucial in linking continuous lattices to topological algebra. Its definition is more in the spirit of the interval and order topologies, and indeed it may be viewed as a mixture of the two. However, it remains asymmetrical—the Lawson topologies on L and L^{op} need not agree. But, even if one is seeking an appropriate Hausdorff topology for continuous lattices, this asymmetry is not at all surprising in view of the examples we have developed. We also show that the new topology is closely related to the earlier topology, because in any complete lattice a set is Scott open iff it is a Lawson-open upper set. (Proposition 1.6). Though the Scott topology determines the underlying partial order, the Lawson topology does not do so in general, however.

In Section 1 it is shown that the Lawson topology on a complete lattice is always quasicompact and T_1. In Section 2 we see that for meet-continuous complete lattices the Lawson topology is Hausdorff if and only if the lattice is continuous. (Hence the asymmetry, because L may be continuous when L^{op} is not.) In fact continuous lattices equipped with the Lawson topology give compact Hausdorff topological semilattices which have a basis of subsemilattices—a most important class of semilattices in topological algebra. This interplay culminates in the Fundamental Theorem 3.4 of Chapter VI, which equates the two classes. (In the exercises the notion of a generalized continuous lattice is introduced in order to characterize precisely when the Lawson topology is Hausdorff for arbitrary complete lattices.)

In Section 3 we show in which way the Lawson topology can be defined in terms of convergence, where the notion of convergence is, once again, given in lattice-theoretical terms. This resumes and concludes a theme which we began to investigate in Section 1 of Chapter II. The last objective of the present chapter is to clarify when the Scott and Lawson topologies have *countable bases*. In Section 4 we prove this happens either for both or for neither, and the situation is described in purely lattice-theoretic terms; in fact, we treat arbitrary infinite cardinalities .

1. THE LAWSON TOPOLOGY

The Scott topology is rather coarse—a fact implicitly used to great advantage in the the previous chapter. For this reason we seldom mentioned closures, which give the entire lower set of a point; similarly, the quasicompactness in the Scott topology of a complete lattice is trivial, since any open cover must cover 0, and the open set containing 0 must be the whole space. To utilize the tools of closures and compactness more efficiently a refinement of the Scott topology must be found.

A guiding idea is the consideration of various topologies on such simple lattices as 2^X. The subbasic open sets of the Scott topology are the sets

$$W_x = \{f \in 2^X : f(x) = 1\}, \text{ for } x \in X.$$

Finite intersections of these are basic open sets, and these can be written as $\uparrow k$, where $k \in 2^X$ is the characteristic function of a finite set.

How do we then arrive at the product topology on 2^X when 2 is given the discrete topology? (This is certainly a natural Hausdorff topology to aim for.) In this topology, all sets $\uparrow k$ are also *closed*; and since any set of the form $\uparrow f$ is an intersection of sets $\uparrow k$, then all sets $\uparrow f$ have to be closed. Indeed, if one considers the Scott-open sets *plus* all sets of the form $2^X \backslash \uparrow f$ as a subbasis for a topology, the Tychonoff topology on 2^X results. This motivates considering on any complete lattice the topology arising if, in addition to all Scott-closed sets, we declare all upper sets of points closed.

We begin by defining an auxiliary topology which we will use more extensively in Chapter V but which serves at the moment only to refine the Scott topology in the way just described.

1.1. DEFINITION. Let L be a lattice. We call the topology generated by the complements $L \backslash \uparrow x$ of principal filters (as subbasic open sets) *the lower topology* and denote it $\omega(L)$. □

1.2. LEMMA. *Let* S *and* T *be complete lattices.*

(i) *A function* $f: S \to T$ *is continuous relative to the lower topologies if it preserves arbitrary infs;*

(ii) *If f is continuous, it preserves infs of filtered sets;*

(iii) *If it is continuous and preserves finite infs, then it preserves arbitrary infs.*

Proof. (i): Suppose that f preserves arbitrary infs. We must show that the inverse image of a subbasic closed set of the form $\uparrow t$ in T is also closed in S. Let $s = \inf f^{-1}(\uparrow t)$. Then

$$f(s) = f(\inf f^{-1}(\uparrow t)) = \inf ff^{-1}(\uparrow t) \geq \inf \uparrow t = t;$$

that is, $s \in f^{-1}(\uparrow t)$. But $f^{-1}(\uparrow t)$ is a filter (since f preserves infs) and thus equals $\uparrow s$. Hence, $f^{-1}(\uparrow t)$ is closed in S.

(ii): Next suppose that f is continuous relative to the lower topologies. Let F be filtered in S and $s = \inf F$. Whenever $x \notin \downarrow s$, then F cannot be contained in $\uparrow x$; hence, it is eventually in $L \setminus \uparrow x$. Thus, F converges to s relative to the lower topology. Then $f(s)$ is in the closure of $f(F)$, which is certainly contained in $\uparrow \inf f(F)$, so $\inf f(F) \leq f(\inf F)$.

Next we claim that f is monotone: If $x \leq y$ in S, then, by the continuity of f, the set $f^{-1}(\uparrow f(x))$ is closed in S and is in particular an upper set. Since it contains x, it must contain y; whence, $f(y) \in \uparrow f(x)$. That proves $f(x) \leq f(y)$.

Therefore, $f(\inf F) \leq f(u)$, for all $u \in F$, and so $f(\inf F) \leq \inf f(F)$. Thus, $\inf f(F) = f(\inf F)$.

(iii): If in addition, f preserves finite infs, then f preserves arbitrary infs (by in O-1.10) □

In words we can say that a ***semilattice*** map between two complete lattices is continuous relative to the lower topologies iff it preserves infs of filtered sets. This is the same as saying that it preserves arbitrary infs, so that the extra assumption of continuity makes the passage from the finite infs to the infinite ones.

Let us note that the lower topology $\omega(L)$ on the complete lattice L is generally coarser than $\sigma(L^{op})$, since a basis of $\omega(L)$ is given by the sets $L \setminus \uparrow F$ for finite $F \subseteq L$. But one should note that not every $\omega(L)$-closed set is necessarily of the form $\uparrow F$ for finite F—even though the collection of all principal filters $\uparrow x$ is closed under arbitrary intersections. However, the closure of a singleton $\{x\}$ is $\uparrow x$, and in particular the lower topology is T_0.

1.3. LEMMA. *If* S *and* T *are complete semilattices, then* $\omega(S \times T)$ *is the product topology of the topologies* $\omega(S)$ *and* $\omega(T)$.

Proof. This is immediate from the following two relations:

$$(S \setminus \uparrow s) \times (T \setminus \uparrow t) = (S \times T) \setminus ((S \times \uparrow t) \cup (\uparrow s \times T)), \text{ and}$$

$$\uparrow(s,t) = (S \times \uparrow t) \cap (\uparrow s \times T). \ \square$$

Note that the situation here is considerably simpler than the corresponding one for the Scott topologies (see II-4.11).

1.4. LEMMA. *If* L *is a complete lattice, then* $(L,\omega(L))$ *is a topological semilattice, that is, the inf operation*

$$(x,y) \mapsto x \wedge y : (L,\omega(L)) \times (L,\omega(L)) \to (L,\omega(L))$$

is continuous.

Proof. The inf operation preserves arbitrary infs hence is continuous as a function $(L\times L,\omega(L\times L))\to(L,\omega(L))$ by 1.2. The assertion follows from 1.3. □

We now proceed to the essential definition:

1.5. DEFINITION. Let L be a complete lattice. Then the common refinement $\sigma(L)\vee\omega(L)$ of the Scott topology and the lower topology is called the *Lawson topology* and is denoted $\lambda(L)$. The *space* $(L,\lambda(L))$ is written ΛL. □

In other words, the Lawson topology has as a ***subbasis*** the sets U, with $U\in\sigma(L)$, together with the sets $L\setminus\uparrow x$, for $x\in L$. The sets $U\setminus\uparrow F$, where $U\in\sigma(L)$ and F is finite in L, form a ***basis*** for $\lambda(L)$. Note that both U and $L\setminus\uparrow F$ satisfy property (S) (see II-1.3 and II-1.4(v-vi)); hence, all $U\setminus\uparrow F$ and all Lawson-open sets satisfy property (S) by II-1.4(vii). All one appears to be able to say beyond this about the structure of Lawson-open sets in general is the following:

1.6. PROPOSITION. *Let* L *be a complete lattice.*

(i) *An upper set U is Lawson open iff it is Scott open;*

(ii) *A lower set is Lawson-closed iff it is closed under sups of directed sets.*

Proof. (i): Since $\sigma(L)\subseteq\lambda(L)$ we have to show that every Lawson open upper set is Scott open. But by the preceding remarks such a set satisfies property (S), and then as an upper set it is Scott open by II-1.4(v).

(ii): The second assertion follows from the first in view of II-1.4(i). □

A picture may help in the visualization of the open sets in the Lawson topology:

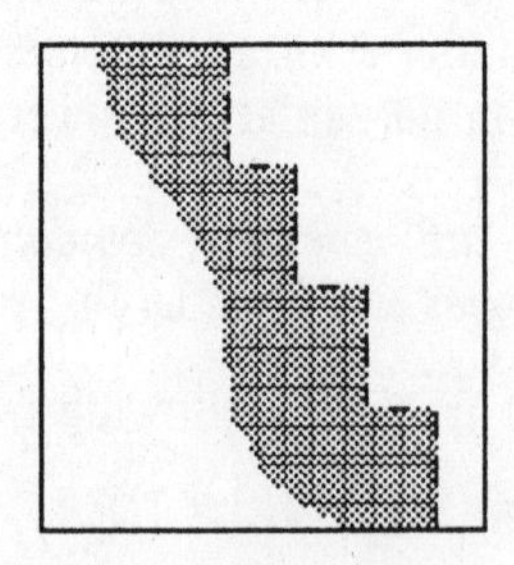

It will be seen that the Lawson topology on the unit square $[0,1]^2$ is just the ordinary Euclidean topology.

The examples in O-4.5 show that for a Lawson-open set U the upper set $\uparrow U$ need not be open and that the Lawson interior of an upper set need not be an upper set. When we talk about meet-continuous lattices, then things improve as we will see in 2.1 below.

1.7. LEMMA. *Let* L *be a complete lattice and* F *a filtered subset. Then* $\inf F = \lim F$ *with respect to* $\lambda(L)$, *and this limit is unique.*

Proof. The sets $L \setminus \uparrow x$ with $x \not\leq \inf F$ are subbasic $\omega(L)$-neighborhoods of $\inf F$, and F is eventually contained in any of them. Hence F converges to $\inf F$ with respect to $\omega(L)$. It also converges to $\inf F$ with respect to $\sigma(L)$ trivially, since every Scott neighborhood of $\inf F$ is an upper set and hence contains F. As a consequence, F converges to $\inf F$ with respect to $\lambda(L)$.

Now let y be any limit of F with respect to $\lambda(L)$. If $u \in F$, then $\downarrow u$ is $\lambda(L)$-closed and F is eventually in $\downarrow u$. Hence $y \leq u$ for all $u \in F$, and so $y \leq \inf F$. Conversely, $\uparrow \inf F$ is $\lambda(L)$-closed and contains F. Thus $y \in \uparrow \inf F$, and this proves $y = \inf F$. □

1.8. THEOREM. *Let* S *and* T *be complete lattices and* $f\colon S \to T$ *a semilattice morphism. Then the following statements are equivalent:*

(1) *f is Lawson-continuous (that is, $\lambda(S)$-$\lambda(T)$-continuous);*

(2) *f preserves arbitrary infs and directed sups;*

(3) *f preserves liminfs.*

Remark. Condition (3) means of course that $f(\underline{\lim}\, x_j) = \underline{\lim}\, f(x_j)$ for all nets $(x_j)_{j \in J}$ on S; compare this with II-2.1(2).

Proof. (2) implies (1): Assume that f preserves arbitrary infs and directed sups. Then f is $\omega(S)$-$\omega(T)$-continuous by 1.2 and $\sigma(S)$-$\sigma(T)$-continuous by II-2.1. Hence f is Lawson continuous.

(1) implies (2): We suppose that f is $\lambda(S)$-$\lambda(T)$-continuous. Let F be a filtered set in S. Then $\inf F = \lim F$ (with respect to $\lambda(S)$) by 1.7. Since f is λ-continuous we have $f(\inf F) = f(\lim F) = \lim f(F)$. The latter is $\inf f(F)$, since $f(F)$ is filtered because f is a semilattice morphism and 1.7 applies once more. Thus, f preserves infs of filtered sets and, hence, infs of arbitrary sets (compare O-1.10). If $U \in \sigma(T) \subseteq \lambda(T)$, then $f^{-1}(U) \in \lambda(S)$. Because $f^{-1}(U)$ is an upper set owing to the monotonicity of f, we have therefore the conclusion $f^{-1}(U) \in \sigma(L)$ by 1.6. Hence f is Scott continuous, and so it preserves directed sups by II-2.1.

(2) implies (3): Immediate from the definition of $\underline{\lim}$ in II-1.1.

(3) implies (2): If D is directed, then $f(\sup D) = f(\underline{\lim}\, D) = \underline{\lim}\, f(D) = \sup f(D)$, since f is order preserving as a semilattice morphism. If F is filtered, then $\inf f(\downarrow x \cap F) = \inf F$ for all $x \in F$; hence,

$$\underline{\lim}\, F = \bigvee_{x \in F} \inf f(\downarrow x \cap F) = \inf F.$$

Thus $f(\inf F) = f(\underline{\lim}\, F) = \underline{\lim}\, f(F) = \inf f(F)$. This shows that f preserves filtered infs. Since f is a semilattice morphism, f also preserves arbitrary infs (see O-1.10). □

1.9. THEOREM. *For a complete lattice* L, *the Lawson topology* λ(L) *is a quasicompact* T_1*-topology.*

Proof. Firstly, for $x \in L$ we have $\{x\} = \downarrow x \cap \uparrow x$. Now, $\downarrow x$ is Scott closed, while $\uparrow x$ is closed in the lower topology. Hence, $\{x\}$ is Lawson closed; that is, λ(L) is a T_1-topology.

To prove that λ(L) is quasicompact, we use the Alexander Lemma: a space is quasicompact if every open cover consisting of ***subbasic*** open sets contains a finite subcover.

Thus assume $\{U_j \in \sigma(L) : j \in J\}$ and $\{L \setminus \uparrow x_k : k \in K\}$ together form a cover of L. Let $x = \sup\{x_k : k \in K\}$. Then

$$\bigcup\{L\setminus\uparrow x_k : k\in K\} = L\setminus\bigcap\{\uparrow x_k : k\in K\} = L\setminus\uparrow x.$$

But $x \notin L\setminus\uparrow x$; therefore, there is a j such that $x \in U_j$. Since U_j is Scott open, there are indices $k_1,\ldots,k_n$ such that $x_{k_1}\vee\ldots\vee x_{k_n} \in U_j$. Then

$$U_j \cup (L\setminus\uparrow x_{k_1}) \cup \ldots \cup (L\setminus\uparrow x_{k_n}) = L,$$

and the proof is complete. □

This compactness theorem shows in a certain sense that we have not refined the Scott topology too much, even though we have improved the T_0-separation property of the Scott topology to the T_1-separation of the Lawson topology. It will be of paramount importance to understand when in fact the Lawson topology is Hausdorff; most of the following observations serve to study this question. In particular the next theorem shows the suitability of the Lawson topology for continuous lattices.

1.10. THEOREM. *For a continuous lattice* L, *the Lawson topology* λ(L) *is a compact Hausdorff topology.*

Proof. After Theorem 1.9 it remains to show that λ(L) is Hausdorff. Suppose that $x \neq y$ in L, and assume that $x \not\leq y$. Then by I-1.6 there is a $u \ll x$ with $u \not\leq y$. Then $\Uparrow u$ is a Scott- (hence, Lawson-) open neighborhood of x (see II-1.5(5)), and $L\setminus\uparrow u$ is an ω(L)- (hence, Lawson-) open neighborhood of y. Clearly these two neighborhoods are disjoint □

A finer analysis of the Lawson topology is possible for meet-continuous lattices. We make it the subject of the *next* section. We close *this* section with a discussion of the relationship between subalgebras and the Lawson topology (in the same spirit as Theorem 1.8 for morphisms).

1.11. THEOREM. *Let* L *be a continuous lattice, and let* S *be a subsemilattice containing* 1. *The following conditions are equivalent:*

(1) S *is closed in the Lawson topology;*

(2) S *is closed with respect to the formation of arbitrary infs and directed sups in* L.

(3) *For all nets* $(x_j)_{j\in J}$ *in* S, *we have* $\underline{\lim}\ x_j \in S$, *where* $\underline{\lim}\ x_j$ *is taken with respect to* L.

Proof. (1) implies (2): By Lemma 1.7 a filtered inf is a $\lambda(L)$-limit; hence S is closed under filtered infs and thus arbitrary infs (see O-1.5). Since all Lawson-open sets satisfy property (S) (see the remarks after Definition 1.5), it follows that directed sets converge to their sups with respect to $\lambda(L)$. Hence S is closed with respect to directed sups.

(2) implies (3): Immediate.

(3) implies (2): Since filtered infs and directed sups are special cases of taking $\underline{\lim}$ of a net, we conclude S is closed with respect to taking directed sups and filtered infs (and hence arbitrary infs).

(2) implies (1): By I-2.7(ii) S itself is a continuous lattice. By Theorem 1.8 the inclusion mapping from S to L is continuous. Since S is compact by Theorem 1.9, its image under inclusion is compact and hence closed since L is Hausdorff (Theorem 1.10). But the image of S is just S with the relative topology. □

EXERCISES

1.12. EXERCISE. Let L be an algebraic lattice.

(i) ΛL is compact zero dimensional.

(ii) The Lawson topology has a basis of open-closed subsemilattices.

(HINT: By 1.9, ΛL is quasicompact. By II-1.15 $\sigma(L)$ has a basis of sets of the form $\uparrow k$ with $k \in K(L)$. But $\uparrow k$ is $\omega(L)$-closed, thus, $\lambda(L)$-closed; further $\Uparrow k$ is $\sigma(L)$-open by II-1.10. If $k \in K(L)$, then $\Uparrow k = \uparrow k$, and so $\uparrow k \in \lambda(L)$; whence, $\uparrow k$ is $\lambda(L)$-open-closed for any compact k. If L is algebraic, then the sets $\uparrow k \setminus (\uparrow k_1 \cup \ldots \cup \uparrow k_n)$, where $k, k_1, \ldots, k_n \in K(L)$, constitute a basis for $\lambda(L)$. All of these sets are open-closed.)

Example O-4.5(2) shows that there are complete lattices L such that ΛL is compact zero dimensional while L is not algebraic. In the next section we note that in the class of meet-continuous lattices this aberration cannot occur.

1.13. EXERCISE. Let X be a locally quasicompact topological space. Let $\Gamma(X)$ denote the lattice of closed sets (O-2.7(3)). (Recall $\Gamma(X)^{\mathrm{op}} \simeq \mathcal{O}(X)$ is a continuous lattice by I-1.7(5).)

(i) $F \ll G$ in $\Gamma(X)^{\mathrm{op}}$ iff there is a quasicompact set Q such that $F \cup Q = X$ and $G \cap Q = \varnothing$.

(ii) The Scott topology on $\Gamma(X)^{\mathrm{op}}$ has as a basis the sets of the form $\{G \in \Gamma(X) : G \cap Q = \varnothing\}$, where Q is a quasicompact subset of X.

(iii) The lower topology $\omega(\Gamma(X)^{op})$ has as a subbasis the sets of the form $\{G\in\Gamma(X) : G\cap U \neq \emptyset\}$, where U is an open subset of X.

(iv) The Lawson topology has as a basis the sets of the form

$$\{G\in\Gamma(X) : G\cap Q = \emptyset \text{ and } G\cap U_k \neq \emptyset,\ k=1,\dots,n\},$$

where Q is quasicompact and $U_1,\dots,U_n$ is a finite collection of open sets.

(HINT: (i) follows from I-1.4(ii). For (ii) use II-1.14(2) and (i) above. Then (iii) and (iv) are straightforward.) □

We now investigate the lower topology a bit more closely.

1.14. EXERCISE. Let L be a complete lattice and $U,V\in\omega(L)$. Set $A = L\setminus U$ and $B = L\setminus V$. Then the following statements are equivalent:

(1) $U \ll V$ in $\omega(L)$.

(2) For all directed sets $D\subseteq L$ with $\sup D\in B$ we have $D\cap A \neq \emptyset$.

(HINT: (1) implies (2): Let D be directed in L with $\sup D\in B$. Then $\{L\setminus\uparrow d : d\in D\}$ is a directed collection in $\omega(L)$ with

$$V = L\setminus B \subseteq L\setminus\uparrow\sup D = \bigcup\{L\setminus\uparrow d : d\in D\}.$$

Hence, (1) implies that there is a $d\in D$ with $U\subseteq L\setminus\uparrow d$; that is, $d\in A$.

(2) implies (1): Let $\{L\setminus\uparrow x : x\in M\}$ be a collection of subbasic open sets covering V. Then $V\subseteq\bigcup\{L\setminus\uparrow x : x\in M\} = L\setminus\uparrow y$ where $y = \sup\{x : x\in M\}$. Thus $\uparrow y\subseteq L\setminus V = B$. By hypothesis $\sup F\in A$ for some finite set $F\subseteq M$ (since $\{\sup F : F \text{ finite}, F\subseteq M\}$ is directed). Thus $\uparrow\sup F\subseteq A$; that is,

$$U = L\setminus A\subseteq L\setminus(\uparrow\sup F) = \bigcup\{L\setminus\uparrow x : x\in F\}.$$

By the generalized version of Alexander's Lemma (I-3.21) we have $U\ll V$.) □

We had many reasons to consider situations where the Scott topology $\sigma(L)$ of a complete lattice itself was a continuous lattice. (See II-1.14, II-4.11, II-4.16). Now we ask when the lower topology $\omega(L)$ is a continuous lattice. We begin by introducing some technical but convenient notation. In fact condition 1.14(2) motivates the following definition:

1.15. DEFINITION. For two upper sets A,B in a complete lattice L we write

$A \,((\, B$ iff for every directed set $D\subseteq L$ with $\sup D\in B$ we have $A\cap D\neq\emptyset$.

We say that "*A shields B*". □

Notice that $A \,((\, B$ implies $A\subseteq B$ (consider singleton sets D!), and that $a\ll b$ is equivalent to $\uparrow a \,((\, \uparrow b$. By 1.14, if A and B are $\omega(L)$-closed, then $A \,((\, B$ iff $L\setminus A \ll L\setminus B$ in $\omega(L)$. An upper set U shields itself (that is, satisfies $U \,((\, U$) iff it is Scott open(II-1.3).

1.16. EXERCISE. *Let* L *be a complete lattice. Then the following conditions are equivalent:*

(1) *$\omega(\mathrm{L})$ is a continuous lattice.*

(2) *For each $\omega(\mathrm{L})$-closed set C and each $y \notin C$ there is a finite set $F \subseteq \mathrm{L}$ with $y \notin \uparrow F$ and $\uparrow F \mathrel{((} C$.*

(3) *Whenever $x \not\leq y$ in L then there is a finite set $F \subseteq \mathrm{L}$ with $y \notin \uparrow F$ and $\uparrow F \mathrel{((} \uparrow x$.*

(4) *For each $\omega(\mathrm{L})$-closed set C, the set $\{x : C \mathrel{((} \uparrow x\}$ is the Scott interior of C, and $\omega(\mathrm{L})$ is continuous.*

(5) *Whenever $x \not\leq y$ in L there is a Scott-open neighborhood of x whose $\omega(\mathrm{L})$-closure does not contain y.*

(HINT: (1) iff (2): (1) says that for each $\mathrm{L} \backslash C \in \omega(\mathrm{L})$ with $y \in \mathrm{L} \backslash C$ there is a basic $\omega(\mathrm{L})$-neighborhood $\mathrm{L} \backslash \uparrow F$ of y with $\mathrm{L} \backslash \uparrow F \ll \mathrm{L} \backslash C$. Hence apply 1.14.

(3) implies (2): Suppose C is a given $\omega(\mathrm{L})$-closed set. Then

$$C = \bigcap \{\uparrow G : G \text{ finite and } C \subseteq \uparrow G\}.$$

Hence there is a finite set G with $C \subseteq \uparrow G$ and $y \notin \uparrow G$. For each $g \in G$ by (3) there is a finite set F_g such that $\uparrow F_g \mathrel{((} \uparrow g$; set $F = \bigcup \{F_g : g \in G\}$. Then F is finite and $y \notin \uparrow F$, while $\uparrow F \mathrel{((} \uparrow G$, hence $\uparrow F \mathrel{((} \uparrow C$.

(1) implies (4): Let U be the Scott interior of C. Then $U \mathrel{((} U$, hence $U \mathrel{((} \uparrow x$ for all $x \in U$. Thus $U \subseteq \{x : C \mathrel{((} \uparrow x\}$. (This holds without any further hypothesis!). In order to show equality it suffices to demonstrate that $\{x : C \mathrel{((} \uparrow x\}$ is Scott open.

Let D be a directed set with $C \mathrel{((} \uparrow \sup D$. This means

$$\mathrm{L} \backslash C \ll \sup \{\mathrm{L} \backslash \uparrow d : d \in D\}$$

in the continuous lattice $\omega(\mathrm{L})$. Hence by I-1.19 there is a $d \in D$ with $\mathrm{L} \backslash C \ll \mathrm{L} \backslash \uparrow d$; that is, $C \mathrel{((} \uparrow d$.

(4) implies (5): Let $x \not\leq y$, then, since $\omega(\mathrm{L})$ is continuous, there is an $\omega(\mathrm{L})$-closed C such that $y \notin C$ and $C \mathrel{((} \uparrow x$. Then $x \in \operatorname{int} C$ by (4), and so we have $\mathrm{cl}_\omega(\mathrm{int}_\sigma(C)) \subseteq C$.

(5) implies (3): Let $x \not\leq y$ and find a $\omega(\mathrm{L})$-closed set C with $y \notin C$ and $x \in \operatorname{int} C$. Then $C \mathrel{((} \uparrow x$. Since $C = \bigcap \{\uparrow F : F \text{ finite and } C \subseteq \uparrow F\}$ we find a finite set F with $y \notin \uparrow F$ and $C \subseteq \uparrow F$. Then $\uparrow F \mathrel{((} \uparrow x$.) $\square$

This last result suggests a new definition. We notice that the condition (3) of 1.16 generalizes condition I-1.6(A_1) which reads in the current terminology as follows:

(A_1) Whenever $x \not\leq y$, then there is a $u \in \mathrm{L}$ with $y \notin \uparrow u$ and $\uparrow u \mathrel{((} \uparrow x$.

We therefore adopt the following convention:

1.17. DEFINITION. A lattice L is called a *generalized continuous lattice,* or shortly a GCL-lattice, iff it is complete and satisfies:

(GA) Whenever $x \not\leq y$, then there is a finite $F \subseteq \mathrm{L}$ with $y \notin \uparrow F$ and $\uparrow F \mathrel{((} \uparrow x$.

For each $x \in L$, let fin(x) denote the set of finite subsets $F \subseteq L$ with $\uparrow F \ll \uparrow x$. □

Clearly, condition (5) of Theorem 1.16 implies that ΛL is a Hausdorff space. Thus Theorem 1.16 also generalizes Theorem 1.10. The final word on this aspect of 1.14 will be given in Section 3 (and its exercises).

1.18. EXERCISE. Find a complete lattice which is GCL but is not continuous (see O-4.5(2)). □

1.19. EXERCISE. For a complete lattice L show that the following is an equivalent condition to be a GCL-lattice:

(GA_1) For each $x \in L$ and each choice function $f \in \mathsf{X}_{F \in \mathrm{fin}(x)} F$ we have $x \leq \sup\{f(F) : F \in \mathrm{fin}(x)\}$. □

1.20. EXERCISE. Let L be a GCL-lattice and X, Y, Z upper sets. Then we have the following statements:

(i) If $X \ll Z$ and $Y \ll Z$, then $X \vee Y \ll Z$, where for short we write $X \vee Y = \{x \vee y : x \in X \text{ and } y \in Y\}$.

(ii) If C is $\omega(L)$-closed and $X \ll C$, then there is a finite set F with $C \subseteq \uparrow F \subseteq X$ and $X \ll \uparrow F \ll C$.

(HINT: (i) is straightforward. (ii) follows from 1.16 for $\omega(L)$-closed X. Find some finite set G with $C \subseteq \uparrow G \subseteq \uparrow X$ and $G \ll X$, and apply the preceding remark to G and C.) □

1.21. EXERCISE. Consider the Lawson topology on an up-complete poset L.

(i) It remains true that an upper set is Scott open iff it is Lawson-open (cf. 1.6);

(ii) If F is a filtered subset which has an inf, then $\inf F = \lim F$ with respect to $\lambda(L)$ (cf. 1.7);

(iii) The Lawson topology on a complete semilattice is quasicompact T_1.

(HINT: Consider S^1(see O-2.12), note that $1 \in K(S^1)$ and that $\{1\}$ is open with respect to $\lambda(S)$.) □

The next exercises relate two traditional lattice topologies, the interval topology and the order topology, to the Lawson topology. The interval topology has for a subbase of closed sets all principal ideals and all principal filters. In order to define the order topology on L, we say a net $(x_j)_{j \in J}$ *order converges to* x iff $x = \liminf x_j = \limsup x_j$. This notion of convergence defines a topology on L, the order topology. (See Section II-1; see also Birkhoff [1967] for further details concerning these topologies.)

1.22. EXERCISE. Let L be a complete lattice.

(i) The interval topology is the join (as topologies) of the upper and lower topology.

(ii) The interval topology is contained in the Lawson topology on L and on L^{op}.

(iii) If the interval topology is Hausdorff, then the interval and the Lawson topologies on L and L^{op} all agree.

(HINT: Use 1.9.) □

1.23. EXERCISE. Let L be a complete lattice.

(i) The open upper sets in the order topology are the Scott-open sets (and dually).

(ii) The Lawson topology on L and on L^{op} is contained in the order topology. □

NOTES

In describing the history of the Lawson topology, it is best to distinguish two viewpoints: that of topological algebra and that of lattice theory. In topological algebra one studies the structure of algebraic structures such as groups, rings, and semigroups which are already equipped with the topology such that the operations are continuous. In this vein compact topological semilattices and lattices have been studied since the fifties by A.D. Wallace and the numerous mathematicians following in his footsteps; we will comment on this piece of history in Chapter VI where we concentrate on compact semilattices. However, we will see in the next section of the present chapter how compact semilattice theory and continuous lattice theory relate (2.13). In lattice theory, on the other hand, one considers lattices and looks for topologies which are naturally defined in terms of the given lattice structure. Typical examples are the topologies $\sigma(L)$ (Section II-1), $\omega(L)$, and $\lambda(L)$; there are, of course, others, but these do not interest us here.

The blending of the topological algebra viewpoint and the lattice theoretical viewpoint, as far as continuous lattices were concerned, was accomplished by K.H. Hofmann and A. Stralka in ATLAS [1976]. What was discovered there was that what had been studied by the topological algebraists under the name of compact Lawson semilattices was in fact the very same thing as continuous lattices (although in ATLAS this discovery is not phrased quite so explicitly; the first paper in print being explicit about this is Lea's note [1976/77]).

The explicit definition of the topology $\lambda(L)$ given here has evolved in the SCS-Seminar since 1976. The name *Lawson topology* was chosen at the First Workshop on Continuous Lattices in April 1977 at Tulane. However, just as one finds the Scott topology defined for special complete lattices such as those of the form $\mathcal{O}(X)$ before Scott's paper in the work of Day and Kelly, the Lawson topology has its precursors on $\mathcal{O}(X)$. Indeed the Lawson topology was considered as early as 1961 by J.M.G. Fell [1962] when X is a locally quasicompact space, although the definition was given in terms of the description of basic open sets which are not on the surface recognizable as yielding the same topology as we see in Exercise 1.13. In fact the topology was introduced on $\Gamma(X)^{op}$ and it was shown by Fell that it was compact Hausdorff and that $(A,B) \mapsto A \cup B$ was continuous. These studies were continued later by Dixmier [1968] who provided more information on this topology.

Theorem 1.8 is new. Theorem 1.9 was first published by Hofmann [1977], with a proof due to D. Scott [SCS-4]. In this line, 1.10 had been known to the SCS-Seminar since 1976. The more sophisticated investigations of generalized continuous lattices contained in Exercises 1.14 through 1.20 are more recent and are due to G. Gierz and J.D. Lawson [198*]; the presentation here is a bit reorganized. Note that Theorem 1.16 once more gives lattice theoretical properties of a complete lattice in terms of properties of one of the canonical topologies on L, this time in terms of the lower topology $\omega(L)$. Familiarity with GCL-lattices is not absolutely necessary for an understanding of the basic theory of continuous lattices, notably since they will turn out to be continuous in the case that meet-continuous lattices are considered. But in Section 3 we will show in the Exercises that the Lawson topology of a complete lattice is Hausdorff iff the lattice is GCL.

2. MEET-CONTINUOUS LATTICES REVISITED

For more detailed information on the developments of the previous section we turn to meet-continuous lattices. Here there is rather more to say on the nature of Lawson-open sets. In discussing various topologies we will use subscripts to distinguish relative to which the closure, the interior, etc. is to be taken.

2.1. PROPOSITION. *In a meet-continuous lattice* L *we have:*
(i) *If* $U \in \lambda(L)$, *then* $\uparrow U \in \sigma(L)$;
(ii) *If* X *is an upper set, then* $\mathrm{int}_\sigma X = \mathrm{int}_\lambda X$;
(iii) *If* X *is a lower set, then* $\mathrm{cl}_\sigma X = \mathrm{cl}_\lambda X$.

Proof. (i): One easily verifies that in a meet-continuous lattice, if a set U has property (S) of II-1.3, then $\uparrow U$ has property (S). Then apply II-1.4(v).
(ii): Trivially $\mathrm{int}_\sigma X \subseteq \mathrm{int}_\lambda X$. By (i), $\mathrm{int}_\lambda X \subseteq \uparrow \mathrm{int}_\lambda X \subseteq \mathrm{int}_\sigma X$.
(iii) is equivalent to (ii) in a straightforward way. □

2.2. PROPOSITION. *Let* S *and* T *be complete lattices such that* $\sigma(T)$ *is a continuous lattice. Then* $\Lambda(S \times T) = \Lambda S \times \Lambda T$.

Proof. By II-4.11 we have $\Sigma(S \times T) = \Sigma S \times \Sigma T$. From 1.3 we recall that $\omega(S \times T)$ is the product topology of $\omega(S)$ and $\omega(T)$. Suppose that ξ_1, ξ_2 are topologies on X and η_1, η_2 topologies on Y. Assume $U_k \in \xi_k$ and $V_k \in \eta_k$, $k=1,2$. Then the relation $(U_1 \cap U_2) \times (V_1 \cap V_2) = (U_1 \times V_1) \cap (U_2 \times V_2)$ shows that basic open sets of $(\xi_1 \vee \xi_2) \times (\eta_1 \vee \eta_2)$ are basic open in $(\xi_1 \times \eta_1) \vee (\xi_2 \times \eta_2)$ and vice versa. We apply this observation with $X = S$, $\xi_1 = \sigma(S)$, $\xi_2 = \omega(S)$ and $Y = T$, $\eta_1 = \sigma(T)$, $\eta_2 = \omega(T)$ and obtain the assertion. □

2.3. THEOREM. *Let* L *be a meet-continuous lattice. Then all the translations* $x \mapsto ax : \Lambda L \to \Lambda L$, *for* $a \in L$, *are continuous. If also* $\sigma(L)$ *is a continuous lattice, then* ΛL *is a topological semilattice.*

Proof. Every map $x \mapsto ax$ is a semilattice morphism preserving arbitrary infs and, since L is meet-continuous, directed sups (O-4.1). Hence it is Lawson-continuous by 1.8. The function $(x,y) \mapsto xy : L \times L \to L$ likewise preserves arbitrary infs and directed sups (O-4.2). Hence by 1.8 it yields a continuous map $\Lambda(L \times L) \to \Lambda L$. If $\sigma(L)$ is continuous, then 2.2 applies and gives $\Lambda(L \times L) = \Lambda L \times \Lambda L$ which yields the desired conclusion. □

2.4. COROLLARY. *In a meet-continuous lattice* L *with continuous* $\sigma(L)$ *the Lawson topology is Hausdorff iff the graph of* $\leq$ *is closed in* $\Lambda L \times \Lambda L$.

Proof. Since ΛL is a topological semilattice by 2.3, then the function $m : \Lambda L \times \Lambda L \to \Lambda L$ defined by $m(x,y) = (x, xy)$ is continuous, and the graph of $\leq$ is m^{-1} of the diagonal in $L \times L$. Thus, if the latter is closed, then so is the graph of $\leq$. Since the diagonal is the intersection of the graphs of $\leq$ and of $\leq^{op}$, the converse is clear. □

We need a technical concept which is related to the Scott interior of an upper set, and which generalizes the idea of the sets $\Uparrow x$.

2.5. DEFINITION. For an upper set X in a complete lattice we write X^0 for the set of all elements u such that whenever $u \leq \sup D$ and D is directed, then $X \cap D \neq \emptyset$.

2.6. REMARK. (i) $(\uparrow x)^0 = \Uparrow x$.

(ii) *For any upper set X we have* $\operatorname{int}_\sigma X \subseteq X^0$.

Proof. (i) is immediate from I-1.1 and 2.5, and (ii) is clear from the definition of Scott-open sets in II-1.3. □

2.7. LEMMA. *Let X and Y be upper sets in a meet-continuous lattice. Then* $(X \cup Y)^0 = X^0 \cup Y^0$.

Proof. We have always $X^0 \cup Y^0 \subseteq (X \cup Y)^0$. Now suppose that $s \in (X \cup Y)^0 \setminus (X^0 \cup Y^0)$. Then there are directed sets D and E with $s \leq \sup D$ and $s \leq \sup E$, but $D \cap X = \emptyset = E \cap Y$. Since L is meet continuous, we have $s \leq (\sup D)(\sup E) = \sup DE$ (see O-4.2); hence the directed set DE meets $(X \cup Y)$. We remark $DE \cap X = \emptyset = DE \cap Y$; hence, $DE \cap (X \cup Y) = \emptyset$, and this is a contradiction. □

As a corollary, we obtain

2.8. LEMMA. *If F is a finite set in a meet-continuous lattice then we have*

$$\operatorname{int}_\sigma \uparrow F \subseteq \bigcup \{\Uparrow x : x \in F\}.$$

Proof. By 2.6(ii) we have $\operatorname{int}_\sigma \uparrow F \subseteq (\uparrow F)^0 = (\bigcup\{\uparrow x : x \in F\})^0$; by 2..7 and induction we observe $(\bigcup\{\uparrow x : x \in F\})^0 = \bigcup\{(\uparrow x)^0 : x \in F\}$. The assertion follows from 2.6(i). □

We now arrive at an important characterization theorem for continuous lattices in the class of meet-continuous lattices in terms of the Hausdorff separation of the Lawson topology.

2.9. THEOREM. L *is a continuous lattice iff it is meet continuous and Hausdorff in its Lawson topology.*

Proof. The first implication is clear from I-2.2 and 1.10.

Conversely, in view of I-1.6, consider two points x and y with $x \not\leq y$. We hope to find a $u \ll x$ such that $u \not\leq y$. Clearly $xy < x$, and it suffices to find a $u \ll x$ such that $u \not\leq xy$. Therefore, we assume that $y < x$.

Since L is Hausdorff, there are disjoint open $\lambda(L)$-neighborhoods V and W of y and x, respectively, and we may assume that V is of the form $U\setminus\uparrow F$ with a Scott-open neighborhood U of y and a finite set F. But since $y<x$, we may also assume that $W\subseteq U$ (for otherwise we replace W by $U\cap W$). Now we claim that $W\subseteq\uparrow F$. For if not, then there would be a $w\in W\setminus\uparrow F\subseteq U\setminus\uparrow F = V$, which is impossible since V and W are disjoint. But then also $\uparrow W\subseteq\uparrow F$. Since $\uparrow W$ is Scott open by 2.1(i), we know that $\mathrm{int}_\sigma\uparrow F$ contains x. By Lemma 2.8 this implies that there is a $u\in F$ such that $x\in\twoheaduparrow u$; that is, $u\ll x$. Since $y\in U\setminus\uparrow F\subseteq L\setminus\uparrow u$ we have $u\not\leq y$ as was desired. □

This theorem reveals the assumption of continuity for a meet-continuous lattice to be a separation property rather than a compactness property. Furthermore, 2.9 together with 2.4 and II-1.14 show that the graph of $\leq$ is closed in $\Lambda L \times \Lambda L$ for a continuous lattice L.

2.10. DEFINITION. We say that a semilattice with a topology *has small (open) semilattices* iff each point has a neighborhood basis of (open) subsemilattices (cf. also VI-3.1 ff.). □

2.11. PROPOSITION. *Let* L *be a meet-continuous lattice. Then the following are equivalent*:

(1) *Each point of* L *has a Scott-neighborhood basis of open filters.*

(2) ΛL *has small open semilattices.*

Proof. (1) implies (2): Let $x\in W\in\lambda(L)$. Then there is a $V\in\sigma(L)$ and a finite set $F\subseteq L$ with $x\in V\setminus\uparrow F\subseteq W$. By (1) there is a filter $U\in\sigma(L)$ with $x\in U\subseteq V$. Then $x\in U\setminus\uparrow F\in\lambda(L)$, with $U\setminus\uparrow F\subseteq W$. One obtains easily that $U\setminus\uparrow F$ is a semilattice.

(2) implies (1): Let $x\in W\in\sigma(L)$. Then there is a subsemilattice $V\in\lambda(L)$ with $x\in V\subseteq W$. Then $x\in\uparrow V$, $\uparrow V$ is a filter, and $\uparrow V\in\sigma(L)$ by 2.1. □

2.12. LEMMA. *Let* L *be a meet-continuous lattice. Then for each* $x\in L$ *we have*

$$\sup\{\inf U : x\in U\in\lambda(L)\} = \sup\{\inf U : x\in U\in\sigma(L)\}.$$

Proof. Since $\sigma(L)\subseteq\lambda(L)$, clearly the left hand side is $\geq$ the right hand side. But if $U\in\lambda(L)$, then $\inf U = \inf\uparrow U$ and $\uparrow U\in\sigma(L)$ by 2.1, hence the reverse inequality holds. □

We are ready for a crucial theorem:

2.13. THEOREM. *Let* L *be a meet-continuous lattice. Then the following are equivalent*:

(1) L *is continuous.*

(2) ΛL *has small open semilattices and* $\sigma(L)$ *is a continuous lattice.*

(3) ΛL *is a compact Hausdorff topological semilattice with small open semilattices.*

(3^1) ΛL *is a compact Hausdorff topological semilattice with small compact semilattices.*

(4) $x = \sup\{\inf U : x \in U \in \lambda(\mathrm{L})\}$, *for all* $x \in$ L.

Proof. By 2.11, condition (2) is equivalent to II-1.14(3), and by 2.12, condition (4) is equivalent to II-1.14(4). Hence (1) iff (2) iff (4).

(1) implies (3): If L is continuous, then ΛL is compact Hausdorff by 1.10. By 2.3 and II-1.14, we know that ΛL is a topological semilattice, and from (2) we know that ΛL has small open semilattices.

(3) implies (3^1) implies (4) is straightforward. □

In Chapter VI-3 we will see that any compact topological unital semilattice with small open semilattices has a continuous lattice as underlying poset, and that its given topology is the Lawson topology. Thus Theorem 2.13 constitutes an essential portion of a fundamental link between continuous lattices and compact semilattices.

EXERCISES

2.14. EXERCISE. Let L be a meet-continuous lattice. Then L is GCL (1.17) iff L is continuous.

(HINT: It was noticed in the context of Definition 1.17 that every continuous lattice is GCL. Every GCL-lattice has a Hausdorff Lawson topology by 1.16(5). Hence Theorem 2.9 applies to give the converse.) □

In this book most of the complete lattices we discuss are in fact meet continuous. For these, there is no difference between the GCL-property and continuity. In particular, for meet-continuous lattices, the conditions of Theorem 1.16 are satisfied iff L is continuous. (Example O-4.5(2), by the way, described a GCL-lattice which was not meet continuous.)

Let us link the sets X^0 introduced in 2.5 with the "shielding" relation used in the proof and statement of 1.16:

2.15. EXERCISE. Let X be an upper set in a complete lattice L.

(i) $X^0 = \{x \in \mathrm{L} : X \,((\, \uparrow x\}$. (See 1.15.)

(ii) If L is a GCL-lattice (1.17) then $X^0 = \mathrm{int}_\sigma X$. (Compare 2.6(ii).) □

2.16. EXERCISE. Let L be a meet-continuous lattice. Then the following statements are equivalent:

(1) L is algebraic;

(2) ΛL has small open semilattices and σ(L) is algebraic;

(3) ΛL is a compact zero dimensional Hausdorff topological semilattice with small open-closed semilattices;

(4) ΛL is a zero dimensional Hausdorff topological semilattice.

(HINT: For (1) iff (2) refer to 2.13 and II-1.15, considering II-1.11 and 2.11. In 1.12 we showed (1) implies (3), and (3) implies (4) is trivial. Suppose (4) and take $x \in L$ and let U be a neighborhood of x. By (4) find an open-compact neighborhood $V \subseteq U$; by continuity of multiplication and the fact V and its complement are closed find a neighborhood W of x in V such that $WV \subseteq V$. Let $W^* = \bigcup\{W^n : n=1,2,...\}$ and note that $W^* \subseteq V$. Now $\inf W^* = \lim \overline{W^*} \in V^- = V$, whence 2.13(4) is satisfied. Thus ΛL has small semilattices by 2.13(2). If W is open-closed in ΛL, then $\uparrow W$ is open-closed, hence is a compact element in $\sigma(L)$; this shows that $\sigma(L)$ is algebraic. Thus we have shown (2).) □

NOTES

As the diagram of "hierarchies" at the end of Chapter I indicates (see end of Section 4 in Chapter I), all of the theory directly relevant for continuous lattices takes place within the class of meet-continuous lattices. (The GCL-lattices which we introduced in the exercises to the previous section are an exception; they do not fit into the hierarchy insofar as those GCL-lattices which are not continuous cannot even be meet continuous (2.14).) To the discussions on the Lawson topology we have added the hypothesis of meet-continuity in the present section and thereby obtained results such as 2.9 and 2.13. Theorem 2.9 is due to Gierz and Lawson [SCS-42], but the proof here is more direct. The equivalence (1) iff (3) in 2.13 is implicit in Hofmann and Stralka [1976] although without identification of the Lawson topology in explicit terms. Theorem 2.3 is new (as is Proposition 2.2); these results use once again the hypothesis that the Scott topology $\sigma(L)$ is a continuous lattice; the reason for this goes back to II-4.11.

3. LIM-INF CONVERGENCE

In this section we resume the theme of Section II-1 where we derived the Scott topology from lim-inf convergence and then discussed the relations between these notions. Here we refine the concept of lim-inf convergence and investigate its relationship to the Lawson topology. Roughly speaking, since $\sigma(L)\subseteq\lambda(L)$, we expect fewer convergent nets to belong to the finer topology $\lambda(L)$. Our first observation is purely lattice theoretical in view of the fact that the lim-inf is a purely lattice theoretical idea (see II-1.1).

3.1. PROPOSITION. *Let* L *be a complete lattice,* $x\in L$ *and* $(x_j)_{j\in J}$ *a net on* L. *Then the following statements are equivalent:*

(1) $x = \underline{\lim}\, y_k$ *for all subnets* $y_k = x_{f(k)}$ *of* $(x_j)_{j\in J}$;

(2) $x = \underline{\lim}\, x_j$ *and* $x\geq\inf y_k$ *for all subnets* y_k *of* x_j;

(3) $x = \underline{\lim}\, x_j$ *and* $x\geq\inf x_{p(j)}$ *for all* $p: J\to J$ *with* $j\leq p(j)$ *for all* $j\in J$.

Proof. (1) implies (2): Trivial by II-1.1.

(2) implies (3): Obvious in view of the fact that for each of the functions p we consider, $(x_{p(j)})_{j\in J}$ is a subnet of $(x_j)_{j\in J}$.

(3) implies (1): Let $(y_k)_{k\in K}$ be a subnet of $(x_j)_{j\in J}$ with $y_k = x_{f(k)}$. For $j\in J$ there is a $k\in K$ such that $k\leq k^*$ implies $j\leq f(k^*)$, hence

$$\inf_{k^*\geq k} y_{k^*}\geq\inf_{j^*\geq j} x_{j^*}.$$

Thus $y = \underline{\lim}\, y_k$ gives $y\geq x$. We have to show $y\leq x$. For this it suffices to prove $\inf_{k\geq k_0} y_k\leq x$ for arbitrarily large k_0. Since $y_k = x_{f(k)}$ is a subnet of x_j, we know that for each $j\in J$ there is a $g(j)\in K$ with $g(j)\geq k_0$ such that $k\geq g(j)$ implies $f(k)\geq j$. We define $p: J\to J$ be $p = fg$; then clearly $p(j) = f(g(j))\geq j$. Thus,

$$\inf_{k\geq k_0} y_k = \inf_{k\geq k_0} x_{f(k)}\leq \inf x_{f(g(j))} = \inf x_{p(j)}\leq x$$

by (3). $\square$

Now we consider the class $\mathcal{LI}$ of all pairs $((x_j)_{j\in J}, x)$ of nets on L and elements in L which satisfy the equivalent conditions of 3.1. According to the discussion in Section II-1 this convergence notion determines a topology $\mathcal{O}(\mathcal{LI})$, and we wish to give this topology an official name.

3.2. DEFINITION. For a complete lattice L, the topology $\mathcal{O}(\mathcal{LI})$ is called the *lim-inf topology* and is written $\xi(L)$. We abbreviate $(L,\xi(L))$ as ΞL. $\square$

We note immediately that, for any directed set D in a complete lattice L, the element $x = \sup D$ and the net $(d)_{d\in D}$ satisfy 3.1(2); whence the pair $((d)_{d\in D}, \sup D)\in\mathcal{LI}$. Thus, if $U\in\xi(L)$ and $\sup D\in U$, then D is eventually in U; that is, U satisfies condition (S) of II-1.3. From II-1.4(v) we then derive immediately a parallel to 1.6:

3.3. PROPOSITION. *Let* L *be a complete lattice.*

(i) *An upper set is* $\xi(\mathrm{L})$*-open if and only if it is Scott open.*

(ii) *A lower set is* $\xi(\mathrm{L})$*-closed iff it is closed under sups of directed sets.* □

The following remark is useful:

3.4. LEMMA. *Let* L *be a complete lattice. A set* $A \subseteq \mathrm{L}$ *is* $\xi(\mathrm{L})$*-closed iff for every ultrafilter* $\mathfrak{F}$ *on* L *with* $A \in \mathfrak{F}$ *one has* $\underline{\lim}\, \mathfrak{F} \in A$*, where we define*

$$\underline{\lim}\, \mathfrak{F} = \sup_{F \in \mathfrak{F}} \inf F.$$

Proof. Suppose that $\mathrm{L} \backslash A \in \xi(\mathrm{L})$ and that $\mathfrak{F}$ is an ultrafilter on L with $A \in \mathfrak{F}$. We define a net on L: let J be the set of all pairs (x,F) with $x \in F \in \mathfrak{F}$ and write $(x,F) \leq (y,G)$ iff $G \subseteq F$, and for $j = (x,F)$ we set $x_j = x$. Then for all $F \in \mathfrak{F}$ we have $\inf F = \inf_{j \geq (x,F)} x_j$ (with an arbitrary $x \in F$), and thus

$$\underline{\lim}\, \mathfrak{F} = \sup_{F \in \mathfrak{F}} \inf F = \sup_{j \in J} \inf_{j^* \geq j} x_{j^*} = \underline{\lim}\, x_j.$$

Since $A \in \mathfrak{F}$, we have $x_j \in A$ for $j \geq (x,A)$, and thus from $\mathrm{L} \backslash A \in \xi(\mathrm{L})$ and the definition of $\xi(\mathrm{L})$ we deduce $\underline{\lim}\, \mathfrak{F} \in A$.

Conversely, suppose that A contains the lim-inf of every ultrafilter $\mathfrak{F}$ with $A \in \mathfrak{F}$. Consider $((x_j)_{j \in J}, x) \in \mathcal{L}\mathfrak{I}$, and assume that x_j is cofinally in A; we must show that $x \in A$. It is no loss of generality to assume that all x_j are in A (otherwise we pass to a subnet of x_j). The family $\{G_j : j \in J\}$, where we set $G_j = \{x_{j^*} : j \leq j^*\}$, is a filter basis on A, and $x = \sup_j \inf G_j$.

Let $\mathfrak{F}$ be any ultrafilter containing the G_j. Clearly $\mathfrak{F}$ contains A. If $y = \underline{\lim}\, \mathfrak{F}$, then $y \in A$ by assumption, and since $G_j \in \mathfrak{F}$, we have $x \leq y$. If we can show $y \leq x$, we are done.

For this we need to show that for arbitrarily small $F_0 \in \mathfrak{F}$ we have indeed $\inf F_0 \leq x$. Let $F_0 \in \mathfrak{F}$ be given. For each $j \in J$, we know $G_j \in \mathfrak{F}$; hence, there is an $F \in \mathfrak{F}$ with $F \subseteq F_0 \cap G_j$, and this F contains an $x_{p(j)}$ with $p(j) \leq j$. Now

$$\inf F_0 \leq \inf x_{p(j)} \leq x$$

by 3.1(3), and this is what had to be shown. □

3.5. LEMMA. *For every complete lattice we have* $\lambda(\mathrm{L}) \subseteq \xi(\mathrm{L})$.

Proof. It suffices to show $\sigma(\mathrm{L}) \subseteq \xi(\mathrm{L})$ and $\omega(\mathrm{L}) \subseteq \xi(\mathrm{L})$. The first inclusion follows from 3.3(i). Secondly, let $x \in \mathrm{L}$; we would like to show that $\uparrow x$ is $\xi(\mathrm{L})$-closed. But if $\mathfrak{F}$ is any ultrafilter with $\uparrow x \in \mathfrak{F}$, then clearly $x \leq \underline{\lim}\, \mathfrak{F}$, and so the assertion follows from 3.4. □

As we observed in Section II-1, if $((x_j)_{j \in J}, x) \in \mathcal{L}\mathfrak{I}$, then $x = \underline{\lim}\, x_j$ with respect to $\xi(\mathrm{L})$; and from 3.4 we know that every ultrafilter $\mathfrak{F}$ on L converges to $\underline{\lim}\, \mathfrak{F}$ with respect to $\xi(\mathrm{L})$. In particular, $\Xi \mathrm{L}$ is quasicompact. Since $\{x\} = \uparrow x \cap \downarrow x$, and both $\uparrow x$ and $\downarrow x$ are $\xi(\mathrm{L})$-closed by 3.5, we have proved:

3.6. PROPOSITION. *For every complete lattice* L*, the lim-inf topology is quasicompact* T_1 *and contains the Lawson topology. If the latter is Hausdorff,* *then* $\lambda(L) = \xi(L)$. □

Note that this result gives a second independent proof of the quasicompactness of ΛL (see 1.9).

3.7. LEMMA. (i) *Let* L *be a complete lattice such that* ΞL *is Hausdorff. Then the graph of* $\leq$ *is closed in* $\Xi L \times \Xi L$.

(ii) *If* $(X,\leq)$ *is a poset and a compact Hausdorff space such that* $\leq$ *has a closed graph, then any closed upper set* C *has a neighborhood basis of open upper sets.*

(iii) *Under the assumption of* (ii), *if* $\mathfrak{F}$ *is any ultrafilter on* X *containing all open upper set neighborhoods of* C, *then* $\underline{\lim}\, \mathfrak{F} \in C$.

Proof. (i): We take an ultrafilter $\mathfrak{F}$ in the graph of $\leq$ and show that its limit (x_1,x_2) (which exists by compactness—see 3.6) satisfies $x_1 \leq x_2$. If we denote with $\pi_k : L \times L \to L$ the first and second projection, $k=1,2$, then $\pi_k(\mathfrak{F})$ is an ultrafilter, since π_k is surjective. Hence, $\pi_k(\mathfrak{F})$ converges to $\underline{\lim}\, \pi_k(\mathfrak{F})$ by the definition of $\xi(L)$; and, since limits are unique, we have $x_k = \underline{\lim}\, \pi_k(\mathfrak{F})$, in view of the definition of the product topology. But if $F \in \mathfrak{F}$, then $(u,v) \in F$ implies $u \leq v$, since $\mathfrak{F}$ is in the graph of $\leq$. Hence, $\inf \pi_1(F) \leq \inf \pi_2(F)$, and thus $x_1 = \underline{\lim}\, \pi_1(\mathfrak{F}) \leq \underline{\lim}\, \pi_2(\mathfrak{F}) = x_2$.

(ii): If A is a closed subset of the compact space X, then $\downarrow A$ is closed. (Indeed if $x = \lim x_j$ with $x_j \in \downarrow A$, then we have $x_j \leq a_j \in A$. Because of compactness of A, a subnet $a_{f(k)}$ of a_j converges to $a \in A$; whence, $(x,a) = \lim (x_{f(k)}, a_{f(k)}) \in \text{graph}(\leq)$, hence $x \in \downarrow A$.) Now if U is an arbitrary open neighborhood of C, set $A = X \setminus U$. Then $V = X \setminus \downarrow A$ is an open upper set by what was just shown and it satisfies $C \subseteq V \subseteq U$.

(iii): If we have $\underline{\lim}\, \mathfrak{F} \notin C$, then we would find an open upper set U such that $\underline{\lim}\, \mathfrak{F} \notin U^-$, $C \subseteq U$ by (ii), since C is the intersection of its closed neighborhoods by compactness of X. But then we would have $X \setminus U^- \in \mathfrak{F}$ and $U \in \mathfrak{F}$ and this would be a contradiction. □

3.8. LEMMA. *Let* L *be a complete lattice such that* ΞL *is Hausdorff. Then for* $x \not\leq y$ *there is a finite set* F *with* $y \notin \uparrow F$ *and* $\uparrow F$ *a* $\sigma(L)$*-neighborhood of* x.

Proof. Let $x \not\leq y$. We will find a finite set $F \subseteq L$ such that $y \notin \uparrow F$ and such that there is an open upper set $U \in \xi(L)$ such that $x \in U \subseteq \uparrow F$. By 3.3 we then have $U \in \sigma(L)$.

By way of contradiction, we now assume that for all finite $F \subseteq L$ with $y \notin \uparrow F$ and all $x \in U = \uparrow U \in \xi(L)$ we have $U \setminus \uparrow F \neq \emptyset$. Then the collection of all of these $U \setminus \uparrow F$ is a filter basis B which is contained in some ultrafilter $\mathfrak{F}$. Let $u = \underline{\lim}\, \mathfrak{F}$, then $u = \lim \mathfrak{F}$ with respect to $\xi(L)$ by 3.4. Since $\mathfrak{F}$ contains all open upper set neighborhoods U of $\uparrow x$, Lemma 3.7(i) and (iii) apply, and we have $u \in \uparrow x$; that is, $x \leq \sup_{E \in \mathfrak{F}} \inf E$. Since $x \not\leq y$, there must be at least one $E \in \mathfrak{F}$ with $y \notin \uparrow \inf E$. If $e = \inf E$, then $\uparrow e$ contains E, whence $\uparrow e \in \mathfrak{F}$. On the other hand, $y \notin \uparrow e$ shows $L \setminus \uparrow e \in B$; hence $L \setminus \uparrow e \in \mathfrak{F}$. This is a contradiction.

□

It may be useful to visualize the organization of the proof by contemplating the following complete distributive lattice, which does not satisfy the hypothesis of 3.8. We define

$$L = \{(s,t) \in [0,1]^2 : (s \neq 0 \text{ and } s \neq 1) \text{ implies } t < 1\}.$$

Take $y = (0,1)$ and $x = (1,1) = 1$. Then $y \ll x$ (in particular $x \not\leq y$), and y is contained in the $\lambda(L)$-closure of every Scott-open neighborhood of x. (Argue that the projection of any such U on the first coordinate must contain $(0,1]$.) If B and $\mathfrak{F}$ are chosen as in 3.8, then $u = \underline{\lim}\ \mathfrak{F}$ agrees with y.

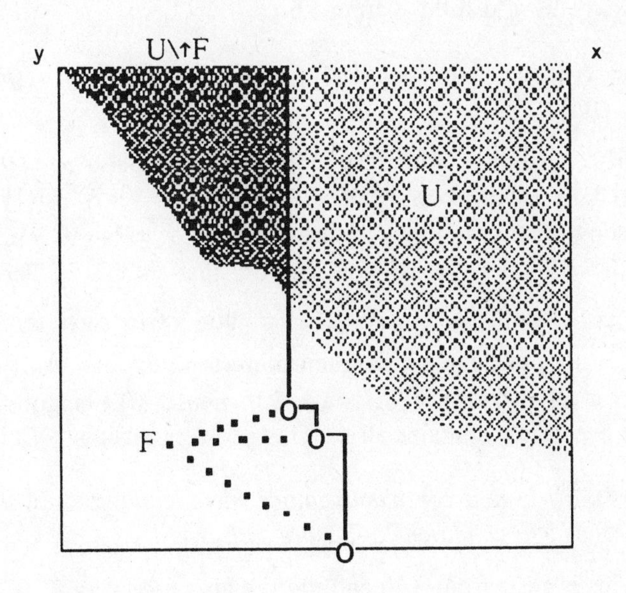

We can now show the following result:

3.9. THEOREM. *Let* L *be a complete lattice. Then the following conditions are equivalent:*

(1) *For all $x \not\leq y$ in* L *there is a* σ(L)*-neighborhood of x and an* ω(L)*-neighborhood of y which are disjoint;*

(2) ΛL *is Hausdorff;*

(3) ΞL *is Hausdorff.*

Moreover, if these conditions are satisfied, then the Lawson topology λ(L) *and the lim-inf topology* ξ(L) *agree.*

Proof. (1) implies (2) is immediate, and (2) implies (3) and the last conclusion are clear from 3.6. If (3) is satisfied, then 3.8 applies; thus $\mathrm{int}_\sigma \uparrow F$ is a $\sigma(\mathrm{L})$-neighborhood of x and $\mathrm{L}\setminus\uparrow F$ is an $\omega(\mathrm{L})$-neighborhood of y which satisfy the requirements of (1). □

We have studied how the Lawson topology compares with the topology associated with $\mathcal{LJ}$ convergence, and we wish to make a few observations on how topological convergence relative to the Lawson topology compares with the $\mathcal{LJ}$ convergence.

Let us denote with $\mathcal{L}_\lambda$ the class of all $((x_j)_{j\in J}, x)$ with $x = \lim x_j$ with respect to $\lambda(\mathrm{L})$. If $((x_j)_{j\in J}, x)\in\mathcal{LJ}$, then $x = \underline{\lim}\, x_j$; and thus $x = \lim x_j$ with respect to $\sigma(\mathrm{L})$ and with respect to $\omega(\mathrm{L})$, hence, with respect to $\lambda(\mathrm{L})$ as well. Thus $\mathcal{LJ}\subseteq\mathcal{L}_\lambda$. (This also follows from 3.6.)

3.10. LEMMA. *If* L *is a continuous lattice and* $x = \lim x_j$ *with respect to* $\lambda(\mathrm{L})$, *then* $((x_j)_{j\in J}, x)\in\mathcal{LJ}$.

Proof. If $x = \lim x_j$ with respect to $\lambda(\mathrm{L})$, then also $x = \lim y_k$ with respect to $\lambda(\mathrm{L})$ for all subnets y_k of x_j. In order to verify 3.1(1), it suffices therefore to show that $x = \underline{\lim}\, x_j$. Since we infer $x = \lim x_j$ with respect to $\sigma(\mathrm{L})$, by II-1.8 we get $x \leq \underline{\lim}\, x_j$. Now suppose that $x < \underline{\lim}\, x_j$. Then by 3.9(1) there is a $U\in\sigma(\mathrm{L})$ with $\bigvee^{\mathrm{up}}_{j\in J} \bigwedge_{i\geq j} x_i = \underline{\lim}\, x_j \in U$, such that the $\omega(\mathrm{L})$-closure K of U does not contain x. Then there is j such that $\bigwedge_{i\geq j} x_i\in U$, that is, $x_i\in U\subseteq K$ for all $i\geq j$. But $\mathrm{L}\setminus K$ is a $\omega(\mathrm{L})$-, hence $\lambda(\mathrm{L})$-neighborhood of x and therefore eventually contains all x_i. This is a contradiction. □

3.11. THEOREM. *If* L *is a continuous lattice, then* $x = \lim x_j$ *with respect to* $\lambda(\mathrm{L})$ *iff the equivalent conditions of* 3.1 *are satisfied.*

Proof. This is clear from 3.10 and the remarks preceding 3.10. □

3.12. COROLLARY. *Let* L *be a continuous lattice. If* $\mathfrak{F}$ *is a filter on* L *and* $x = \lim \mathfrak{F}$ *with respect to* $\lambda(\mathrm{L})$, *then* $x = \underline{\lim}\, \mathfrak{F}$. *In particular, if* $\mathfrak{F}$ *is an ultrafilter, then* $\mathfrak{F}$ *converges to* $\underline{\lim}\, \mathfrak{F}$ *with respect to* $\lambda(\mathrm{L})$.

Proof. Construct a net x_j, with $j = (x,F)$ and $x\in F\in\mathfrak{F}$ as in 3.4; then we find $x = \lim x_j$ with respect to $\lambda(\mathrm{L})$. By 3.11 we have

$$x = \underline{\lim}\, x_j = \sup_j \inf_{k\geq j} x_k.$$

Since $\inf_{k\geq(x,F)} x_k = \inf F$, we note $\underline{\lim}\, x_j = \underline{\lim}\, \mathfrak{F}$. This proves the first assertion. If $\mathfrak{F}$ is an ultrafilter, then $x = \lim \mathfrak{F}$ exists since $\Lambda\mathrm{L}$ is compact Hausdorff by 1.10. The final assertion then follows from the preceding. □

For meet-continuous lattices we have a converse to 3.11.

3.13. THEOREM. *Let* L *be a meet-continuous lattice. Then the following conditions are equivalent:*

(1) $x = \lim x_j$ *with respect to* λ(L) *iff the equivalent conditions of* 3.1 *are satisfied. (Equivalently: topological convergence relative to the Lawson topology is lim-inf convergence of all subnets.)*

(2) L *is a continuous lattice.*

Proof. (2) implies (1): From 3.11 above.

(1) implies (2): We suppose that for each net x_j with $x = \lim x_j$ with respect to λ(L) we know that $x = \underline{\lim}\, x_j$, and we prove 2.13(4); this will accomplish the proof by 2.13. Now let $x \in$ L be given. We define a net as follows: let J be the set of all pairs (u,U) with $u \in U$ and $x \in U \in \lambda$(L), and write $(u,U) \leq (v,V)$ iff $V \subseteq U$. For $j = (u,U)$ we set $x_j = u$. Clearly $x = \lim x_j$ with respect to λ(L). Hence, by hypothesis, we know that

$$x = \underline{\lim}\, x_j = \sup_j \inf_{j \leq i} x_i = \sup_{x \in U \in \lambda(L)} \inf U.$$

This indeed yields 2.13(4). □

This theorem should be viewed as a parallel to II-1.8, and it therefore completes that circle of ideas. The results of Sections 2 and 3 indicate that the Hausdorff separation of ΛL is essential for a truly satisfactory utilization of the Lawson topology. For meet-continuous lattices we know that ΛL is Hausdorff iff L is a continuous lattice (2.9). In the absence of meet-continuity the Hausdorff separation of ΛL is a rather subtle condition as might be anticipated by the relatively complicated proof of 3.9. We will discuss the lattice-theoretical aspects of this condition in the Exercise 3.15.

EXERCISES

3.14. EXERCISE. If L is a complete lattice with Hausdorff Lawson topology and if S is a subsemilattice of L, then the following statements are equivalent:

(1) S is closed under arbitrary infs and directed sups (that is, is a subalgebra).

(2) S is closed relative to the Lawson topology.

Remark. If L is continuous, the Lawson topology is Hausdorff.

(HINT: (1) implies (2) follows by showing (1) implies S is closed in ΞL and then using 3.6. (2) implies (1) follows as in 1.11.) □

The discussions in the exercises for Section 1 concerning GCL-lattices prepare us for a final elucidation of the Hausdorff separation of the Lawson topology in terms of lattice-theoretical properties. We have the following result which, in view of the various equivalent conditions of 1.16 characterizing generalized continuous lattices (GCL-lattices), is conclusive and satisfactory:

3.15. EXERCISE. For a complete lattice L, the following conditions are equivalent:

(1) ΛL is Hausdorff;

(2) L is a generalized continuous lattice.

(HINT: Condition 1.16(5) and condition 3.9(1) express one and the same thing.) □

The result of 3.15 generalizes Theorem 2.9; the link is given by 2.14 which says that every meet-continuous GCL-lattice is continuous. The proof of 3.15 is no easy matter in view of what went into 1.16 and 3.9. The proof of 2.9 is much simpler (although not in itself trivial either).

Before we produce additional equivalent conditions, we elaborate further on the lim-inf convergence of ultrafilters.

3.16. EXERCISE. Let L be a complete lattice and $\mathfrak{F}$ an ultrafilter on L.

(i) The set of ω(L)-cluster points of $\mathfrak{F}$ is $\uparrow(\underline{\lim}\ \mathfrak{F})$;

(ii) A set $A \subseteq$ L is ω(L)-closed iff it is an upper set and for every ultrafilter $\mathfrak{F}$ on L with $A \in \mathfrak{F}$ one has $\underline{\lim}\ \mathfrak{F} \in A$ iff it is an upper set and is λ(L)-closed.

(HINT: (i): Let C be the set of cluster points of $\mathfrak{F}$. Then C is the intersection of all ω(L)-closures F^- of sets $F \in \mathfrak{F}$, and each such closure is contained in $\uparrow(\inf F)$. Hence, $C \subseteq \underline{\lim}\ \mathfrak{F}$ (for any filter $\mathfrak{F}$).

Conversely let $\underline{\lim}\ \mathfrak{F} \leq y$ and take $F \in \mathfrak{F}$. If $y \notin F^-$ (closure with respect to ω(L)), then there is a finite set E with $F^- \subseteq \uparrow E$ and $y \notin \uparrow E$. From $F \subseteq \uparrow E$ we obtain $\uparrow E \in \mathfrak{F}$. Since $\mathfrak{F}$ is an ultrafilter, there is $e \in E$ with $\uparrow e \in \mathfrak{F}$. Whence $e \leq \underline{\lim}\ \mathfrak{F}$, contradicting $e \not\leq y$. Thus $y \in F^-$ for all $F \in \mathfrak{F}$; that is, $y \in C$.

(ii): If $A = A^-$ and $\mathfrak{F}$ is an ultrafilter with $A \in \mathfrak{F}$, then all cluster-points of $\mathfrak{F}$ are in A, hence $\underline{\lim}\ \mathfrak{F} \in A$. Conversely suppose $A = \uparrow A$ and that $\underline{\lim}\ \mathfrak{F} \in A$ for every ultrafilter with $A \in \mathfrak{F}$. If $y \in A^-$, there is an ultrafilter $\mathfrak{F}$ with $A \in \mathfrak{F}$ and $y = \omega$(L)-lim $\mathfrak{F}$. By (i) we have $y \geq \underline{\lim}\ \mathfrak{F}$. From $\underline{\lim}\ \mathfrak{F} \in A = \uparrow A$ we obtain $y \in A$. If $A = \uparrow A$ is ω(L)-closed, then it is trivially λ(L)-closed. Conversely, if A is λ(L)-closed it is ξ(L)-closed (3.6); hence, every ultrafilter $\mathfrak{F}$ with $A \in \mathfrak{F}$ yields $\underline{\lim}\ \mathfrak{F} \in A$ by 3.4.) □

The characterization theorems for the Hausdorff separation of the Lawson topology may be amplified as follows:

3.17. EXERCISE. A complete lattice L is GCL iff the following condition is satisfied:

(UF) For all ultrafilters $\mathfrak{F}$ on L the set of σ(L)-cluster points of $\mathfrak{F}$ is $\downarrow(\underline{\lim}\ \mathfrak{F})$.

Remark. Note that $\downarrow(\underline{\lim}\ \mathfrak{F})$ is always contained in the set of σ(L)-cluster points of $\mathfrak{F}$ (see II-1.8).

(HINT: (UF) implies 3.15(1): By (UF) and 3.16(i), the only λ(L)-cluster point of an ultrafilter $\mathfrak{F}$ is $\underline{\lim}\ \mathfrak{F}$.

3.9(1) implies (UF): If $x \not\leq \underline{\lim}\, \mathcal{F}$, then x has a $\sigma(\mathrm{L})$-open neighborhood U whose $\omega(\mathrm{L})$-closure C does not contain $\underline{\lim}\, \mathcal{F}$. By 3.16(ii) we know $C \notin \mathcal{F}$, whence $U \notin \mathcal{F}$. Since $\mathcal{F}$ is an ultrafilter, $\mathrm{L} \setminus U \in \mathcal{F}$. Thus x is not a $\sigma(\mathrm{L})$-cluster point of $\mathcal{F}$.) □

3.18. EXERCISE. The class of GCL-lattices is closed under the formation of subalgebras, products, homomorphic images (see I-2.6) and surjective images of monotone, Lawson-continuous maps.

(HINT: Products: the identity map $\Lambda(\mathbf{X}_j \mathrm{L}_j) \to \mathbf{X}_j \Lambda(\mathrm{L}_j)$ is continuous. Subalgebras: the inclusion $\mathrm{S} \to \mathrm{T}$ of a subalgebra is continuous (1.8). Homomorphic images: If $\Lambda\mathrm{S}$ is Hausdorff, then $\Lambda(\mathrm{S} \times \mathrm{S})$ is Hausdorff (since $\Lambda(\mathrm{S}) \times \Lambda(\mathrm{S})$ is). For any homomorphism $f : \mathrm{S} \to \mathrm{T}$ the kernel $\ker f = (f \times f)^{-1}(\text{diagonal } (\mathrm{T} \times \mathrm{T}))$ is a subalgebra of $\mathrm{S} \times \mathrm{S}$: hence it is closed in $\Lambda(\mathrm{S} \times \mathrm{S})$ by 3.14; but $\Lambda(\mathrm{S} \times \mathrm{S}) = \Lambda(\mathrm{S}) \times \Lambda(\mathrm{S})$ by compactness. Thus the quotient topology on $\mathrm{S}/\ker f$ is compact. We note with the aid of 3.4 that the quotient topology and $\xi(\mathrm{S}/\ker f)$ agree. Thus $\xi(\mathrm{S}/\ker f)$ is Hausdorff, hence agrees with $\lambda(\mathrm{S}/\ker f)$. If f is surjective, then $\mathrm{S}/\ker f$ and T are isomorphic lattices, whence $\Lambda(\mathrm{T})$ is Hausdorff.

Surjective images of Lawson-continuous monotone maps: show that $\ker f$ is closed on $\mathrm{S} \times \mathrm{S}$ by considering ultrafilters on $\mathrm{S} \times \mathrm{S}$ containing $\ker f$. Then proceed as before.) □

3.19. EXERCISE. A complete lattice L is GCL iff the following equivalent conditions are satisfied:

(SM) The sup-morphism $(I \mapsto \sup I) : \mathrm{Id}\, \mathrm{L} \to \mathrm{L}$ is $\omega(\mathrm{L})$-continuous;

(SM$'$) The sup-morphism $(I \mapsto \sup I) : \mathrm{Id}\, \mathrm{L} \to \mathrm{L}$ is $\lambda(\mathrm{L})$-continuous.

(HINT: (SM) implies GCL: Id L is algebraic, hence GCL; thus use 3.18.
GCL implies (SM): Show that for each $x \in \mathrm{L}$ the set

$$U_x = \{I \in \mathrm{Id}\, \mathrm{L} : x \not\leq \sup I\}$$

is $\omega(\mathrm{Id}\, \mathrm{L})$-open. Indeed if $x \not\leq \sup I$, find a finite $F \subseteq \mathrm{L} \setminus \downarrow \sup I$ with $\uparrow F \ll \uparrow x$ by GCL. Set $V = \{J \in \mathrm{Id}\, \mathrm{L} : f \notin J \text{ for all } f \in F\}$. Then $I \in V$, and $J \in V$ implies $x \not\leq \sup J$. For $x \leq \sup J$ would imply $\uparrow F \cap J \neq \emptyset$; that is, $\downarrow f \subseteq I$ for some $f \in F$.) □

In the following exercise we develop for the topology $\omega(\mathrm{L})$ analogs of earlier derived results for $\sigma(\mathrm{L})$. A subset X of a complete lattice L is said to have a property (Ω) if the following condition is satisfied:

(Ω) If $x \in X$ and $x = \underline{\lim}\, \mathcal{F}$ for an ultrafilter $\mathcal{F}$, then $X \in \mathcal{F}$.

Compare this with II-1.3(S).

3.20. EXERCISE. (i) A subset of L is ω-open iff it is a lower set satisfying property (Ω) (compare II-1.4(v));

(ii) Every upper set satisfies property (Ω) (compare II-1.4(vi));

(iii) If an entire collection of subsets satisfies property (Ω), then every subset in the topology it generates also satisfies (Ω) (compare II-1.4(vii));

(iv) In the Lawson topology on a complete lattice, the lower open sets are precisely the ω-open sets (compare 1.6);

(vi) The interval topology on L is the join of the ω-topology and its dual.

(HINT: Use 3.16.) □

The next three exercises develop basic properties of hypercontinuous lattices, analogs of continuous lattices. For more details, see Gierz and Lawson [1979].

3.21. EXERCISE. Let L be a complete lattice. We define a relation $\prec$ on L by $x \prec y$ iff whenever the intersection of a nonempty collection of upper sets is contained in $\uparrow y$, then the intersection of finitely many is contained in $\uparrow x$.

(i) Show that the following statements are equivalent:

(1) $x \prec y$;

(2) If the intersection of a nonempty collection of sets open in the upper topology $\upsilon(L)$ (see II-1.17 and the preceding definition) is contained in $\uparrow y$, then the intersection of finitely many of them is contained in $\uparrow x$;

(3) If $T \subseteq L$, $T \neq \emptyset$, and $\bigcap\{L \setminus \downarrow t : t \in T\} \subseteq \uparrow y$, then there exists a finite subset $F \subseteq T$ such that $\bigcap\{L \setminus \downarrow t : t \in F\} \subseteq \uparrow x$;

(4) $y \in \operatorname{int}_\upsilon \uparrow x$, where the interior is taken in the upper topology.

(ii) Show that the relation $\prec$ is an auxiliary relation and that $x \prec y$ implies $x \ll y$. □

3.22. EXERCISE. Let L be a complete lattice. The lattice L is called a *hypercontinuous lattice* if the relation $\prec$ is approximating (that is, if for all $y \in L$, $y = \sup\{x : x \prec y\}$).

(i) Show that a hypercontinuous lattice is continuous.

(ii) Show that $x \prec y$ iff $x \ll y$ in a hypercontinuous lattice.

(iii) Show that L is hypercontinuous iff for all $y \in L$:

$$y = \sup\{\inf U : y \in U \text{ and } U \text{ is open in the upper topology}\}. \quad \square$$

The next exercise gives various equivalent formulations for the notion of a hypercontinuous lattice.

3.23. EXERCISE. Let L be a continuous lattice. The following statements are equivalent:

(1) L is hypercontinuous;

(2) The Scott topology is the upper topology ($\sigma(\mathrm{L}) = \upsilon(\mathrm{L})$);

(3) The Lawson topology is the interval topology ($\lambda(\mathrm{L}) = \omega(\mathrm{L}) \vee \upsilon(\mathrm{L})$);

(4) The Lawson topologies on L and L^{op} agree (in this case we say they are *linked*);

(5) L^{op} is a generalized continuous lattice and the Lawson topologies are linked;

(6) The interval topology is Hausdorff;

(7) For every ultrafilter $\mathfrak{F}$ on L,

$$\sup\{\inf F : F \in \mathfrak{F}\} = \inf\{\sup F : F \in \mathfrak{F}\}. \quad \square$$

NOTES

With the exception of Theorems 3.11 and 3.13, the results of this section are due to Gierz and Lawson [198*]. This applies in particular to the key results 3.6 and 3.15. Exercise 3.18 is parallel to I-2.7, while 3.19 is parallel to O-3.15, O-4.2(1) and I-2.1.

4. BASES AND WEIGHTS

Recall that if X is a topological space, then the weight $w(X)$ is a cardinal defined by the equation

$$w(X) = \min\{\text{card}\,\mathfrak{B} : \mathfrak{B} \text{ is a basis of the topology } \mathcal{O}(X)\}.$$

In particular, X satisfies the second axiom of countability iff $w(X) \leq \aleph_0$.

Since with a complete lattice L we have associated at least two topological spaces, ΣL and ΛL, we also obtain two cardinals $w(\Sigma L)$ and $w(\Lambda L)$. For continuous lattices these are *equal* as we now show. It will in fact be useful to characterize this unique cardinal through the immediate lattice-theoretical data pertaining to L. For this purpose we formulate the following definition:

4.1. DEFINITION. Let L be a continuous lattice. A subset $B \subseteq L$ is called a *basis* of L iff

(i) B is a sup-subsemilattice of L containing 0;
(ii) $x = \sup(\downarrow x \cap B)$ for all $x \in L$. □

Condition (i) means that $0 \in B$ and $B \vee B \subseteq B$; that is, B is closed under finite sups. From I-4.3 and I-4.4 we recognize immediately that for an algebraic lattice L the subset K(L) is a basis; conversely, if K(L) is a basis of a complete lattice L, then L is algebraic. A basis is of course not uniquely determined in general. It is therefore useful to have several ways of recognizing a basis.

4.2. PROPOSITION. *Let B be a sup-subsemilattice of a continuous lattice* L. *Then the following conditions are equivalent:*

(1) *B is a basis of* L;
(2) *Whenever $y \not\leq x$, there is a $b \in B$ with $b \not\leq x$ and $b \ll y$;*
(3) *Whenever $x \ll y$, there is a $b \in B$ with $x \leq b \ll y$;*
(4) $K(L) \subseteq B$, *and whenever $y \not\leq x$, there is a $b \in B$ with $b \not\leq x$ and $b \leq y$;*
(5) *Whenever $y \not\leq x$, there is a $b \in B$ with $b \not\leq x$ and $b \leq y$.*

Proof. It is evident from Definition 4.1, that (1) iff (2). By I-1.6, clearly (3) implies (1). If (1) is satisfied and $x \ll y$, then $\downarrow x \cap B$ is a sup-subsemilattice because B is a sup-semilattice as is $\downarrow x$ as well (see I-1.2(ii)). Since $x \ll y$ and since $y = \sup(\downarrow y \cap B)$ by (1), we find a $b \in \downarrow y \cap B$ with $x \leq b$ by I-1.1. Hence, (1) implies (3); thus, the first three statements are equivalent. Evidently (3) implies (4) (recall I-1.2(i)), and (4) implies (5). We claim (5) implies (2). Let $y \not\leq x$. By I-1.6(A_1) there is a $y^* \ll y$ with $y^* \not\leq x$. By (5) there is a $b \in B$ with $b \not\leq x$ and $b \leq y^*$. From I-1.2(ii) we have $b \ll y$. □

Notice the corollary that, for an algebraic lattice L, the subset K(L) is the *unique* smallest basis. Further we remark that condition (5) allows us to introduce the concept of a basis for an arbitrary (complete) lattice.

4.3. PROPOSITION. *Let* L *be a continuous lattice and B a basis. Then the function $r_B = (J \mapsto \sup J)$:* Id $B \to$ L *is a surjective map preserving arbitrary infs and arbitrary sups whose domain* Id B *is an algebraic lattice with*

$$\mathrm{K}(\mathrm{Id}\, B) = \{\downarrow b : b \in B\} \cong B.$$

The lower adjoint of r_B is $x \mapsto \;\Downarrow x \cap B$.

Proof. We have a function $f = (J \mapsto \downarrow J) : \mathrm{Id}\, B \to \mathrm{Id}\, \mathrm{L}$ since r_B is surjective by 4.1. This function clearly preserves arbitrary unions; that is, arbitrary sups. By O-1.5 we know $r_B = r_{\mathrm{L}} f$, and r_{L} preserves arbitrary sups by I-2.1(5). Thus r_B preserves arbitrary sups.

Next we show that the function $x \mapsto \;\Downarrow x \cap B : \mathrm{L} \to \mathrm{Id}\, B$ is a lower adjoint of r_B. Indeed if $x \in \mathrm{L}$ and $J \in \mathrm{Id}\, B$, then $\Downarrow x \cap B \subseteq J$ iff $\Downarrow x \subseteq \downarrow J$ iff $x \leq \sup \downarrow J = \sup J$ by I-2.1 and O-1.5. Thus r_B is an upper adjoint and thus preserves arbitrary infs by O-3.3.

The remainder about the structure of Id B follows from I-4.12. □

4.4. PROPOSITION. *Let* L *be a continuous lattice. Then the following statements are equivalent:*

(1) L *is algebraic;*

(2) L *has a unique smallest basis;*

(3) L *has a minimal basis.*

Moreover, if these conditions are satisfied, the unique smallest basis is K(L).

Proof. As we observed above, (1) implies (2) follows from 4.2. Trivially (2) implies (3).

(3) implies (1): Let B be a minimal basis of L and let $J \in \mathrm{Id}\, B$. Set $x = \sup J$. We know $\Downarrow x \cap B \subseteq J$ from 4.3. Define $C = (B \setminus J) \cup (\Downarrow x \cap B)$. Then C is a sup-subsemilattice with $C \subseteq B$.

Let $y \in \mathrm{L}$. If $y \not\leq x$, then $\sup(\Downarrow y \cap C) = \sup(\Downarrow y \cap B)$, since $\Downarrow y \cap C$ is cofinal in $\Downarrow y \cap B$. Hence, $\sup(\Downarrow y \cap C) = y$ in this case. Now let $y \leq x$ If $b \in \;\Downarrow y \cap B$, then $b \in \;\Downarrow x \cap B$ by I-1.2. Hence, $\Downarrow y \cap B \subseteq \;\Downarrow y \cap C \subseteq \;\Downarrow y \cap B$; whence, $\sup(\Downarrow y \cap C) = \sup(\Downarrow y \cap B) = y$ in this case, too. Thus, C is a basis by 4.1. By the minimality of B we conclude $C = B$, which means that $J = \;\Downarrow x \cap B$. We have shown that the lower adjoint of $r_B : \mathrm{Id}\, B \to \mathrm{L}$ is surjective; hence, r_B is injective by O-3.7. Thus by 4.3 it is an isomorphism, and since the lattice Id B is algebraic by 4.3, we have shown (1). □

We have to accept the fact in the light of 4.4 that, for continuous lattices which are not algebraic, there are no minimal bases. So there seems little hope to have a "canonical" basis in this case. While there is no minimal basis in general, at least the set of *cardinals* of bases of the continuous lattice has a minimum.

4.5. DEFINITION. Let L be a continuous lattice. The cardinal

$$w(\mathrm{L}) = \min\{\operatorname{card} B : B \text{ is a basis of } \mathrm{L}\}$$

is called the *weight* of the lattice L. □

By 4.4 we know that for any algebraic lattice $w(\mathrm{L}) = \operatorname{card} \mathrm{K}(\mathrm{L})$.

4.6. PROPOSITION. *Let* X *be a topological* T_0*-space whose topology* $\mathcal{O}(\mathrm{X})$ *is a continuous lattice. Then*:

(i) *If* X *is infinite,* $w(\mathrm{X}) = w(\mathcal{O}(\mathrm{X}))$;

(ii) *If* X *is finite,* $w(\mathrm{X}) = \operatorname{card} \mathrm{X} \leq \operatorname{card} \mathcal{O}(\mathrm{X}) = w(\mathcal{O}(\mathrm{X}))$.

Remark. See in this conection I-1.7(5), I-1.8, II-4.10, II-4.11 and Chapter V, Section 5 below.

Proof. (i): Any basis of the continuous lattice $\mathcal{O}(\mathrm{X})$ is clearly a basis of the topological space X; whence $w(\mathrm{X}) \leq w(\mathcal{O}(\mathrm{X}))$. Now let $\mathfrak{B}$ be a basis of the topological space X with card $\mathfrak{B} = w(\mathrm{X})$. If X is infinite, then we may assume that $\mathfrak{B}$ is closed under finite unions and contains $\varnothing$; that is, is a sup-subsemilattice of $\mathcal{O}(\mathrm{X})$ with zero. Now let $U, V \in \mathcal{O}(\mathrm{X})$ with $V \not\subseteq U$. Then there is a $v \in V \setminus U$. Since $\mathfrak{B}$ is a basis, we find a $W \in \mathfrak{B}$ with $v \in W \subseteq V$. But $v \notin U$ implies $W \not\subseteq U$. By 4.2(5) we know that $\mathfrak{B}$ is a basis of the lattice $\mathcal{O}(\mathrm{X})$; whence, $w(\mathcal{O}(\mathrm{X})) \leq \operatorname{card} \mathfrak{B} = w(\mathrm{X})$.

(ii) If X is finite, then every point has a smallest neighborhood; whence, $w(\mathrm{X}) = \operatorname{card} \mathrm{X}$, since X is a T_0-space. Further $\mathcal{O}(\mathrm{X})$ is finite; it therefore follows that $w(\mathcal{O}(\mathrm{X})) = \operatorname{card} \mathcal{O}(\mathrm{X})$. □

If, in 4.1 in the definition of a basis of a continuous lattice, we had only demanded condition (ii) to be satisfied, we would have arrived at the equality $w(\mathrm{X}) = w(\mathcal{O}(\mathrm{X}))$ in the finite case, too.

4.7. THEOREM. *For any continuous lattice* L *one has*

$$w(\mathrm{L}) = w(\Sigma\mathrm{L}) = w(\Lambda\mathrm{L}) = w(\operatorname{Id} B)$$

for any basis B *of* L *with card* $B = w(\mathrm{L})$. *Further* $w(\mathrm{L}) \leq w(\sigma(\mathrm{L}))$, *and equality holds if* L *is infinite.*

Remark. The weights $w(\Lambda\mathrm{L})$ and $w(\Sigma\mathrm{L})$ are weights of topological spaces in the traditional sense recalled in the introductory remarks of this section; the other weights are lattice-theoretical weights in the sense of 4.5.

Proof. $w(\Sigma\mathrm{L}) \leq w(\Lambda\mathrm{L})$: Let $\mathfrak{B}$ be a basis of $\Lambda\mathrm{L}$ of cardinality $w(\Lambda\mathrm{L})$, then $\{\uparrow U : U \in \mathfrak{B}\}$ is a basis of $\Sigma\mathrm{L}$ by 2.1(i). Hence,

$$w(\Sigma\mathrm{L}) \leq \operatorname{card}\{\uparrow U : U \in \mathfrak{B}\} \leq w(\Lambda\mathrm{L}).$$

$w(\Lambda\mathrm{L}) \leq w(\mathrm{L})$: Let B be a basis of L with card $B = w(\mathrm{L})$. If L is finite (whence, $\Lambda\mathrm{L}$ is discrete), then $B = \mathrm{L}$; and the assertion is clear. Now let L be infinite. For $b_0, b_1, \ldots, b_n \in B$ set

$$W(b_0,b_1,\ldots,b_n) = \Uparrow b_0\setminus(\uparrow b_1\cup\ldots\cup\uparrow b_n),$$

and $W(b) = \Uparrow b$ for $b\in B$. Then

$$\text{card}\{W(b_0,\ldots,b_n) : b_0,\ldots,b_n\in B,\ n=0,1,2,\ldots\} = \text{card}\ B = w(\mathrm{L}).$$

But we claim that the $W(b_0,\ldots,b_n)$ form a basis of $\Lambda\mathrm{L}$.

Indeed let $U\in\lambda(\mathrm{L})$, and take $x\in U$. Then there is a $\lambda(\mathrm{L})$-neighborhood $\Uparrow x_0\setminus(\uparrow x_1\cup\ldots\cup\uparrow x_n)\subseteq U$ of x (see II-1.10(i) and the remarks following 1.5). Thus $x_0\ll x$, and hence, by 4.2(3) we find a $b_0\in B$ with $x_0\leq b_0\ll x$; that is, $x\in\Uparrow b_0\subseteq\Uparrow x_0$. For each $k=1,\ldots,n$ we have $x_k\not\leq x$; hence, by 4.2(2) there is a $b_k\ll x_k$ with $b_k\not\leq x$. Therefore

$$x\in W(b_0,b_1,\ldots,b_n) = \Uparrow b_0\setminus(\uparrow b_1\cup\ldots\cup\uparrow b_n)\subseteq \Uparrow x_0\setminus(\uparrow x_1\cup\ldots\cup\uparrow x_n)\subseteq U.$$

It follows that $w(\Lambda\mathrm{L})\leq w(\mathrm{L})$.

$w(\mathrm{L})\leq w(\Sigma\mathrm{L})$: Let $\mathfrak{B}$ be a basis of $\Sigma\mathrm{L}$ of cardinality $w(\Sigma\mathrm{L})$. Let $B_0 = \{\inf U : U\in\mathfrak{B}\}$. Then card $B_0\leq$ card $\mathfrak{B} = w(\Sigma\mathrm{L})$. Let B be the set of finite sups of elements of B_0. Then B is a sup-semilattice. If L is finite, $B_0 = \mathrm{L} = B$. If L is infinite, then card B = card B_0. We claim that B is a basis of L.

Let $x\ll y$. Then $y\in\Uparrow x\in\sigma(\mathrm{L})$; hence, there is a $U\in\mathfrak{B}$ with $y\in U\subseteq\Uparrow x$. But then $x\leq\inf U\ll y$ and $\inf U\in B_0$. Thus, by 4.2(3), we know that B is a basis. Hence $w(\mathrm{L})\leq$ card B = card $B_0\leq w(\Sigma\mathrm{L})$.

$w(\mathrm{L}) = w(\mathrm{Id}\ B)$ for any basis B of L with card $B = w(\mathrm{L})$: We have $w(\mathrm{Id}\ B)$ = card $\mathrm{K}(\mathrm{Id}\ B)$ (since Id B is algebraic) = card $B = w(\mathrm{L})$ (by 4.3).

$w(\mathrm{L})\leq w(\sigma(\mathrm{L}))$: If L is finite, then $\sigma(\mathrm{L})$ consists of all upper sets; whence $w(\sigma(\mathrm{L}))\geq$ card L. But $w(\Sigma\mathrm{L})$ = card L since $\uparrow x$ is the smallest $\sigma(\mathrm{L})$-neighborhood of each $x\in\mathrm{L}$. Thus $w(\mathrm{L})\leq w(\sigma(\mathrm{L}))$. If L is infinite, then we find $\sigma(\mathrm{L}) = \mathcal{O}(\Sigma\mathrm{L})$ (see the paragraph preceding II-2.1). Hence 4.6 proves the assertion $w(\mathrm{L}) = w(\Sigma\mathrm{L}) = w(\sigma(\mathrm{L}))$. □

After Theorem 4.7 we may say "the weight of a continuous lattice" or "the weight of the Scott topology" without ambiguity.

4.8. COROLLARY. *Every continuous lattice is the quotient of an algebraic (and even an arithmetic) lattice of equal weight by a quotient map preserving all sups and infs.*

Proof. If L is finite, everything is trivial. If L is infinite, let B be a basis of cardinality $w(\mathrm{L})$. We may assume that B is a sublattice, since adding all finite meets does not raise the cardinality. Then Id B is an arithmetic lattice by 4.3 and I-4.7. By Theorem 4.7 we have $w(\mathrm{Id}\ B) = w(\mathrm{L})$; and, by 4.3, L is a quotient of Id B. □

4.9. COROLLARY. *Passing to the Scott topology on an infinite continuous lattice does not raise weights. Specifically, if* L *is an infinite continuous lattice, then* $w(\mathrm{L}) = w(\sigma^n(\mathrm{L}))$ *for* $n=1,2,3,\ldots$. □

4.10. COROLLARY. *Let* L *be a continuous lattice. Then* L *has a countable basis iff* ΛL *is a compact metric space.* □

We now calculate the weights of function spaces (see II-4.1).

4.11. THEOREM. *Let* X *be a* T_0*-space such that* $\mathcal{O}(X)$ *is a continuous lattice and let* L *be a nonsingleton continuous lattice. If at least one of X or* L *is infinite, then* $w([X,\Sigma L]) = \max\{w(X),w(L)\}$.

Proof. We know that $\Sigma 2$ is a retract of ΣL, since L is nonsingleton (see II-3.2(iii)). Hence $\mathcal{O}(X) \simeq [X,\Sigma 2]$ is a retract of $[X,\Sigma L]$ in *UPS*; whence, $\Sigma\mathcal{O}(X)$ is a retract in ***TOP*** of $\Sigma[X,\Sigma L]$. Thus $w(X) \leq w(\mathcal{O}(X))$ (by 4.6) $= w(\Sigma\mathcal{O}(X))$ (by 4.7)$\leq w(\Sigma[X,\Sigma L]) = w([X,\Sigma L])$ (by 4.7).

Further, L is a retract of $[X,\Sigma L]$ in ***UPS***, hence ΣL is a retract of $\Sigma[X,\Sigma L]$ in ***TOP*** (see proof of II-4.10). Thus $w(L) = w(\Sigma L)$ (by 4.6) $\leq w(\Sigma[X,\Sigma L]) = w([X,\Sigma L])$ (by 4.7). Thus we have shown that

$$w([X,\Sigma L]) \geq \max\{w(X),w(L)\}.$$

Now we let $\mathfrak{B}$ be a basis of X of cardinality $w(X)$ and B a basis of L of cardinality $w(L)$. For $U \in \mathfrak{B}$ and $b \in B$ we denote with $c_{U,b} : X \to L$ the function which takes the value b on U and the value 0 elsewhere, a continuous function. Then we have

$$\begin{aligned} \operatorname{card}\{c_{U,b} : (U,b) \in \mathfrak{B} \times B\} &= \operatorname{card}\mathfrak{B} \operatorname{card} B \\ &= w(X)w(L) \\ &= \max\{w(X),w(L)\}, \end{aligned}$$

since at least one of $w(X)$ or $w(L)$ is infinite.

Now consider $f,g \in [X,\Sigma L]$ with $g \not\leq f$; we are preparing to show that there is a $c_{U,b} \leq g$ with $c_{U,b} \not\leq f$; then by 4.2(4), the set of finite sups of functions $c_{U,b}$ is a basis of $[X,\Sigma L]$; this will show that

$$w([X,\Sigma L]) \leq \operatorname{card}\{\text{finite sups of } c_{U,b}\} = \operatorname{card}\{c_{U,b}\} = \max\{w(X),w(L)\}.$$

Now $g \not\leq f$ implies the existence of some $x \in X$ with $g(x) \not\leq f(x)$. Since B is a basis, by 4.2(2) we find a $b \in B$ such that $b \not\leq f(x)$ and $b \ll g(x)$. It follows that $g(x) \in \Uparrow b \in \sigma(L)$. Since g is Scott continuous, there is a $U \in \mathfrak{B}$ with $x \in U$ and $g(U) \subseteq \Uparrow b$. This implies that $c_{U,b} \leq g$. But $c_{U,b}(x) = b \not\leq f(x)$, hence $c_{U,b} \not\leq f$ which we had to show. □

4.12. COROLLARY. *Let* S *and* T *be continuous lattices of which at least one is infinite. Then we have* $w([S \to L]) = \max\{w(S),w(T)\}$.

Proof. We have $[S \to L] = [\Sigma S,\Sigma T]$ (see II-4.1 ff.). Hence $w([S \to L]) = w([\Sigma S,\Sigma T]) = \max\{w(\Sigma S),w(\Sigma T)\}$ (by 4.11) $= \max\{w(S),w(T)\}$ by 4.7. □

4.13. COROLLARY. *Let* L *be an infinite continuous lattice. Then we have* $w([L \to L]) = w(L)$. □

In other words, forming the space of self-maps does not raise weights for infinite continuous lattices. This will become relevant in sections VI-3 and VI-4. We will now discuss how the other basic constructions of I-2.7 fare with respect to weights.

4.14. PROPOSITION. (i) *Let* $\{L_j : j \in J\}$ *be a family of nonsingleton continuous lattices. Then*

$$w(\mathsf{X}_j L_j) = \Sigma_j w(L_j) = \max(\{\text{card } J, \sup\{w(L_j) : j \in J\}),$$

if at least one of J *or* L_j *is infinite, and*

$$w(\mathsf{X}_j L_j) = \mathsf{X}_j w(L_j)$$

if everything in sight is finite.

(ii) *If* L *is a continuous lattice and* S *a subalgebra of* L, *then*

$$w(S) \leq w(L).$$

(iii) *If* L *is a continuous lattice and* S *a homomorphic image of* L, *then*

$$w(S) \leq w(L).$$

Remark. For the notions of subalgebra and homomorphic image see I-2.6.

Proof. (i): We may assume that one of J or L_j, for $j \in J$, is infinite. Let B_j be a basis of L_j of cardinality $w(L_j)$. Then the set of all $(b_j)_{j \in J}$ with $b_j \in B_j$ and all but a finite number of the b_j equal to 0 is a basis of $\mathsf{X}_j L_j$ of cardinality $\Sigma_j w(L_j)$. Hence $w(\mathsf{X}_j L_j) \leq \Sigma_j w(L_j)$. Since each factor is a retract with respect to the Scott topologies, say, the reverse inequality is clear from 4.7.

(ii): We consider the Lawson topologies on S and L. The inclusion map $\Lambda S \to \Lambda L$ is continuous by 1.8. Since both topologies are compact Hausdorff (see 1.10), this function is an embedding. Hence $w(\Lambda S) \leq w(\Lambda L)$. Then $w(S) \leq w(L)$ by 4.7.

(iii): Let $g : L \to S$ be a surjective homomorphism and take a basis B of L of cardinality $w(L)$. Now let $y \not\leq x$ in S. Set $J = g^{-1}(\downarrow x)$ and $u = \min g^{-1}(\uparrow y)$. Then J is a Scott-closed subset of L since g is Scott continuous and $\downarrow x$ is Scott closed in S. Moreover, $u \notin J$ for otherwise $y = g(u) \in g(J) = \downarrow x$. But $u = \sup(\twoheaddownarrow u \cap B)$ because B is a basis of L. Now there is a $b \in B$ with $b \ll u$ such that $b \notin J = g^{-1}(\downarrow x)$. Then $g(b) \notin \downarrow x$; that is, $g(b) \not\leq x$. Further $g(b) \leq g(u) = y$. This shows that $g(B)$ is a basis of S by 4.2(5). It then follows that $w(S) \leq \text{card L} = w(L)$. □

Parts (i) and (ii) of 4.14 together will enable us to calculate the weights of projective limits in the category of continuous lattices and maps preserving infs and directed sups (see IV-3.25). We next observe the existence of a basis-type set with particular properties:

4.15. LEMMA. *A continuous lattice* L *contains a subset* B *with card* $B = w(L)$ *such that* $b = \inf \twoheaduparrow b$ *for all* $b \in B$. *Moreover,* $x = \sup(\twoheaddownarrow x \cap B)$ *for all* $x \in L$.

Proof. The assertion is clear if L is finite.

Assume that L is infinite. Then the lattice $\sigma(L)$ has a basis $\mathfrak{B}$ of cardinality $w(L)$ by 4.7. Set $B = \{\inf U : U \in \mathfrak{B}\}$. We have $x = \sup(\downarrow x \cap B)$ for all $x \in L$ by II-1.14(4). It follows that the set B^* of finite sups of elements in B is a basis of L. Hence, since L is infinite, card B = card $B^* \geq w(L)$. By definition we have card $B \leq$ card $\mathfrak{B} = w(L)$. Hence, card $B = w(L)$. □

We introduce another standard cardinal invariant for topological spaces and apply it to continuous lattices. The density of a topological space X is the minimum of all cardinals card D, where D is a dense subset.

4.16. DEFINITION. Let L be a complete lattice. Then the *density* of L is the cardinal $d(L) = \min\{\text{card } D: D \text{ is a dense subset of } \Lambda L\}$.

4.17. PROPOSITION. *Let* L *be a continuous lattice. Then* $w(L) \leq 2^{d(L)}$,*and equality can, but need not occur.*

Proof. Let B be a subset of cardinality $w(L)$ which satisfies the conditions of Lemma 4.15. Let D be a dense subset of ΛL of cardinality $d(L)$. For $b \in B$, the set $\Uparrow b \cap D$ is dense in the $\lambda(L)$-open set $\Uparrow b$. We claim that in fact $\inf(\Uparrow b \cap D) = b$. For if not, then the element $b_0 = \inf(\Uparrow b \cap D)$ satisfies $b < b_0$. Then $\emptyset \neq \Uparrow b \setminus \uparrow b_0 \in \lambda(L)$, and thus there is a $d \in (\Uparrow b \setminus \uparrow b_0) \cap D$. On the other hand, by the definition of b_0 we would have to have $b_0 \leq d$, and this is a contradiction. Now we know that the function $b \mapsto \Uparrow b_0 \cap D : B \to 2^D$ is injective. Hence, we conclude

$$w(L) = \text{card } B \leq \text{card } 2^D \leq 2^{\text{card } D} = \text{card } 2^{d(L)}.$$

This establishes the asserted inequality.

We remark next that equality may be attained: let X be an arbitrary infinite set, and let βX be the Stone-Cech compactification of the discrete space X. We let $L = \mathcal{O}(\beta X)$ and we know that L is a continuous lattice (see I-1.7(5)); in fact, it is arithmetic with K(L) = lattice of compact open subsets of $\beta X \cong 2^X$. Thus,

$$w(L) = \text{card } K(L) \text{ (by 4.4)} = \text{card } 2^X = 2^{\text{card } X}.$$

If we let F denote the set of all finite subsets of $X \subseteq \beta X$, then it is a straightforward exercise to show that F is dense in L with respect to $\lambda(L)$. Thus $d(L) \leq \text{card } F = \text{card } X$; whence,

$$w(L) \leq 2^{d(L)} \text{ (by what was shown above)} \leq 2^{\text{card } X} = w(L).$$

Hence $w(L) = 2^{d(L)}$ in this example, and $d(L)$ can be any infinite cardinal.

In order to show that the inequality $w(L) < 2^{d(L)}$ can occur, we have many possibilities to choose from. If Ω is the first uncountable ordinal and $L = [1,\Omega]^{op}$, then L is algebraic and K(L) = L. Whence,

$$w(L) = \text{card } K(L) = \text{card } L = \aleph_1;$$

but since no countable subset of $[1,\Omega]$ can be dense in the interval topology (which agrees with the Lawson topology), we know $\aleph_1 \leq d(L) \leq \text{card } L = \aleph_1$. Thus $w(L) = \aleph_1 < 2^{\aleph_1} = 2^{d(L)}$.

One can also show easily that the standard Cantor set C provides an example $L = C$ with

$$w(L) = \text{card } K(L) = \aleph_0 < 2^{\aleph_0} = 2^{d(L)}. \quad \square$$

EXERCISES

4.18. EXERCISE. For each cardinal $\boldsymbol{m} \geq x_0$ let $\boldsymbol{CL}_{\boldsymbol{m}}$ denote the full subcategory of $\boldsymbol{CL}$ with objects L with $w(L) \leq \boldsymbol{m}$. Show that this subcategory is closed under taking limits of diagrams of cardinality at most $\boldsymbol{m}$ and under all finite colimits. What can be said of the formation of function spaces $[L \to M]$? $\square$

4.19. EXERCISE. Let L be a continuous lattice and B a basis. For the elements of B we define $a \prec b$ iff $a \ll_L b$. An element $I \in \text{Id } B$ is called a *Dedekind cut*, if whenever $a \in I$, then $a \prec b$ for some $b \in L$. We have:

(i) B is a sup-semilattice with identity and a relation $\prec$ satisfying the following conditions for all u,x,y,z:

(1) $x \prec y$ implies $x \leq y$
(2) $u \leq x \prec y \leq z$ imply $u \prec z$.
(3) $x \prec z$ and $y \prec z$ imply $x \vee y \prec z$.
(4) $0 \prec x$.
(5) $x \prec z$ and $x \neq z$ together imply $x \prec y \prec z$ and $x \neq y$ for some y.

(ii) The function $x \mapsto (\downarrow\!\!\!\downarrow x \cap B) : L \to \text{Id } B$ (4.3) is an order isomorphism from L onto the subset of Dedekind cuts of B.

(HINT: (1)-(4) are immediate, and (5) follows from the interpolation property of $\ll$ and 4.1(ii).

The injectivity of the function $x \mapsto (\downarrow\!\!\!\downarrow x \cap B)$ follows from 4.3.

Let I be a Dedekind cut. Set $x = \sup_L I$. Then $x = \sup_L \downarrow I$, and so $\downarrow\!\!\!\downarrow x \subseteq \downarrow I$ by I-2.1(2). Hence $\downarrow\!\!\!\downarrow x \cap B \subseteq I$; by (5) we have $I \subseteq \downarrow\!\!\!\downarrow x \cap B$. Thus every Dedekind cut is of the form $\downarrow\!\!\!\downarrow x \cap B$.) $\square$

4.20. EXERCISE. Let S be a sup-semilattice with identity and an auxiliary relation $\prec$ satisfying 4.19, (1)–(5).

(i) The poset L of its Dedekind cuts is a continuous lattice, and the function $x \mapsto s_\prec(x) : S \to L$ a surjection of S onto a basis of L. Moreover, if $x,y \in S$, then $x \prec y$ implies $s_\prec(x) \ll s_\prec(y)$ in L.

(ii) Suppose that S also satisfies the following condition:

(6) For all $x,y \in S$ the relation $x \not\leq y$ implies the existence of a $z \in S$ with $z \not\leq y$ and $z \prec x$.

Then the function $x \mapsto s_{\prec}(x) : S \to L$ is an isomorphism onto a basis of L, and $s_{\prec}(x) \ll s_{\prec}(y)$ implies $x \prec y$ for $x, y \in S$.

(iii) If S arises as a basis from a continuous lattice as in 4.19, then (6) is satisfied.

(HINT: (i): Id S is an algebraic lattice by I-4.12. Define $p : \text{Id } S \to \text{Id } S$ by $p(I) = s_{\prec}(I)$ (note that p is well-defined!); then im p = L, and p is a projection (even: kernel operator) preserving directed sups. Hence L = im p is a continuous lattice by I-2.14. The interpolation property shows that for each Dedekind cut I and $x \in I$ we have $s_{\prec}(x) \ll I$ in L. Thus $\{s_{\prec}(x) : x \in S\}$ is a basis of L, and $x \prec y$ implies $s_{\prec}(x) \ll s_{\prec}(y)$.) □

Thus the sup-semilattices $(S, \prec)$ together with an auxiliary relation satisfying (1)–(6) are precisely the bases of continuous lattices with the relation being induced from the way-below relation.

4.21. EXERCISE. Let L be a continuous lattice of countable weight. Then the set IRR L is a G_δ in ΛL. (In particular, IRR L is a Borel set.)

(HINT: Any locally compact space which is countable at infinity is a countable union of compact subspaces; hence, any continuous T_2 image of a locally compact space which is countable at infinity is an F_σ (that is, a countable union of closed subspaces). By I-3.40, L\IRR L is a continuous image of an open subset of $\Lambda L \times \Lambda L$; and, if ΛL is second countable, the assertion follows from the previous remark.) □

PROBLEM. Develop a complete theory of weights for arbitrary complete lattices (cf. 4.2) and also for up-complete semilattices and up-complete posets. □

NOTES

The material in this section is largely due to Hofmann. Bases for continuous lattices were considered by Scott and by Ershov [1972]. A forerunner of 4.17 for algebraic lattices was given by Hofmann, Mislove and Stralka [1974]. Exercises 4.19 and 4.20 contain material which is due to M. Smyth (see [SCS-4] and Smyth [1978]); obviously we have here a variation of the theme of auxiliary relations in I-1.9 ff. These results provide an axiomatic characterization of sup-semilattices which occur as bases of continuous lattices. Proposition 4.21 is a lattice theoretical version of a result of Dixmier's [1968].

CHAPTER IV

Morphisms and Functors

With the exception of certain developments in Chapter II, notably Sections 2 and 4, we largely refrained from using category-theoretic language (even when we used its tools in the context of Galois connections). Inevitably, we have to consider various types of functions between continuous lattices, and this is a natural point in our study to use the framework of category theory.

In Section 1 we discuss duality based on the formalism of Galois connections. We discuss the categories ***INF*** and ***SUP***, whose objects are complete lattices (in both cases) and whose morphisms are functions preserving arbitrary infs (respectively, sups); these categories are dual (1.3). We saw as early as I-2.6 ff., that maps preserving infs *and* directed sups play an important role in our theory. This leads us to consider the subcategory $\boldsymbol{INF}^{\uparrow}$ of ***INF***. Its dual under the ***INF—SUP*** duality is denoted $\boldsymbol{SUP}^0$; its morphisms are precisely characterized in 1.4(1)–(2), but as a category in itself, $\boldsymbol{SUP}^0$ plays a minor role. More important, however, are the full subcategories $\boldsymbol{AL} \subseteq \boldsymbol{CL} \subseteq \boldsymbol{INF}^{\uparrow}$ and $\boldsymbol{AL}^{\mathrm{op}} \subseteq \boldsymbol{CL}^{\mathrm{op}} \subseteq \boldsymbol{SUP}^0$, which consist of algebraic and continuous lattices, respectively. We thus have a duality between ***CL*** and $\boldsymbol{CL}^{\mathrm{op}}$ (1.10) and between ***AL*** and $\boldsymbol{AL}^{\mathrm{op}}$; the latter extends to the very useful duality between ***AL*** and the category ***S*** of semilattices with identity and semilattice morphisms preserving the identity (1.15). Further duality theorems involving distributivity and prime elements will be given at the end of the first section, but this context will not be fully developed before Chapter V. In view of the fact that certain other categories of complete, continuous or algebraic lattices had been introduced in Chapter II-2.2, we survey the relevant categories in a diagram on the next page. The exercises for this section contain the interesting self-duality of continuous posets and continuous semilattices.

In Section 2 we introduce at long last the "character theory" which is appropriate for continuous lattices. Indeed, we show that the homomorphisms (in the sense of I-2.6) of a continuous lattice into the unit interval *separate points.* While this is (modulo results of III-2) a result on compact semilattices, we present here a lattice-theoretic approach which is new in this context, but which has related forerunners in the work of Raney on completely distributive lattices in the 1950's. The principal results in the section are more general in that they apply to complete lattices which are not necessarily continuous (2.12, 2.16, 2.20).

In Section 3 we present a general description of a limiting process introduced by Scott which produces continuous lattices L which are naturally isomorphic to their own function spaces [L→L]. We describe the functorial setting to the extent that it is necessary and convenient; we analyze the concept of projective limits in the relevant categories (such as $\boldsymbol{INF}^{\uparrow}$, ***CL*** and ***AL***).

In Section 4 we establish not only the result indicated above, but also some newer results which, for example, give us in a functorial fashion a continuous lattice L which is naturally isomorphic to its own ideal lattice Id L (or, in another application, to the lattice of all Lawson-open lower sets).

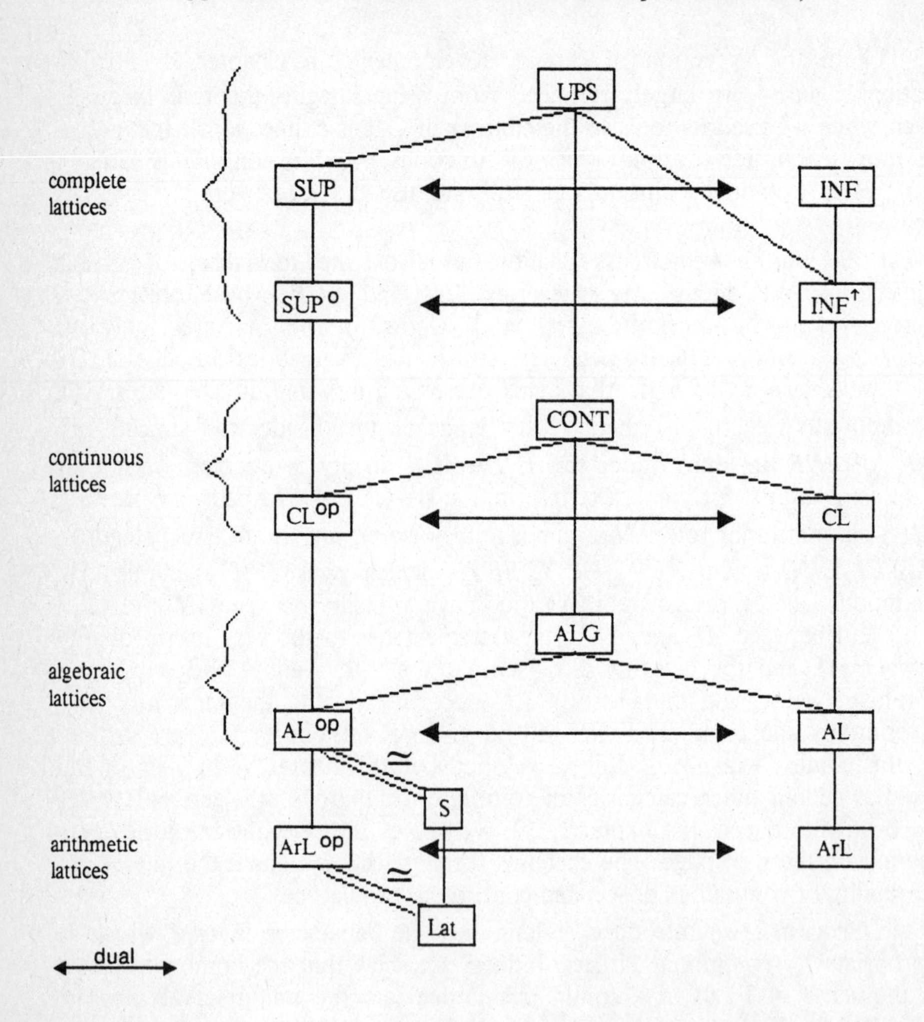

1. DUALITY THEORY

1.1. DEFINITION. The categories ***INF*** and ***SUP*** have the same class of objects, namely, the class of all complete lattices. The morphisms of ***INF*** preserve arbitrary infs, the morphisms of ***SUP*** preserve arbitrary sups. Sometimes ***INF*** is called the category of *all complete inf-semilattices* and ***SUP*** the category of *all complete sup-semilatices*. Evidently, ***INF***$\cap$***SUP*** is the category of *all complete lattices and complete lattice morphisms.* □

It is perhaps noteworthy that the role of ***INF***$\cap$***SUP*** is relatively secondary in our framework. The supply of morphisms in this category is too restricted. A noteworthy exception is the sup-morphism $r = (I \mapsto \sup I) : \mathrm{Id}\, L \to L$, which is in ***INF***$\cap$***SUP*** for a continuous lattice L (see O-3.15, 4.2, and I-2.1).

We recall from Chapter O, Section 3 the concept of adjoint functions between partially ordered sets. In particular, given a pair of complete lattices S and T, by O-3.5, to each map $g : S \to T$ in ***INF*** there corresponds a unique lower adjoint $D(g) : T \to S$ in ***SUP***, and to each map $d : T \to S$ in ***SUP*** there corresponds a unique upper adjoint $G(d) : S \to T$ in ***INF***. In order to define a functor, we have to define D and G on objects; thus for any complete lattice C we write $D(C) = C$ and $G(C) = C$.

1.2. LEMMA. *The assignments* $D :$ ***INF***$\to$***SUP***^{op} *and* $G :$ ***SUP***$\to$***INF***^{op} *are functors (that is, D and G are contravariant functors.).*

Proof. The (lower or upper) adjoint of an identity map of a complete lattice is the identity map (O-3.7). That the composition of upper (resp., lower) adjoints is again the upper (resp., lower) adjoint of the composition is well known in category theory and is immediate from the definition O-3.1; indeed if we have $g_1 : S_1 \to S_2$ and $g_2 : S_2 \to S_3$ then $g_2 g_1(s_1) \geq s_3$ iff $g_1(s_1) \geq D(g_2)s_3$ iff $s_1 \geq D(g_1)D(g_2)s_3$ on one hand, but also $g_2 g_1(s_1) \geq s_3$ iff $s_1 \geq D(g_2 g_1)$ on the other. Thus $D(g_2 g_1) = D(g_1)D(g_2)$. The assignment G is treated analogously. Thus D and G are (contravariant) functors. □

1.3. THEOREM. (INF-SUP DUALITY). *The categories* ***INF*** *and* ***SUP*** *are dual under the functors D and G given through the Galois connection of functions. Specifically, D and G preserve objects (that is, the "dual" of a complete lattice is itself under this duality). Moreover,* $GD(g) = g$ *and* $DG(d) = d$ *for all g in* ***INF*** *for all d in* ***SUP***.

Proof. This is trivial: by definition D and G preserve objects, and the identities $GD(g) = g$ and $DG(d) = d$ are clear from the adjunction. □

This simple duality nevertheless is quite useful as a basis and a guide to the invention of other duality theories. Examples are the self-duality of the category of continuous semilattices and the duality between unital semilattices and algebraic lattices which we discuss later (see also I-4.12).

Our first task is to investigate how the functors D and G translate certain preservation properties of morphisms.

1.4. THEOREM. *Let* S *and* T *be complete lattices and* $g : S \to T$ *the upper adjoint of* $d : T \to S$ (*that is,* $d = D(g)$, $g = G(d)$). *Then the following statements are equivalent:*

(1) *g preserves directed sups* (*that is, g is Scott continuous,* see II-2.2).

(2) *If $U \subseteq T$ is any Scott-open set in* T, *then* $\uparrow d(U)$ *is Scott open in* S.

These conditions imply:

(3) *d preserves $\ll$ (that is, if $t \ll t^*$ in* T *then $d(t) \ll d(t^*)$ in* S),

and if T *is continuous, then all three conditions are equivalent.*

Proof. (1) implies (2): Let U be Scott open in T. In order to show that $\uparrow d(U)$ is Scott open in S, we take a directed set $D \subseteq S$ with sup $D \in \uparrow d(U)$ and we show that $D \cap \uparrow d(U) \neq \emptyset$. Now, sup $D \in \uparrow d(U)$ implies $d(u) \leq \sup D$ for some $u \in U$. Since $t \leq gd(t)$ for all $t \in T$ (O-3.6(2)), we know $g(\sup D) \geq gd(u) \geq u \in U$. Hence $g(\sup D) \in U$. But $g(\sup D) = \sup g(D)$ by hypothesis (1), and $g(D)$ is directed since g preserves order. Since U is Scott open there is an $x \in D$ with $g(x) \in U$, and thus $dg(x) \in d(U)$. But $dg(x) \leq x$ (by O-3.6(2)), so $x \in \uparrow d(U)$. Thus $D \cap \uparrow d(U) \neq \emptyset$.

(2) implies (1): Let D be a directed set in S with $s = \sup D$. We have always $g(s) \geq \sup g(D)$; hence we must show $g(s) \leq \sup g(D)$. We proceed by contradiction. Assume that $g(s) \notin \downarrow \sup g(D)$. With $U = L \setminus \downarrow \sup g(D)$, we have $g(s) \in U$, $\sup g(D) \notin U$, and U is Scott open. By hypothesis (2) we know that $\uparrow d(U)$ is Scott open in S. We note $dg(s) \in d(U)$ and $dg(s) \leq s$ (O-3.6(2)), hence $s \in \uparrow d(U)$. So, since $\uparrow d(U)$ is Scott open, we have an $x \in D$ with $x \in \uparrow d(U)$, that is, $x \geq d(u)$ for some $u \in U$. But then in view of O-3.6(2) once more, $u \leq gd(u) \leq g(x) \leq \sup g(D)$; whence $\sup g(D) \in U$, and this is the desired contradiction.

(1) implies (3): Suppose $t \ll t^*$ in T and let $D \subseteq S$ be directed with $d(t^*) \leq \sup D$. By O-3.1 this means $t^* \leq g(\sup D)$, but $g(\sup D) = \sup g(D)$ by (1). Now there is an $s \in D$ with $t \leq g(s)$ by I-1.1. Thus $d(t) \leq s$ by O-3.1, whence $t \ll t^*$.

(3) implies (1) for continuous T: Let D be an up-directed set in S with $s = \sup D$. We have to show that $g(s) \leq \sup g(D)$. Take an arbitrary $t \ll g(s)$. By hypothesis (3) we have $d(t) \ll dg(s) \leq s = \sup D$ (O-3.6(2)). Thus by I-1.2(ii) and I-1.1 we find an $x \in D$ with $d(t) \leq x$, which is equivalent to $t \leq g(x)$ by the definition of adjunction O-3.1. Thus $t \leq \sup g(D)$. Now assume that T is continuous. Since $t \in \twoheaddownarrow g(s)$ was arbitrary, then we have $g(s) = \sup \twoheaddownarrow g(s)$; the claim follows. □

In several of the following results we have occasion to say that a map is *open*, meaning of course that it maps open sets to open sets.

1.5. REMARK. *Condition* (2) *in* 1.4 *implies*:

(2′) *d is relatively open onto its image with respect to the Scott topology on* T *and the topology on d(T) induced by the Scott topology of* S.

Furthermore condition (2′) *implies*:

(2″) *The corestriction d : T→d (T) is open with respect to the Scott topologies.*

Proof. (2) implies (2′): It suffices to observe that $d(U) = d(\mathrm{T})\cap\uparrow d(U)$. The left is always included in the right; so suppose that $t\in d(\mathrm{T})$ and $t\geq d(u)$ for $u\in U$. Now $t = d(v) = d(v)\vee d(u) = d(v\vee u)$. But $U = \uparrow U$; so $v\vee u\in U$, and thus, $t\in d(U)$.

(2′) implies (2″): Let $U\subseteq\mathrm{T}$ be Scott open and let $D\subseteq d(\mathrm{T})$ be directed with $\sup_{d(\mathrm{T})} D\in d(U)$. Since d is a lower adjoint, d preserves all sups, and so $\sup_{\mathrm{S}} D = \sup_{d(\mathrm{T})} D\in d(U)$. Now, (2′) implies $d(U) = V\cap d(\mathrm{T})$ for some $V\subseteq\mathrm{S}$ Scott open, so $\sup_{\mathrm{S}} D\in V$. Hence $D\cap V\neq\emptyset$, and since $D\subseteq d(\mathrm{T})$, we conclude $D\cap d(U) = D\cap(V\cap d(\mathrm{T}))\neq\emptyset$. □

In general, (2′) does not imply (2): If $d : I\to I^2$ is the embedding $d(t) = (t,1)$, then $V = d(]1/2,1])$ is relatively Scott open, since it is of the form $d(I)\cap(]1/2,1]\times I)$; but $\uparrow V = V$ is not Scott open in I^2.

On the other hand, if d is surjective (that is, g is injective by O-3.7) then $\uparrow d(U) = d(U)$, and thus (2″) implies (2). Hence we have:

1.6. COROLLARY. *If g : S→T in **INF** is upper adjoint to d : T→S in **SUP** and if g is injective (equivalently, d is surjective* (see O-3.7)), *then g preserves directed sups iff d is Scott open.* □

1.7. COROLLARY. *Let* L *be a complete lattice and k : L→L a kernel operator. Then the following statements are equivalent*:

(1) *k preserves directed sups*;

(2) *For each Scott-open set U of k(L) the set $\uparrow U$ is Scott open in* L.

These conditions imply

(3) *For $x,x^*\in k(\mathrm{L})$, we have $x\ll_{k(\mathrm{L})}x^*$ iff $x\ll_{\mathrm{L}}x^*$.*

Moreover, if k(L) is continuous, then all three conditions are equivalent. If L *is continuous and* (1) *holds, then k(L) is continuous, too.*

Proof. By O-3.10, the corestriction $k^\circ : \mathrm{L}\to k(\mathrm{L})$ of k is upper adjoint to the inclusion $k_\circ : k(\mathrm{L})\to\mathrm{L}$. Since $k_\circ$ preserves sups, then (1) holds iff k° preserves directed sups. Then Theorem 1.4 applies to give the equivalence of (1) and (2). We have always that $x\ll_{\mathrm{L}}x^*$ implies $x\ll_{k(\mathrm{L})}x^*$. Hence 1.4 shows that (3) follows from the other two conditions and is in fact equivalent if k(L) is continuous. Finally, if L is continuous and (1) is satisfied, then $k(L)$ is continuous by I-2.10. □

1.8. COROLLARY. *Let* L *be a complete lattice and* $c : L \to L$ *a closure operator* (O-3.8). *Then the following statements are equivalent:*

(1) $c(L)$ *is closed in* L *under directed sups.*

(2) *The corestriction* $c^\circ : L \to c(L)$ *is Scott open.*

These conditions imply

(3) $c(x) \ll_{c(L)} c(x^*)$ *for all* $x \ll_L x^*$ *in* L.

Moreover, if $c(L)$ *is continuous, then all three conditions are equivalent.* □

Remark. By I-2.15, $c(L)$ is continuous if L is continuous and c preserves infs. By O-3.12(ii), the lattice $c(L)$ is always complete.

Proof. Condition (1) is equivalent to saying that the inclusion map $c_\circ : c(L) \to L$ preserves directed sups. Hence the corollary follows from 1.4 and 1.6. □

In order to reformulate Theorem 1.4 in terms of duality we require suitable categories.

1.9. DEFINITION. We introduce the following subcategories:

INF$^\uparrow$ has as objects all complete lattices and as morphisms maps preserving arbitrary infs and directed sups (that is, ***INF***-maps that are also Scott continuous).

SUP^0 has as objects all complete lattices and as morphisms all ***SUP***-morphisms d where for each Scott-open U in the domain of d the set $\uparrow d(U)$ is Scott open in the range.

CL is the full subcategory of ***INF***$^\uparrow$ with objects all continuous lattices.

CL^{op} is the full subcategory of ***SUP***^{0} with objects all continuous lattices.

We call ***CL*** the category of *continuous lattices* and ***CL***^{op} the *dual category of continuous lattices.* □

Evidently ***INF***$^\uparrow$ = ***INF***$\cap$***UPS***. Notice that we view ***CL***^{op} as a concrete category of functions between continuous lattices, and the ^{op}-notation is justified by the next theorem.

1.10. THEOREM. (i) (***INF***$^\uparrow$ – ***SUP***^{0} DUALITY). *The categories* ***INF***$^\uparrow$ *and* ***SUP***^{0} *are dual under the adjoint functors D and G.*

(ii) (*CL*−*CL*^{op} DUALITY). *The categories* ***CL*** *and* $\boldsymbol{CL}^{\mathrm{op}}$ *are dual under the adjoint functors D and G, and a map between continuous lattices is in* $\boldsymbol{CL}^{\mathrm{op}}$ *iff it preserves all sups and the relation* $\ll$. □

Let us now determine how the morphisms of CL^{op} treat algebraic lattices, which, as we know from Section I-4, play an important role in our theory. Our discussion leads us to another important duality theory.

1.11. PROPOSITION. *Let* $d : \mathrm{T} \to \mathrm{S}$ *be a* ***SUP****-map. Then* (1) *below implies* (2); *and if* T *is algebraic, then the two statements are equivalent*:

(1) *d preserves* $\ll$;

(2) $d(\mathrm{K(T)}) \subseteq \mathrm{K(S)}$.

Proof. (1) implies (2): If $c \in \mathrm{K(T)}$, then $c \ll c$, and thus $d(c) \ll d(c)$ by (1), hence $d(c) \in \mathrm{K(S)}$.

Now suppose that T is algebraic.

(2) implies (1): Suppose that $t \ll t^*$ in T. By I-4.5 there is a compact c with $t \leq c \leq t^*$. Then $d(t) \leq d(c) \leq d(t^*)$ and $d(c) \in \mathrm{K(S)}$ by (2), Hence $d(t) \ll d(t^*)$ (I-4.5). □

1.12. COROLLARY. *Let* $g : \mathrm{S} \to \mathrm{T}$ *be an* ***INF****-map and suppose that* T *is algebraic. Then the following are equivalent*:

(1) *g preserves directed sups.*

(2) $\mathrm{K}(g) = (c \mapsto \inf g^{-1}(\uparrow c)) : \mathrm{K(T)} \to \mathrm{K(S)}$ *is a well-defined sup-semilattice morphism.*

Proof. Notice that the lower adjoint $d = D(g)$ is the mapping such that $t \mapsto \inf g^{-1}(\uparrow t)$ by O-3.4. Thus condition (2) is equivalent to

(2′) $d(\mathrm{K(T)}) \subseteq \mathrm{K(S)}$;

since d preserves sups and K(L) is a sup-semilattice for each complete lattice L by I-4.3. But the equivalence of (1) and (2′) follows from 1.4 and 1.11. □

In order to express this last fact in a systematic way, we introduce further categories (compare, however, II-2.2!).

1.13. DEFINITION. We define the following categories:

S is the category of all unital semilattices (that is, commutative idempotent monoids) with maps preserving the semilattice operation and the identity.

AL and ***ArL*** are the full subcategories of ***CL*** consisting of all algebraic and of all arithmetic lattices, respectively.

$\boldsymbol{AL}^{\mathrm{op}}$ and $\boldsymbol{ArL}^{\mathrm{op}}$ are the full subcategories of $\boldsymbol{CL}^{\mathrm{op}}$ consisting of $\boldsymbol{AL}$- and of $\boldsymbol{ArL}$-objects, respectively. □

1.14. COROLLARY. (i) *The assignment which associates with each algebraic lattice* L *the sup-semilattice* K(L) *and with each morphism* $g : \mathrm{S} \to \mathrm{T}$ *in* $\boldsymbol{AL}$ *the map* $\mathrm{K}(g) = D(g) \mid \mathrm{K}(\mathrm{T}) : \mathrm{K}(\mathrm{T}) \to \mathrm{K}(\mathrm{S})$ *is a functor* $\boldsymbol{AL} \to \boldsymbol{S}^{\mathrm{op}}$.

(ii) *The assignment* $\mathrm{L} \mapsto \mathrm{K}(\mathrm{L})$ *and* $d \mapsto d \mid \mathrm{K}(\mathrm{T}) : \mathrm{K}(\mathrm{T}) \to \mathrm{K}(\mathrm{S})$ *for maps* $d : \mathrm{T} \to \mathrm{S}$ *in* $\boldsymbol{AL}^{\mathrm{op}}$ *is a functor* $\boldsymbol{AL}^{\mathrm{op}} \to \boldsymbol{S}$. □

The reader should note that the functor K has an "inverse" and, in fact, defines a duality between $\boldsymbol{AL}$ and $\boldsymbol{S}$. Let us make the following remark: we have specified the category $\boldsymbol{S}$ as a variety defined in terms of one binary operation and one nullary (constant) operation (namely, the semilattice multiplication and the identity), but we have not spoken of an order. Each semilattice may be considered either as an inf-semilattice relative to the order given by $x \leq y$ iff $x = xy$ or alternatively as a sup-semilattice relative to the opposite order given by $x \leq y$ iff $y = xy$ (see O-1.8). Conversely, every inf-semilattice with maximal element and every sup-semilattice with minimal element is an object in $\boldsymbol{S}$. In our present context it is a bit more convenient to think of a unital semilattice as a sup-semilattice with minimal element, but we recall that this is purely a matter of convenience and has no theoretical significance.

If S is a sup-semilattice with minimal element, then Id S is in $\boldsymbol{AL}$ according to I-4.12, and the principal ideal embedding $x \mapsto {\downarrow}x : \mathrm{S} \to \mathrm{Id}\ \mathrm{S}$ induces an isomorphism between S and K(Id S). Conversely, we recall from I-4.12 that, if L is an algebraic lattice, then each $x \in \mathrm{L}$ yields an ideal ${\downarrow}x \cap \mathrm{K}(\mathrm{L})$ of the sup-semilattice K(L), and that the function $x \mapsto {\downarrow}x \cap \mathrm{K}(\mathrm{L}) : \mathrm{L} \to \mathrm{Id}(\mathrm{K}(\mathrm{L}))$ is an isomorphism. Let us observe that for each sup-semilattice morphism $f : \mathrm{S} \to \mathrm{T}$ preserving minimal elements we induce a morphism $\mathrm{Id}(f) = (J \mapsto f^{-1}(J)) : \mathrm{Id}(\mathrm{T}) \to \mathrm{Id}(\mathrm{S})$ and that this morphism preserves arbitrary intersections and unions of up-directed families and, hence, is an $\boldsymbol{AL}$-morphism. Evidently, $\mathrm{Id} : \boldsymbol{S} \to \boldsymbol{AL}^{\mathrm{op}}$ is a functor, and what we have observed amounts to the following theorem:

1.15. THEOREM. ($\boldsymbol{AL}-\boldsymbol{S}$ DUALITY). (i) *The categories* $\boldsymbol{S}$ *of unital semilattices and* $\boldsymbol{AL}$ *of algebraic lattices (with maps preserving infs and directed sups) are dual. If one considers* $\boldsymbol{S}$ *as a category of sup-semilattices with minimal elements (and semilattice morphisms preserving minimal elements), then the duality is established through the functors* $\mathrm{Id} : \boldsymbol{S} \to \boldsymbol{AL}^{\mathrm{op}}$ *and* $\mathrm{K} : \boldsymbol{AL} \to \boldsymbol{S}^{\mathrm{op}}$.

(ii) *Under this duality, the full subcategory* $\boldsymbol{S}_{\mathrm{lat}}$ *of unital lattices with identity preserving semilattice maps is placed into duality with the category* $\boldsymbol{ArL}$ *of arithmetic lattices.*

Proof. Only the last assertion is not yet proved, but it is immediate from the fact that an algebraic lattice L is arithmetic iff K(L) is a lattice with smallest element (see I-4.7). □

Notice that 1.15 also says that the categories S of unital semilattices and the dual category AL^{op} of algebraic lattices are equivalent.

1.16. DEFINITION. Let ***Lat*** denote the category of all unital lattices and all lattice homomorphisms preserving the unit element.

1.17. THEOREM. *The category* $\boldsymbol{ArL}^{op}$ *of all arithmetic lattices and all maps preserving finite infs and arbitrary sups and respecting the way-below relation is equivalent to the category* ***Lat***.

Proof. If S and T are algebraic lattices, then for each $d : S \to T$ in $\boldsymbol{AL}^{op}$ we have a commutative diagram

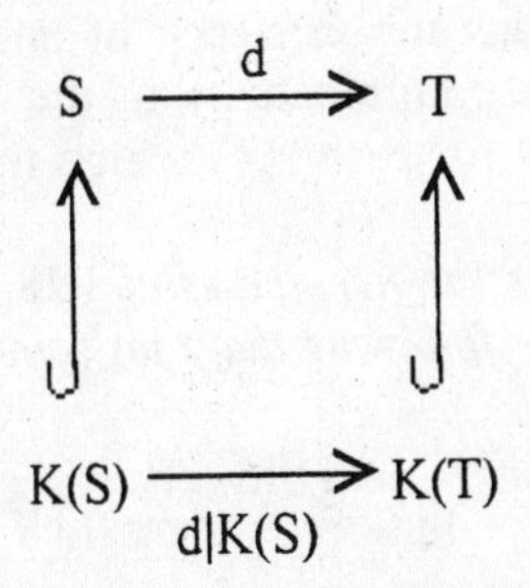

and by a remark following 1.14, $d \mapsto d \mid K(S)$: $\boldsymbol{AL}^{op}(S,T) \to S(K(S),K(T))$ is a bijection. If now S and T are arithmetic, then K(S) and K(T) are lattices. If further d preserves finite infs, then $d \mid K(S)$ is in ***Lat***. Conversely, suppose that $d \mid K(S)$ preserves finite infs. Let s, $s^* \in S$. Then $D = \downarrow s \cap K(S)$ and $D^* = \downarrow s^* \cap K(S)$ are directed sets in S with $s = \sup D$ and $s^* = \sup D^*$. We have $ss^* = \sup DD^*$, since S is meet continuous (see O-4.2, I-2.2 and I-4.5). Then $d(ss^*) = d(\sup DD^*) = \sup d(DD^*)$ (since d preserves arbitrary sups) $= \sup d(D) \sup d(D^*)$ (since $d \mid K(S$ preserves finite infs and T is meet continuous) $= d(\sup D)d(\sup D^*)$ (since d preserves sups) $= d(s)d(s^*)$. Thus d preserves finite infs. We have shown that $d \mapsto d \mid K(S)$ establishes a bijection between $\boldsymbol{ArL}^{op}(S,T)$ and ***Lat***(K(S),K(T)). □

We encounter in the previous theorem a lower adjoint d which in addition preserves finite infs, that is, which was a lattice homomorphism. It is worth our while to ask systematically whether or not this additional property of a right adjoint is recognizable by looking at its left adjoint. The following propositions provide the tools and are also of independent interest:

1.18. PROPOSITION. *Let* S *and* T *be lattices with minimal and maximal elements. Let* $g : S \to T$ *be upper adjoint to* $d : T \to S$. *Then the* ***AL****-morphism* $\mathrm{Id}(g) : \mathrm{Id}\, S \to \mathrm{Id}\, T$ *is given by* $\mathrm{Id}\, g(I) = d^{-1}(I) = \downarrow g(I)$, *and we have a commutative diagram (with the principal ideal embeddings as vertical arrows)*

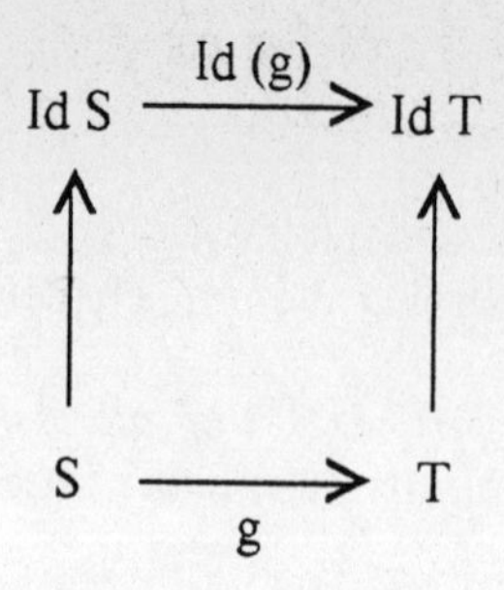

Proof. First we have to show that $d^{-1}(I) = \downarrow g(I)$ for all $I \in \mathrm{Id}(S)$. But $t \in d^{-1}(I)$ means $d(t) \in I$ which is equivalent to the existence of an $i \in I$ with $d(t) \leq i$. By O-3.1 this means the existence of an $i \in I$ with $t \leq g(i)$ which says precisely that $t \in \downarrow g(I)$. The commutativity of the diagram is a consequence of the relation $\downarrow g(s) = \downarrow g(\downarrow s)$ for each $s \in S$, which follows from O-1.11. □

1.19. PROPOSITION. *Under the hypotheses of* 1.18, *the lower adjoint* $D(\mathrm{Id}(g))$ *is given by* $J \mapsto \downarrow d(J)$ *and the following diagram commutes*

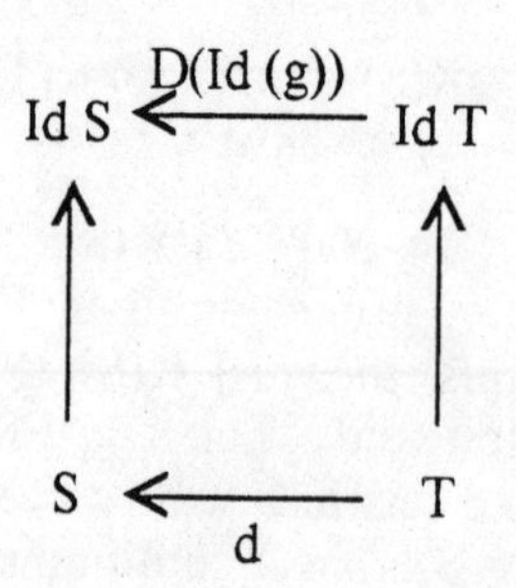

Proof. First we must identify the right adjoint of Id(g): But $d^{-1}(I) \supseteq J$ means $I \supseteq d(J)$ which is equivalent to $I \supseteq \downarrow d(J)$ since I is an ideal. (Notice that $\downarrow d(I)$ is always an ideal since d preserves sups.) As before the commutativity of the diagram follows from O-1.11. □

We complement these observations by the following:

1.20. PROPOSITION. *Assume that the hypotheses of* 1.18 *are satisfied, and in addition that* S *and* T *are complete lattices. then the following diagram commutes*:

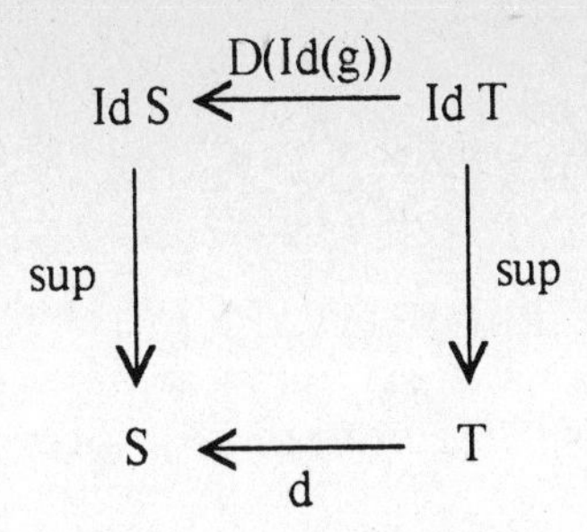

Moreover $g \in \boldsymbol{INF}^{\uparrow}$, iff the following diagram commutes also:

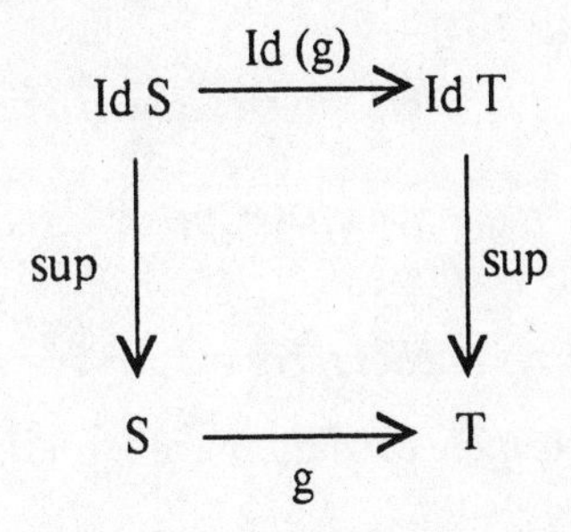

Proof. Exercise (see Hofmann and Stralka [1976]). □

We need the following sharpening of Remark I-3.16:

1.21. LEMMA. *Let* S *be a poset, and* $I \in \mathrm{Id}\ \mathrm{S}$. *Then the following statements are equivalent*:

(1) $\mathrm{S} \setminus I$ *is filtered.*

(2) $I \in$ PRIME Id S (*see* I-3.11).

Remark. If S is a semilattice, then (1) is equivalent to $\mathrm{S} \setminus I$ is a filter.

Proof. (1) implies (2): Let $I_1, I_2 \in \mathrm{Id}\ \mathrm{S}$. If neither I_1 nor I_2 is contained in I, then we find elements $i_n \in I_n \setminus I$, $n=1,2$, and by (1) we find a $j \leq i_n$, $n=1,2$, with $j \in \mathrm{S} \setminus I$. But then also $j \in I_n$, $n=1,2$, since I_n is an ideal, and so $j \in (I_1 \cap I_2) \setminus I$. The result follows by contraposition.

(2) implies (1): Suppose that $s, s^* \in \mathrm{S} \setminus I$. Then $\downarrow s, \downarrow s^* \not\subseteq I$, whence $\downarrow s \cap \downarrow s^* \not\subseteq I$ by (2). Hence there is an element $s^{**} \in \mathrm{S} \setminus I$ with $s^{**} \leq s, s^*$.

The justification of the remark is clear. □

We generalize definition I-3.17 a little bit:

1.22. DEFINITION. An ideal in a poset is called a *prime ideal* iff it satisfies the equivalent conditions of 1.21. □

1.23. REMARK. (i) *If* L *is a semilattice, and* $s \in L$, *then* $\downarrow s$ *is a prime ideal of* L *iff* s *is a prime element of* L *as defined in* I-3.11.

(ii) *The principal ideal map* $p = (x \mapsto \downarrow x) : L \to \mathrm{Id}\, L$ *satisfies*

$$p^{-1}(\mathrm{PRIME}\ \mathrm{Id}\ L) = \mathrm{PRIME}\ L.$$

This implies $p(\mathrm{PRIME}\ L) \subseteq \mathrm{PRIME}\ \mathrm{Id}\ L$. □

We now derive from Theorem I-3.14 in conjunction with I-4.12 the following fact:

1.24. PROPOSITION. *Let* S *be a sup-semilattice with minimal element. Then the following statements are equivalent:*

(1) Id S *is distributive;*

(2) *Every ideal of* S *is the intersection of prime ideals.* □

1.25. PROPOSITION. *Under the hypotheses of* 1.24, *the conditions there are also equivalent to the following (expressed in terms of the sup-semilattice* S *alone):*

(3) $x \leq s \vee s^*$ *always implies there exist* $t \leq s$ *and* $t^* \leq s^*$ *with* $x = t \vee t^*$.

Furthermore, if S *is a lattice, these conditions are equivalent to:*

(4) S *is distributive.*

Proof. Exercise (see Hofmann, Mislove and Stralka [1974]. See also I-3.29.) □

After this preparation we have the following information on lower adjoints d which are also lattice morphisms:

1.26. THEOREM. *Let* S *and* T *be unital lattices and suppose that* $g : S \to T$ *is an upper adjoint to* $d : T \to S$. *Then each of the following conditions implies the next.*

(1) d *is a lattice morphism (that is, preserves finite infs);*

(2) *For each filter* F *of* S, *the set* $d^{-1}(F)$ *is a filter of* T;

(3) $\mathrm{Id}(g)$ *(that is, the function* $J \mapsto \downarrow g(J)$*) preserves prime ideals;*

(4) $g(\mathrm{PRIME}\ S) \subseteq \mathrm{PRIME}\ T$ *(that is,* g *preserves primes).*

If S *is distributive* (3) *implies* (1), *and if each element of* S *is an inf of primes, then all four conditions are equivalent.*

Remark. If S is a distributive continuous lattice, then PRIME S order-generates S by I-3.13.

Proof. (2) is an immediate consequence of (1).

(2) implies (3): If we abbreviate $r = \mathrm{Id}(g)$, then we have $r(I) = d^{-1}(I)$ by 1.18 and 1.19. This is equivalent to $d^{-1}(S \setminus I) = T \setminus r(I)$. If I is a

prime ideal, then $S \setminus I$ is a filter, and then $T \setminus r(I)$ is a filter by (2). This means that $r(I)$ is a prime ideal.

(3) implies (4): Let $p \in \mathrm{PRIME}\ S$. Then $\downarrow p$ is a prime ideal by 1.23. Hence $\downarrow g(\downarrow p)$ is a prime ideal by (3). But $\downarrow g(\downarrow p) = \downarrow g(p)$ by O-1.11, and so $g(p)$ is prime by 1.23.

Now suppose that S is distributive. Then every ideal I of S is an intersection of prime ideals by 1.24. In particular every principal ideal is the intersection of prime ideals.

(3) implies (1): Suppose $t,u \in T$; then we always have $d(tu) \leq d(t)d(u)$. To prove the converse, we note that $\downarrow d(tu)$ is the intersection of prime ideals, and so $d(t)d(u) \leq d(tu)$ if $d(t)d(u) \in I$ for every prime ideal of S containing $d(tu)$. Let I be such a prime ideal. Then $tu \in d^{-1}(I) = r(I)$, and $r(I)$ is a prime ideal by (3). Hence $t \in r(I)$ or $u \in r(I)$, whence $d(t) \in d(r(I)) \subseteq I$ or $d(u) \in d(r(I)) \subseteq I$. In either case, $d(t)d(u) \in I$, as we had to show.

Finally suppose that every element of S is an inf of primes. This implies that every principal ideal of S is the intersection of principal ideals generated by primes, and these are prime ideals by 1.23. Hence the proof of (3) implies (1) applies with $I = \downarrow p$ for some prime p of S, since $d^{-1}(I) = d^{-1}(\downarrow p) = \downarrow g(\downarrow p)$ (by 1.18) $= \downarrow g(p)$ (by O-1.11) is prime by (4) and 1.23. □

It appears on the surface that we proved a more general statement in so far as for (3) implies (1) we really used only that the prime ideals separate the points from ideals. But this property in itself is sufficient to make S distributive.

1.27. COROLLARY. (i) *Let* S *be a continuous Heyting algebra and* T *a complete lattice. Then the mapping* $g \mapsto D(g) : \boldsymbol{INF}(S,T) \to \boldsymbol{SUP}(T,S)$ *induces a bijection from the subset of all prime preserving maps onto the subset of all lattice homomorphisms in* $\boldsymbol{SUP}(T,S)$.

(ii) *If* T *is continuous, then the same map induces a bijection from the set of all prime preserving morphisms in* $\boldsymbol{CL}(S,T)$ *onto the subset of all lattice homomorphisms in* $\boldsymbol{CL}^{op}(T,S)$. □

1.28. COROLLARY. *The category of all continuous* [*algebraic*] *Heyting algebras with prime preserving* **CL***-maps* [**AL***-maps*] *is dual to the category of all continuous* [*algebraic*] *Heyting algebras with lattice homomorphisms preserving arbitrary sups and respecting the way-below relation* [*compact elements*]. □

EXERCISES

1.29. EXERCISE. Let S and T be (up-complete) posets and (g,d) with $g : S \to T$ and $d : T \to S$ an adjunction between S and T. Then the conclusions of Theorem 1.4 hold. If S is an upper semilattice, then conclusions of 1.5 hold, too.

(HINT: Check the proofs of 1.4 and 1.5.) □

1.30. EXERCISE. If T is an (up-complete) poset and $S \subseteq T$ such that some function $d: T \to S$ is lower adjoint to the inclusion $S \to T$ (that is, S is the image of T of some closure operator of T (see O-3.10)). Then S is up-closed in T iff d is Scott open.

(HINT: Check the proof of 1.6.) □

1.31. EXERCISE. Corollary 1.7 holds for (up-complete) posets L, with the possible exception of the very last sentence. In the case of a complete semilattice, 1.7 holds in its entirety.

(HINT: Check the proof of 1.7 and recall Section I-2. Recall that we do not know whether I-2.10, 2.14 hold for continuous posets or semilattices.) □

1.32. EXERCISE. Corollary 1.8 persists for (up-complete) posets. □

1.33. EXERCISE. Let $\boldsymbol{U}_{\mathrm{up}}$ be the category of all up-complete posets and Scott-continuous upper adjoints, and let $\boldsymbol{U}_{\mathrm{low}}$ be the category of up-complete posets with lower adjoints d such that $\uparrow d(U)$ is Scott open for every Scott-open U in its domain.

(i) $\boldsymbol{U}_{\mathrm{up}}$ and $\boldsymbol{U}_{\mathrm{low}}$ are dual via Galois connections.

(ii) If T is a continuous poset, and S is up-complete, then $d \in \boldsymbol{U}_{\mathrm{low}}(S,T)$ iff d is a lower adjoint such that $s \ll s^*$ implies $d(s) \ll d(s^*)$.

1.34. EXERCISE. Let (g,d) be a Galois connection between up-complete posets S and T of which T is algebraic (I-4.28). Then the three conditions 1.11(1), (2) and 1.12(1) are equivalent. □

1.35. EXERCISE. Let $\boldsymbol{A}_{\mathrm{up}}$ be the category of algebraic posets with Scott-continuous upper adjoint maps and $\boldsymbol{P}$ the category of posets and monotone maps $f: S \to T$ such that $f^{-1}(I)$ is directed for every ideal I of T. Then $\boldsymbol{A}_{\mathrm{up}}$ and $\boldsymbol{P}$ are dual under the functors

$$\mathrm{Id}: \boldsymbol{P} \to \boldsymbol{A}_{\mathrm{up}},\ \mathrm{Id}(f)(I) = f^{-1}(I) \text{ and } \mathrm{K}: \boldsymbol{A}_{\mathrm{up}} \to \boldsymbol{P},\ \mathrm{K}(f)(c) = \min f^{-1}(\uparrow c).$$

(HINT: Verify the steps leading to Theorem 1.15.) □

1.36. EXERCISE. Furnish the proof of 1.20. □

1.37. EXERCISE. Furnish the proof of 1.23. □

1.38. EXERCISE. Furnish the proof of 1.25 (compare Exercise I-3.29.) □

1.39. DEFINITION. The following categories are defined:

CPoset$_0$ is the category whose objects are continuous posets and whose morphisms are Scott-continuous maps f such that $f^{-1}(U)$ is empty or a filter for any open filter U.

CSem$_0$ is the subcategory whose objects are continuous semilattices (see I-1.26) and whose morphisms are also semilattice morphisms.

For the poset $\mathbf{CPoset}_0(\mathrm{S},\mathrm{T})$ with its pointwise partial order we write $\langle \mathrm{S}\to\mathrm{T}\rangle_0$. We will write $\langle \mathrm{S}\to\mathrm{T}\rangle$ for the set of all $f\in \mathbf{CPoset}_0(\mathrm{S},\mathrm{T})$ which are *cofinal*, that is, which satisfy $\mathrm{T} = \downarrow f(\mathrm{S})$.

> ***CPoset*** and ***CSem*** are the full subcategories of $\mathbf{CPoset}_0$ and $\mathbf{CSem}_0$ with cofinal morphisms. □

We note that the set $\langle \mathrm{S}\to\mathrm{T}\rangle$ can be characterized succinctly as the set of all $f: \mathrm{S}\to\mathrm{T}$ such that $f^{-1}(U)$ is a nonempty open filter if U is a nonempty open filter. It is also the case that $\langle \mathrm{S}\to\mathrm{T}\rangle_0$ and $\langle \mathrm{S}\to\mathrm{T}\rangle$ are up-complete posets.

1.40. EXERCISE. The function $f\mapsto f^{-1}(1)$: $\langle \mathrm{S}\to 2\rangle \to \mathcal{O}\mathrm{Filt}\,(\mathrm{S})$ into the continuous poset of open filters (see I-3.32) is an isomorphism whose inverse is

$$U\mapsto \chi_U : \mathcal{O}\mathrm{Filt}\,(\mathrm{S})\to\langle \mathrm{S}\to 2\rangle,$$

where χ_U is the characteristic function of U. □

1.41. EXERCISE. Set $\langle \mathrm{S}\rangle = \langle \mathrm{S}\to 2\rangle$.

(i) Then $\langle\bullet\rangle$: $\mathbf{CPoset}\to\mathbf{CPoset}^{\mathrm{op}}$ is a functor which maps ***CSem*** into $\mathbf{CSem}^{\mathrm{op}}$; here we take as usual $f(\varphi) = \varphi\circ f$. If we identify the functor $\langle\bullet\rangle$ with the functor $\mathcal{O}\mathrm{Filt}$ (which we may!), then $\mathcal{O}\mathrm{Filt}(f)(V) = f^{-1}(V)$.

(ii) The natural function η_{S} : $\mathrm{S}\to\langle\langle \mathrm{S}\rangle\rangle$, where $\eta_{\mathrm{S}}(s)(\varphi) = \varphi(s)$, corresponds to the natural function

$$\alpha_{\mathrm{S}} = (s\mapsto \{U\in\mathcal{O}\mathrm{Filt}\,(\mathrm{S}) : s\in U\}) : \mathrm{S}\to\mathcal{O}\mathrm{Filt}\,(\mathcal{O}\mathrm{Filt}\,(\mathrm{S})).\ \square$$

We now come to an important duality theorem due to Lawson [1979a].

1.42. EXERCISE. (i) For each continuous poset S the maps η_{S} : $\mathrm{S}\to\langle\langle \mathrm{S}\rangle\rangle$ and α_{S} : $\mathrm{S}\to\mathcal{O}\mathrm{Filt}\,(\mathcal{O}\mathrm{Filt}\,(\mathrm{S}))$ are isomorphisms.

(ii) The categories ***CPoset*** and ***CSem*** are self dual under the functor $\langle\bullet\rangle$ (or, equivalently, $\mathcal{O}\mathrm{Filt}$).

(HINT: We outline the proof:

$\alpha = \alpha_{\mathrm{S}}$ is injective: Let $s\not\leq s^*$ in S; then there is an open filter U with $s\in U$ but $s^*\notin U$ by I-3.31. Hence $\alpha(s^*)\not\leq\alpha(s)$.

α is surjective: Let $\mathcal{U}$ be an open filter in $\mathcal{O}\mathrm{Filt}$ (S). Since $\mathcal{U}$ is open, for each $V\in\mathcal{U}$ there is a $U = U_V\in\mathcal{U}$ with $U\ll V$ (see I-3.32). Hence, by I-3.32, there is an $x_V\in V$ such that $U\subseteq\uparrow x_V$. In particular, for all $W\in\mathcal{U}$ with $W\subseteq U$ we have $x_V\leq x_W$. Thus $(x_V)_{V\in\mathcal{U}}$ is a directed net (see O-1.2); hence $x = \sup x_U$ exists in the up-complete poset S. Since $x_V\leq x$ and $x_V\in V$ for all $V\in\mathcal{U}$, then $x\in V$ for all $V\in\mathcal{U}$, and thus $x\in\bigcap\mathcal{U}$. Conversely, if $f\in\bigcap\mathcal{U}$, then $x_V\leq f$ since $\bigcap\mathcal{U}\subseteq U_V\subseteq\uparrow x_V$. Thus $x = \sup x_V\leq f$. We have shown that $x = \min\bigcap\mathcal{U}$ exists in S.

It is clear that $\mathcal{U}\subseteq\alpha(x)$. Conversely, let $W\in\alpha(x)$, that is, $x\in W\in\mathcal{O}\mathrm{Filt}$ (S). In particular $W\in\sigma(\mathrm{S})$; hence there is a $V\in\mathcal{U}$ such that $x_V\in W$ by II-1.16(iv). But then $U_V\subseteq\uparrow x_V\subseteq W$, and it follows that $W\in\mathcal{U}$ since $\mathcal{U}$ is a filter, hence an upper set. We have shown $\alpha(x) = \mathfrak{U}$.) □

1.43. EXERCISE. (i) If S is an algebraic poset (I-4.28) (resp., semilattice), then ⟨S⟩ is an algebraic poset (resp., semilattice) with $K(\mathcal{O}\mathrm{Filt}(S)) = \{\uparrow k : k \in K(S)\}$; and this latter poset is isomorphic to the dual of S in the ***AL-S***-duality, if S is a lattice.

(ii) Thus, the categories ***APoset*** and ***ASem*** of algebraic posets and semilattices are self-dual under the Lawson duality. □

1.44. EXERCISE. Let L_1 and L_2 be complete lattices. The poset $\mathbf{SUP}(L_1,L_2{}^{op})^{op}$ is called the *tensor product* of L_1 and L_2 and is written $L_1 \otimes L_2$.

(i) $L_1 \otimes L_2 = \mathbf{SUP}(L_1,L_2{}^{op})^{op} \simeq \mathbf{INF}(L_2{}^{op},L_1) = \mathbf{SUP}(L_2,L_1{}^{op})^{op} = L_2 \otimes L_1$.

(ii) Is the tensor product associative?

(iii) Show that this tensor product classifies the bimorphisms in ***SUP*** in the sense that

$$\mathbf{SUP}(L_1 \otimes L_2, L_3) \simeq \mathbf{SUP}(L_1, \mathbf{SUP}(L_2,L_3)). \ \square$$

1.45. EXERCISE. (i) L_1 and L_2 are complete lattices. For a subset $G \subseteq L_1 \times L_2$ with $(0,1), (1,0) \in G$ the following statements are equivalent:

(1) $G = \{(x_1,x_2) \in L_1 \times L_2 : x_2 \leq f(x_1)\}$ for some $f \in L_1 \otimes L_2$.

(2) If $X \leq G$, then $(\inf pr_1 X, \sup pr_2 X), (\sup pr_1 X, \inf pr_2 X) \in G$.

(3) If $X_1 \times X_2 \subseteq G$, then $\downarrow(\sup X_1, \sup X_2) \subseteq G$.

(4) G is Scott closed in $L_1 \times L_2$, and if $(x_1,x_2), (y_1,y_2) \in G$, then $(x_1 \wedge y_1, x_2 \vee y_2), (x_1 \vee y_1, x_2 \wedge y_2) \in G$.

(ii) In particular, the poset $\mathcal{G}(L_1,L_2)$ of all Scott-closed subsets $G \subseteq L_1 \times L_2$ satisfying (1)–(4) is isomorphic to $L_1 \otimes L_2$.

(iii) The tensor product $L_1 \otimes L_2$ is a continuous lattice iff L_1 and L_2 are continuous lattices.

(HINT: For these and more details we refer to Bandelt [1979]).

NOTES

The general idea of Galois connections is a classical theme in lattice theory almost since its inception, as is exemplified by the paper of O. Ore [1944]. In the Notes for Section 3 of Chapter O we have given some references surveying the background of Theorem 1.3. The results of Theorem 1.4 (and of 1.5, 1.6) are new, but the equivalence of (1) and (3) was first proved by Hofmann and Stralka [1976]. The duality of ***CL*** and $\boldsymbol{CL}^{\mathrm{op}}$ originated from the same source; this result turned out to be rather useful in dealing with homomorphisms between continuous lattices (Hofmann and Mislove [1975], [1976] and [1977], Hofmann, Mislove and Stralka [1973] and [1975], Hofmann and Stralka [1976]).

The duality described in Theorem 1.15 was extensively discussed by Hofmann, Mislove and Stralka [1973], [1974] and [1975], although in the form of a Pontryagin duality for semilattices it was introduced by Austin [1963]. For further references check Hofmann, Mislove and Stralka [1974]. Propositions 1.18, 1.19 and 1.20 come from Hofmann and Stralka [1976]. Except for some embellishments, Theorem 1.26 occurs in a paper of Hofmann and Lawson [1978]. The duality theorem in Excercise 1.42 is due to Lawson [1979a]; its power is not yet fully exploited since the attention to continuous and algebraic posets and semilattices is fairly recent. The material on tensor products in the exercise is due to Bandelt [1979], where one can find further references on the topic of tensor products of continuous lattices.

2. MORPHISMS INTO CHAINS

For lattice-ordered structures L the morphisms L→M into a complete chain M play a role which is analogous to that of characters in the theory of groups. In this line it is of great importance to know when the morphisms into the unit interval I = [0,1] separate the points. After the developments in Chapters I–III we are primarily interested in $INF^{\uparrow}$-morphisms, that is, morphisms which preserve arbitrary infs and directed sups (see I-2.6 ff., II-2.1 ff. and III-1.8).

The duality which we introduced in the preceding section, notably Theorem 1.4, allows us to reduce the question of surjective $INF^{\uparrow}$-morphisms L→C with C a chain to a question on SUP^0-embeddings C→L (see 1.10); these latter ones allow by 1.4(3) the simpler description of being sup-preserving and respecting the relation ≪, since complete chains are always continuous lattices. Therefore, we are led to consider subsets in a complete lattice S which are totally ordered with respect to the way-below relation ≪. In order to maintain a framework of suitable generality we formulate the discussion in terms of an arbitrary auxiliary order ≺ on L (see I-1.9). We soon notice, however, that the strong interpolation property (I-1.15) plays a central role.

2.1. DEFINITION. Let L be a complete lattice. A binary relation on L which satisfies conditions (i), (ii) and (iv) (but not necessarily (iii)) of I-1.9 will be called an *extra order* on L.

As in I-1.11 we call an extra order ≺ *approximating* provided that for all $x \in$L we have $x = \sup s_{\prec}(x)$, where $s_{\prec}$: L→(lower sets of L) is the monotone map $s_{\prec}(x) = \{y \in \mathrm{L} : y \prec x\}$.

A subset $C \subseteq$L is *a strict chain* (or, more accurately, *a ≺-strict chain*, if specification is needed) iff $x,y \in C$ implies $x \prec y$ or $x = y$ or $y \prec x$ in L. □

All singletons are examples of strict chains, and for each $x \in$L the set $\{0,x\}$ is a strict chain by I-1.9(iv). The axiom of choice allows us to find a much greater variety of strict chains, however.

2.2. LEMMA. *Let* L *be a complete lattice and* C_0 *a strict chain. Then the collection* Γ *of all strict chains* C *in* L *with* $C_0 \subseteq C$ *is inductive with respect to* ⊆. *Consequently, every strict chain is contained in a maximal one.*

Proof. It is clear that Γ is inductive, and so the result follows by Zorn's Lemma. □

We now concentrate our attention on maximal strict chains.

2.3. LEMMA. *Let* C *be a maximal strict chain in a complete lattice* L. *For a subset* $X \subseteq C$, *set* $s = \sup_{\mathrm{L}} X$. *Then* $\sup_C X$ *exists, and*

(i) $\sup_C X = \min(\uparrow s \cap C)$;

(ii) *if* $s \in C$, *then* $s = \sup_C X$;
(iii) if $s \notin C$, *then* $s < \sup_C X$.

Proof. We have the two cases to consider: that $s \in C$ and that $s \notin C$. If $s \in C$, then $s = \sup_C X$ and evidently $s = \min(\uparrow s \cap C)$. Now suppose that $s \notin C$. Then for any $c \in C$ with $c < s$ there is some $d \in C$ with $c < d < s$. Since C is strict, we have $c \prec d$, whence $c \prec s$ by I-1.9(ii). Since C is maximal strict, the chain $C \cup \{s\}$ is no longer strict; after what was just said this can only be the case if there is some $c^* \in C$ with $s < c^*$ but not $s \prec c^*$. Suppose for a moment that there were a $d^* \in C$ with $s < d^* < c^*$. Then, since C is strict, we could conclude that $d^* \prec c^*$ and thus $s \prec c^*$ because of I-1.9(ii), which is not the case. But this means that $c^* = \min(\uparrow s \cap C)$. □

2.4. PROPOSITION. *Every maximal strict chain in a complete lattice is a complete chain (in its own right).*

Remark. Observe that we do not and cannot claim that a maximal strict chain is either sup-closed or inf-closed in L.

Proof. By 2.3 every subset of a maximal strict chain C has sup in C. By O-2.2 this suffices. □

In the following study of the structure of maximal strict chains we often refer to the *strong interpolation property* for $\prec$, which we recall for reference from I-1.15:

(SI) $x \prec z$ and $x \neq z$ together imply $(\exists y)(x \prec y \prec z$ and $x \neq y)$.

2.5. PROPOSITION. *Let C be a maximal strict chain in a complete lattice* L. *Then*
(i) $0 \in C$; *and*
(ii) *if* $\max C \prec 1$, *then* $1 \prec 1$ *and* $\max C = 1$.
(iii) *If* $\prec$ *satisfies the strong interpolation property, then we have for* $x, z \in C$

(SI_C) $x \prec z$ *and* $x \neq z$ *together imply* $(\exists y \in C)(x \prec y \prec z$ *and* $x \neq y)$.

Proof. (i): For any $s \in L$ one has $0 \prec s$ by I-1.9(iv); thus, (i) follows from the maximality of C.

(ii): Suppose $\max C \prec 1$. If $\max C < 1$, then $C \cup \{1\}$ would be a larger strict chain, which would contradict the maximality of C. Hence, $\max C = 1$, and thus $1 \prec 1$.

(iii): Let $x \prec z$ with $x \neq z$. If $[x,z]_C$ contains an element $y \neq x, z$, then $x \prec y \prec z$, since C is strict. If, however, $[x,z]_C = \{x,z\}$, then we apply (SI) to find a $y \in S$ with $x \neq y$ and $x \prec y \prec z$; then $C \cup \{y\}$ is a strict chain, and by maximality of C we conclude $y \in C$. □

We say that a strict chain C *satisfies the interpolation property* if it satisfies condition (SI_C).

The trouble with maximal strict chains is that, despite their completeness in their own right, they are not in general sup-closed in L which is what we need for a SUP^0-embedding. If, for example, L is the square I^2, then the chain $([0,1[\times\{0\})\cup\{(1,1)\}$ is maximal strict but not sup-closed. We therefore need a modification procedure. The following lemmas will prove useful in showing that the modified chains we construct have desirable properties.

2.6. LEMMA. *Let* L *be a complete lattice equipped with an extra order* $\prec$. *Let* C *be a strict chain in* L *which satisfies the interpolation property. If* $p,q\in C$ and $p<q$, *then there exists a* $y\in$L *such that* $p<y\prec q$ and $y = \sup_L\{x\in C\colon x\prec y\}$.

Proof. Since C is strict, we have $p\prec q$. By (SI_C) there exists $w\in C$ such that $p\prec w\prec q$ and $p \neq w$. Again by (SI_C) there exists $x_1\in C$ such that $p\prec x_1\prec w$ and $p\neq x_1$. Inductively choose $x_n\in$C such that $x_{n-1}\prec x_n\prec w$. Let $y = \sup_L\{x_n\}$. Then $x_n\prec x_{n+1}\leq y$ implies $\sup_L\{x\in C : x\prec y\}\geq\sup_L\{x_n\} = y$. Finally $p<x_1\leq y\leq w\prec q$, that is $p<y\prec q$. □

We call a chain $C\sqsubseteq$L *sup-closed* if $X\leq C$ implies $\sup_L X = \sup_C X$.

2.7. LEMMA. *For a sup-closed strict chain* C *in a complete latice* L *with an extra order* $\prec$ *the following statements are equivalent:*

(1) *C satisfies the interpolation property* (SI_C).

(2) $c = \sup_L\{x\in C : x\prec c\}$, *for all* $c\in C$.

Proof. (1) implies (2): Let $c\in C$ and let $d = \sup_L\{x\in C : x\prec c\}$. Then $d\in C$ since C is sup-closed. If $d \neq c$, then $d<c$. Since C is strict, we have $d\prec c$. By (SI_C) there exists $y\in C$ such that $d<y\prec c$. This contradicts the definition of d.

(2) implies (1): If $x\prec z$ in C and $x \neq z$ and z not $\prec z$, then from $z = \sup_L\{y\in C : y\prec z\}$ we can find a $y\in C$ such that $x<y\prec z$. Since C is strict, we conclude $x\prec y$. □

The next proposition involves an important construction which allows one to obtain sup-closed chains from given chains.

2.8. PROPOSITION. (CHAIN MODIFICATION LEMMA). *Let* C *be a strict chain in a complete lattice* L *equipped with an extra order* $\prec$. *Define* $D = D(C)$ *by*

$$y\in D \text{ iff } y = \sup_L\{x\in C : x\prec y\}.$$

We then have:

(i) *D is a strict chain which is sup-closed in L.*

(ii) *If C satisfies the interpolation property, then D does also (and hence the equivalent condition* (2) *of* 2.7).

(iii) *If C satisfies the interpolation property, then* $a,b\in C$ *and* $a<b$ *imply that there exists a* $d\in D$ *such that* $a<d\prec b$.

Proof. (i): Let us first show that D is a strict chain. For this it suffices to show that if $a,b\in D$ and $a\not\leq b$, then $b\prec a$. Since $a = \sup_L\{x\in C : x\prec a\}$, there exists $x\in C$ such that $x\prec a$ and $x\not\leq b$. If $y\in C$ and $y\prec b$ then $y\prec x$ (since C is a strict chain and $x \leq y \prec b$ is impossible). As y was arbitrary, we conclude from $b \in D$ that $b \leq x \prec a$. Thus $b \prec a$.

The verification that D is sup-closed in L is routine (the sup of sups is a sup).

(ii): Now suppose C satisfies (SI_C). Let $p = \sup_L\{b \in D : b \prec d\}$, where $d \in D$. We show the assumption $p < d$ leads to a contradiction. If $p < d$, then for some $c \in C$ we have $c \prec d$ and $c \not\leq p$. If $c \prec c$, then $c = \sup_L\{x \in C : x \prec c\}$. Hence $c \in D$, and thus $c \leq p$, which is impossible. If not $c \prec c$, then $c < d$. Since $d \in D$, we have $c < c^* \prec d$ for some $c^* \in C$. By Lemma 2.6 there exists $y \in$ L such that $c < y \prec c^*$ such that $y \in D$. Then $c < y \leq p$, a contradiction. Hence $p = d$.

(iii): The last assertion follows from 2.6. □

For most purposes one can take maximal strict chains and modify them as in 2.8 to obtain strict and sup-closed chains. One may, however, actually obtain ***maximal*** strict chains which are sup-closed by a little additional work. We take a brief detour to present this construction. The next example shows that if a maximal strict chain is modified as in 2.8, it no longer need be maximal strict.

2.9. EXAMPLE. We consider the chain T $= \{0, 1/2, 2/3, \ldots, (n-1)/n, \ldots, 1\}$ and the chain $3 = \{0,1,2\}$ in their natural orders. On the set S $=$ T$\times 3$ we consider the binary relation $\leq$ given by $(t,n) \leq (t^*,n^*)$ iff $(n,n^*) = (0,2)$ or ($t \leq t^*$ and $n \leq n^*$).

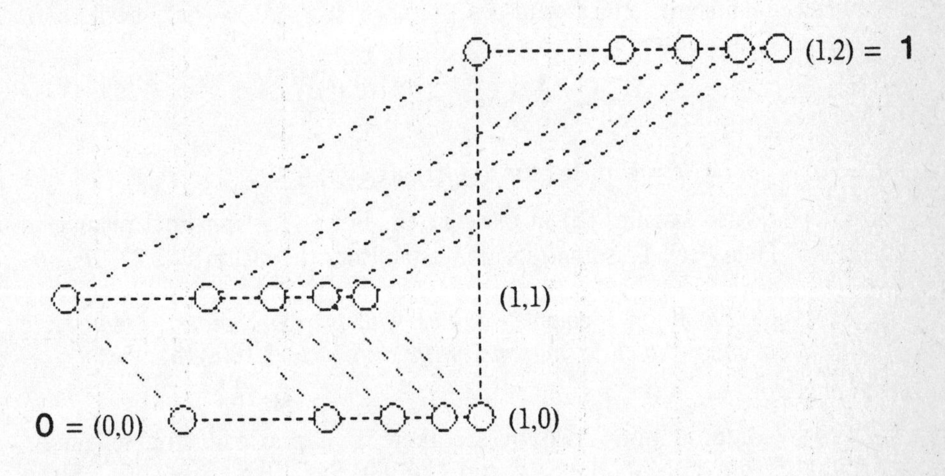

Then it is verified straightforwardly that $\leq$ is the partial order of a continuous lattice. The relations $(1,0) \ll (1,1)$ and $(1,1) \ll (1,2)$ fail, but $(1,0) \ll (1,2)$ holds. Consider the strict chain $C = ((\mathrm{T}\setminus\{1\}) \times \{0\}) \cup \{(1,1)\}$; then C is maximal strict and $D(C) = \mathrm{T} \times \{0\}$, but $D(C) \cup \{(1,2)\}$ is a strict chain; whence, $D(C)$ is not maximal. □

2.10. LEMMA. *Let* L *be a complete lattice and* C_0 *a chain satisfying the following properties:*

(a) C_0 *is strict.*

(b) $c = \sup_L(s_{\prec}(c) \cap C_0)$ *for all* $c \in C_0$.

Then C_0 *is contained in a chain which is maximal with respect to* (a) *and* (b), *and every such chain is sup-closed and satisfies the interpolation property* (SI_C).

Proof. We consider the collection $\mathcal{S}$ of all chains C containing C_0 and satisfying (a) and (b) with C in place of C_0. This collection is inductive with respect to $\subseteq$; indeed let $\{C_j : j \in J\}$ be a tower in $\mathcal{S}$ and C its union. Then $c \in C$ implies $c \in C_j$ for some j, and thus

$$c = \sup_L(s_{\prec}(c) \cap C_j) \leq \sup_L(s_{\prec}(c) \cap C) \leq c;$$

that C is strict was shown in the proof of 2.2. Hence by Zorn's Lemma there are maximal members in $\mathcal{S}$. Let now C be one of them; we claim that C is sup-closed in L. Assume not. Then there is a subset $X \subseteq C$ such that if $t = \sup_L X$, then $t \notin C$. Since C contains 0 (by maximality), $X \neq \emptyset$.

We claim that $C \cup \{t\}$ satisfies (a) and (b). If $c \in C$ and $c < t$, then there is an $x \in X$ with $c < x$. Since C is strict, $c \prec x$, and thus $c \prec t$. If $t < c$, then since $C \cup \{t\}$ is a chain and since $c = \sup_L\{y \in C : y \prec c\}$, we have $t < y \prec c$ for some $y \in C$. Hence $t \prec c$. Therefore $C \cup \{t\}$ is a strict chain. We note that $x \in X$ implies the existence of some $x^* \in X$ with $x < x^*$ since $t \notin X$. As $x \prec x^*$, because of strictness of C, we observe

$$X \subseteq s_{\prec}(t) \cap C \subseteq s_{\prec}(t) \cap (C \cup \{t\}).$$

Hence

$$t = \sup_L X \leq \sup_L(s_{\prec}(t) \cap (C \cup \{t\})) \leq t.$$

Hence, $C \cup \{t\}$ also satisfies (b) in place of C_0. But C was maximal relative to (a) and (b). Thus $t \in C$. This contradiction establishes the claim by 2.7. □

2.11. LEMMA. *Let* L *be a complete lattice with an extra order* $\prec$ *satisfying* (SI) *and* C *a chain which is maximal with respect to* 2.10(a),(b). *Then* C *is maximal strict.*

Proof. Suppose not. Then by 2.2 there is a maximal strict chain M containing C with $y \in M \setminus C$. Let $t = \sup_L(\downarrow y \cap C)$. By 2.10 we have $t \in C$. Since M is a strict chain, $t \neq y$ implies $t \prec y$. By 2.5(iii) M satisfies the interpolation property.

We apply the Chain Modification Lemma (2.8) to M to obtain $D = D(M)$. Since $C \subseteq M$ and C satisfies 2.10(b), we conclude $C \subseteq D$. By 2.8, D is a strict chain satisfying 2.10(b). The maximality of C implies $C = D$. Again by 2.8 there exists $d \in D$ such that $t < d \prec y$. Thus $d \in \downarrow y \cap D = \downarrow y \cap C$, and so $d \leq t$ by definition of t, a contradiction. □

Thus we know that if there are strict chains with 2.10(b) at all, then there are maximal strict chains which are sup-closed. This is where the construction $D(C)$ for maximal strict chains C comes in. In any case, we have the following result:

2.12. THEOREM. *Let* L *be a complete lattice equipped with an extra order* $\prec$ *satisfying* (SI). *If* $x \neq 0$ *in* L, *then there is a maximal* $\prec$*-strict chain* M *which is sup-closed in* L *and contains an element* $m \neq 0$ *with* $m \leq x$.

Proof. Step 1. $C_1 = \{0,x\}$ is a strict chain by I-1.9(iv).

Step 2. C_1 is contained in a maximal strict chain C_2 by 2.2.

Step 3. Apply the Chain Modification Lemma 2.8 to the chain C_2 to obtain $C_3 = D(C_2)$. By 2.8, C_3 is a sup-closed strict chain satisfying the interpolation property and condition 2.10(b) (since C_2 satisfies it by 2.5(iii)).

Step 4. By 2.10, the chain C_3 is contained in a strict chain C_4 which is maximal with respect to (a) and (b) of 2.10, and by 2.10 is sup-closed. By 2.11, the chain C_4 is maximal strict.

We notice that in step 3 we have $0 < d \prec x$ for some $d \in C_3$ by 2.8. Thus the proof is complete if we set $M = C_4$ and $m = d$. □

For some additional information we need sharper hypotheses:

2.13. PROPOSITION. *Assume the hypotheses of Theorem* 2.12 *and suppose that* y *is an arbitrary element of* L *for which there is a* $u \prec x$ *with* $u \not\leq y$. *Then the chain* M *of* 2.12 *can be found so that, in addition,* $m \not\leq y$.

Proof. We only modify step 1 by setting $C_1 = \{0,u,x\}$, and we proceed for the remainder as in the proof of Theorem 2.12. If $u < x$, then again by 2.8 the d chosen in 2.12 can be picked larger than u. If $u = x$, then we have $x \in C_3$ and, hence, we choose $m = x$. □

2.14. COROLLARY. *Let* L *be a complete lattice and* $\prec$ *an extra relation which satisfies the strong interpolation property and which is approximating* (I-1.11). *Then for two elements with* $x \not\leq y$ *there is a sup-closed (maximal) strict chain* M *with an element* m *satisfying* $y \not\geq m \leq x$.

Proof. Immediate from 2.13. □

The principal application of these results concerns the way-below relation.

2.15. LEMMA. *Let* L *be a complete lattice and* C *a sup-closed,* $\ll$*-strict chain satisfying* (SI_C). *Then* $g = (x \mapsto \max(\downarrow x \cap C)) : \mathrm{L} \to C$ *is an* $\boldsymbol{INF}^{\uparrow}$*-morphism from* L *onto the complete chain* C. *If* $m \in C$ *is such that* $y \not\geq m \leq x$, *then* $g(y) < m \leq g(x)$.

Proof. Since C is sup-closed and $\ll$-strict, then the embedding $d : C \to \mathrm{L}$ is a $\boldsymbol{SUP}^0$-map (see 1.4 and 1.9 and note that $\ll_{\mathrm{L}}$ induces $\ll_C$ by (SI_C)).

Hence its upper adjoint g is an $INF^{\uparrow}$-map by 1.10. The remainder follows from the definition of g. □

2.16. THEOREM. *Let* L *is a complete lattice in which the way-below relation* $\ll$ *has the strong interpolation property. If* $0 \neq x \in$ L, *then there is an* $INF^{\uparrow}$-*quotient map* $g : \mathrm{L} \to \mathrm{M}$ *onto a complete chain* M *such that:*

(i) $g(x) \neq 0$;

(ii) *For every factorization* $\mathrm{L} \to^{h} C \to^{f} \mathrm{M}$ *of* g *in* $INF^{\uparrow}$ *with* C *a chain and* h *surjective, the map* f *is an isomorphism.*

Remark. We could say more shortly in place of (ii) that g is a *maximal chain quotient.*

Proof. By Theorem 2.12 we find a sup-closed maximal strict chain M and an $m \in \mathrm{M}$ with $0 \neq m \leq x$. Then $g = (s \mapsto \max(\downarrow s \cap \mathrm{M})) : \mathrm{L} \to \mathrm{M}$ is the desired morphism by 2.15. The remainder follows from the maximality of M via the duality 1.10. □

Theorem 2.16 gives an important characterization of continuous lattices:

2.17. THEOREM. *For a complete lattice* L, *the following conditions are equivalent:*

(1) L *is a continuous lattice;*

(2) *There is an* $INF^{\uparrow}$-*embedding of* L *into a product of complete chains.*

Proof. (2) implies (1): I-1.3(1), I-2.7(i) and (ii).

(1) implies (2): L is continuous iff $\ll$ is approximating (I-1.6). Then, by 2.14 and 2.15, the $INF^{\uparrow}$-quotients from L onto chains separate the points. The usual method for forming an embedding now applies. □

We round this topic off by observing that, for a complete chain, the ***CL***-morphisms into the unit interval $\mathbf{I} = [0,1]$ separate the points.

2.18. PROPOSITION. *Every chain allows an embedding into a cube (that is, a lattice* $\mathbf{I}^{X}$*) which preserves all existing sups and infs.*

Proof. Let C be the given chain. We proceed in steps in order to show that the maps $C \to \mathbf{I}$ preserving arbitrary infs and sups separate the points of C.

Step 1. Every chain allows an embedding into a complete chain which preserves infs and sups. (This is, of course, well known even for lattices; let us briefly indicate a proof for the special case: We let $\mathcal{C} \subseteq \mathrm{Id}\ C$ for a chain C be the set of all ideals J such that J has a maximum only if $c = \max J$ has a successor c^* in C. Then define $j : C \to \mathcal{C}$ by $j(x) = \downarrow x \setminus \{x\}$. Show that $\mathcal{C}$ is closed under arbitrary unions, hence is a complete chain. Clearly j preserves order. Now let $X \subseteq C$ and set $c = \inf X$, $d = \sup X$ (if they exist). If $c \in X$, then $j(c) = \min j(X)$ and if $d \in X$ then $j(d) = \max j(X)$. Thus assume that $c \notin X$, $d \notin X$. Show that $\downarrow X \in \mathcal{C}$ and $\downarrow X = \bigcup\{\downarrow x \setminus \{x\} : x \in X\}$ which will show that $j(d) = \sup j(X)$. Show that $\bigcap j(X) = \downarrow \inf X = \downarrow c \notin \mathcal{C}$ (if $c \notin X$), whence

$\inf j(X) = \downarrow c\backslash\{c\} = j(c)$.)

Step 2. Assume from now on that C is complete and that $x<y$ in C. Define $f : C\to[x,y]_C$ by $f(c) = (c\wedge y)\vee x$. Then f is a retraction preserving arbitrary sups and infs. Thus, it suffices to find a map $g : C\to\mathbf{I}$ preserving sups and infs such that $g(0_C) = 0$ and $g(1_C) = 1$.

Step 3. Suppose that $\mathrm{K}(C) \neq \{\min C\}$. Let $0 \neq k\in\mathrm{K}(C)$ and define $g : C\to[0,1]$ by $g(\uparrow k) = \{1\}$ and $g(C\backslash\uparrow k) = \{0\}$.

Step 4. From now on we may assume that $\mathrm{K}(C) = \{\min C\}$. This is the only part that requires a bit of work. Since $\mathrm{K}(C)$ is degenerate the relation $<$ agrees with $\ll$ with the exception of $0\ll 0$, where $0<0$ fails. By recursion, we find a strict chain $D\subseteq C$ which is order isomorphic to the poset Q of dyadic rationals in $]0,1[$: indeed $<$ satisfies the interpolation property, and thus we define a function $h : Q\to C$ by letting $h(1/2)$ be any point with $0<h(1/2)<1$, by letting $h(1/4)<h(1/2)<h(3/4)<1$, and by proceeding recursively. Let I be the sup-closure of D in X, that is, $I = \{\sup_C X : X\subseteq D\}$. Then I is isomorphic to the unit interval $\mathbf{I}$: the function h extends to a function $h : \mathbf{I}\to C$ through $h(r) = \sup_C(\downarrow r\cap Q)$, which gives an order isomorphism from $\mathbf{I}$ onto I. Now I is a sup-closed strict chain in C; thus the inclusion map $I\to C$ belongs to $\boldsymbol{SUP}^0$, and the left adjoint $g = (c \mapsto \max(\downarrow c\cap I)) : C\to I$, therefore, belongs to $\boldsymbol{INF}^{\uparrow}$ by 1.10. But since the domain of g is a chain, it preserves arbitrary sups, as it preserves directed sups. Clearly, $g(0_C) = 0_I$ and $g(1_C) = 1_I$. Thus g is the desired map. □

Notice that 2.18 says that the *complete* lattice homomorphisms of any chain into $\mathbf{I}$ separate the points. This gives us the following consequence:

2.19. THEOREM. *For a poset* L *the following statements are equivalent:*

(1) L *is a continuous lattice;*

(2) L *is isomorphic to a subset of a cube which is closed under arbitrary infs and directed sups;*

(3) L *is a complete lattice and* $\boldsymbol{INF}^{\uparrow}(\mathrm{L},\mathbf{I})$ *separates the points of* L.

Proof. (1) implies (2): By Theorem 2.17, L allows the $\boldsymbol{INF}^{\uparrow}$-embedding into a product $\mathsf{X}_{j\in J}\, C_j$ of complete chains. By 2.18, every chain, and thus every product of chains allows an $\boldsymbol{INF}^{\uparrow}$-embedding into a cube.

(2) implies (1): This follows directly from 2.17.

(3) iff (2): Clear. □

This theorem—which in the introduction to this book we used as a first, preliminary definition of a continuous lattice—allows us to consider very concrete representations of continuous lattices; but we should recall that the representation of a continuous lattice as a substructure of a cube in itself does not tell us too much about its structure, although for many questions the existence of enough $\boldsymbol{CL}$-morphisms into $\mathbf{I}$ is of vital importance. The elements of $\boldsymbol{CL}(\mathrm{L},\mathbf{I})$ play the role of *characters.*

Let us remark also that theorem 2.16 can be sharpened in one direction.

2.20. PROPOSITION. *Let* L *be a complete lattice in which* $\ll$ *satisfies the strong interpolation property. Suppose that* $x \not\leq y$ *in* L. *Then the following statements are equivalent:*

(1) *There is a map* $f \in \boldsymbol{INF}^{\uparrow}(\mathrm{L}, \mathrm{I})$ *with* $f(y) < f(x)$.

(2) *There is a* $u \in \mathrm{L}$ *with* $u \ll x$ *and* $u \not\leq y$.

Proof. (1) implies (2): Let $d : \mathrm{I} \to \mathrm{L}$ be the lower adjoint of f. Then let $v = (f(x)+f(y))/2$, and $u = d(v)$. Now $v < f(x)$ implies $v \ll f(x)$ by I-1.3(1); whence, $u = d(v) \ll df(x)$ (by 1.4(3)) $\leq x$ (by O-3.6). If we had $u \leq y$, then $v \leq fd(v)$ (by O-3.6) $= f(u) \leq f(y)$, a contradiction. Thus $u \not\leq y$.

(2) implies (1): By 2.14 and 2.15 we have an $\boldsymbol{INF}^{\uparrow}$-morphism onto a complete chain separating y and x, then 2.18 proves (1). □

EXERCISES

2.21. EXERCISE. Let L be a complete lattice, $\prec$ an auxiliary relation. Let $\prec\!\bullet$ be the auxiliary relation satisfying the strong interpolation property which is derived from $\prec$ according to I-1.24. If $x \neq 0$ in L, then there is a maximal $\prec\!\bullet$-strict chain C which is sup-closed in L and contains an element $m \neq 0$ with $m \leq x$. □

PROBLEM. Discuss the properties of C with respect to the auxiliary order $\prec$.

2.22. EXERCISE. Let L be a complete lattice. We recall that for an auxiliary order $\prec$ on L (see I-1.9 ff.) we denote with $\prec\!\bullet$ according to I-1.24 the associated auxiliary order with the strong interpolation property. By III-4.19, 20, the set $\mathrm{Id}_{\prec\bullet}\mathrm{L}$ of $\prec$-Dedekind cuts of L is a continuous lattice.

Next the image L$'$ of the sup-map $(I \mapsto \sup I) : \mathrm{Id}_{\prec\bullet}\mathrm{L} \to \mathrm{L}$ is the image of the Scott-continuous kernel operator $k = (x \mapsto \sup\{y \in \mathrm{L} : y \prec\!\bullet x\}) : \mathrm{L} \to \mathrm{L}$.

Now let $\prec \; = \; \ll$. The sup-map $\mathrm{Id}_{\ll\bullet}\mathrm{L} \to \mathrm{L}$ is injective and L$'$ is a continuous lattice. If $f : \mathrm{L} \to \mathrm{M}$ is any $\boldsymbol{INF}^{\uparrow}$-morphism into a continuous lattice, then $f = (f \mid \mathrm{L}')k$.

(HINT: The first statements are clear. Consider $\prec \; = \; \ll$: Show $\{x : x \ll\!\bullet \sup I\} = I$ for $I \in \mathrm{Id}_{\ll\bullet}\,\mathrm{L}$, which gives injectivity of the sup-map. Then L$'$ is a continuous lattice. If f is given, consider the lower adjoint $d : \mathrm{M} \to \mathrm{L}$ and use 1.4 to show that d factors through L$'$.) □

Note that this Exercise gives an explicit construction of a left reflection of the category $\boldsymbol{INF}^{\uparrow}$ into $\boldsymbol{CL}$ with front adjunction $k : \mathrm{L} \to \mathrm{L}'$.

In the following we discuss some results concerning the lattices of kernel and closure operators. We recall from O-3.13 that the lattice of all closure operators of a complete lattice is isomorphic to the opposite of the lattice of all inf-closed subsets, and from I-2.9 that in the case of a continuous lattice the lattice of all Scott-continuous closure operators is isomorphic to the opposite of the lattice of all subalgebras.

2.23. EXERCISE. For any lattice L let Ker L (resp., Clos L) denote the poset of all kernel (resp., closure) operators of L. If L is complete, we have two functions $C, N : (L \to L) \to (L \to L)$ (cf. I-2.16) given by:

$$C(f)(x) = \inf(\uparrow x \cap \{y : f(y) \leq y\}) \text{ and } N(f)(x) = \sup(\downarrow x \cap \{y : f(y) \geq y\}).$$

Then:

(i) C is a closure operator and N is a kernel operator;

(ii) im C = Clos L and im N = Ker L;

(iii) $C(f) = \min\{c \in \text{Clos L} : f \leq c\}$ and $N(f) = \max\{k \in \text{Ker L} : k \leq f\}$;

(iv) Clos L is inf-closed and Ker L is sup-closed (O-3.12);

(v) Clos L is closed under directed sups and Ker L under filtered infs;

(vi) If L is a continuous lattice, then Clos L is a continuous lattice. □

2.24. EXERCISE. For a complete lattice L, let ker L$\subseteq$Ker L be the poset of Scott-continuous kernel operators and clos L$\subseteq$Clos L the poset of Scott-continuous closure operators. Let L be a continuous lattice. Then

(i) clos L is a continuous lattice.
(ii) The following statements are equivalent:

  (1) ker L is a continuous lattice;

  (2) ker L is an algebraic lattice;

  (3) ker L is an algebraic lattice all of whose ***CL***-quotients are algebraic;

  (4) L is an algebraic lattice all of whose ***CL***-quotients are algebraic;

  (5) L is an algebraic lattice and K(L) contains no order dense chains.

(HINT: (i): We know that ΛL is a compact semilattice with small semilattices (III-2.13). If X and Y are subalgebras (that is, closed subsemilattices: see III-1.11) then so is XY, and if S is the semilattice of all subalgebras with respect to $(X, Y) \mapsto XY$, then the partial order of this semilattice is reverse containment, and thus S $\cong$ clos L by I-2.9. It suffices therefore to show that S is a compact topological semilattice with small semilattices, and that the topology induced on S from the standard topology on the set of compact subsets of the compact space ΛL agrees with the Lawson topology. (See VI-3.4)

(ii): For a proof we refer to Hofmann and Mislove [1977].) □

2.25. EXERCISE. Let L be a complete lattice and cong$^-$ L the lattice of all congruences on L which are subalgebras of L in the sense of I-2.6, 2.8, 2.11. Then (ker L)$^{\text{op}} \cong$ cong$^-$ L.

(HINT: See I-2.11.) □

As a consequence of 2.25, the results of 2.24(ii) provide information on the lattice of closed congruences on a continuous lattice.

2.26. EXERCISE. Let L be a continuous lattice. Then the poset ker L of Scott-continuous kernel operators (see 2.24, 2.25) is a complete lattice on which the way-below relation has the strong interpolation property.

(HINT: We refer to Hofmann and Mislove [1977].) □

2.27. EXERCISE. (i) Let L be a continuous lattice. If R is a ***CL***-congruence of L with more than one class, then there is a surjective map f : cong$^-$ L→C from the lattice of closed congruences on a complete chain such that $f(R)$ = min C<max C = f(L×L) and that f preserves arbitrary sups and infs of filtered sets.

(ii) There is a morphism cong$^-$ L→I^X preserving arbitrary sups and filtered infs into a cube such that the only element mapped to the top is the congruence maximal L×L.

(HINT: Apply 2.26 and Proposition 2.20.) □

2.28. EXERCISE. Use Theorem 2.19 to give a new proof of the fact that every continuous lattice carries a compact Hausdorff topology such that the inf operation is jointly continuous and that every point has a basis of open semilattice neighborhoods.

2.29. EXERCISE. Let L be a complete lattice and define $x \lhd y$ iff whenever $y \leq \sup X$ then $x \leq x^*$ for some $x^* \in X$. [Note that for $\mathcal{M} = 2^X$ in I-2.22 we have $\lhd = \dashv$.] Then:

(i) $\lhd$ is an extra order which is not an auxiliary order;
(ii) $x \lhd y$ implies $x \ll y$;
(iii) The relation $\lhd$ is approximating if L is completely distributive;
(iv) Under these circumstances, $\lhd$ satisfies the interpolation property. □

If $\lhd$ is approximating, then the developments of this section apply and show the existence of maximal $\lhd$-strict chains according to Theorem 2.12 and Corollary 2.14. Lemma 2.15 applies with $\lhd$ in place of $\ll$ and yields a morphism g : L→C in ***INF∩SUP***. This allows us to conclude a parallel theorem to 2.19:

2.30. EXERCISE. For a poset L the following statements are equivalent:

(1) L is a completely distributive lattice;

(2) L is isomorphic to a subset of some cube (that is, a lattice I^X) which is closed under *arbitrary* infs and *arbitrary* sups;

(3) L is a complete lattice and (***INF∩SUP***)(L,I) separates the points of L.

(HINT: This Theorem is due to Raney, [1952] and was reported in Chapter I following I-3.15.) □

2.31. EXERCISE. A completely distributive lattice is hypercontinuous.

(HINT: Use 2.30(2) above and III-3.23(4).) □

NOTES

The general idea of constructing morphisms into chains by using (maximal) complete strict chains relative to suitable auxiliary relations dates back some twenty years; forerunners are to be found in Raney's classical paper on completely distributive lattices [1952], and the closest to what we do here is Bruns' treatment of this technique [1961]. However, none of these papers exactly applies to the situation of continuous lattices which we cover here. In this context a first indication was given by Hofmann and Stralka [1976], but the argument was found to contain a gap [SCS-4] which was patched by Scott; a fairly complete elaboration of the techniques presented here were given in Hofmann [SCS-5]; supplements were provided by Carruth [SCS-6,7]. Some results are new including Example 2.8 and the construction of maximal strict chains which are sup-closed (Theorem 2.12).

Theorem 2.19 is a core result of the entire theory of continuous lattices. In a certain form, this theorem is due to Lawson, who showed in [1969] that a compact topological semilattice has enough continuous semilattice homomorphisms into the interval to separate the points if and only if it has small semilattices. Theorem III-1.8, Theorem III-2.13, and Theorem VI-3.4 below show that Lawson's result is equivalent to 2.19. The construction of the left reflection $\mathbf{INF}^{\uparrow} \to \mathbf{CL}$ in 2.22 was first noted in Gierz, Hofmann, Keimel and Mislove [SCS-12]. The results in 2.24 are from Hofmann and Mislove [1977]; in the equivalence of (4) and (5) in 2.24(ii) was proved in Hofmann, Mislove and Stralka, [1973]. The result in 2.27 is new. Exercise 2.29 retrieves the studies of Raney, [1952], Bruns, [1961], Papert, [1959].

We remark that the methods to construct maximal strict chains require the interpolation property in its strong form (SI) (see 2.5), and this is the only place where the interpolation property (INT)(see I-1.15) is not sufficient.

3. PROJECTIVE LIMITS AND FUNCTORS WHICH PRESERVE THEM

One of the original motivations for considering continuous lattices has much to do with the construction of continuous lattices L which are naturally isomorphic to their own function spaces [L→L] (see II-2.5); indeed, such lattices provide set-theoretical models for the λ-calculus of Church and Curry. In the next section we discuss the general principle underlying the construction of these lattices; in the present section, however, we prepare for this discussion by providing the main ingredients.

First we present a thorough investigation of projective limits in the category of complete lattices and maps preserving arbitrary infs and directed sups, and then we take a close look at functors which preserve projective limits (but not necessarily arbitrary limits) in this category and its relevant subcategories. It turns out that projective limits in the categories at hand have many features which are not at all apparent from purely categorical considerations. These special characteristics are responsible for the existence of several relevant functors which do preserve projective limits—while they do not even preserve products in general, let alone arbitrary limits. Our task is to describe manageable criteria which allow us to test concrete functors for the preservation properties vis-à-vis projective limits.

In a review of projective limits it is just as easy and efficient to recall the concept of a limit in general; in fact, the notation is in many ways simpler if we adopt the category-theoretical conventions from the start. In what follows a *small category* is one whose class of morphisms (and objects) is a ***set***. In general, we regard categories like ***CL*** as forming a ***proper class***, but—as is usually the convention—the hom-sets ***CL***(S,T) are sets and not proper classes.

3.1. DEFINITION. Let $\boldsymbol{C}$ be an arbitrary category. A *diagram* in $\boldsymbol{C}$ is simply a functor $D : \boldsymbol{I} \to \boldsymbol{C}$ from a small category $\boldsymbol{I}$.

For each object L of $\boldsymbol{C}$ we can associate a functor $|L| : \boldsymbol{I} \to \boldsymbol{C}$ which takes any object i of $\boldsymbol{I}$ to the fixed object L of $\boldsymbol{C}$ and any arrow $i \to j$ to the identity map of L. This is merely a device to introduce in simple terms the concept of a *cone over a diagram* $D : \boldsymbol{I} \to \boldsymbol{C}$ *with vertex* L. Such a cone is by definition a natural transformation $g : |L| \to D$. Explicitly, this means that for every object i of $\boldsymbol{I}$ we have a $\boldsymbol{C}$-morphism $g_i : L \to D(i)$ such that for any arrow $a : i \to j$ in $\boldsymbol{I}$ the diagram

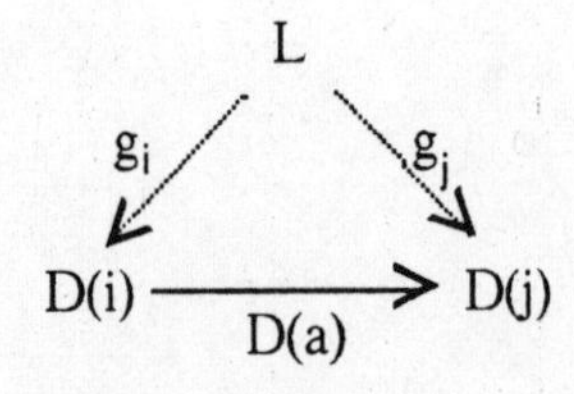

commutes.

A cone $g : |\mathrm{L}| \to D$ over a diagram D is called a *limit cone* provided it has the following universal property: whenever a cone $h : |\mathrm{H}| \to D$ over the same diagram D is given, then there is a unique C-morphism $h^* : \mathrm{H} \to \mathrm{L}$ such that $g|h^*| = h$, where $|h^*| : |\mathrm{H}| \to |\mathrm{L}|$ is the constant natural transformation with $|h^*|_i = h^* : \mathrm{H} \to \mathrm{L}$ for all objects i of $\boldsymbol{I}$. In explicit terms, this means that for each object i of $\boldsymbol{I}$ the diagram

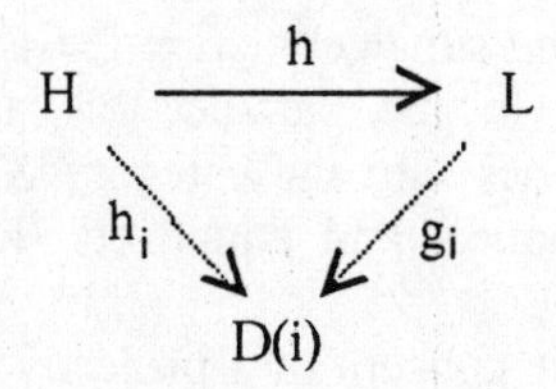

commutes.

The vertex L of a limit cone of a diagram D is called the *limit* of D and is denoted $\lim D$, and the natural transformation $g : |\lim D| \to D$ is called the *limit natural transformation.* In the same vein, the maps $g_i : \lim D \to D(i)$ are called the *limit maps.* □

Limits—if they exist—are unique up to an isomorphism. Our interest here is in a special kind of limit, called *projective limit,* but before we make its definition precise we record some special limits to exemplify the concept. Suppose that I is an arbitrary ***set***. Then we may consider I as a category, whose objects are the elements of I and whose only maps are the identity maps of these objects. (These morphisms have to be there by the definition of a category, and we allow no others. Such categories are called *discrete categories.*) A diagram $D : I \to C$ in a category C is then nothing but a family of objects $\{D(i) : i \in I\}$—indeed each family of objects indexed by a set can be described as a diagram in this fashion. If this diagram has a limit, it is called the *product* of the family, and is written $\Pi_{i \in I} D(i)$. The limit maps $\pi_j : \Pi_{i \in I} D(i) \to D(j)$ are called *projections.*

Another simple type of limit is the *equalizer* of a pair of maps $f, g : \mathrm{L} \to \mathrm{M}$ in C. The equalizer is an object E in C together with a unique map $e : \mathrm{E} \to \mathrm{L}$ such that $fe = ge$ and that for any morphism $h : \mathrm{H} \to \mathrm{L}$ with $fh = gh$ there is a unique map $h^* : \mathrm{H} \to \mathrm{E}$ with $h = eh^*$. It should be clear how this fits into the general scheme of limits: let $\boldsymbol{I}$ be the category

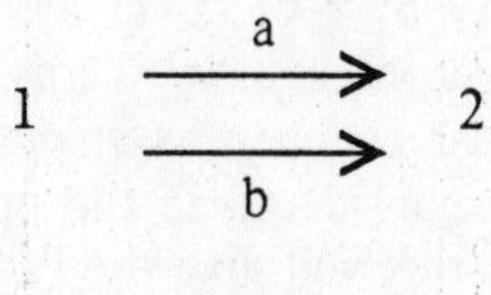

$$1 \overset{a}{\underset{b}{\rightrightarrows}} 2$$

(plus the identity maps of 1 and 2), and let $D : \boldsymbol{I} \to C$ be the diagram given by $D(1) = \mathrm{L}$, $D(2) = \mathrm{M}$, $D(a) = f$ and $D(b) = g$. Then $\lim D = \mathrm{E}$, and the limit maps are $g_1 = e$, $g_2 = fe = ge$.

Not only are products and equalizers good examples of limits, they suffice for the construction of arbitrary limits as Freyd has shown (see, e.g., Mac Lane [1971], p. 109). Indeed, it follows that in a category every diagram has a limit if every family has a product and every pair of maps with the same domain and codomain has an equalizer. Such categories are called *complete*.

Most of our categories such as ***INF***, ***SUP***, ***CL***, ***AL*** are complete, since the cartesian products are products in the sense of limits and the equalizer of two maps $f,g : L \to M$ is just the subalgebra $E = \{x \in L : f(x) = g(x)\}$ with $e : E \to L$ the inclusion morphism. (In the parlance of category theory, the forgetful functor from these categories into the category ***SET*** of all sets and functions preserves and creates products and equalizers (hence limits). See, e.g., Mac Lane [1971], p. 108).

Now we indroduce the concept of a projective limit:

3.2. DEFINITION. Let I be a partially ordered set. We may consider I as a small category in the following fashion: the elements of I are the objects, and for two objects i and j there is one and only one arrow $i \to j$ whenever $i \leq j$ (see remarks following O-3.1).

A *projective system* in a category C is a diagram $D : I^{op} \to C$ whose domain is a poset I which is, in addition, directed. The limit, lim D, of a projective system is called a *projective limit*, and the limit cone over a projective system is called a *projective limit cone*. □

Let us take stock of what a projective system is in terms of objects and maps. It is a family of objects $D(j)$ indexed by the elements j of a directed set I, and a system of maps $g_{ij} : D(j) \to D(i)$ for every pair i,j of elements in I with $i \leq j$ such that the following relations are satisfied for all $i \leq j \leq k$ in I:

(i) $g_{ii} = 1_{D(i)}$;

(ii) $g_{ij} g_{jk} = g_{ik}$.

If the system has a limit, lim D, then there are limit maps $g_i : \lim D \to D_i$ such that, for all $j \leq k$ in I,

(iii) $g_{jk} g_k = g_j$.

The functorial definition of a projective system automatically takes care of all these conditions.

Just as in the theory of ordered sets, most of the elementary concepts in category theory have a dual (see O-1.7) or an opposite. The opposite category C^{op} is obtained from a category by reversal of arrows; how this is done formally is explained in any source on category theory, e.g., in Mac Lane [1971], p. 33. The introduction of dual concepts is then simple: if any concept is generally introduced, it can be considered in the opposite category and interpreted in the original category; this will give the "co-concept". ***Example***: a *co-cone under a diagram* $D : I \to C$ in the category C is a natural transformation $g : D \to |L|$, i.e., a system of maps $g_i : D_i \to L$ with the commuting relations dual to those of the cone. The dual of a limit cone is the *colimit cone*; its *co-vertex* is called a *colimit* of the system.

Sometimes, for reasons of historical priority, variations to this nomenclature exist. Thus, a co-projective system is called a *direct system*, a co-projective limit is called a *direct limit*. The direct limit of a direct system D will never-theless be written colim D. The reader who is not already familiar with the practice of rote dualization should give the explicit definitions of these dual concepts as an exercise.

The formal dualization by reversal of arrows is a convenient routine—but it may not always have much concrete significance in a given category. However, we saw in the discussions of the first section of this chapter that the category ***INF*** of complete lattices with inf-preserving maps was equivalent to the opposite category $\boldsymbol{SUP}^{\mathrm{op}}$ of the category ***SUP*** of all complete lattices with sup-preserving maps; thus, $\boldsymbol{SUP}^{\mathrm{op}}$ in this case has a concrete meaning. The functors implementing the equivalence were given by the Galois connection (cf. 1.3). For the purposes of our present discussion it will be useful to have some explicit notation.

3.3. NOTATION. If $g : \mathrm{S}\to\mathrm{T}$ is a map in ***INF*** we write $\hat{g}$ in place of $D(g)$ (see remarks following 1.1). Thus $\hat{}\ : \boldsymbol{INF}\to\boldsymbol{SUP}^{\mathrm{op}}$ is an isomorphism of categories. The subcategory $\boldsymbol{INF}^{\uparrow}$ is mapped isomorphically onto $(\boldsymbol{SUP}^{0})^{\mathrm{op}}$, the full subcategory ***CL*** of $\boldsymbol{INF}^{\uparrow}$ is mapped isomorphically onto $\boldsymbol{CL}^{\mathrm{op}}$ and the subcategory ***AL*** onto $\boldsymbol{AL}^{\mathrm{op}}$, respectively. □

It is fairly clear that for any projective system $D : J^{\mathrm{op}}\to\boldsymbol{INF}^{\uparrow}$ the limit $\lim D$ serves at the same time as colimit of the direct system $\hat{D}: J\to\boldsymbol{SUP}^{0}$ given by $\hat{D}(i\to j) = (g_{ij})\hat{}$. This is simply a consequence of duality. The limit maps $g_j : \lim D\to D(j)$ dualize to colimit maps $\hat{g}_j : \hat{D}(j)\to\operatorname{colim}\hat{D} = \lim D$, where we recall $\hat{D}(j) = D(j)$. However, a sufficiently careful analysis of the colimit maps $\hat{g}_j$ will reveal more than one would expect from arrow-theoretical generalities. We now undertake such an analysis.

3.4. PROPOSITION. *Let $D : J^{\mathrm{op}}\to\boldsymbol{INF}^{\uparrow}$ be a projective system in $\boldsymbol{INF}^{\uparrow}$, and let $g_j : \lim D\to \mathrm{L}_j$ denote its limit cone, where $\mathrm{L}_j = D(j)$. Then the colimit cone $\hat{g}_j : \mathrm{L}_j\to\lim D$ is determined for all $i,j\in J$ by the formula:*

(1) $g_j\hat{g}_i = \sup\{g_{jk}\hat{g}_{ik} : i,j\leq k \text{ in } J\}$; *moreover we have*

(2) $\sup \hat{g}_j g_j = 1_{\lim D}$.

Proof. Fix i in J and denote for any j in J the cofinal subset $\{k\in J: i,j\leq k\}$ by J^{ij}. Then we have an ***UPS***-morphism $g_{jk}\hat{g}_{ik} : \mathrm{L}_i\to\mathrm{L}_j$ for $k\in J^{ij}$. We claim that $(g_{jk}\hat{g}_{ik})_{k\in J^{ij}}$ is a monotone net in $[\mathrm{L}_i\to\mathrm{L}_j]$ (cf. O-1.2). Indeed let $k\leq k^*$ in J^{ij}, then $g_{jk^*}\hat{g}_{ik^*} = (g_{jk}g_{kk^*})(\hat{g}_{kk^*}\hat{g}_{ik})$ (see 3.2(ii) above!) $\geq g_{jk}\hat{g}_{ik}$, since $g_{kk^*}\hat{g}_{kk^*}\geq 1$ by O-3.6. In the complete lattice $[\mathrm{L}_i\to\mathrm{L}_j]$ the directed supremum $\sup\{g_{jk}\hat{g}_{ik} : k\in J^{ij}\}$ exists, and we call it $f_j : L_i \to L_j$.

Suppose $j \leq j'$. We claim $f_j = g_{jj'} f_{j'}$. For a proof calculate as follows:

$$\begin{aligned} g_{jj'} f_{j'}(x) &= g_{jj'}(\sup\{g_{j'k}\hat{g}_{ik}(x) : k\in J^{ij'}\}) \\ &= \sup\{g_{jj'}g_{j'k}\hat{g}_{ik}(x) : k\in J^{ij'}\} \\ &= \sup\{g_{jj}\hat{g}_{ik}(x) : k\in J^{ij'}\} \\ &= f_j(x). \end{aligned}$$

(The second equation follows because $g_{jj'}$ is Scott continuous and the net $(g_{j'k}\hat{g}_{ik}(x))_{k\in J^{ij'}}$ is directed). Remark then that the infinite diagram

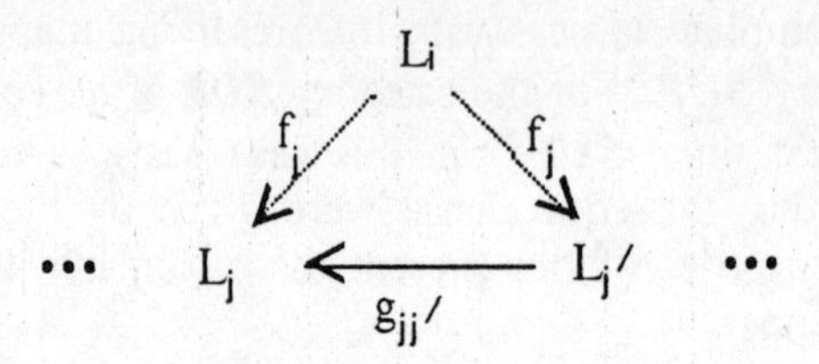

is a cone in ***UPS*** over the projective system $D : J^{op} \to$ ***INF***$^\uparrow$.

Now $g_k : \lim D \to L_k$, where $k\in J$, is a limit cone in ***INF***$^\uparrow$. The natural inclusion functor ***INF***$^\uparrow \to$ ***UPS*** preserves products and equalizers, because in both categories the products are the cartesian products and equalizers are created by ***SET***-equalizers for any pair of ***INF***-maps. Thus, the inclusion functor preserves projective limits; in consequence we know that the system of maps $g_k : \lim D \to L_k$ is a limit cone also in ***UPS***. Then, by the universal property of the limit in ***UPS*** (see 3.1 above), there are unique ***UPS***-mappings $g_i' : L_i \to \lim D$ with $f_j = g_j g_i'$ for all $i\in J$.

As the last step we claim $g_i' = \hat{g}_i$; once this claim is established we are done since then $g_j\hat{g}_i = g_j g_i' = f_j = \sup\{g_{jk}\hat{g}_{ik} : k\in J^{ij}\}$ by definition of f_j.

In order to prove the claim we calculate

$$g_i g_i' = f_i = \sup\{g_{ik}\hat{g}_{ik} : k\in J^{ii}\} \geq 1_{L_i},$$

since $g_{ik}\hat{g}_{ik} \geq 1_{L_i}$ by O-3.6. On the other hand, for all $j\in J$ we have

$$\begin{aligned} g_j g_i' g_i(x) &= f_j g_i(x) \\ &= \sup\{g_{jk}\hat{g}_{ik}g_i(x) : k\in J^{ij}\} \\ &= \sup\{g_{jk}\hat{g}_{ik}g_{ik}g_k(x) : k\in J^{ij}\} \\ &\leq \sup\{g_{jk}g_k(x) : k\in J^{ij}\} \\ &= \sup\{g_j(x) : k\in J^{ij}\} \\ &= g_j(x), \end{aligned}$$

where the inequality in the middle follows because $\hat{g}_{ik}g_{ik} \leq 1_{L_k}$ by O-3.6. Thus $g_j g_i' g_i(x) \leq g_j(x)$ for all $j\in J$, and since the limit maps g_j separate the points of $\lim D$ we have $g_i' g_i \leq 1_{\lim D}$. The two relations $g_i g_i' \geq 1$ and $g_i' g_i \leq 1$ together show $g_i' = \hat{g}_i$ by O-3.6.

Finally, we prove equation (2). We fix an arbitrary $j \in J$ and calculate $g_j(\sup_i \hat{g}_i g_i)$. Now an argument similar to that in the beginning of this proof shows that $(\hat{g}_i g_j)_{i \in J}$ is directed, and so we may asume that $j \leq i$ for each i we consider. Then, $g_j \in \boldsymbol{UPS}$ implies that

$$
\begin{aligned}
g_j(\sup_i \hat{g}_i g_j) &= \sup_i g_j \hat{g}_i g_i \\
&= \sup_i \sup_{k \in J^{ij}} g_{jk} \hat{g}_{ik} g_i \\
&= \sup \{ g_{jk} \hat{g}_{ik} g_i : i,j \in J,\ i,j \leq k \} \\
&= \sup \{ g_{ji} g_{ik} \hat{g}_{ik} g_{ik} g_k : i,j \in J,\ i,j \leq k \} \text{ (since } j \leq i) \\
&= \sup \{ g_{ji} \hat{g}_{ik} g_k : i,j \in J,\ i,j \leq k \} \\
&= g_j,
\end{aligned}
$$

where the next to last equation follows because $g_{ik} \hat{g}_{ik} g_{ik} = \hat{g}_{ik}$, by O-3.6.The limit maps g_j separate the points of lim D, and so we find that $\sup_i \hat{g}_i g_i = 1_{\lim D}$ as was asserted. □

Some comments are in order. In the first place, the immediate data given in the projective system D are the g_{ij} and then also, by duality, the $\hat{g}_{ij}$. Hence, the right hand side of relation (1) is expressed in terms of the given data. The limit maps g_i are available once the limit is calculated, and they separate the points of lim D. Hence for any given $i \in J$, if the left hand side of (1) is known for all $j \in J$, then $\hat{g}_i$ is known. Therefore, (1) gives an explicit way to calculate the colimit maps $\hat{g}_i$ (cf. O-3.7(3)). We know from O-3.6(2) that $\hat{g}_j g_j \leq 1$; equation (2) tells us that "in the limit" equality holds irrespective of the surjectivity of g_j.

Secondly, the dual maps $\hat{g}_{ij}$ and $\hat{g}_i$ exist whenever a projective system in the bigger category ***INF*** is given. However, in the proof we needed to know that all g_{ij} preserved directed sups. Moreover, in order that compositions such as $g_{jk} \hat{g}_{ik}$ or $g_j \hat{g}_i$ are meaningful at all in any of the categories which are of interest to us, the functions g_{jk} and g_j had better be in ***UPS*** (since then the compositions in question are still in ***UPS***). It is then clear why—in the calculations of 3.4 in particular—we have already left the purely arrow-theoretical domain.

We now investigate the properties of the colimit maps $\hat{g}_i : L_i \to \lim D$ further.

3.5. PROPOSITION. Suppose the same data as in 3.4 are given. If

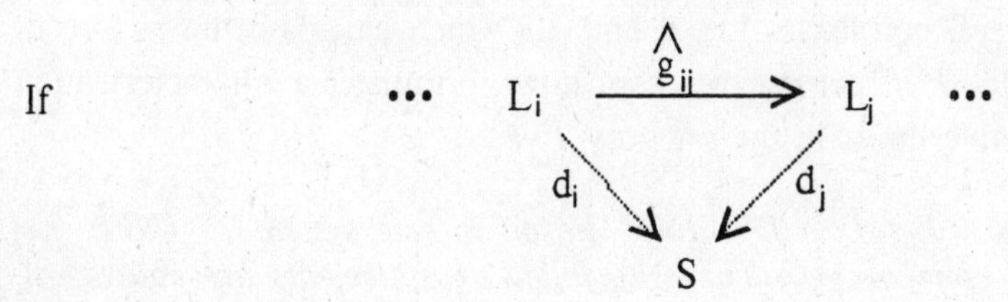

is any co-cone in ***UPS*** under the direct system D, then there is a unique function $d : \lim D \to S$ in ***UPS*** such that $d_i = d\hat{g}_i$ for all $i \in J$. Moreover, the function d is given by the formula:

$$(3)\quad d(x) = \sup\{d_j g_j(x) : j \in J\}.$$

Proof. Define d by equation (3). We first note that d is in ***UPS***, because all d_j and g_j are in ***UPS*** and $[L \to S]$ is closed under sups.

Now let $i \in J$ and $x \in L_i$. Then we calculate:

$$\begin{aligned} d\hat{g}_i(x) &= \sup\{d_j g_j \hat{g}_i(x) : j \in J\} \text{ (by (3))} \\ &= \sup_j \{d_j \sup\{g_{jk}\hat{g}_{ik}(x) : k \in J^{ij}\}\} \text{ (by (1) in 3.4)} \\ &= \sup\{d_j g_{jk}\hat{g}_{ik}(x) : j \in J,\ k \in J^{ij}\} \text{ (since } d_j \in \boldsymbol{UPS}). \end{aligned}$$

But $j \leq k$ implies $d_j = d_k\hat{g}_{jk}$, and so $d_j g_{jk} = d_k\hat{g}_{jk}g_{jk} \leq d_k$, since $\hat{g}_{jk}g_{jk} \leq 1$ by O-3.6. Therefore, $d_j g_{jk}\hat{g}_{ik} \leq d_k\hat{g}_{ik} = d_i$, whence $d\hat{g}_i(x) \leq d_i(x)$.

In order to show the other inequality, we first observe that for a fixed i and any $i \leq k$ we always have $d_i = d_k\hat{g}_{ik}$; thus we can write

$$d_i(x) = \sup\{d_k g_{kk}\hat{g}_{ik}(x) : k \in J \text{ with } i \leq k\}.$$

If we form the sup over the larger index set $\{(j,k) \in J \times J : i \leq k \text{ and } j \leq k\}$ we possibly enlarge the sup; whence

$$\begin{aligned} d_i(x) &\leq \sup\{d_j g_{jk}\hat{g}_{ik}(x) : j,k \in J : i,j \leq k\} \\ &= \sup_j (d_j \sup_{j,i \leq k} g_{jk}\hat{g}_{ik}(x)) \\ &= \sup_j d_j g_j \hat{g}_i(x) \qquad \text{(by 3.4(1))} \\ &= d\hat{g}_i(x) \qquad \text{(by (3))}. \end{aligned}$$

This shows that $d_i = d\hat{g}_i$ for all $i \in J$.

In order to show the uniqueness of d, we consider a second ***UPS***-mapping $d^* : \lim D \to S$ with $d\hat{g}_i = d^*\hat{g}_i$ for all i. Then also $d\hat{g}_i g_i = d^*\hat{g}_i g_i$ for all $i \in J$, and so $d = d \sup_i \hat{g}_i g_i$ (by (2) of 3.4) $= \sup_i d\hat{g}_i g_i$ (since $d \in$ ***UPS***) $= \sup_i d^*\hat{g}_i g_i = \ldots = d^*$ (for the same reasons). Hence, d is unique, and this completes the proof. □

We note that the equation (2) of 3.4 is a special case of (3) above which obtains for $d_j = \hat{g}_j$ and $d = 1_L$. However, relation (2) was needed in the proof of Proposition 3.5 in order to show uniqueness of d. Thus it is not possible to derive (2) as a special case from (3).

There are several corollaries to 3.4 and 3.5 which elucidate the nature of projective limits in $\boldsymbol{INF}^{\uparrow}$. First we can now formulate a characterization theorem for projective limits in the category $\boldsymbol{INF}^{\uparrow}$:

3.6. PROPOSITION. *Let $D : J^{op} \to \boldsymbol{INF}^{\uparrow}$ be an inverse system in $\boldsymbol{INF}^{\uparrow}$. Let $g_j : L \to D(j)$ be a cone over D. Then the following statements are equivalent:*

(1) $g_j : L \to D(j)$ *is a limit cone over* D *in* $\boldsymbol{INF}^{\uparrow}$;

(2) $\hat{g}_j : D(j) \to L$ *is a colimit cone under* $\hat{D}$ *in* $\boldsymbol{UPS}$.

Proof. (2) implies (1): Since all maps $g_{ik} = D(i \to k)$ and g_1 are in $\boldsymbol{INF}^{\uparrow}$, then all maps $\hat{g}_{ik}$ and $\hat{g}_i$ are in $\boldsymbol{SUP}^0$ by 1.10. Then L is, in particular, a colimit of $\hat{D}$ in $\boldsymbol{SUP}^0$. But then (1) follows by simple dualizing.

(1) implies (2): This is the content of 3.5. □

We observe that the universal property expressed in condition (2) holds in the category $\boldsymbol{UPS}$ which contains substantially more morphisms than the category $\boldsymbol{SUP}^0$ which is the precise dual of $\boldsymbol{INF}^{\uparrow}$.

Now let us suppose that we have a projective system $D : J^{op} \to \boldsymbol{INF}^{\uparrow}$, and that we are given a morphism $g : L \to \lim D$ in $\boldsymbol{INF}^{\uparrow}$. We write $g_i' = g_i g$, where $g_i : \lim D \to L_i$ is the limit map. With the information we have, it is now possible to characterize precisely the circumstances under which g is injective, respectively, surjective in terms of the functions g_i' alone. In particular, this gives criteria for g to be an isomorphism. This will become important when we explore when a functor preserves projective limits.

3.7. PROPOSITION. *The following statements are equivalent:*

(1) g *is injective*;

(2) $\hat{g}g = 1_L$;

(3) $\sup \hat{g}_i' g_i' = 1_L$.

3.8. PROPOSITION. *The following statements are equivalent:*

(1) g *is surjective*;

(2) $\operatorname{im} g_j \subseteq \operatorname{im} g_j'$ *for all* $j \in J$;

(3) $g_j' \hat{g}_j' = g_j \hat{g}_j$ *for all* $j \in J$.

Remark. Note that (3) is equivalent to the following by 3.4(1):

(3′) $g_j' \hat{g}_j' = \sup \{ g_{jk} \hat{g}_{jk} : j \leq k \}$.

Proof of Proposition 3.7. (1) iff (2) by O-3.7.

(2) implies (3): $\sup \hat{g}_j' g_j' = \sup \hat{g}\hat{g}_j g_j' = \hat{g}(\sup \hat{g}_j g_j')$ (since $\hat{g} \in \boldsymbol{UPS}$) $= \hat{g}(\sup \hat{g}_j g_j g) = \hat{g}(\sup \hat{g}_j g_j) g$ (since sup is calculated pointwise) $= \hat{g}g$ (by 3.4(2)) $= 1_L$ by (2).

(3) implies (2): $\hat{g}g = (\sup \hat{g}_j' g_j) g$ (by 3.5(3) with $d_j = \hat{g}_j'$ and $d = \hat{g}$) $= \sup \hat{g}_j' g_j g = \sup \hat{g}_j' g_j' = 1_L$ (by (3)). □

Proof of Proposition 3.8. (1) implies (2): $\operatorname{im} g_j' = g_j g(L) = \operatorname{im} g_j$ if g is surjective.

(2) implies (3): Suppose that $y = g_j\hat{g}_j(x)$ with $x,y \in L_j$. Then y is in im g_j, and thus in im g_j' by (2). Thus $y = g_j'(z)$ for some $z \in L$, and so $y = g_j'(z) = g_j'\hat{g}_j'g_j'(z)$ (by O-3.6(3)) $= g_j'\hat{g}_j'(y) = g_j'\hat{g}_j'g_j\hat{g}_j(x) = g_j'\hat{g}\hat{g}_jg_j\hat{g}_j(x)$ (since $g_j' = g_jg$) $= g_j'\hat{g}\hat{g}_j(x)$ (by O-3.6(3) again) $= g_j'\hat{g}_j'(x)$. Thus (3) is proved since x was arbitrary.

(3) implies (1): Now, $1_{\lim D} = \sup \hat{g}_jg_j$ (by 3.4(2)) $= \sup \hat{g}_jg_j\hat{g}_jg_j$ (by O-3.6(3)) $= \sup \hat{g}_jg_j'\hat{g}_j'g_j$ (by (3)) $= \sup \hat{g}_jg_jg\hat{g}\hat{g}_jg_j = g\hat{g}$, since $\sup \hat{g}_jg_j = 1_{L_j}$ and $\sup g\hat{g}\hat{g}_jg_j(s) = g\hat{g}\sup \hat{g}_jg_j(s) = g\hat{g}(s)$ by 3.4(2) and since $(h,x) \mapsto h(x)$ is Scott continuous (cf. II-2.9 or II-2.14(ii)). □

It is not always easy in concrete cases to decide whether the surjectivity of all bonding maps g_{jk} of a projective system entails the surjectivity of the limit maps g_j. The answer is positive if the category in question is based on compact spaces and continuous maps. In the situation of the category $\boldsymbol{INF}^{\uparrow}$, however, the situation is extremely simple.

3.9. PROPOSITION. *Let D be a projective system in $\boldsymbol{INF}^{\uparrow}$ such that all maps in the image of D are surjective. Then the limit maps $g_j : \lim D \to D(j)$ are surjective.*

Proof. If all g_{jk} are surjective, then $g_{jk}\hat{g}_{jk} = 1$ for all $k \in J^{ij}$ by O-3.7. But then $g_j\hat{g}_j = 1$ by 3.4(1). This implies that g_j is surjective by O-3.7. □

We have discussed projective limits in a category in general and then in the category $\boldsymbol{INF}^{\uparrow}$ in particular. Now we discuss the preservation of projective limits by functors between categories in general, and then by self-functors of $\boldsymbol{INF}^{\uparrow} \to \boldsymbol{INF}^{\uparrow}$ in particular.

3.10. DEFINITION. Let $F : \boldsymbol{A} \to \boldsymbol{B}$ be a functor between complete categories and let $D : \boldsymbol{J} \to \boldsymbol{A}$ be a diagram in $\boldsymbol{A}$. Let $g_j : \lim D \to D(j)$ be the limit cone over D in $\boldsymbol{A}$. Then $Fg_j : F(\lim D) \to FD(j)$ is a cone over the diagram $FD : \boldsymbol{J} \to \boldsymbol{B}$ in $\boldsymbol{B}$. Now let $h_j : \lim FD \to FD(j)$ be the limit cone over the diagram FD in $\boldsymbol{B}$. By the universal property of the limit (3.1.) there is a unique map $f : F(\lim D) \to \lim FD$ such that $h_jf = Fg_j$ for all j. We say that F *preserves the limit of D* iff f is an isomorphism. In general, F is said to *preserve limits* iff f is an isomorphism for all diagrams D in $\boldsymbol{A}$ and we say that F *preserves projective limits* iff f is an isomorphism for all projective systems D in $\boldsymbol{A}$. □

There are numerous functors occurring in nature which preserve projective limits but do not preserve limits. Cech cohomology on compact spaces with values in the opposite category of graded modules is one of the better known examples, and we will see that most of the functors which interest us here fall into the same category. We recall (cf. pp. 207,208) that a functor which preserves limits must preserve products and equalizers, and by a theorem of Freyd any functor preserving products and equalizers preserves arbitrary limits. But since many of the functors which we will discuss on $\boldsymbol{INF}^{\uparrow}$ will not preserve products, it will be important to have a criterion for functors on $\boldsymbol{INF}^{\uparrow}$ preserving projective limits. We then consider self-functors on $\boldsymbol{INF}^{\uparrow}$.

3.11. THEOREM. *Let $F : \boldsymbol{INF}^{\uparrow} \to \boldsymbol{INF}^{\uparrow}$ be a self-functor of $\boldsymbol{INF}^{\uparrow}$ and let $D : J^{\mathrm{op}} \to \boldsymbol{INF}^{\uparrow}$ a projective system in $\boldsymbol{INF}^{\uparrow}$. Denote with $g_j : \lim D \to D(j)$ is the limit cone of D and $h_j : \lim FD \to FD(j)$ is the limit cone of FD. Then the following statements are equivalent:*

(1) *F preserves the limit of D.*

(2) (i) $(Fg_j)(Fg_j)\hat{} = h_j \hat{h}_j$ *for all $j \in J$, and*

(ii) $\sup_j (Fg_j)\hat{}(Fg_j) = 1_{\lim FD}$.

Proof. We consider the diagram

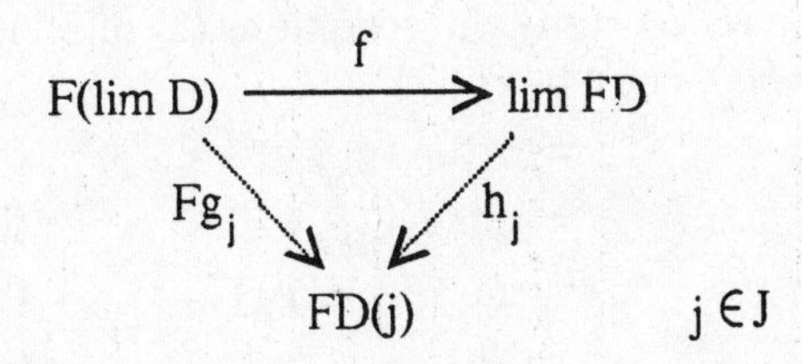

and we find ourselves in the situation discussed in Propositions 3.7 and 3.8, which give a characterization for the injectivity and surjectivity of f, respectively. We recognize that condition 2(ii) is that of injectivity and that of 2(i) is that of surjectivity of f. Consequently, the theorem follows from 3.7 and 3.8. □

Information on certain special cases can be helpful for some applications.

3.12. COROLLARY. *Let $F : \boldsymbol{INF}^{\uparrow} \to \boldsymbol{INF}^{\uparrow}$ be a self functor of $\boldsymbol{INF}^{\uparrow}$ which maps surjective maps to surjective maps. If D is a projective system with surjective bonding maps $g_{jk} = D(k \to j)$ then F preserves the limit of D if*

$$\sup_j (Fg_j)\hat{}(Fg_j) = 1_{\lim FD}.$$

Proof. The map f in the diagram of the proof of 3.11 is surjective by 3.9. Thus only the injectivity of f is in question. but as we observed in the proof of 3.11, this injectivity follows from 2(ii) of 3.11. □

We should remark that both 3.11 and 3.12 remain valid verbatim if F is a functor $A \to A$ on a full and complete subcategory of $\boldsymbol{INF}^{\uparrow}$. As $\boldsymbol{CL}$ and $\boldsymbol{AL}$ illustrate, such sub categories are of interest.

We can now derive useful sufficient conditions for the preservation of projective limits; indeed it is the following corollary which we will use in the applications.

3.13. COROLLARY. *Suppose that the functor $F: A \to A$ on a complete and full subcategory of $INF^{\uparrow}$ satisfies the following conditions:*

For each complete lattice L in A there is a complete lattice C containing FL where FL is closed in C under directed sups and a function

$$\pi : [L \to L] \to C^{FL}$$

such that for each M$\in A$:

(A1) *π is Scott continuous,*

(A2) $\pi(\hat{g}g) = (Fg)\hat{}(Fg)$ *for all* $g \in A(L,M)$,

(A3) $\pi(g\hat{g}) = (Fg)(Fg)\hat{}$ *for all* $g \in A(M,L)$.

Then F preserves projective limits and the injectivity and surjectivity of morphisms.

Remark. To be completely precise, we should say equality holds in (A2) and (A3) when the codomain of the left-hand side is restricted to FL.

Proof. We have to verify the conditions (2) of 3.11.

First 2(i):
$$\begin{aligned} (Fg_j)(Fg_j)\hat{} &= \pi(g_j\hat{g}_j) && \text{(by A3)}\\ &= \pi(\sup\{g_{jk}\hat{g}_{jk} : k \in J^{ij}\}) && \text{(by 3.4)}\\ &= \sup \pi(g_{jk}\hat{g}_{jk}) && \text{(by A1)}\\ &= \sup (Fg_{jk})(Fg_{jk})\hat{} && \text{(by A3)}\\ &= h_j\hat{h}_j && \text{(by 3.4(1)).} \end{aligned}$$

Next 2(ii):
$$\begin{aligned} \sup_j (Fg_j)\hat{}(Fg_j) &= \\ &= \sup_j \pi(\hat{g}_j g_j) && \text{(by A2)}\\ &= \pi(\sup \hat{g}_j g_j) && \text{(by A1)}\\ &= \pi(1_{\lim D}) && \text{(by 3.4(2))}\\ &= \pi(1 \circ \hat{1})\\ &= F(1) \circ F(1)\hat{} && \text{(by A3)}\\ &= 1 \circ 1 = 1. \end{aligned}$$

If g is injective in A, then $\hat{g}g = 1$ by O-3.7. Then $(Fg)\hat{}(Fg) = \pi(\hat{g}g)$ (A2) $= \pi(1) = 1$ (by what we just saw). Hence Fg is injective by O-3.7 again. That F preserves surjectives is shown in a similar way using A3. □

Remark. Frequently π will have range $[FL \to FL] \subseteq C^{FL}$.

The conditions given in 3.13 which secure projective limit preservation for a functor may appear to be technical. However, a second look shows that they are just a valid expression for the statement "F preserves adjunction". Strictly speaking, this statement would mean $(Fg)\hat{} = F\hat{g}$; but since $\hat{g}$ is in SUP^0 and not in general in $INF^{\uparrow}$ (let alone in A), $F\hat{g}$ the right hand side of this equation does not even make sense. Let us suppose for the moment that $F: A \to A$ extends to a functor $F: UPS \to UPS$ (which we denote with the same letter). Then the equation $(Fg)\hat{} = F\hat{g}$ is meaningful for $g \in INF \cap UPS = INF^{\uparrow}$, and if it is satisfied, then conditions A2 and A3 follow if we set $\pi(f) = Ff$.

Condition A1 would follow if we were given the information that the functor F preserves sups of directed nets $f_j : \mathrm{L} \to \mathrm{S}$ in ***UPS***. Thus, from this point of view, the conditions in 3.13 are not unnatural, and the applications we have in mind for 3.13 will confirm this claim. On the other hand, many of the functors $F : \mathbf{INF}^{\uparrow} \to \mathbf{INF}^{\uparrow}$ which we will consider have no extension to ***UPS*** simply because Fg is expressed in terms of g and $\hat{g}$ (see 3.18 below for an example), and that precludes the desired extension.

The following variant of 3.13 may be useful:

3.14. PROPOSITION. *Suppose that the functor $F : \mathbf{A} \to \mathbf{A}$ on a complete and full subcategory of $\mathbf{INF}^{\uparrow}$ satisfies the following condition:*

For each L *in* ***A****, there is a Scott-continuous function $\pi : [\mathrm{L} \to \mathrm{L}] \to \mathrm{C}^{F\mathrm{L}}$ with a complete lattice $\mathrm{C} \supseteq F\mathrm{L}$ such that π satisfies* A2 *of* 3.13. *Then F preserves the limit of any projective system D whose bonding maps g_{ij} are surjective.*

Proof. This is a corollary of 3.12 and the proof of 3.13. □

Now we investigate the preservation property of various functors which are of interest in continuous lattice theory.

The first type arises from function space constructions such as we considered in Chapter II-2. We know that the assignment

$$(\mathrm{S},\mathrm{T}) \mapsto [\mathrm{S} \to \mathrm{T}] : \mathbf{UPS}^{\mathrm{op}} \times \mathbf{UPS} \to \mathbf{UPS}$$

is a functor, where $[g \to h](\varphi) = h\varphi g$. Normally we cannot expect to obtain a covariant functor in both arguments from this functor which is inherently contravariant in the first argument. However, the ***INF-SUP*** duality allows us to produce a covariant functor in ***both*** arguments on a smaller domain by setting

$$B : \mathbf{INF}^{\uparrow} \times \mathbf{UPS} \to \mathbf{UPS},$$
$$B(\mathrm{S},\mathrm{T}) = [\mathrm{S} \to \mathrm{T}],$$
$$B(g,h)(\varphi) = h\varphi\hat{g}.$$

3.15. LEMMA. *Let $g : \mathrm{S} \to \mathrm{S}'$ and $h : \mathrm{T} \to \mathrm{T}'$ be in $\mathbf{INF}^{\uparrow}$. Then the morphism $B(g,h) : [\mathrm{S} \to \mathrm{T}] \to [\mathrm{S}' \to \mathrm{T}']$ has an adjoint given by*

$$B(g,h)^{\wedge} = B(\hat{g},\hat{h}).$$

In particular, $B(g,h) \in \mathbf{INF} \cap \mathbf{UPS} = \mathbf{INF}^{\uparrow}$.

Proof. $B(\hat{g},\hat{h})B(g,h)(\varphi) = \hat{h}h\varphi\hat{g}g \leq \varphi$ since $\hat{g}g \leq 1$ and $\hat{h}h \leq 1$ by O-3.6. Similarly $B(g,h)B(\hat{g},\hat{h}) \geq 1$. The assertion follows from O-3.6 and O-3.3. □

By this Lemma we know B induces a functor $B : \mathbf{INF}^{\uparrow} \times \mathbf{INF}^{\uparrow} \to \mathbf{INF}^{\uparrow}$. The following remark points out special properties of this functor which are similar to the conditions in 3.13.

3.16. LEMMA. *Maintaining the same notation, we have*

(i) $B(g,h)^{\wedge}B(g,h) = [\hat{g}g \rightarrow \hat{h}h]$;

(ii) $B(g,h)\,B(g,h)^{\wedge} = [g\hat{g} \rightarrow h\hat{h}]$.

Moreover, the function $(g,h) \mapsto [g \rightarrow h] : [S' \rightarrow S] \times [T \rightarrow T'] \rightarrow [[S \rightarrow T] \rightarrow [S' \rightarrow T']]$ *preserves directed sups.*

Proof. Straightforward. □

The functor $B : \boldsymbol{INF}^{\uparrow} \times \boldsymbol{INF}^{\uparrow} \rightarrow \boldsymbol{INF}^{\uparrow}$ will give rise to a self functor of $\boldsymbol{INF}^{\uparrow}$ as soon as we specify a functor $\boldsymbol{INF}^{\uparrow} \rightarrow \boldsymbol{INF}^{\uparrow} \times \boldsymbol{INF}^{\uparrow}$ which we can apply before B. There are at least the following three natural cases:

(a) The diagonal embedding $\Delta = (S \mapsto (S,S)) : \boldsymbol{INF}^{\uparrow} \rightarrow \boldsymbol{INF}^{\uparrow} \times \boldsymbol{INF}^{\uparrow}$;

(b) The functor $L_A = (S \mapsto (S,A)) : \boldsymbol{INF}^{\uparrow} \rightarrow \boldsymbol{INF}^{\uparrow} \times \boldsymbol{INF}^{\uparrow}$, for a fixed complete lattice A;

(c) The functor $R_A = (S \mapsto (A,S)) : \boldsymbol{INF}^{\uparrow} \rightarrow \boldsymbol{INF}^{\uparrow} \times \boldsymbol{INF}^{\uparrow}$, for a fixed complete lattice A.

In the above we have indicated the mapping behavior of the functors on the objects of $\boldsymbol{INF}^{\uparrow}$ but not on the maps; the intention is, however, obvious.

3.17. LEMMA. *The composite functors* $B\Delta$, BL_A, $BR_A : \boldsymbol{INF}^{\uparrow} \rightarrow \boldsymbol{INF}^{\uparrow}$ *satisfy the conditions of* Corollary 3.13, *hence preserve projective limits, injective morphisms and surjective morphisms.*

Proof. Consider first the case of $B\Delta$. Define $\pi : [L \rightarrow L] \rightarrow [B\Delta L \rightarrow B\Delta L]$ by $\pi(g) = [\hat{g} \rightarrow g]$. Then π preserves directed sups by 3.16. From 3.16 we also obtain the information

$$\pi(\hat{g}g) = B(g,g)^{\wedge}B(g,g) \text{ and } \pi(g\hat{g}) = B(g,g)B(g,g)^{\wedge}.$$

This shows that the hypotheses A1, A2, A3 are satisfied for $B\Delta$.

Next we turn to the functor BL_A. We define $\pi : [L \rightarrow L] \rightarrow [BL_A \rightarrow BL_A]$ by $\pi(g) = [\hat{g} \rightarrow 1_A]$. The proof proceeds as in the previous case. The situation with R_A is analogous. □

We fix the notation for the two most important cases.

3.18. DEFINITION. The functor $B\Delta : \boldsymbol{INF}^{\uparrow} \rightarrow \boldsymbol{INF}^{\uparrow}$ is *the function-space functor* Funct which associates with a complete lattice L the function space Funct(L) $= [L \rightarrow L]$ of all Scott-continuous functions from L to itself. If $g : S \rightarrow T$ is an $\boldsymbol{INF}^{\uparrow}$-morphism, then $\text{Funct}(g)(\varphi) = g\varphi\hat{g}$.

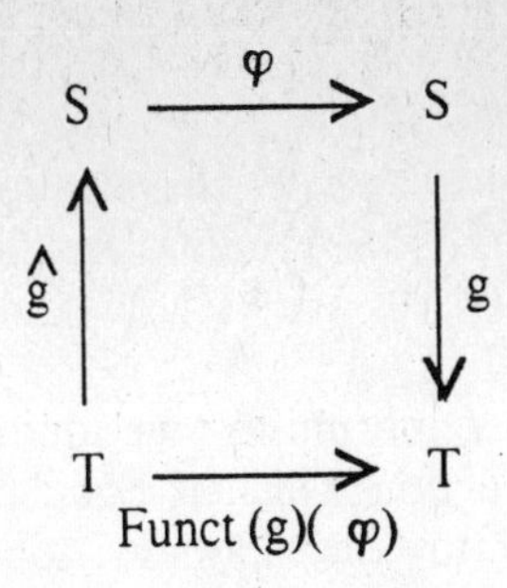

If A is any fixed complete lattice, we denote the functor BL_A by [?→A].

We summarize:

3.19. THEOREM. *The function space functor Funct:* $\boldsymbol{INF}^{\uparrow} \to \boldsymbol{INF}^{\uparrow}$ *preserves projective limits, injectivity and surjectivity of morphisms. The functor* [?→A] : $\boldsymbol{INF}^{\uparrow} \to \boldsymbol{INF}^{\uparrow}$ *has the same property for any complete lattice* A. □

An analogous statement is true for the functor [A→?]; in the exercises we generalize this case (cf. 3.27).

3.20. COROLLARY. *The functor* σ : $\boldsymbol{INF}^{\uparrow} \to \boldsymbol{INF}^{\uparrow}$ *which associates with a complete lattice its Scott topology* σ(L) *and with a morphism* g : S→T *the function* $U \mapsto \hat{g}^{-1}(U) : \sigma(S) \to \sigma(T)$ *preserves projective limits, injectivity and surjectivity of morphisms.*

Proof. There is a natural isomorphism

$$f \mapsto f^{-1}(1) : [L \to 2] \to \sigma(L)$$

and the map $B(g, 1_2)$ corresponds to the map $U \mapsto \hat{g}^{-1}(U)$ under this isomorphism. The result is now a special case of the second part of Theorem 3.19. □

In order to give another example we consider on a complete lattice L and the lattice $P(L)$ of all Lawson-closed upper sets A of L with the order given by $A \leq A'$ iff $A' \subseteq A$. In particular, the finite inf operation on $P(L)$ is set theoretical union, and arbitrary sups are intersections. We note that the lattice $P(L)$ is isomorphic to the lattice of Lawson open lower sets in the natural order of containment. As we shall observe, however, the consideration of the closed upper sets is more convenient. It is easy to make P into a contravariant functor on $\boldsymbol{INF}^{\uparrow}$; but we need a covariant functor. This appears to be possible only on a subcategory on which the upper set of a Lawson-closed set is Lawson closed; this we know to be true on continuous lattices.

Thus we want to verify first that $P(L)$ is a continuous lattice if L is continuous. Let A be a Lawson-closed subset of L. Then $\uparrow A = \pi_2((\text{graph} \leq) \cap (A \times L))$ is also Lawson closed, since the graph of $\leq$ is closed by III-2.9 and III-2.4 in view of II-1.14. (See also VI-1.14 below.) Thus the function $k : \Gamma(L)^{op} \to \Gamma(L)^{op}$ on the lattice of Lawson closed sets given by $k(A) = \uparrow A$ is a kernel (!) operator, whose image is $P(L)$. But k is Scott continuous since $A \mapsto \uparrow A$ preserves intersections of filter bases of compact sets. Now $\Gamma(L)^{op} \cong O(\Lambda L)$ is a continuous lattice since ΛL is a compact Hausdorff space (see III-1.10 and I-1.7(5)). Hence $P(L) = \text{im } k$ is a continuous lattice by I-2.10.

If $g : S \to T$ is a *CL*-morphism, we define $P(g)(A) = \uparrow g(A)$. Then $P(g) : P(S) \to P(T)$ is a function which is Scott continuous, since $g : S \to T$ is continuous with respect to the Lawson topologies and thus preserves intersections of filterbases of compact sets. We now claim that $P(g)$ has a lower adjoint $P(g)\hat{}$ given by $P(g)\hat{}(B) = g^{-1}(B)$. Indeed $B \leq P(g)(A)$ means $\uparrow g(A) \subseteq B$, and if B is an upper set, this is equivalent to $g(A) \subseteq B$, which means $A \subseteq g^{-1}(B)$; but reinterpreted in $P(A)$ this relation means $g^{-1}(B) \leq A$, and this shows $P(g)\hat{}(B) = g^{-1}(B)$.

In particular, this means that $P(g)$ preserves arbitrary infs as an upper adjoint, and thus $P(g)$ is a *CL*-morphism. We have shown that $P : \textbf{CL} \to \textbf{CL}$ is a functor.

3.21. DEFINITION. The functor $P : \textbf{CL} \to \textbf{CL}$ is called the *power set functor.* □

There is a slightly better information available about $P(g)\hat{}$.

Remark. $P(g)\hat{}(B) = g^{-1}(B) = \uparrow \hat{g}(B)$ *for all* $B \in P(T)$.

Proof. Let $a \in g^{-1}(B)$. Then $g(a) = b \in B$. By the definition of a lower adjoint we have $\hat{g}(b) \leq a$, and so $a \in \uparrow \hat{g}(B)$. Conversely, if $a \in \uparrow \hat{g}(B)$, then $\hat{g}(b) \leq a$ for some $b \in B$. Thus $b \leq g(a)$ and so $a \in g^{-1}(\uparrow b) \subseteq g^{-1}(B)$. □

In particular, if $g : S \to T$ is a *CL*-morphism, then for $A \in P(S)$ and $B \in P(T)$ we calculate (recalling O-1.11) as follows:

(i) $P(g)\hat{}P(g)(A) = P(g)\hat{}(\uparrow g(A)) = \uparrow \hat{g}(\uparrow g(A)) = \uparrow \hat{g}g(A)$;

(ii) $P(g)P(g)\hat{}(B) = P(g)(\uparrow \hat{g}(B)) = \uparrow g(\uparrow \hat{g}(B)) = \uparrow g\hat{g}(B)$.

For any L in *CL* we consider the complete lattice $C \supseteq PL$ of all upper sets of L and define $\pi : [L \to L] \to C^{PL}$ by $\pi(f)(A) = \uparrow \varphi(A)$. Then (i) and (ii) show that A2 and A3 of 3.13 are satisfied. We show that (A1) is satisfied, too. Let

$\varphi = \sup \varphi_j$ in $[L \to L]$ for a directed net φ_j. We must show that $\bigcap_i \uparrow \varphi_i(A) = \uparrow \varphi(A)$ for all $A \in P(L)$. Now $x \in \bigcap_j \uparrow \varphi_j(A)$ iff for each j there is an $a_j \in A$ with $\varphi_j(a_j) \leq x$, that is, $a_j \in \varphi_j^{-1}(\downarrow x)$, and this holds iff $A \cap \varphi_j^{-1}(\downarrow x) \neq \emptyset$ for all j. Now $\downarrow x$ is a Scott-closed set, and so $\varphi_j^{-1}(\downarrow x)$ is Scott closed, since φ_j is Scott continuous. Thus $\varphi_j^{-1}(\downarrow x)$ is Lawson closed by III-1.6(ii), and so is $A \cap \varphi_j^{-1}(\downarrow x)$. Now for any j we have $\varphi_j \leq \varphi_k$ eventually, and this implies $\varphi_k^{-1}(\downarrow x) \subseteq \varphi_j^{-1}(\downarrow x)$ eventually. Therefore the $A \cap \varphi_j^{-1}(\downarrow x)$ from a filter basis of

Lawson-compact sets which then must have a nonempty intersection $A \cap \bigcap_j \varphi_j^{-1}(\downarrow x)$. We finally claim that $\bigcap_j \varphi_j^{-1}(\downarrow x) = \varphi^{-1}(\downarrow x)$: indeed $y \in \bigcap_j \varphi_j^{-1}(\downarrow x)$ iff $\varphi_j(y) \leq x$ for all j iff $\varphi(y) = \sup \varphi_j(y) \leq x$ iff $y \in \varphi^{-1}(\downarrow x)$. Thus $A \cap \varphi^{-1}(\downarrow x) \neq \emptyset$, and this means that $\varphi(a) \leq x$ for some $a \in A$, which is tantamount to $x \in \uparrow\varphi(A)$. Since $\varphi_j \leq \varphi$ implies $\uparrow\varphi(A) \subseteq \uparrow\varphi_j(A)$, we proved the claim $\bigcap_j \uparrow\varphi_j(A) = \uparrow\varphi(A)$. Thus $\sup \pi(\varphi_j) = \pi(\sup \varphi_j)$. In view of 3.13 we have shown the following result:

3.22. THEOREM. *The power set functor* P : ***CL***$\rightarrow$***CL*** *preserves projective limits and injectivity and surjectivity of morphisms.* □

As a next example we consider the *ideal functor* Id : $\boldsymbol{INF}^{\uparrow} \rightarrow \boldsymbol{INF}^{\uparrow}$ of 1.18 and 1.19. We recall that for a complete lattice L we let Id L be the complete lattice of ideals of L, and if g : S$\rightarrow$T is in $\boldsymbol{INF}^{\uparrow}$, then $(\mathrm{Id}\ g)(I) = \downarrow g(I)$. In order to apply 3.13 for a complete lattice L we define $C = \{A \subseteq L : A = \downarrow A\}$ and $\pi : [L \rightarrow L] \rightarrow C^{\mathrm{Id}\ L}$ by $\pi(\varphi)(I) = \downarrow\varphi(I)$. Since directed sups in Id L are simply directed unions, then clearly π is Scott continuous. By 1.18 and 1.19 we calculate $(\mathrm{Id}\ g)^{\wedge}(\mathrm{Id}\ g)(I) = \downarrow\hat{g}(\downarrow g(I)) = \downarrow\hat{g}g(I)$ (cf. O-1.11). Similarly we have $(\mathrm{Id}\ g)(\mathrm{Id}\ g)^{\wedge}(I) = \downarrow g(\downarrow\hat{g}(I)) = \downarrow g\hat{g}(I)$. Thus A1, 2, 3 of 3.13 are satisfied and we obtain:

3.23. THEOREM. *The ideal functor* Id : $\boldsymbol{INF}^{\uparrow} \rightarrow \boldsymbol{INF}^{\uparrow}$ *preserves projective limits and injectivity and surjectivity of morphisms*

Remark. The ideal functor maps $\boldsymbol{INF}^{\uparrow}$ into the full subcategory ***ArL*** of arithmetic lattices. □

The discussion of these self-functors would remain somewhat incomplete without some words of explanation to what extent each of them "enlarges" the size of an object. In speaking of the size here we make reference to the weight of a complete lattice whose various aspects were analyzed in Chapter III, Section 4. We also recall that we introduced the concept of weight for continuous lattices only.

3.24. PROPOSITION. *Let* L *be a continuous lattice. If* L *is infinite, then:*

(i) $w(\mathrm{Funct}\ L) = w(L)$;

(ii) $w([L \rightarrow A]) = \max(w(L), w(A))$;

(ii′) $w(\sigma(L)) = w(L)$;

(iii) $w(P(L)) = w(L)$;

(iv) $w(\mathrm{Id}\ L) = \mathrm{card}\ L$.

If L *is finite, then* $w(F(L)) < \aleph_0$ *for all of the functors* F *considered in* (i)–(iv).

Proof. (i), (ii) and (ii$'$) are consequences of III-4.12 and III-4.13. By III-4.7 we know that $w(\lambda(\mathrm{L})) = w(\Lambda\mathrm{L}) = w(\mathrm{L})$. But $P(\mathrm{L})$ is isomorphic to a subalgebra of $\lambda(\mathrm{L})$, hence by III-4.14(ii) we have $w(P(\mathrm{L})) \leq w(\mathrm{L})$. Since $P(\mathrm{L})$ is the image of $\lambda(\mathrm{L})$ under a Scott-continuous kernel operator (see discussion preceding 3.21!), we have $w(P(\mathrm{L})) \leq w(\lambda(\mathrm{L})) = w(\mathrm{L})$ by III-4.14(iii). Thus (iii) follows. By III-4.4 and I-4.12 we know $w(\mathrm{Id\ L}) = \mathrm{card}(\mathrm{K}(\mathrm{Id\ L})) = \mathrm{card\ L}$, since $\mathrm{K}(\mathrm{Id\ L}) \cong \mathrm{L}$. This proves (iv). The statement concerning the finite case is clear. □

We note that the functors Funct, σ, and P do not "enlarge" L, while Id does, in general.

If L is a continuous lattice which is the projective limit of continuous lattices L_j, we need to know the weight of L in terms of the weights of the L_j. This is the place to record the relevant information.

3.25. PROPOSITION. *Let $D : J^{\mathrm{op}} \to \mathbf{CL}$ be a projective system of continuous lattices, and assume that J is infinite. Then*

$$w(\lim D) \leq \max\{\mathrm{card}\, J, \sup\{w(D(j)) : j \in J\}\}.$$

Remark. The assumption that J be infinite is no loss of generality: If J is finite, then $k = \max J$ exists and $\lim D = D(k)$.

Proof. First let $g_j : \lim D \to D(j)$ be the limit maps. Then $x \mapsto (g_j(x))_{j \in J}$: $\lim D \to \Pi_{j \in J} D(j)$ is a $\mathbf{CL}$-embedding, whence $w(\lim D) \leq w(\Pi_{j \in J} D(j))$ by III-4.14(ii). But by III-4.14(i) we have

$$w(\Pi_{j \in J} D(j)) = \max\{\mathrm{card}\, J, \sup\{w(D(j)) : j \in J\}\}.$$

This proves the proposition. □

EXERCISES

3.26. EXERCISE. The full subcategories in $\mathbf{INF}^{\uparrow}$ of continuous and algebraic lattices are complete subcategories. What is the situation with arithmetic lattices? □

3.27. EXERCISE. Let X be a T_0-space Then $[X,?] : \mathbf{INF}^{\uparrow} \to \mathbf{INF}^{\uparrow}$ is a functor given by $[X,\mathrm{L}] = \mathbf{TOP}(X, \Sigma\mathrm{L})$ with the pointwise order on objects, and for maps by $[X,g] : [X,\mathrm{S}] \to [X,\mathrm{T}]$, where $[X,g](\varphi) = g\varphi$ with $g \in \mathbf{INF}^{\uparrow}$. Then

(i) $[X,g]^{\wedge} = [X,g^{\wedge}]$.

(ii) The functor extends to a functor $[X,?] : \mathbf{UPS} \to \mathbf{UPS}$ such that the map $\varphi \mapsto [X,\varphi] : [\mathrm{S} \to \mathrm{T}] \to [[X,\mathrm{S}] \to [X,\mathrm{T}]]$ is Scott continuous.

(iii) The hypotheses of 3.13 are satisfied.

(iv) The functor $[X,?] : \mathbf{INF}^{\uparrow} \to \mathbf{INF}^{\uparrow}$ preserves projective limits and injectivity and surjectivity of morphisms.

(v) If $\mathcal{O}(X)$ is a continuous lattice, the function maps $\mathbf{CL}$ into itself, and it maps $\mathbf{AL}$ into itself.

Remark. This functor generalizes the functor BR_A = [A→?] considered in 3.17, since [A→L] = [ΣA,ΣL]. □

3.28. EXERCISE. With the notation introduced for the powerset functor in 3.21 we have the following results:

(i) For a continuous lattice $PL \cong \omega(L)$, where $\omega(L)$ is the lower topology.

Remark. See III-3.20(iv). This brings us back to the exercises in Chapter III, where the lower topology $\omega(L)$ was discussed in its own right. In particular we can derive from (i) above, III-1.6, 1.16, 1.17, and 2.14 a new proof of the continuity of PL.

(ii) The power set functor $P : \mathbf{CL} \to \mathbf{CL}$ maps the full subcategory $\mathbf{AL}$ of algebraic lattices into itself.

(HINT: (i): The isomorphism is given by $A \mapsto L \setminus A$: indeed we have to observe that every Lawson-closed upper set is $\omega(L)$-closed. If $A \in PL$ and $x \notin A$, then we find some $d \ll a$ with $d \not\leq x$; whence, A is covered by the Scott-(hence: Lawson-) open upper set $\uparrow d$. Since A is compact, finitely many of these suffice to cover. Thus, there is a finite set $F = \{d_1, \ldots, d_n\}$ such that $A \subseteq \uparrow F$ and $x \notin \uparrow F$. A is then the intersection of the $\omega(L)$-closed sets $\uparrow F$ and, hence, is $\omega(L)$-closed.

(ii): If $A \in PL$, then $A \in K(PL)$ iff there is a finite subset $F \subseteq K(L)$ such that $A = \uparrow F$: as in part (i) we note that A is the intersection of all $\uparrow G$ with $G \subseteq L$ finite and $A \subseteq \uparrow G$. This is a filter basis, and since $A \in K(PL)$ there is a finite $F \subseteq L$ with $A = \uparrow F$. If $x \in F$ were not compact, then $y \ll x$ would imply $y < x$, and then $\uparrow y \cup \uparrow (F \setminus \{x\})$, $y \ll x$ would be a filter basis in Lawson-closed upper sets whose intersection is A, but none of whose members was equal to A, in contradiction to the fact that A is compact in PL. It then follows easily that PL is algebraic if L is algebraic.) □

NOTES

The importance of projective limits for continuous lattices and their applications was first pointed out by Scott in [1972], where limits of projective systems whose index domain was the natural numbers and whose maps were all surjective were utilized. The results of 3.4 and 3.5 in this special case were established in that paper; however, the treatment of arbitrary projective limits given here is new. The question of which functors preserve projective limits is treated systematically here for the first time, although several authors recognized earlier how it connects with Scott's constructions. That construction and generalizations of it are the topic of the next section. The power-domain functor has been discussed by Smyth [1978]. The preservation of projective limits indexed by the natural numbers and sufficient conditions for such preservation have recently been considered by Smyth and Plotkin [1979].

4. FIXED-POINT CONSTRUCTIONS FOR FUNCTORS

Scott's construction of continuous lattices L which are isomorphic to their own function space [L→L] is a special case of the general construction to be discussed in this section; the first part is entirely functorial, and the second part, in which we study applications to the categories $\boldsymbol{INF}^{\uparrow}$, ***CL***, ***AL***, relies on the results prepared in the previous section.

In order to illustrate the basic idea, we return briefly to a fixed-point theorem for posets and summarize some of the ideas going into its well-known proof. To this end suppose that A is a down-complete poset and $F : A \to A$ a self-map. We require two properties:

(i) F is monotone;
(ii) F preserves filtered infs.

Now look at the subset A_F of all $x \in A$ such that $x \geq Fx$. We assume that A_F is non empty (if A has an identity, then it is necessarily in A_F). For $x \in A_F$ inductively construct the sequence

$$x \geq Fx \geq F^2x \geq F^3x \geq \ldots,$$

where the monotonicity of the sequence follows from property (i) of F. Since A is down-complete, we can form $\inf_n F^n x$, which we denote by $\tilde{F}x$.

From property (ii) we calculate:

$$F\tilde{F}x = F(\inf_n F^n x) = \inf_n F^{n+1}x = \inf_n F^n x = \tilde{F}x.$$

Thus, $\tilde{F}x$ is in fact a fixed point of F and is of course contained in that subset $A_F^{\circ} \subseteq A_F$ which consists of all the fixed points. Note that the restriction and corestriction $\tilde{F} : A_F \to A_F^{\circ}$ is a retraction.

In order to strike the proper analogy, remember that every poset A may be considered a category with $x \leq y$ being tantamount to $x \to y$. In this reading, a filtered net $(x_j)_{j \in J}$ is just a projective system, and the existence of infs of filtered sets means the existence of projective limits. A function $F : A \to A$ satisfying (i) is simply a ***functor***, and, if (ii) is satisfied, then the functor ***preserves projective limits.*** This, then, is the way we want to generalize the fixed-point construction to arbitrary categories and later apply it to categories like $\boldsymbol{INF}^{\uparrow}$ or ***CL*** with the kind of functors preserving projective limits we saw in the previous section. It remains to be seen, however, how we should generalize the definitions of the subsets A_F and A_F°.

4.1. DEFINITION. We will say that a category A is *pro-complete* iff every projective system (see 3.2) has a limit—in short iff projective limits exist. A functor between pro-complete categories will be called *pro-continuous* iff it preserves projective limits (see 3.10). □

4.2. CONSTRUCTION. Let $F : A \to A$ be a self-functor of a pro-complete category A. (We will eventually assume that F is pro-continuous.) We assume that $p : FL \to L$ is an arbitrary morphism from FL to L for some object L. We denote with $\tilde{F}$L the projective limit of the following inverse system in A:

(1)
$$L \xleftarrow{\;p\;} FL \xleftarrow{\;Fp\;} F^2L \xleftarrow{\;F^2p\;} F^3L \xleftarrow{\;F^3p\;} \cdots$$

It is to be understood that the full inverse system contains all finite compositions of the morphisms listed in diagram (1).

Let $p' : \tilde{F}L \rightarrow L$ be the limit map from the limit to the first term of the sequence. We apply the functor F to diagram (1) together with its limit cone. By 3.2 there is a unique natural map

(2)
$$\tilde{p} : F\tilde{F}L \rightarrow \tilde{F}L$$

such that the following diagram commutes:

(3)
$$\begin{array}{ccccccccc}
 & & FL & \longleftarrow & F(FL) & \xleftarrow{\;F(Fp)\;} & F(F^2L) & \xleftarrow{\;F(F^2p)\;} \cdots & F\tilde{F}L \\
 & & \| & & \| & & \| & & \downarrow \tilde{p} \\
L & \xleftarrow{\;p\;} & FL & \xleftarrow{\;Fp\;} & F^2L & \xleftarrow{\;F^2p\;} & F^3L & \xleftarrow{\;F^3p\;} \cdots & \tilde{F}L
\end{array}$$

We have $p(Fp') = p'\tilde{p}$, which means that the following diagram commutes:

(4)
$$\begin{array}{ccc}
FL & \xleftarrow{\;Fp'\;} & F\tilde{F}L \\
p \downarrow & & \downarrow \tilde{p} \\
L & \xleftarrow{\;p'\;} & \tilde{F}L
\end{array}$$

If F is pro-continuous (i.e., preserves projective limits), then $\tilde{p}$ is an isomorphism. □

We have now associated with a morphism $p : FL \rightarrow L$ a new morphism $\tilde{p} : F(\tilde{F}L) \rightarrow \tilde{F}L$ in a natural fashion. We must discuss in what way this process is functorial. The guiding idea is to consider $p : FL \rightarrow L$ as an "algebra" (which could be an object of a suitable category of algebras, called *comma categories* in the literature, but whose formalism we do not need to enter into here) to which we associate a new "algebra" $\tilde{p} : F(\tilde{F}L) \rightarrow \tilde{F}L$. This new algebra is more special if F is pro-continuous as then $\tilde{p}$ is an isomorphism. This now is the idea that generalizes the formation of the subset A_F of a poset A.

4.3. DEFINITION. Let $F : A \to A$ be a self-functor of a category A. An *F-algebra* is a pair (L,p) consisting of an object L of A together with an A-morphism $p : FL \to L$. If (S,p) and (T,q) are F-algebras, then a *morphism of F-algebras* $f : (S,p) \to (T,q)$ is an A-map $f : S \to T$ such that the following diagram commutes:

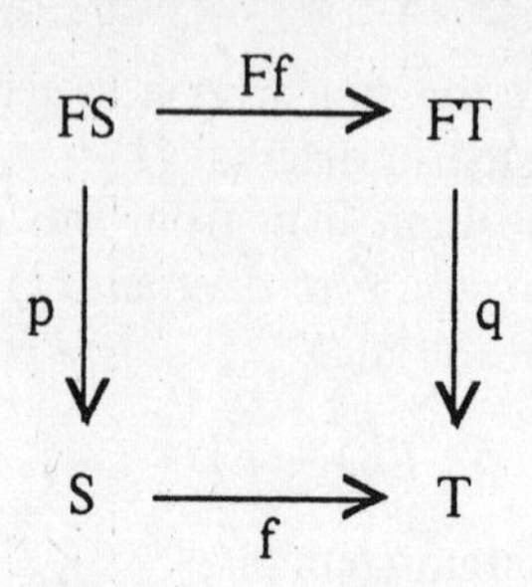

The class of all F-algebras together with the class of F-algebra mor-phisms clearly forms a category which we call the *category of F-algebras* and denote by A_F. The full subcategory of all F-algebras (L,p) for which p is an isomorphism will be denoted $A_F^{\,o}$. □

At this point $A_F^{\,o}$ may very well be an empty category. In Construction 4.2 we have associated with each F-algebra (L,p) an F-algebra $(\tilde{F}L,\tilde{p})$ which in fact is in $A_F^{\,o}$ if F is pro-continuous. We would like to know, of course, to what extent the assignment $(L,p) \mapsto (\tilde{F}L,\tilde{p})$ is functorial.

4.4. LEMMA. *Let A be a pro-complete category and F a self-functor on A. Then for each F-algebra morphism $f : (S,p) \to (T,q)$, there is an F-algebra morphism $\tilde{F}f : (\tilde{F}S,\tilde{p}) \to (\tilde{F}T,\tilde{q})$. Furthermore, the map p' of* 4.2 *is an F-algebra morphism $p' : (\tilde{F}L,\tilde{p}) \to (L,p)$ such that the following diagram commutes:*

(1)
$$\begin{array}{ccc} (\tilde{F}S,\tilde{p}) & \xrightarrow{\tilde{F}f} & (\tilde{F}T,\tilde{q}) \\ \downarrow{\scriptstyle p'} & & \downarrow{\scriptstyle q'} \\ (S,p) & \xrightarrow[f]{} & (T,q) \end{array}$$

Proof. The existence of $\tilde{F}f$ follows immediately from the properties of the limit; it is the unique map which makes the following diagram commute:

(2)

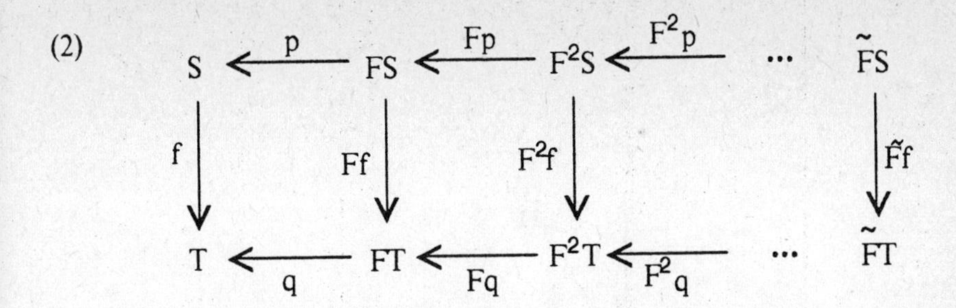

in particular, this proves $fp' = q'(\tilde{F}f)$, which shows the commutativity of (1).

Applying F to diagram (2) gives rise to a three-dimensional commutative diagram

(3)

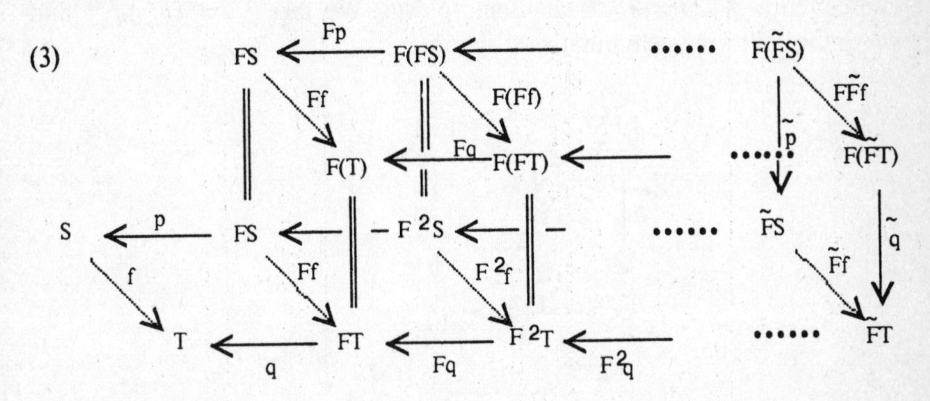

The commutativity of the right-most facet is the statement that $\tilde{F}f$ is an F-algebra morphism. The commutativity of diagram (4) in 4.2 shows that p' is an F-algebra morphism. □

4.5. COROLLARY. *The assignments* $(\mathrm{L},p) \mapsto (\tilde{F}\mathrm{L},\tilde{p})$ *and* $f \mapsto \tilde{F}f$ *determine a self-functor* $\Phi : A_F \to A_F$ *of the category of F-algebras.* □

If F preserves projective limits, then each $\Phi(p)$ is contained in the subcategory A_F°. Let $\Phi^{\circ} : A_F \to A_F^{\circ}$ denote the corestriction of this functor. We now settle the question to what extent our construction is universal.

4.6. THEOREM. *Let* $F : A \to A$ *be a pro-continuous self-functor of a pro-complete category. Then the functor* $\Phi^{\circ} : A_F \to A_F^{\circ}$ *is right adjoint to the inclusion functor.*

Remark. We reformulate in explicit terms what the assertion means: Suppose $f : (\mathrm{S},q) \to (\mathrm{L},p)$ is a morphism in A_F where $q : F\mathrm{S} \to \mathrm{S}$ is an isomorphism and where $p : F\mathrm{L} \to \mathrm{L}$ is arbitrary. Then there is a unique $f_0 : (\mathrm{S},q) \to (\tilde{F}\mathrm{L},\tilde{p})$ such that $f = p'f_0$; that is, there is a commutative diagram

(1)

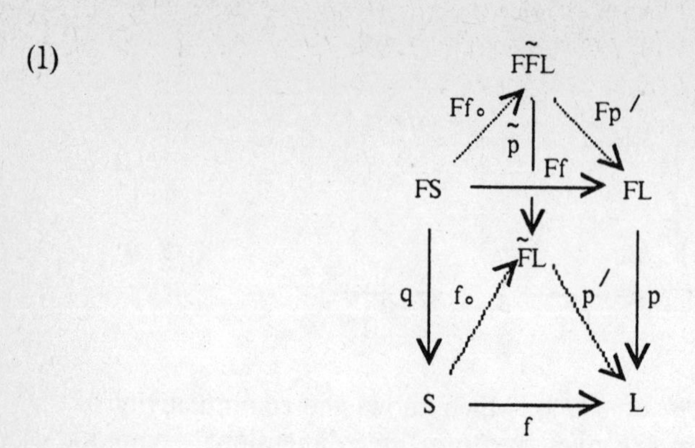

Proof. If $q : FS \to S$ is an isomorphism, then so is $F^n q : F^{n+1}S \to F^n S$. As a consequence $q' : \tilde{F}S \to S$ is an isomorphism. We take $f_0 = (\tilde{F}f)q'^{-1}$ and observe the following commutative diagram:

(2)

$$\begin{array}{ccccc}
\tilde{F}FS & \xrightarrow{\tilde{F}Ff} & & \tilde{F}FL & \\
\downarrow \tilde{q} & \searrow Fq' & & \downarrow \tilde{p} & \searrow Fp' \\
 & & FS & \longrightarrow & FL \\
\tilde{F}S & \xrightarrow{\tilde{F}f} & \downarrow q & \tilde{F}L & \downarrow p \\
 & \searrow q' & & \searrow p' & \\
 & & S & \xrightarrow{f} & L
\end{array}$$

This proves the ***existence*** of the required morphism f_0.

In order to establish ***uniqueness***, we assume that $f = p'g$ in A_F with a map $g : (S,q) \to (\tilde{F}L,p)$. Thus we have a commutative diagram

(3)

$$\begin{array}{ccccccccccc}
FS & \overset{Fg}{\underset{Ff_0}{\rightrightarrows}} & F\tilde{F}L & \xrightarrow{Fp^{(n)}} & F^{n+1}L \cdots & F^3L & \xrightarrow{F^2p} & F^2L & \xrightarrow{Fp} & FL \\
\downarrow q & & \downarrow \tilde{p} & & \downarrow & \downarrow F^2p & & \downarrow Fp & & \downarrow p \\
S & \overset{g}{\underset{f_0}{\rightrightarrows}} & \tilde{F}L & \xrightarrow[p^{(n)}]{} & F^nL & \cdots\ F^2L & \longrightarrow & FL & \xrightarrow[p]{} & L
\end{array}$$

where $p^n : \tilde{F}L \to F^nL$ is the limit map. Then $p^{(1)}g = p^{(1)}f_0$. Since $\tilde{p}$ is an isomorphism, we observe

$$p^{(2)} = (F^2p)F(p^{(2)})\tilde{p}^{-1} = F(p^{(1)})\tilde{p}^{-1}.$$

Thus

$$\begin{aligned} p^{(2)}f_0 &= F(p^{(1)})\tilde{p}^{-1}f_0 = F(p^{(1)})(Ff_0)q^{-1} \\ &= F(p^{(1)}f_0)q^{-1} = F(p^{(1)}g)q^{-1} \\ &= \ldots = p^{(2)}g. \end{aligned}$$

Now we attack $p^{(3)} = (F^3p)(Fp^{(3)})\tilde{p}^{-1} = F(p^{(2)})p^{-1}$ and calculate

$$p^{(3)}f_0 = \ldots = F(p^{(2)}f_0) = F(p^{(2)}g) = \ldots = p^{(3)}g.$$

Continuing by induction, this diagram chasing yields the information

$$p^{(n)}f_0 = p^{(n)}g \text{ for } n = 1,2,3,\ldots$$

By the uniqueness in the universal property of the limit $\tilde{F}L = \lim F^nL$ we now conclude $f_0 = g$. This completes the proof of the theorem. □

Before we apply the general construction to the special categories we are working with, we observe that for some functors $F : A \to A$ there is in fact at least one functor from A into the category A_F of F-algebras. This together with the functor of 4.5 and 4.6 gives a functorial method to associate with any A-object L an F-algebra $(\tilde{F}L,\tilde{p})$ for which p is an isormorphism.

4.7. OBSERVATION. Let $F : A \to A$ be a self-functor of a category and suppose that there is a natural transformation $p_L : FL \to L$. Then the assignment $L \mapsto (L,p_L)$ is a functor $A \to A_F$.

Proof. We only need to recall that by the definition of a natural transformation, for each A morphism f: S→T, the diagram

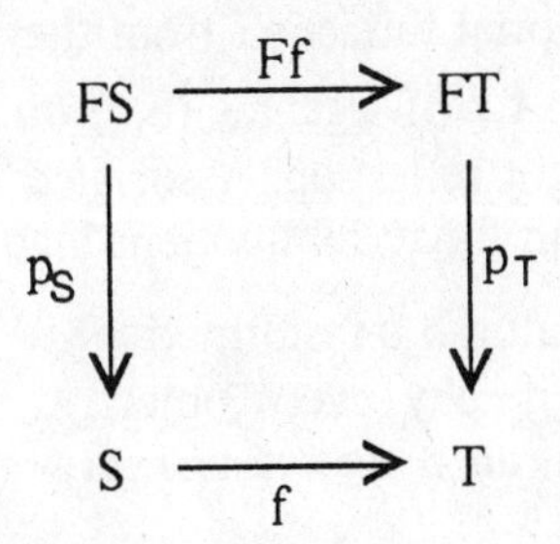

commutes. The rest is clear. □

We proved that the functor $\Phi : A_F \to A_F^{\circ}$ was universal. If the functor $L \mapsto (L,p_L)$ which we just noted were also universal, then in fact we could speak of a universal construction $L \mapsto (\tilde{F}L,\tilde{p}_L)$. However, the universality of Φ is that of a right adjoint; thus $L \mapsto (L,p_L)$ would have to be a right adjoint in order to compose. A quick inspection of what this means will identify this as a rare occurrence unless F preserves arbitrary limits—which is not the case for the functors we consider here. This observation is independent of the particular nature of the natural transformation p_L. It is therefore not to be expected that the construction $L \mapsto (\tilde{F}L, \tilde{p}_L)$ is universal in the sense of being an adjoint functor. We simply record:

4.8. OBSERVATION. If F is a pro-continuous self-functor of a pro-complete category A and if there is a natural transformation $p_L : FL \to L$, then there is a functor $L \mapsto (\tilde{F}L, \tilde{p}_L) : A \to A_F^{\circ}$ from A to the category of F-algebras (S,q) with q an isomorphism. □

At this point we specialize to the categories $\boldsymbol{INF}^{\uparrow}$, $\boldsymbol{CL}$ and $\boldsymbol{AL}$ which we have treated in Section 3. We consider a complete subcategory A of $\boldsymbol{INF}^{\uparrow}$ and a pro-continuous self-functor F. Corollary 3.13 describes sufficient conditions for F to be pro-continuous. For any complete lattice L in A and each morphism $p : FL \to L$—that is, for each F-algebra (L,p)—we create an F-algebra $(\tilde{F}L, \tilde{p})$ and an F-algebra morphism $p' : (\tilde{F}L, \tilde{p}) \to (L,p)$. Recall that this is a morphism $p' : \tilde{F}L \to L$ compatible with $\tilde{p}$ and p. If $p : FL \to L$ is surjective, and if F preserves surjectivity (as is guaranteed by the conditions of 3.13), then all maps of the projective system (1) in 4.2 are surjective. From 3.9 we know that the limit maps are surjective, and this says in particular that p' is surjective. We can then say that $p' : (\tilde{F}L, \tilde{p}) \to (L,p)$ is a *quotient of F-algebras.* It will serve a good purpose for the applications to summarize:

4.9. SCHOLIUM. Let A be a complete subcategory of $\boldsymbol{INF}^{\uparrow}$ (such as, $\boldsymbol{CL}$, $\boldsymbol{AL}$). Let $F : A \to A$ be a pro-continuous self-functor preserving surjectivity (and any functor satisfying the hypotheses of 3.13 has these properties). Then we have the following conclusions:

(i) There is a functorial retraction from the category A_F of F-algebras in A to the full subcategory A_F° of algebras (S,q) on which q is an isomorphism, and this retraction is a right reflection. Associated with a given F-algebra (L,p) is an F-algebra $(\tilde{F}L, \tilde{p})$ with a natural quotient map $p' : (\tilde{F}L, \tilde{p}) \to (L,p)$.

(ii) If there is a natural transformation $p_L : FL \to L$, then there is a ***functorial*** construction whereby every object L of A is a quotient of the underlying A-object $\tilde{F}L$ of an F-algebra $(\tilde{F}L, \tilde{p})$ with isomorphism p. □

Now we apply this scholium to the following functors:

(a) Funct : $\boldsymbol{INF}^{\uparrow} \to \boldsymbol{INF}^{\uparrow}$, the function space functor (3.18);
(b) $P : \boldsymbol{CL} \to \boldsymbol{CL}$, the power set functor (3.21);
(c) Id : $\boldsymbol{INF}^{\uparrow} \to \boldsymbol{INF}^{\uparrow}$, the ideal functor (3.23);
(d) $\sigma : \boldsymbol{INF}^{\uparrow} \to \boldsymbol{INF}^{\uparrow}$, the Scott-topology functor (3.20).

In the case of (a), (b), (c) there are natural surjective transformations $FL \to L$.

4.10. LEMMA. (i) *For a complete lattice* L *let* $z_L = (f \mapsto f(0))$: Funct L→L. *Then* z_L *is a natural surjective* $INF^{\uparrow}$*-morphism with lower adjoint is the map* $x \mapsto \mathrm{const}_x$, *where* const_x *is the constant function with value* x.

(ii) *For a continuous lattice* L *let* $p_L = (A \mapsto \inf A) : PL \to L$. *Then* p_L *is a natural surjective* **CL**-*morphism whose lower adjoint is* $x \mapsto \uparrow x$.

(iii) *For a continuous lattice* L *let* $r_L = (I \mapsto \sup I) : \mathrm{Id}\ L \to L$. *Then* r_L *is a natural surjective* **CL**-*map whose lower adjoint is* $x \mapsto \Downarrow x$.

Proof. (i): For $f \in [L \to L]$ we have $\mathrm{const}_x \leq f$ iff $x \leq f(0)$ since f is monotone; whence, $f(0) = \min f(L)$. In particular, z_L preserves infs by O-3.3. Since sups in $[L \to L]$ are computed pointwise, z_L preserves sups. Surjectivity and naturality are clear.

(ii): For $A \in PL$ we have $\uparrow x \leq A$ in PL; that is, $A \subseteq \uparrow x$ iff $x \leq \inf A = p_L(A)$. As an upper adjoint p_L preserves arbitrary infs. We now show that p_L preserves directed sups. For this purpose let $\{A_j\}$ be a filter basis of Lawson-closed upper sets. We must show that $p_L(\sup_j A_j) = \sup_j p_L(A_j)$, that is, $\inf(\bigcap_j A_j) = \sup_j \inf A_j$. Since the left side always dominates the right, we must show the inequality $\leq$. For this purpose let $x \ll \inf(\bigcap_j A_j)$ be arbitrary; since L is a continuous lattice it now suffices to show $x \leq \sup_j \inf A_j$. This will be accomplished if we exhibit an index k such that $x \leq \inf A_k$. Now $x \ll \inf(\bigcap_j A_j)$ implies that the intersection $\bigcap_j A_j$ of the filterbasis of Lawson-compact sets A_j is contained in the open set $\Uparrow x$. Hence there must exist a member A_k of this filterbasis which is contained in the open set $\uparrow x$. But then $\inf A_k \geq x$ as we had to show. Clearly p_L is surjective and natural.

(iii): The assertions concerning r_L were shown in I-2.1 and 1.18 above. □

For the functor σ there is no natural transformation $\sigma(L) \to L$.

4.11. REMARK. *The maps* z_L *and* r_L *of* 4.10(i) *and* (iii) *are complete lattice morphisms, that is, preserve arbitrary infs and sups.*

Remark. Recall that the preservation of arbitrary infs in the case of r_L requires that L be a continuous lattice!

Proof. In the case of z_L, this is immediate, since sups are calculated pointwise; in the case of r_L we refer to I-2.1. □

We notice that the lower adjoint $x \mapsto \mathrm{const}_x$ of z_L is likewise a complete lattice map. By 1.10 it preserves the way-below relation. It is also true (see 4.16 below) that z_L preserves the way-below relation.

We now summarize the Scholium in the particular case of the four functors Funct, P, Id and σ. The list, of course, is in no way exhaustive. It is, however, somewhat representative of the situation. In each case we record to which extent the functor F "increases" L in terms of weight.

4.12. THEOREM. (i) *If* $\boldsymbol{INF}^{\uparrow}_{\mathrm{Funct}}$ *denotes the category of all function space algebras* (L,*p*), $p : [\mathrm{L}\to\mathrm{L}]\to\mathrm{L}$ *in* $\boldsymbol{INF}^{\uparrow}$ *and* $\boldsymbol{INF}^{\uparrow}{}_{\mathrm{Funct}}{}^{\circ}$ *the full subcategory of algebras* (L,*p*) *with p an isomorphism, then there is a functorial retraction (indeed right reflection)* $(\mathrm{L},p) \mapsto (\mathrm{Funct}^{\sim}\mathrm{L},\tilde{p})$ *from the former to the latter category such that there is a natural quotient of Funct-algebras* $(\mathrm{Funct}^{\sim}\mathrm{L},\tilde{p}) \to (\mathrm{L},p)$.

(ii) *There is a functorial construction associating with any complete lattice* L *in* $\boldsymbol{INF}^{\uparrow}$ *a complete lattice* $\mathrm{Funct}^{\sim}\mathrm{L}$ *such that* L *is a quotient of* $\mathrm{Funct}^{\sim}\mathrm{L}$ *and such that* $\mathrm{Funct}^{\sim}\mathrm{L}$ *is naturally isomorphic to its own function space. If* L *is continuous (resp., algebraic) so is* $\mathrm{Funct}^{\sim}\mathrm{L}$.

(iii) *For* card L$>$1 *we have* $w(\mathrm{Funct}^{\sim}\mathrm{L}) = \max(\aleph_0, w(\mathrm{L}))$.

Proof. The Scholium together with 3.19 proves everything with the exception of the statement on the weights. By induction, from 3.24(i) we derive $w(\mathrm{Funct}^n\mathrm{L}) = w(\mathrm{L})$ and so $w(\mathrm{Funct}^{\sim}\mathrm{L}) = w(\mathrm{L})$ by 3.25, if L is infinite. If L is finite, then $w(\mathrm{Funct}^n\mathrm{L}) < \aleph_0$ by induction and 3.24(i). The assertion $w(\mathrm{Funct}^{\sim}\mathrm{L}) = \aleph_0$ then follows from 3.25 and card L$<$card Funct L for 1$<$card L$<\aleph_0$. □

4.13. THEOREM. (i) *If* $\boldsymbol{CL}_P$ *denotes the category of all power-set algebras* (L,*p*) *where* $p : P\mathrm{L}\to\mathrm{L}$ *in* $\boldsymbol{CL}$, *and* $\boldsymbol{CL}_P{}^{\circ}$ *the full subcategory of all* (L,*p*) *with p an isomorphism, then there is a functorial retraction (indeed right reflection)* $(\mathrm{L},p) \mapsto (\tilde{P}\mathrm{L},\tilde{p}) : \boldsymbol{CL}_P \to \boldsymbol{CL}_P{}^{\circ}$ *such that there is a natural map of power-set algebras* $(\tilde{P}\mathrm{L},\tilde{p}) \to (\mathrm{L},p)$.

(ii) *There is a functorial construction which associates with any continuous lattice* L *a continuous lattice* $\tilde{P}\mathrm{L}$ *such that* L *is a quotient of* $\tilde{P}\mathrm{L}$ *and* $\tilde{P}\mathrm{L}$ *is naturally isomorphic to its own power set space. If* L *is algebraic, then so is* $\tilde{P}\mathrm{L}$.

(iii) *We have* $w(\tilde{P}\mathrm{L}) = \max(\aleph_0, w(\mathrm{L}))$.

Proof. The proof follows from the Scholium 4.9, Theorem 3.22, Lemma 4.10(ii), and from 3.24 and 3.25 (in the same way as the weight statement was proved in 4.12). □

4.14. THEOREM. (i) *Using our standard notation, we find a functorial retraction (indeed right reflection)* $(\mathrm{L},p) \mapsto (\mathrm{Id}^{\sim}\mathrm{L},\tilde{p})$ *from the category of all ideal algebras* (L,*p*), $p : \mathrm{Id}\,\mathrm{L}\to\mathrm{L}$ *in* $\boldsymbol{INF}^{\uparrow}$ *to the full subcategory of all ideal algebras* (L,*p*) *in which p is an isomorphism. There is a natural quotient* $(\mathrm{Id}^{\sim}\mathrm{L},\tilde{p}) \to (\mathrm{L},p)$ *of ideal algebras.*

(ii) *On the proper subcategory* $\boldsymbol{CL}$ *of continuous lattices, there is a functorial construction which associates with any continuous lattice* L *an arithmetic lattice* $\mathrm{Id}^{\sim}\mathrm{L}$ *such that* L *is a* $\boldsymbol{CL}$*-quotient of* $\mathrm{Id}^{\sim}\mathrm{L}$ *and* $\mathrm{Id}^{\sim}\mathrm{L}$ *is isomorphic to its own ideal lattice.*

(iii) *For non-singleton continuous lattices we have*

$$\aleph_0 \leq w(\mathrm{Id}^{\sim}\mathrm{L}) \leq \exp^{\aleph_0}(\mathrm{card}\,\mathrm{L}),$$

where $\exp \boldsymbol{m} = 2^{\boldsymbol{m}}$ *for all cardinals* $\boldsymbol{m}$ *and* $\exp^{\aleph_0} \boldsymbol{m} = \sup_n \exp^n \boldsymbol{m}$.

Proof. The proof proceeds as before. As far as the weights are concerned, we know from 3.24 that $w(\mathrm{Id}\ \mathrm{L}) = \mathrm{card}\ \mathrm{L}$ for all complete lattices L. But $\mathrm{card}\ \mathrm{Id}\ \mathrm{L} \leq \exp \mathrm{card}\ \mathrm{L}$, whence $w(\mathrm{Id}^n \mathrm{L}) \leq \exp^n \mathrm{card}\ \mathrm{L}$. We then obtain the relation $w(\mathrm{Id}\ \mathrm{L}) \leq \sup \exp^n \mathrm{card}\ \mathrm{L}$ from 3.25. □

4.15. PROPOSITION. (i) *There is a right reflection* $(\mathrm{L},p) \mapsto (\tilde{\sigma}(\mathrm{L}),\tilde{p})$ *from the category* $\boldsymbol{INF}^{\uparrow}_{\sigma}$ *of Scott-topology algebras* (L,p) *with* $p : \sigma(\mathrm{L}) \to \mathrm{L}$ *to the full subcategory of all algebras* (L,p) *for which* p *is an isomorphism.*

(ii) *There is a natural map* $(\tilde{\sigma}(\mathrm{L}),\tilde{p}) \to (\mathrm{L},p)$ *in* $\boldsymbol{INF}^{\uparrow}_{\sigma}$.

(iii) *For continuous* L *we have* $w(\tilde{\sigma}(\mathrm{L})) = \max\{\aleph_0, w(\mathrm{L})\}$.

Proof. The proof is immediate from Scholium 4.9, 3.20, 3.24 and 3.25. □

The significance of function space algebras (L,p) with $p : [\mathrm{L}\to\mathrm{L}] \to \mathrm{L}$ an isomorphism lies in the fact that every element in such an algebra may be identified with a Scott continuous function $\mathrm{L}\to\mathrm{L}$, and every Scott-continuous self-function of L is so obtained. Theorem 4.12 shows that such algebras exist in abundance; in fact, every continuous lattice is a quotient of one of these.

In the same spirit we may interpret power set algebras (L,p), $p : P\mathrm{L}\to\mathrm{L}$ [resp., ideal algebras (L,p), $p : \mathrm{Id}\ \mathrm{L}\to\mathrm{L}$], in ***CL*** for which p is an isomorphism as continuous lattices on which every element may be canonically identified with a Lawson-closed upper set [resp., an ideal], and all such sets occur in this fashion.

In the case of a function space algebra (L,p) with p an isomorphism the formula $x(y) = z$ can be given a meaning for x,y,z in L, and in the case of a power set or ideal algebra (L,p) with p an isomorphism the relation $x \in y$ can be made meaningful for x and y in L. We will say more in the exercises.

EXERCISES

4.16. EXERCISE. The natural map $z_{\mathrm{L}} : \mathrm{Funct}\ \mathrm{L}\to\mathrm{L}$ given by $z_{\mathrm{L}}(f) = f(0)$ has a Scott continuous upper adjoint. In particular, z_{L} preserves $\ll$.

(HINT: For $a \in \mathrm{L}$ define $u_{\mathrm{L}}(a) \in [\mathrm{L}\to\mathrm{L}]$ by

$$u_{\mathrm{L}}(a)(x) = a, \quad \text{if } x = 0$$
$$= 1, \quad \text{if } x > 0.$$

Then $f \leq u_{\mathrm{L}}(a)$ for $f \in \mathrm{Funct}\ \mathrm{L}$ iff $f(0) \leq a$. Scott continuity is straightforward.) □

We generalize this last result in the next exercise.

4.17. EXERCISE. For $x \in L$ we define the evaluation map $ev_x : [L \to A] \to A$ by $ev_x(f) = f(x)$ for complete lattices L and A. The following are equivalent:

(1) ev_x has a lower adjoint.

(2) $\inf f_j(x) = (\inf f_j)(x)$ for any family f_j in $[L \to A]$.

(3) $x \in K(L)$.

Moreover, if these conditions are satisfied, the lower adjoint $m_x : A \to [L \to A]$ given by $m_x(a) = [x \Rightarrow a]$ (see II-2.16).

(HINT: (1) and (2) are equivalent by O-3.5. The implication (3)⇒(1) is readily verified: $m_x(a) \leq f$ iff $m_x(a)(y) \leq f(y)$ for all y iff $a \leq f(y)$ for all y with $x \ll y$, and since $x \ll x$ by (3) and f is monotone this is the case iff $a \leq f(x)$, that is, $a \leq ev_x(f)$. Remains (1) iff (3). Because of the presence of constant functions in $[L \to A]$, the function ev is clearly surjective, and thus its lower adjoint m_x is given by $m_x(a) = \min ev_x^{-1}(\{a\}) = \min \{g \in [L \to A] : g(x) = a\}$ (O-3.7). If $z \ll x$, define $c_z = [z \Rightarrow a]$ (II-2.16). Then $c_z \in [L \to A]$ with $c_z(x) = a$, and thus $m_x \leq c_z$. Since L is continuous $x_a = \sup\{z : z \ll x\}$ and thus, since $m_x(a)(x) = a$ and m_x is monotone, $m_x(a) = \inf \{c_z : z \ll x\}$. We conclude from this relation that

$$\begin{aligned} m_x(a)(y) &= a, && \text{if } x \leq y; \\ &= 0, && \text{otherwise.} \end{aligned}$$

Since $m_x(a)$ is Scott continuous, this implies that $\uparrow x$ is open, i.e., $x \in K(L)$.) □

4.18. EXERCISE. Let (L,p) be a function space algebra with $p : \text{Funct } L \to L$ an isomorphism and L a continuous lattice. Then:

(i) There is a binary operation $(x,y) \mapsto x(y) : L \times L \to L$ which is given by

$$x(y) = p^{-1}(x)(y) = (ev_y \circ p^{-1})(x)$$

which is Scott continuous in both arguments.

(ii) It is Lawson continuous in x for a given fixed y iff $y \in K(L)$.

(iii) It is Lawson continuous in y for a given x iff there is an element $x^\wedge$ in L such that

$$x^\wedge(x(y)) \vee y = y = x(x^\wedge(y)) \wedge y \text{ for all } y \in L.$$

(HINT: Since p is an isomorphism and the function $(f,x) \mapsto f(x) : [L \to L] \times L \to L$ is Scott continuous by II-2.14(ii), then the binary operation is Scott continuous. For fixed y it is Lawson continuous in x iff ev_y is Lawson continuous, since p is Lawson continuous as is p^{-1}. This is the case iff y is compact by 4.17 above. For fixed x it is Lawson continuous in y iff $p^{-1}(x)$ preserves arbitrary infs. This is the case iff it has a lower adjoint by O-3.3, 3.4. This is the case iff there is an $x \in L$ such that $x^\wedge(x(y)) \leq y$ and $x(x^\wedge(y)) \geq y$ for all y by O-3.6 and the definition of the binary operation.) □

4.19. EXERCISE. Under the hypotheses of 4.18, the space L is a topological monoid relative to the operation $(x,y) \mapsto x \circ y$ that is given by $x \circ y = p(p^{-1}(x) \circ p^{-1}(y))$. Moreover one has the identity

$$(x \circ y)(z) = x(y(z)).$$

(HINT: The space [L→L] is a topological monoid under composition by II-4.23. Since p is an isomorphism and the operation "∘" on L is just transported composition, the first assertion follows. The second is straightforward from the definitions.) □

The two preceding exercises show that every function space algebra (L,p) in which p is an isomorphism is a topological monoid relative to the Scott topology in such a way that it acts on itself (although not generally by translation!) in a Scott-continuous fashion; the action is such that every Scott-continuous self-function is realized by the action of precisely one monoid element. Let us express this in a definition:

4.20. DEFINITION. A continuous lattice L will be called a *self-acting monoid* if there is a topological monoid multiplication $(x,y) \mapsto x \circ y$ on L and a monoid action on itself $(x,y) \mapsto x(y)$ which is Scott continuous in both arguments such that for each $f \in$[L→L] there is a unique $x \in$L such that $f(y) = x(y)$ for all y. □

We recall that the term monoid action means the validity of the equations $e(x) = x$ for the identity e of the monoid and of $(x \circ y)(z) = x(y(z))$. We have observed the following result; its converse is also true.

4.21. EXERCISE. There is a bijection between the objects of $\boldsymbol{CL}_{\mathrm{Funct}}{}^{\circ}$ (function space algebras (L,p) with p an isomorphism) and self-acting monoids. □

4.22. EXERCISE. A morphism f : S→T *of self-acting monoids* is a $\boldsymbol{CL}$-morphism which satisfies $f(x(y)) = f(x)(f(y))$ and $f(x \circ y) = f(x) \circ f(y)$. Prove that the second relation will follow from the first. □

4.23. EXERCISE. If f : (S,p)→(T,q) is a morphism in $\boldsymbol{CL}_{\mathrm{Funct}}{}^{\circ}$ and if f is injective (that is, if (S,p) is a "subalgebra" of (T,q)), then f : S→T is an injective morphism of self-acting monoids (that is, S is a self-acting submonoid of T).

(HINT: We have to verify the relation $f\varphi = f\varphi f^{\wedge} f$ for all $\varphi \in$[L→L]; if f is injective, $f^{\wedge} f = 1$ by O-3.7.) □

In more general terms, the proof for the preceding proposition shows that we need the relation $f\varphi = f\varphi f^{\wedge} f$ for all $\varphi \in$[L→L] in order to conclude that a $\boldsymbol{CL}_{\mathrm{Funct}}{}^{\circ}$-morphism f induces a morphism of self-acting monoids. Since every $\varphi \in$[L→L] is the sup of functions of the form $[x \Rightarrow a]$ (II-2.16), it suffices that $f[x \Rightarrow a] = f[x \Rightarrow a]f^{\wedge} f$ holds for all a and x. If f is non-constant, this holds for all a and x iff $x \ll y$ is equivalent to $x \ll f^{\wedge} fy$. If we fix y and remember that $y = \{\sup x : x \ll y\}$, then this condition evidently implies (1) $\hat{f} f \geq 1$. The converse is always true, thus $\hat{f}f = 1$, that is, the injectivity of f is a necessary and sufficient condition for f to induce a morphism of self-acting monoids.

NOTES

The fundamental construction of 4.2 was introduced in the special case of the function space functor F = Funct in Scott [1972]. The objective was to construct the function space algebras which we obtained in 4.12; they serve as set-theoretical models for the λ-calculus of Church and Curry. Exercises 4.18–4.23 contribute in this direction. The investigation of the functorial framework of the fixed point construction for functors, independently from the development here has also been discussed by Smyth and Plotkin. They have the concept of F-algebras in their set-up, and instead of arbitrary projective limits and their preservation through functors they concentrate on limits of systems indexed by the natural numbers. The universality Theorem 4.6 is new as is, in its particular form, Scholium 4.9 which we apply to special situations. The function space and the construction of continuous lattices which are isomorphic to their self-function space is Scott's idea; Smyth treated the power set functor along the same lines. The ideal functor considerations of 4.14 are new, as are the results of 4.15 concerning the Scott-topology functor. There is a slight difference as far as the latter is concerned: there is no canonical construction starting from a continuous lattice and producing a continuous lattice which is naturally isomorphic to its own Scott topology; for all three of the preceding functors, the construction yields precisely this result.

CHAPTER V

Spectral Theory of Continuous Lattices

Spectral theory plays an important and well-known role in such areas as the theory of commutative rings, lattices, and of C^*-algebras, for example. The general idea is to define a notion of "prime element" (more often: ideal element) and then to endow the set of these primes with a topology. This topological space is called the "spectrum" of the structure. One then seeks to find how algebraic properties of the original structure are reflected in the topological properties of the spectrum; in addition, it is often possible to obtain a representation of the given structure in a concrete and natural fashion from the spectrum.

By means of the spectral theory of this chapter we associate with every complete lattice L a topological space, denoted by Spec L, and a representation $L \to \mathcal{O}(\mathrm{Spec}\ L)$ of the given lattice into the lattice of open subsets of the spectrum. Frequently one reduces the spectral theory in other mathematical contexts (such as those listed above) to this lattice-theoretical spectral theory by considering a distinguished lattice of subobjects and identifying the spectrum of this lattice with the spectrum of the original structure in a natural way. Since the lattice of open sets of a topological space is a complete Heyting algebra, it should be noted that a spectral representation can be an *isomorphism* only if L itself is a complete Heyting algebra.

The chapter begins with an important lemma (frequently referred to as "THE LEMMA") which plays a vital role in the spectral theory of continuous lattices. It states that a "finitely prime" element of a continuous lattice is also "compactly prime" with respect to the Lawson topology. Section 2 then resumes the theme of order generation begun in Section I-3, where it was shown that in a continuous lattice the set of irreducible elements is order generating in the sense that every element is an inf of irreducibles (see I-3.9 ff.). The investigation is expanded here to *topologically generating* sets—subsets for which the whole lattice is the smallest *closed* subsemilattice containing the set. We show that the closure of the set of non-identity irreducibles is the unique smallest closed order-generating subset of a continuous lattice as well as being the unique smallest closed topologically generating subset of the lattice. In particular, for a distributive continuous lattice, the closure of the set of non-identity primes is the unique smallest closed order-generating subset.

This line is further pursued in Section 3, where we identify the closure of the primes in a distributive continuous lattice as being exactly the pseudoprimes of I-3, or, as they are also recognized here, the set of weak primes. Analogous results are also obtained for the weak irreducibles in a general continuous lattice.

In Section 4 begins the principal topic of the chapter, the spectral theory of complete Heyting algebras in which the primes order generate. We give the set of non-identity primes the hull-kernel topology and call this space the spectrum; the given lattice is isomorphic to the lattice of open subsets of this space. In this fashion we record the duality between the category of complete Heyting algebras with points and lattice morphisms preserving arbitrary sups on one hand, and sober spaces and continuous maps on the other. This prepares the way for the specific spectral theory of continuous lattices discussed in Section 5: the spectrum of a continuous lattice is locally quasicompact and sober, and all locally quasicompact sober spaces are so obtained. The category of continuous Heyting algebras is dual to the category of locally quasicompact sober spaces and continuous maps.

We collect a good deal of supplementary information in the exercises. Several important applications of the spectral theory of continuous lattices will be given in Chapter VII, when more information on the topological algebra of continuous lattices will be available.

1. THE LEMMA

Let us recall that an element p of a lattice L is called *irreducible*, if the relation $a \wedge b = p$ always implies $a = p$ or $b = p$. The element p is called *prime*, if $a \wedge b \leq p$ always implies $a \leq p$ or $b \leq p$. (See Chapter I, Section 3, especially Definitions I-3.5 and I-3.11.)

These definitions can be rephrased in the following way: p is irreducible [prime], if $\inf F = p$ [$\inf F \leq p$] implies $p \in F$ [$p \in \uparrow F$] for every finite nonempty subset F of L. In the presence of a topology on L, one could define p to be *strongly irreducible* [*strongly prime*], if $\inf K = p$ [$\inf K \leq p$] implies $p \in K$ [$p \in \uparrow K$] for every nonempty ***compact*** subset K of L. Since compactness is often a kind of substitute for finiteness, one may conjecture that these strengthened notions of irreducibility and primality are in reality identical with the first ones. We shall prove this conjecture in the case of *continuous* lattices. The following LEMMA is crucial, and we give two versions:

1.1. THE LEMMA. *Let* L *and* M *be complete lattices and let* $i : \mathrm{L} \to \mathrm{M}$ *be Scott continuous. Suppose* p *is a prime element of* M *and the subset* $A \subseteq \mathrm{L}$ *is such that* $\inf i(A) \leq p$. *Then there is an ultrafilter* $\mathcal{U}$ *on* A *with* $i(\underline{\lim}\ \mathcal{U}) \leq p$.

Proof. Let A and p satisfy the hypotheses and define $\mathfrak{I}$ to be the set of all subsets B of A with $\inf i(B) \not\leq p$. Then we can assert:

(i) $A \notin \mathfrak{I}$;
(ii) if $C \subseteq B \in \mathfrak{I}$, then $C \in \mathfrak{I}$;
(iii) if $B \in \mathfrak{I}$ and $C \in \mathfrak{I}$, then $B \cup C \in \mathfrak{I}$.

We can easily prove (ii), because if $C \subseteq B \in \mathfrak{I}$, then $\inf i(C) \geq \inf i(B)$. For (iii), note that if $B \in \mathfrak{I}$ and $C \in \mathfrak{I}$, then $\inf i(B) \not\leq p$ and $\inf i(C) \not\leq p$, whence we see that $\inf i(B \cup C) = \inf i(B) \wedge \inf i(C) \not\leq p$ by the primality of p. Thus, $\mathfrak{I}$ is a proper ideal of subsets of A. But then there is an ultrafilter $\mathfrak{U}$ on A disjoint from $\mathfrak{I}$ (by I-3.19 and Remark). The latter means that $\inf i(B) \leq p$ for all $B \in \mathfrak{U}$. We conclude that

$$i(\underline{\lim}\ \mathfrak{U}) = i(\sup\{\inf B : B \in \mathfrak{U}\}) = \sup\{i(\inf B) : B \in \mathfrak{U}\}$$
$$\leq \sup\{\inf i(B) : B \in \mathfrak{U}\} \leq p,$$

where we have used the fact that i preserves directed sups. □

1.1′. THE LEMMA. *Let* L *and* M *be complete lattices and let* $i : \mathrm{L} \to \mathrm{M}$ *be Scott continuous and preserve arbitrary* infs. *Suppose* p *is an irreducible element of* M *and the subset* $A \subseteq \mathrm{L}$ *is such that* $\inf i(A) = p$. *Then there is an ultrafilter* $\mathfrak{U}$ *on* A *with* $i(\underline{\lim}\ \mathfrak{U}) = p$.

Proof. The proof is the same as that of 1.1, if one replaces everywhere $\not\leq$ by $>$, and $\leq$ by $=$. (One needs that i preserves arbitrary infs to replace the first $\leq$ by $=$ in the last equation.) □

If the set A contains $\underline{\lim}\ \mathfrak{U}$ for every ultrafilter $\mathfrak{U}$ on A, we may conclude that under the hypotheses of 1.1 [resp. 1.1′] that we have $\inf i(A) \leq p$ [$\inf i(A) = p$] implies $p \in \uparrow i(A)$ [$p \in i(A)$, respectively]. Since in a continuous lattice $\underline{\lim}\ \mathfrak{U}$ is the topological limit of $\mathfrak{U}$ with respect to the Lawson topology (III-3.12), we have proved:

1.2. COROLLARY. *Let* L *be continuous and* M *be complete. If* $i : \mathrm{L} \to \mathrm{M}$ *is Scott continuous and* $K \subseteq \mathrm{L}$ *is compact in the Lawson topology, then*

(i) *If* p *is prime in* M, *then* $\inf i(K) \leq p$ *implies* $p \in \uparrow i(K)$;

(ii) *If* p *is irreducible in* M *and, in addition,* i *preserves arbitrary infs, then* $\inf i(K) = p$ *implies* $p \in i(K)$. □

If we specialize (1.1) and (1.1′) to the case where $\mathrm{L} = \mathrm{M}$ and i is the identity map, then we obtain:

1.3. COROLLARY. *Let* L *be a complete lattice.*

(i) *If* p *is prime in* L, *then on every subset* A *of* L *with* $\inf A \leq p$ *there is an ultrafilter* $\mathfrak{U}$ *such that* $\underline{\lim}\ \mathfrak{U} \leq p$;

(ii) *If* p *is irreducible in* L, *then on every subset* A *of* L *with* $\inf A = p$ *there is an ultrafilter* $\mathfrak{U}$ *such that* $\underline{\lim}\ \mathfrak{U} = p$. □

If L is a continuous lattice, then the set of all lim-infs of ultrafilters on A is just the closure A^- of A with respect to the Lawson topology (III-3.12).

Thus 1.3 implies:

1.4. COROLLARY. *Let* L *be a continuous lattice.*

(i) *If p is prime in* L, *then* $\inf A \leq p$ *implies* $p \in \uparrow(A^-)$ *for every nonempty subset A of* L.

(ii) *If p is irreducible in* L, *then* $\inf A = p$ *implies* $p \in A^-$ *for every nonempty subset A of* L. □

We can now state the result that we promised.

1.5. THEOREM. *Let* L *be a continuous lattice. Then for every nonempty subset K of* L *which is compact in the Lawson topology we have:*

(i) *If p is prime in* L, *then* $\inf K \leq p$ *implies* $p \in \uparrow K$;

(ii) *If p is irreducible in* L, *then* $\inf K = p$ *implies* $p \in K$. □

EXERCISES

1.6. EXERCISE. Give an independent proof of Corollary 1.2(i) under the following weaker primality condition on p: For every nonempty finite subset F of L, $\inf i(F) \leq p$ implies $p \in \uparrow i(F)$. □

1.7. EXERCISE. Let K be a compact convex subset of a locally convex topological vector space, and denote by Con(K) the co-continuous lattice of all closed convex subsets of K (see Example I-1.22). Prove that if $A \subseteq K$ has the property that its closed convex hull is equal to K, then A^- contains all extreme points of K (see I-3.39). □

1.8. EXERCISE. Let X be a compact space. Show that for every closed prime ideal I of the ring C(X) of real or complex continuous functions on X, there is an element $x \in$ X such that $I = \{f \in C(X) : f(x) = 0\}$.

(HINT: Consider X as a subset of $\mathcal{O}$(X) via the embedding $x \mapsto X \setminus \{x\}$ and use for i the map $U \mapsto \{f \in C(X) : f(x) = 0 \text{ for all } x \notin U\}$ from the lattice L = $\mathcal{O}$(X) of all open subsets of X to the lattice M of all closed ideals of C(X).) □

We next complete a theme begun in Exercise I-3.42, where we started to investigate the relationship between completely distributive lattices and continuous posets.

1.9. EXERCISE. Let L be a completely distributive lattice (I-2.4, 2.5, 3.15, 3.42) and $p \neq 0$ a coprime. Then $⇊p \cap P$ is directed, where P denotes the set of non-zero coprimes.

(HINT: Let q, r be two co-primes with $0 \neq q$ and $r \ll p$. Then $⇈q \cap ⇈r$ is a Scott- (hence, Lawson-) open neighborhood of $\uparrow p$. But in a completely distributive lattice we have $\lambda(L) = \lambda(L^{op})$. (This follows immediately from the fact that L is embedded under an ***INF$\cap$SUP*** map into $[0,1]^X$ for some X: see IV-

2.30, 2.31; alternatively refer to Proposition VII-2.9.) From I-3.42 we know then that $p = \sup(\mathop{\downarrow\!\!\downarrow} p \cap P)$, whence $(\mathop{\downarrow\!\!\downarrow} p \cap P)^{-} \cap \uparrow p \neq \emptyset$, with closure taken with respect to $\lambda(L^{op}) = \lambda(L)$ by an application of 1.4 to L^{op}. Then we find $(\mathop{\downarrow\!\!\downarrow} p \cap P) \cap \mathop{\uparrow\!\!\uparrow} q \cap \mathop{\uparrow\!\!\uparrow} r \neq \emptyset$, since $\mathop{\uparrow\!\!\uparrow} q \cap \mathop{\uparrow\!\!\uparrow} r$ is an open neighborhood of $\uparrow p$.) □

1.10. EXERCISE. Establish from 1.9 the following theorem.

THEOREM. *If* L *is a completely distributive lattice, then the poset* P *of non-zero coprimes in the induced order is a continuous poset. Dually, if* Q *is the set of primes* $p<1$, *then* $(Q,\geq)$ *is a continuous poset.*

(HINT: Take I-3.42 and note that (1) above shows that $\mathop{\downarrow\!\!\downarrow}_P p$ is directed for $p \in P$. For the dual statement apply the preceding result to L^{op}.) □

1.11. EXERCISE. If L is completely distributive and P the poset of non-zero coprimes, and if U is an open filter in P, then $\uparrow U$ is an open filter in L.

(HINT: Let $v \in \uparrow U$; then $u \leq v$ with $u \in U$; then by I-3.42 and 1.9 above, there is a $u^* \in \mathop{\downarrow\!\!\downarrow} u \cap P$ with $u^* \in U$; but then $u^* \ll v$.) □

We can now throw additional light on the Lawson duality which we discussed in IV-1.39 ff.

1.12. EXERCISE. Let L be a completely distributive lattice. Let Q be the continuous poset of non-zero coprimes with the partial order induced from $\geq$ and let P be the continuous poset of nonzero primes with the partial order induced from $\leq$. Then P and Q are duals of each other in the sense of Lawson duality of IV-1.42.

(HINT: Let U be an open filter of P. Then $\uparrow U$ is a Scott-open upper set in L (see 1.11). Moreover, $I = L \setminus \uparrow U$ is an ideal. Let $p_U = \max I$. Then $p_U \in Q$. Conversely if $p \in Q$, let $U_p = P \setminus \downarrow p$; then U_p is an open filter in P. The maps $U \mapsto p_U : \mathrm{OFilt}\, P \to Q$ and $p \mapsto U_p : Q \to \mathrm{OFilt}\, P$ are inverses of each other. Hence $Q \cong \mathrm{OFilt}\, P$. In the light of Lawson duality, this gives the assertion.) □

1.13. EXERCISE. (i) Let $i : L \to M$ be a morphism in $\boldsymbol{INF}^{\uparrow}$. If $p<1$ is a prime in M and $x \in L$ satisfies $i(x) \leq p$, then there is an irreducible $q \in L$ with $x \leq q$ and $i(q) \leq p$.

(ii) The conclusion also holds for L an up-complete semilattice (O-2.11) and M any poset (where $p \in M$ is prime iff $M \setminus \downarrow p$ is a filter in M), provided that $i^{-1}(U)$ is an open filter for any open filter U of M.

(HINT: Pick q maximal in $\uparrow x \setminus i^{-1}(M \setminus \downarrow p)$.) □

NOTES

THE LEMMA (1.1) is an abstract version of the so-called Jónsson Lemma which plays an important role in universal algebra (Jónsson [1967]). Its relevance in continuous lattices in the form of 1.2, 1.4, 1.5 was discovered by Gierz and Keimel [1976]. From that latter paper we have also drawn the exercises; 1.7 and 1.8 are well-known theorems in analysis. The same paper contains more material on the uses of THE LEMMA; in particular one obtains a characterization of the extreme points in the dual unit ball of semicontinuous function spaces originally due to Cunningham and Roy [1974]. The results on irreducibles contained in 1.4 and 1.5 are from Hofmann and Lawson [1976/77]. The duality of continuous posets and completely distributive lattices in Exercises 1.9 – 1.12 is due to Lawson [198*]. Finally, Exercise 1.13 is from Hofmann, Keimel, and Watkins [SCS-51-52].

2. ORDER GENERATION AND TOPOLOGICAL GENERATION

A subset X of a lattice L has been called *order generating* (I-3.8), if every element of L is the inf of a subset of X. It has been proved (I-3.10) that, in a continuous lattice L, the set IRR L of all irreducible elements of L is order generating. In an algebraic lattice, the set Irr L of completely irreducible elements is the (unique) smallest order generating subset (I-4.23). But in general a continuous lattice does not have any minimal order-generating subset. In the unit interval [0,1] every order-dense subset is order generating, but there is no minimal subset of this type. For this reason we restrict our attention for the moment to order-generating sets which are closed with respect to the Lawson topology. As a consequence of The Lemma of Section 1 we obtain:

2.1. THEOREM. *Among the order-generating subsets of a continuous lattice* L *which are closed with respect to the Lawson topology there is a unique smallest one: the closure* $(\mathrm{IRR}\ L\backslash\{1\})^-$ *of the set of irreducible elements* <1 *in* L.

Proof. By I-3.10, $(\mathrm{IRR}\ L\backslash\{1\})^-$ is order generating. Let X be any Lawson-closed order-generating subset of L. Then 1.5(ii) implies that X contains every irreducible element; whence, $(\mathrm{IRR}\ L\backslash\{1\})^-\subseteq X$. □

In a topological semilattice another notion of generation is natural:

2.2. DEFINITION. A subset X of a topological semilattice L is said to be *topologically generating* if the smallest closed subsemilattice of L containing X and 1 is L itself. □

As a matter of convention in a continuous lattice topological generation is always understood with respect to the Lawson topology. In compact semilattices, and in particular in continuous lattices, topological generation is weaker than order generation:

2.3. PROPOSITION. *Let* L *be a continuous lattice. Then every order-generating subset of* L *is topologically generating with respect to the Lawson topology.*

Proof. Let X be an order generating subset and T the smallest closed subsemilattice of L containing X and 1. Then T is closed under arbitrary infs by virtue of III-1.11(2). Thus for all $x\in L$, we have $x = \inf(\uparrow x\cap X)\in T$, and consequently $L = T$. □

The previous proposition remains true for arbitrary compact semilattices (see VI-2.9 below) and, in fact, for a complete lattice with a compact topology in which $\uparrow x$ and $\downarrow x$ are always closed (cf. O-4.4 and VI-1.3 below). The converse of Proposition 2.3, however, is false. In the unit square $[0,1]\times[0,1]$ with the usual order, the set $[0,1[\times[0,1[$ is topologically but not order generating.

How does topological generation work in continuous lattices? Let X be any subset of a continuous lattice L. If we assign to every filter $\mathfrak{F}$ on X the element $\underline{\lim}\ \mathfrak{F}$ of L, we obtain a map preserving arbitrary infs and directed sups from the lattice F = Filt 2^X of all filters on X into L. (This is nothing but the theorem that the lattice F is the free continuous lattice on the set X (I-4.17)). By III-1.8 and III-3.12, the image of this map is the smallest ***closed*** subsemilattice containing X. In other words, we obtain the smallest closed subsemilattice containing X by taking the set of all directed sups in the set of all infs of subsets of X. In particular, L is topologically generated by X if every element of L is a directed sup of infs of subsets of X. This remark leads us to the:

2.4. PROPOSITION. *Let* L *be a continuous lattice.*

(i) *A subset X is topologically generating if and only if its closure X^- with respect to the Lawson topology is order generating;*

(ii) *Among the Lawson-closed topologically generating subsets of* L *there is a (unique) smallest one: The closure* (IRR L$\setminus\{1\})^-$ *of the set of irreducible elements* <1.

Proof. (i): From 2.3 and 2.2, we conclude "if". In order to prove "only if" we let Y be the set of all infs of subsets of X^-. We use III-1.11 to show that Y is a closed subsemilattice. Firstly, Y is clearly closed under the formation of all infs. We must show that for every directed subset $D\subseteq Y$, we have $\sup D \in Y$. For this purpose set $s = \sup D$ and $t = \inf(\uparrow s\cap X^-) \in Y$; then $s\leq t$, and we must now show $t\leq s$. In view of I-1.6 it suffices to show $w\leq s$ for any $w\ll t$: Now $d = \inf(\uparrow d\cap X^-)$ since $d\in Y$, and so

$$\bigcap_{d\in D}(\uparrow d\cap X^-) = (\bigcap_{d\in D}\uparrow d)\cap X^-) = \uparrow s\cap X^-\subseteq\uparrow t.$$

Thus the intersection of the filterbasis of the λ(L)-compact sets $\uparrow d\cap X^-$ is contained in the open neighborhood $\Uparrow w$ of $\uparrow t$. Hence there is some $c\in D$ with $\uparrow c\cap X^-\subseteq\Uparrow w$, whence $c = \inf(\uparrow c\cap X^-)\geq w$, and since $c\leq s$ we have indeed $w\leq s$.

If we now assume that X is topologically generating, then $Y\cup\{1\} = $ L. This means that X^- is order generating.

(ii): Use (i) and 2.1 above. □

EXERCISES

2.5. EXERCISE. Let L be an algebraic lattice endowed with its Lawson topology.

(i) Show that (Irr L$)^-$ is the smallest closed order-generating and topologically generating set, where Irr L denotes the set of completely irreducible elements (see I-4.23).

(ii) If Y is a topologically generating subset of L, show that every compact element of L is the inf of a subset of Y. □

2.6. EXERCISE. Let L be a continuous lattice and Id L its ideal lattice. Consider the adjunction $i = (x \mapsto \downarrow x)$ and $r = (I \mapsto \sup I)$ between L and Id L as in O-3.15.

(i) If X is order generating in L, show that $i(X)$ is topologically generating in Id L.

(ii) If Y is topologically generating in Id L, show that $r(Y)$ is order generating in L.

(HINT: Use I-2.1 and 2.5(ii) above.) □

2.7. EXERCISE. Let L be a complete lattice for which IRR L is order generating. Show that among the subsets of L which are order generating and closed with respect to the lim-inf topology (see Section III-3) there is a smallest one. (Compare Theorem 2.1.). □

(HINT: Use III-3.4 and1.3(iii) above.) □

2.8. EXERCISE. Let L and M be complete lattices and $f: \mathrm{L} \to \mathrm{M}$ a surjective ***UPS***-map preserving finite infs. Show that IRR M$\subseteq f$(IRR L). □

(HINT: For $p \in$ IRR M pick q maximal in $f^{-1}(p)$. Show that q is irreducible.) □

NOTES

R. Jamison [1974] gave a slightly restricted version of Theorem 2.1 in his dissertation. His interest was in abstract theories of convexities. In his context subsemilattices were "convex" subsets, and from this viewpoint irreducibles become "extreme points". Theorem 2.1 then becomes an analogue to the Krein-Milman theorem: the closed convex hull of the extreme points (that is, the smallest closed subsemilattice containing the irreducibles) is all of L, and conversely any closed set whose closed convex hull is all of L contains the extreme points.

The treatment given in this section is essentially that of Hofmann and Lawson [1976/77].

3. WEAK IRREDUCIBLES AND WEAKLY PRIME ELEMENTS

Throughout this section L denotes a continuous lattice in the Lawson topology. In the preceding section we characterized the closure of the set of irreducible elements as being the smallest closed order-generating subset of L. In this section we characterize the individual elements of (IRR L)$^-$. The distributive case is of particular interest in this regard; here (IRR L)$^-$(= (PRIME L)$^-$) consists precisely of the pseudoprime elements already introduced in Chapter I. (Indeed the notion of a pseudoprime element arose in the context of characterizing those distributive continuous lattices for which the set of prime elements is closed.)

3.1. DEFINITION. An element p of L is called *weakly irreducible* if for any finite family $X_1,X_2,...,X_n$ of subsets of L, the relation $p\in\mathrm{int}(X_1X_2\cdots X_n)$ implies $p\in X_k^-$ for some k. We call p *weakly prime* if for any finite family $X_1,X_2,...,X_n$ of subsets of L, the relation $p\in\mathrm{int}\uparrow(X_1X_2\cdots X_n)$ implies $p\in(\uparrow X_k)^-$ for some k. We denote with WIRR L and WPRIME L the sets of weakly irreducible and weakly prime elements of L, respectively. □

In the definition, the notation $X_1X_2\cdots X_n$ stands for the pointwise product of the sets. Recall that an element p of L is called *pseudoprime* if p is the sup of a prime ideal of L. (see Definition I-3.23.) The set of pseudoprimes is denoted by ΨPRIME L. One easily verifies that every irreducible element is weakly irreducible and that every prime element is weakly prime; that is, we have IRR L⊆WIRR L, and PRIME L⊆WPRIME L. We already have seen that PRIME L⊆ΨPRIME L in I-3.23 ff.

3.2. LEMMA. *The sets* WIRR L *and* WPRIME L *are closed.*

Proof. As the proofs in the two cases are similar, we only consider W = WIRR L. Let $p\in W^-$. We want to show $p\in W$. So we take an arbitrary finite family $X_1,...,X_n$ of subsets of L such that $p\in\mathrm{int}\ X_1\cdots X_n$. Pick any net $(p_j)_{j\in J}$ in W such that $p = \lim_j p_j$. Then we can find a j_0 such that $p_j\in\mathrm{int}\ X_1\cdots X_n$ for all $j\geq j_0$. As all the p_j are weakly irreducible, we conclude that $p_j\in X_1^-\cup...\cup X_n^-$ for all $j\geq j_0$. But this implies $p = \lim_j p_j\in X_1^-\cup...\cup X_n^-$; that is, p is weakly irreducible. □

Definition 3.1 makes sense in every semilattice endowed with a topology and Lemma 3.2 remains true in general. But in order to show that the elements in the closure of IRR L are *exactly* the weakly irreducible elements, we have to use our standing hypothesis that L is a continuous lattice:

3.3. PROPOSITION. *In a continuous lattice* L, *we have* $(\mathrm{IRR\,L})^- = \mathrm{WIRR\,L}$.

Proof. Suppose that there is an element $p \in \mathrm{L} \setminus (\mathrm{IRR\,L})^-$. Then every $x \in (\mathrm{IRR\,L})^-$ has a neighborhood $U(x)$ not containing p. By III-2.13, we may suppose that all the $U(x)$ are closed subsemilattices of L and since L is compact Hausdorff, hence regular, we may assume these are closed. As the interiors of the $U(x)$ cover the compact space $(\mathrm{IRR\,L})^-$, we find a finite subcover of compact subsemilattices $X_1,\ldots,X_n$ which do not contain p but where $\mathrm{IRR\,L} \subseteq X_1 \cup \ldots \cup X_n$.

By adjoining the identity to each of the X_k if necessary, we can assert that the pointwise product $X_1 \cdots X_n$ is a closed subsemilattice of L containing IRR L. As IRR L is order generating in L (I-3.10), we conclude that $\mathrm{L} = X_1 \cdots X_n$. In particular, $p \in \mathrm{int}\, X_1 \cdots X_n$. As $p \notin X_k$ for each k, we conclude that $p \notin \mathrm{WIRR\,L}$. □

We do not have an order-theoretical characterization of weakly irreducible elements. However, for weakly prime elements such a characterization is possible. Note that condition (2) in the following theorem is the primality condition of I-3.24(2), and that condition (3) is analogous to 1.5.

3.4. PROPOSITION. *For an element* p *of a continuous lattice* L *the following conditions are equivalent:*

(1) *p is weakly prime;*

(2) *For any nonempty finite set of elements* $x_1,\ldots,x_n$ *in* L, *the relation* $x_1 \wedge \cdots \wedge x_n \ll p$ *implies* $x_k \leq p$ *for some* k;

(3) *For any nonempty compact subset* K *of* L, *the relation* $\inf K \ll p$ *implies* $p \in \uparrow K$.

Proof. (3) implies (2): Immediate.

(2) implies (1): Suppose that $p \in \mathrm{int} \uparrow (X_1 \cdots X_n)$. Then there is an $x \ll p$ with $x \in \uparrow(X_1 \cdots X_n)$ by the definition of the Lawson topology (see II-1.5(5) and III-1.6(i)). Thus $x \geq x_1 \wedge \ldots \wedge x_n$ with suitable $x_k \in X_k$. Now (2) implies $x_k \leq p$ for some k; whence, we find $p \in \uparrow X_k \subseteq (\uparrow X_k)^-$.

(1) implies (3): Assume that $x \not\leq p$ for all $x \in K$. As L is a continuous lattice, for each $x \in K$ we find an element $u(x) \ll x$ with $u(x) \not\leq p$. As $\uparrow u(x)$ is a neighborhood of x, the compactness of K implies the existence of finitely many elements $x_1,\ldots,x_n$ in K such that $K \subseteq \uparrow u(x_1) \cup \ldots \cup \uparrow u(x_n)$. From the relation $\inf K \ll p$ we conclude $p \in \mathrm{int} \uparrow \inf K \subseteq \mathrm{int} \uparrow (u(x_1) \wedge \ldots \wedge u(x_n))$; whence, $p \in (\uparrow u(x_k))^- = \uparrow u(x_k)$ for some k by (1). But this is a contradiction to the choice of $u(x) \not\leq p$. □

In the preceding theorem condition (2) cannot be strengthened to $x_1 \wedge ... \wedge x_n \ll p$ implies $x_k \ll p$ for some k. Indeed, in the unit square $L = [0,1] \times [0,1]$ the element $p = (1,1)$ is prime; but for $x_1 = (1,0)$ and $x_2 = (0,1)$, we have $x_1 \wedge x_2 \ll p$ and neither $x_1 \ll p$ nor $x_2 \ll p$.

In I-3.24 it had been shown that every pseudoprime element satisfies condition (2) of the preceding proposition. Thus:

3.5. COROLLARY. ΨPRIME L$\subseteq$WPRIME L *and if* L *is distributive, then equality holds.* □

Condition (3) in Proposition 3.4 allows us to conclude that every weakly prime element is weakly irreducible:

3.6. PROPOSITION. WPRIME L $\subseteq$WIRR L.

Proof. Let p be weakly prime. For every $t \ll p$, we have

$$t = \inf(\uparrow t \cap \text{WIRR L}),$$

as IRR L$\subseteq$WIRR L and as IRR L is order generating (I-3.10). Because we know $\uparrow t \cap$WIRR L is compact by 3.2, we can find an element $p_t \in$WIRR L such that $t \leq p_t \leq p$ by 3.4(3). Because in the Lawson topology $p = \lim(t)_{t \ll p}$, we conclude that $p = \lim(p_t)_{t \ll p}$. As $p_t \in$WIRR L and as WIRR L is closed, we get $p \in$WIRR L. □

The containment relations between the different kinds of irreducible and prime elements in a *continuous lattice* are summarized in the following diagram:

$$\begin{array}{ccccccc} \text{Irr L} & \subseteq & \text{IRR L} & \subseteq & (\text{IRR L})^- & = & \text{WIRR L} \\ & & \cup\!\shortparallel & & & & \cup\!\shortparallel \\ \{1\} & \subseteq & \text{PRIME L} & \subseteq & \Psi\text{PRIME L} & \subseteq & \text{WPRIME L} \end{array}$$

In the distributive case one has IRR L = PRIME L (I-3.12) and ΨPRIME L = WPRIME L (I-3.24 and 3.4). Thus, the containment diagram simplifies considerably for *distributive continuous lattices* L (continuous cHa's):

$$\begin{array}{ccccccccc} \text{Irr L} & \subseteq & \text{IRR L} & \subseteq & (\text{IRR L})^- & = & & = & \text{WIRR L} \\ & & \| & & \| & & & & \| \\ \{1\} & \subseteq & \text{PRIME L} & \subseteq & (\text{PRIME L})^- & = & \Psi\text{PRIME L} & = & \text{WPRIME L} \end{array}$$

In particular, the pseudoprimes are exactly the elements in the closure of the set of primes. As we had characterized in I-3.27 and I-4.8 the distributive continuous lattices and, in particular, the algebraic lattices for which every pseudoprime element is prime, we obtain the following criterion for the closedness of PRIME L:

3.7. PROPOSITION. (i) *In a distributive continuous lattice* L *the set* PRIME L *of prime elements of* L *is closed iff the relation* $\ll$ *on* L *is multiplicative* (*that is,* $a \ll x$ *and* $a \ll y$ *always imply* $a \ll x \wedge y$).

(ii) *In a distributive algebraic lattice* L *the set* PRIME L *of prime elements is closed iff* L *is arithmetic* (*that is, if the compact elements form a sublattice of* L). □

In the nondistributive case we do not know similar characterizations of those continuous lattices L in which IRR L or PRIME L is closed.

3.8. EXAMPLES. (1) The left-hand figure below indicates a distributive continuous lattice L with a pseudoprime element p which is not prime. Thus, PRIME L is not closed; indeed, PRIME L is just $L \setminus \{p\}$ in this example. The prime ideal P with $p = \sup P$ is $P = \{x : x < p\}$.

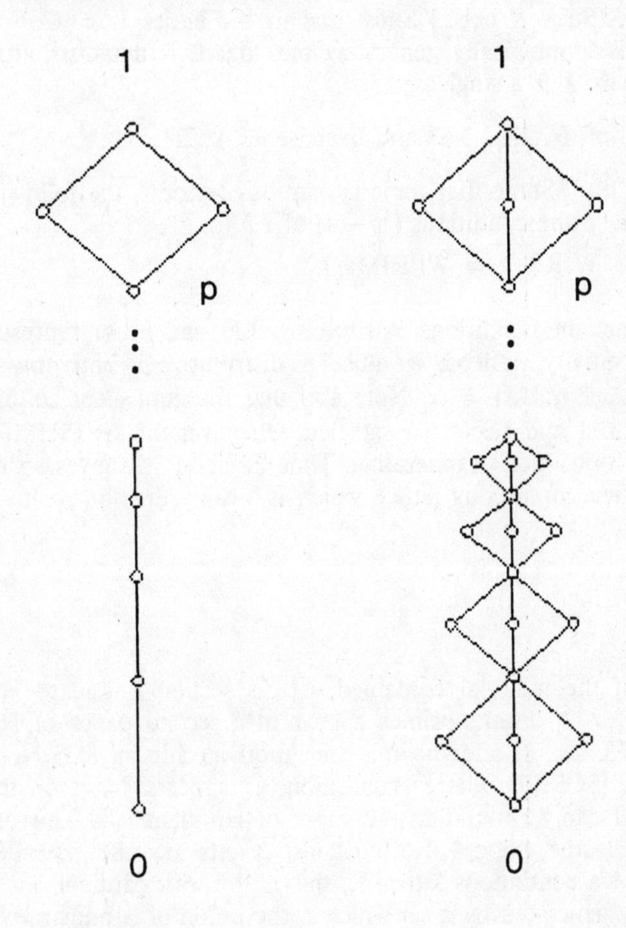

(2) The right-hand figure above indicates a nondistributive continuous lattice with no prime element except 1 but with a pseudoprime element p different from 1. Again $P = \{x : x \ll p\}$ is a prime ideal with sup $P = p$. □

EXERCISES

3.9. EXERCISE. Let L be a continuous lattice in which the set PRIME L of prime elements is topologically generating. Show that L is distributive if and only if the set $T = \{x \in L : x = \inf (\uparrow x \cap \text{PRIME } L)\}$ is a sublattice of L. □

(HINT: See Hofmann and Lawson [1976/77].) □

3.10. EXERCISE. Let K be a compact convex set in a locally convex topological vector space with the property that the set of extreme points is dense in K. (Such K exist.) Show that in the lattice $L = \text{Con}(K)^{\text{op}}$ the set PRIME L is topologically generating but that L is not distributive except in the case where K is a singleton.

(HINT: cf. I-1.22, I-3.39 and Exercise 1.7.) □

3.11. EXERCISE. Show that for a continuous lattice L the following condition is equivalent to the conditions (1)–(4) of I-3.41:

(5) WIRR L = WPRIME L.

Note that the conditions in Exercises 3.11 and I-3.41 represent a kind of weak distributivity; indeed, a lattice is distributive if and only if $a_1 \wedge a_2 \leq x$ implies $(a_1 \vee x) \wedge (a_2 \vee x) = x$. Note also that the equivalent conditions (1)–(5) in Exercise 3.11 and I-3.41 are satisfied, whenever the set PRIME L of prime elements is topologically generating. Thus Exercise 3.10 gives an example of a nondistributive continuous lattice which is weakly distributive in the sense of 3.11.

NOTES

Most of the material contained in this section is due to Hofmann and Lawson [1976/77]. Pseudoprimes appear in a second paper of Hofmann and Lawson [1978, sec. 8] after having been motivated in an SCS-Note of Keimel and Mislove [SCS-19], where Proposition 3.7 appears. More on the subject of Exercise 3.11 can be found in the paper of Hofmann and Lawson mentioned above. The same paper also contains results on the so-called spanning dimension of a continuous lattice L: this is the least cardinal α such that S is topologically generated by a set which is the union of α nonempty chains. The relation of this and similar dimensional concepts with width and breadth of L and IRR L are discussed there in detail. These notions of breadth and width in compact semilattices have also been studied by Baker and Stralka [1970] and Lea [1972].

4. SOBER SPACES AND COMPLETE LATTICES

In this section we assign to every complete lattice a sober topological space and to every topological space a certain distributive complete lattice. These assignments are functorial and establish a dual equivalence between the category of all sober spaces and a certain category of distributive complete lattices. This material concerns complete Heyting algebras in general and is only auxiliary for the following two sections, where we specialize to distributive continuous lattices..

On the level of objects, the essential results can be summarized as follows: a complete lattice L can be represented as the lattice of open subsets of a topological space iff L is order generated by its prime elements. In particular, these lattices are complete Heyting algebras. A topological space X is homeomorphic to the space Spec L of nonunit prime elements endowed with the hull-kernel topology of some lattice L iff X is sober.

4.1. DEFINITION. For a complete lattice L we denote by Spec L the set of all prime elements of L different from 1; that is,

$$\mathrm{Spec}\, L = \mathrm{PRIME}\, L \setminus \{1\}.$$

For every $a \in L$, let

$$\nabla_L(a) = \{p \in \mathrm{Spec}\, L : a \leq p\} = \uparrow a \cap \mathrm{Spec}\, L$$

be the *hull* of a, and let

$$\Delta_L(a) = \mathrm{Spec}\, L \setminus \nabla_L(a) = \mathrm{Spec}\, L \setminus \uparrow a$$

be the complement of the hull of a in Spec L. □

4.2. PROPOSITION. *We have for all* $X \subseteq L$ *and all finite* $F \subseteq L$:

(i) $\nabla_L(0) = \mathrm{Spec}\, L$ *and* $\nabla_L(1) = \emptyset$;

(ii) $\Delta_L(0) = \emptyset$ *and* $\Delta_L(1) = \mathrm{Spec}\, L$;

(iii) $\bigcap\{\nabla_L(a) : a \in X\} = \nabla_L(\sup X)$;

(iv) $\bigcup\{\Delta_L(a) : a \in X\} = \Delta_L(\sup X)$;

(v) $\bigcup\{\nabla_L(a) : a \in F\} = \nabla_L(\inf F)$;

(vi) $\bigcap\{\Delta_L(a) : a \in F\} = \Delta_L(\inf F)$.

Proof. (i) and (ii) are clear. For (iii) it suffices to remark that the condition $p \in \bigcap\{\nabla_L(a) : a \in X\}$ means $p \geq a$ for all $a \in X$; this in turn is equivalent to saying that $p \geq \sup X$; that is, $p \in \nabla_L(\sup X)$. (iv) is similar. For (v) we note that $p \in \nabla_L(\inf F)$ means $p \geq \inf F$; as F is finite and as p is prime, this is equivalent to $p \geq a$ for some $a \in F$, that is, $p \in \bigcup\{\nabla_L(a) : a \in F\}$. (vi) is similar. □

From 4.2 we conclude that the sets of the form $\nabla_L(a)$ for $a \in L$ are exactly the closed sets and the sets of the form $\Delta_L(a)$ for $a \in L$ are exactly the open sets of one and the same topology on Spec L.

4.3. DEFINITION. The *hull-kernel topology* on Spec L is defined to be the topology the open sets of which are the sets of the form $\Delta_L(a)$ for $a \in L$. □

From now on, Spec L will always be endowed with this topology; that is, $\mathcal{O}(\text{Spec L}) = \{\Delta_L(a) : a \in L\}$. Recall that the lower topology $\omega(L)$ on L is defined to have the principal filters $\uparrow a$ for $a \in L$ as subbasic closed sets (III-1.1). As $\nabla_L(a) = \uparrow a \cap \text{Spec L}$ and as these sets are the closed sets of the hull-kernel topology, we conclude that the hull-kernel topology on Spec L is the subspace topology induced from the lower topology $\omega(L)$ on L. It follows that the hull-kernel topology is coarser than the topology induced from the Lawson topology on L (see III-1.5).

Remark. The assignment $a \mapsto \Delta_L(a) : L \to \mathcal{O}(\text{Spec L})$ is surjective and preserves arbitrary sups and finite infs (4.2); the map is bijective if and only if Spec L is order generating in L. Indeed, since $X = \nabla_L(\inf X)$ iff X is a hull-kernel closed subset of Spec L, then ∇_L is injective iff $a = \inf \nabla_L(a)$ for all $a \in L$ iff Spec L is order generating. In this case, the inverse of Δ_L is given by

$$U \mapsto \inf(\text{Spec L} \setminus U) : \mathcal{O}(\text{Spec L}) \to L.$$

It is important to recall at this point the notion of a sober space (see I-3.36 and also the discussion preceding II-1.12).

4.4. PROPOSITION. *For every complete lattice* L, *the space* Spec L *is sober.*

Proof. It is obvious that $\{p\}^- = \nabla_L(p)$ for every $p \in \text{Spec L}$. Now, if p, q are elements in Spec L with $\{p\}^- = \{q\}^-$, then, $p \leq q$ and $q \leq p$. Thus, we have proved Spec L is a T_0-space.

Next let A be any nonempty irreducible closed subset of Spec L. We show that $A = \{p\}^-$ for some $p \in \text{Spec L}$. Since A is closed, $A = \nabla_L(a)$ for some $a \in L$. Let $p = \inf \nabla_L(a)$. Then $p \neq 1$ and $\nabla_L(a) = \nabla_L(p)$. We wish to show that p is prime.

Suppose that $b \wedge c \leq p$. If $x \in A$, then $b \wedge c \leq p \leq x$. Since x is prime either $b \leq x$ or $c \leq x$; that is, $x \in \uparrow b \cup \uparrow c$. Thus, $A \subseteq \nabla_L(b) \cup \nabla_L(c)$. Since A is irreducible, either $A \subseteq \uparrow b$ or $A \subseteq \uparrow c$. Hence, either $b \leq p$ or $c \leq p$. □

We now have assigned to every complete lattice L a topological space Spec L. In order to make this assignment functorial, we consider a map $\varphi : L \to M$ of complete lattices *preserving arbitrary sups and finite infs* (cf. O-3.24). (Note that the preservation of finite infs includes the property $\varphi(1) = 1$.) The upper adjoint (recall O-3.5) $\tau = (y \mapsto \max \varphi^{-1}(\downarrow y)) : M \to L$ has the fundamental property

$$\tau(b) \geq a \text{ iff } b \geq \varphi(a) \text{ for all } a \in \mathrm{L} \text{ and all } b \in \mathrm{M},$$

which is equivalent to

$$\tau^{-1}(\uparrow a) = \uparrow\varphi(a) \text{ for all } a \in \mathrm{L}.$$

We note:

4.5. LEMMA. *τ maps* Spec M *into* Spec L.

Proof. By IV-1.26, τ(PRIME M))$\subset$PRIME L. Let $p \in$ Spec M. We have $\tau(p) \neq 1$; indeed $\tau(p) = 1$ would imply $p \geq \varphi(1) = 1$; that is, $p = 1$. Hence τ(Spec M)$\subset$Spec L. □

We therefore denote by

$$\text{Spec } \varphi : \text{Spec M} \to \text{Spec L}$$

the restriction and corestriction of $\tau : \mathrm{M} \to \mathrm{L}$. The following formula clearly implies that Spec φ is continuous with respect to the hull-kernel topologies on Spec M and Spec L:

4.6. LEMMA. $(\text{Spec } \varphi)^{-1}(\Delta_{\mathrm{L}}(a)) = \Delta_{\mathrm{M}}(\varphi(a))$ *for all* $a \in \mathrm{L}$.

Proof. $(\text{Spec } \varphi)^{-1}(\Delta_{\mathrm{L}}(a)) = \tau^{-1}(\text{Spec L}\setminus\uparrow a) \cap \text{Spec M} = \text{Spec M}\setminus\tau^{-1}(\uparrow a)$ $= \text{Spec M}\setminus\uparrow\varphi(a) = \Delta_{\mathrm{M}}(\varphi(a))$. □

If φ has an upper adjoint τ and φ' an upper adjoint τ', then $\tau' \circ \tau$ is the upper adjoint of $\varphi \circ \varphi'$. Thus Spec $(\varphi \circ \varphi') =$ Spec $\varphi' \circ$ Spec φ. This and the preceding remarks show that we have indeed a functor

$$\text{Spec} : \boldsymbol{SUP}^{\wedge} \to \boldsymbol{TOP}^{\mathrm{op}}$$

Here ***TOP*** is the category of all topological T_0-spaces and all continuous maps (cf. II-2, Remarks preceding 2.11) and $\boldsymbol{SUP}^{\wedge} = \boldsymbol{SUP} \cap \boldsymbol{Lat}$ is the category of all complete lattices and all maps between them preserving arbitrary sups and finite infs (see IV-1, notably 1.1 and 1.26).

There is an obvious functor the other way around

$$\mathcal{O} : \boldsymbol{TOP}^{\mathrm{op}} \to \boldsymbol{SUP}^{\wedge}$$

where to every topological space X we assign its lattice $\mathcal{O}$(X) of open subsets, and for every continuous function f: X→Y we assign the function

$$\mathcal{O}(f) = (U \mapsto f^{-1}(U)) : \mathcal{O}(\mathrm{Y}) \to \mathcal{O}(\mathrm{X}),$$

which clearly preserves arbitrary unions and finite intersections and, hence, is a morphism in the category $\boldsymbol{SUP}^{\wedge}$.

For an arbitrary topological space X, we now consider the set Spec $\mathcal{O}$(X) of all prime elements $U \neq$ X of the lattice $\mathcal{O}$(X) endowed with the hull-kernel topology. Since for every $x \in$X the open set $\mathrm{X}\setminus\{x\}^-$ is prime in $\mathcal{O}$(X), we may define a function

$$\xi_X = (x \mapsto X\setminus\{x\}^-) : X \to \operatorname{Spec} \mathcal{O}(X)$$

Note that $U = \xi_X^{-1}(\Delta_{\mathcal{O}(X)}(U))$ for every open subset U of X; in particular, we find that $\xi_X : X \to \operatorname{Spec} \mathcal{O}(X)$ is continuous. Indeed,

$$\begin{aligned} x \in \xi_X^{-1}(\Delta_{\mathcal{O}(X)}(U)) \text{ iff } & \xi_X(x) \in \Delta_{\mathcal{O}(X)}(U) \\ \text{iff } & U \not\subseteq \xi_X(x) = X\setminus\{x\}^- \\ \text{iff } & x \in U. \end{aligned}$$

Note also that $\xi_X(U) = \Delta_{\mathcal{O}(X)}(U) \cap \xi_X(X)$ for every open subset U of X; in particular, ξ_X is an open map onto its image. Furthermore ξ_X is injective iff $\{x\}^- = \{y\}^-$ implies $x = y$; that is, iff X is a T_0-space. And in this case ξ_X is an embedding. Finally, ξ_X is surjective iff every prime element of $\mathcal{O}(X)$ can be written in the form $X\setminus\{x\}^-$ for some $x \in X$, which is equivalent to saying that every irreducible closed subset of X is of the form $\{x\}^-$ for some $x \in X$. We thus conclude that ξ_X is bijective iff X is a sober space, and that in this case ξ_X is a homeomorphism.

Now we formulate the main result of this section. In the following ***SOB*** denotes the full subcategory of ***TOP*** whose objects are the sober topological spaces, and $\boldsymbol{HEYT_0}$ the full subcategory of $\boldsymbol{SUP^\wedge}$ whose objects are the complete lattices in which the prime elements are order generating. Note that the category $\boldsymbol{HEYT_0}$ is a full subcategory of ***HEYT***, the category of all complete Heyting algebras with maps from $\boldsymbol{SUP^\wedge}$ (see II-2, Remarks preceding II-2.11).

4.7. PROPOSITION. (i) *The functor*

$$\operatorname{Spec} : \boldsymbol{SUP^\wedge} \to \boldsymbol{TOP^{op}}$$

is left adjoint to the functor

$$\mathcal{O} : \boldsymbol{TOP^{op}} \to \boldsymbol{SUP^\wedge}.$$

Front and back adjunctions are given by

$$\Delta_L : L \to \mathcal{O}(\operatorname{Spec} L) \text{ and } \xi_X : X \to \operatorname{Spec} \mathcal{O}(X)$$

where $\Delta_L(a) = \operatorname{Spec} L \setminus \uparrow a$ *and* $\xi_X(x) = X\setminus\{x\}^-$ *for any complete lattice* L *and any topological space* X. *Moreover,*

$$\Delta_{\mathcal{O}(X)} : \mathcal{O}(X) \to \mathcal{O}(\operatorname{Spec} \mathcal{O}(X)) \text{ and } \xi_{\operatorname{Spec} L} : \operatorname{Spec} L \to \operatorname{Spec} \mathcal{O}(\operatorname{Spec} L)$$

are isomorphisms.

(ii) *The categories* ***SOB*** *and* $\boldsymbol{HEYT_0}$ *are dual under the restrictions of the functors* Spec *and* $\mathcal{O}$.

(iii) *The functors*

$$\mathcal{O}\,\mathrm{Spec} : \boldsymbol{SUP}^{\wedge} \to \boldsymbol{HEYT}_0 \text{ and } \mathrm{Spec}\,\mathcal{O} : \boldsymbol{TOP} \to \boldsymbol{SOB}$$

are reflections. In particular, Δ_L *is an isomorphism iff* $L \in \boldsymbol{HEYT}_0$*, and* ξ_X *a homeomorphism iff* X *is sober.*

Remark. For any space X, its sober reflection $X^s = \mathrm{Spec}\,\mathcal{O}(X)$ is called the *sobrification* of X and the natural map $\xi_X : X \to X^s$ is called the *sobrification map.*

Proof. We first prove the naturality of Δ_L and ξ_X. The commutativity of the diagram

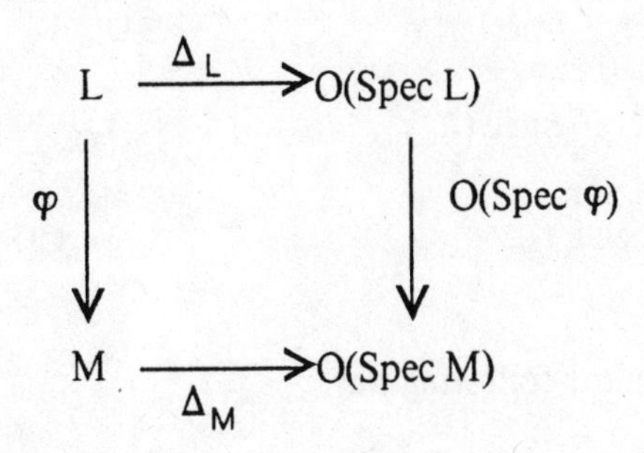

follows from the fact that

$$(\mathcal{O}\mathrm{Spec}\,\varphi)(\Delta_L(a)) = (\mathrm{Spec}\,\varphi)^{-1}(\Delta_L(a)) = \Delta_M(\varphi(a))$$

for every $a \in L$ (see 4.6). The commutativity of

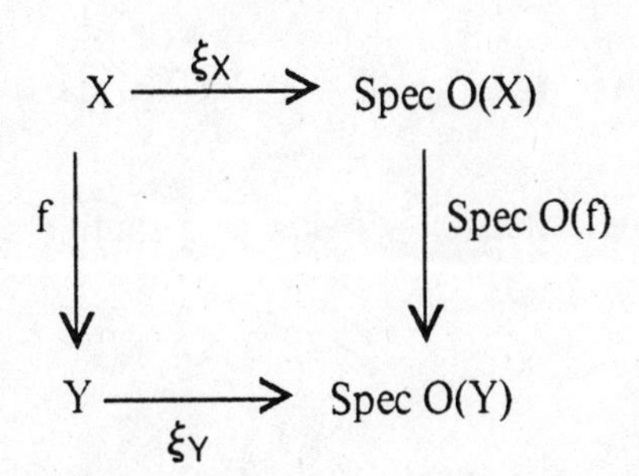

can be seen in the following way:

$$\begin{aligned}(\mathrm{Spec}\,\mathcal{O}(f))(\xi_X(x)) &= (\mathrm{Spec}\,\mathcal{O}(f))(X\backslash\{x\}^-)\\ &= \bigcup \mathcal{O}(f)^{-1}(\downarrow(X\backslash\{x\}^-))\\ &= \bigcup\{U \in \mathcal{O}(Y) : \mathcal{O}(f)(U) \subseteq X\backslash\{x\}^-\}\\ &= \bigcup\{U \in \mathcal{O}(Y) : x \notin f^{-1}(U)\}\\ &= \bigcup\{U \in \mathcal{O}(Y) : f(x) \notin U\}\end{aligned}$$

$$= Y\backslash\{f(x)\}^-$$
$$= \xi_Y(f(x)),$$

for all $x \in X$.

The fact that $\Delta_{O(X)} : O(X) \to O(\mathrm{Spec}\ O(X))$ is a lattice isomorphism follows from the comments before 4.2 and 4.7(i) as does the fact that the mapping $\xi_{\mathrm{Spec\ L}} : \mathrm{Spec\ L} \to \mathrm{Spec}\ O(\mathrm{Spec\ L})$ is a homeomorphism.

In order to show the adjointness of the functors Spec and O, we prove the commutativity of the following diagrams:

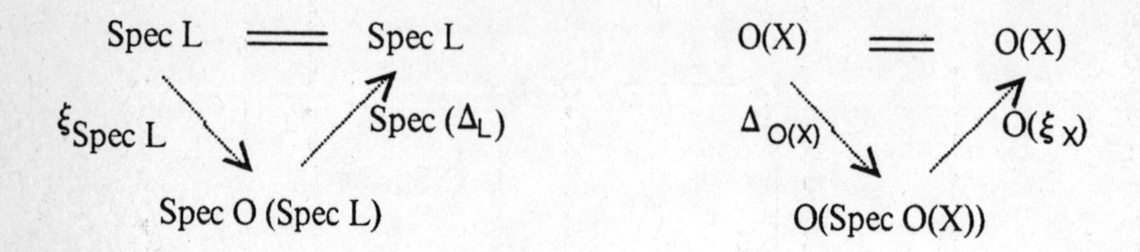

This can be done in the following way:

$$\begin{aligned}(\mathrm{Spec}\ \Delta_L)(\xi_{\mathrm{Spec\ L}}(p)) &= (\mathrm{Spec}\ \Delta_L)(\mathrm{Spec\ L}\backslash\{p\}^-)\\ &= (\mathrm{Spec}\ \Delta_L)(\Delta_L(p))\\ &= \sup \Delta_L^{-1}(\downarrow\Delta_L(p))\\ &= p\end{aligned}$$

for every $p \in \mathrm{Spec\ L}$, and

$$O(\xi_X)(\Delta_{O(X)}(U)) = \xi_X^{-1}(\Delta_{O(X)}(U)) = U$$

for every $U \in \Delta(S)$.

The remaining assertions now follow in a routine manner. □

EXERCISES

4.8. EXERCISE. (i) Let L be a complete lattice, $X \subseteq L$. The following statements are equivalent:

(1) X is sober with respect to the relative lower topology.

(2) For $x \in L$, if $x = \underline{\lim}\ \mathcal{F}$ for some ultrafilter $\mathcal{F}$ on $\uparrow x \cap X$, then $x \in X$.

(ii) Show that the set Spec L satisfies condition (2) in any complete lattice. This gives an alternative argument that Spec L is sober.

(HINT: Use Exercise III-3.16.) □

The following gives an alternate approach to constructing the sobrification of a space.

4.9. EXERCISE. Let X be a topological space. Define a set X^s by

$$X^s = \{A \subseteq X : A \text{ is closed, irreducible, and nonempty}\}.$$

Topologize X^s by open sets $U^s = \{A \in X^s : A \cap U \neq \emptyset\}$ for each open set U of X. If we let $j : X \to X^s$ be the map $x \mapsto \{x\}^-$, show that (X^s, j) is—up to homeomorphism—the sobrification of X. □

4.10. EXERCISE. Let L be a complete lattice. A filter $F \subseteq L$ is said to be *completely prime* if $\sup A \in F$ and $A \neq \emptyset$ always imply $A \cap F \neq \emptyset$. Show that the following statements are equivalent for a proper filter F:

(1) F is completely prime.

(2) F is Scott open and prime.

(3) There exists a prime $p \neq 1$ such that $F = L \setminus \downarrow p$. □

The following exercise gives an alternate approach to defining Spec L.

4.11. EXERCISE. Let L be a complete lattice. Define

$$\text{Spec } L = \{F \subseteq L : F \text{ is a proper completely prime filter}\}.$$

Topologize Spec L by taking as open sets all sets of the form

$$\$(x) = \{F \in \text{Spec } L : x \in F\}.$$

Show that all sets of the form $\$(x)$ for $x \in L$ form a topology on Spec L, and show that this definition is equivalent to the one given in the text.

(HINT: Use part (3) of 4.10 to set up the equivalence.) □

NOTES

The material contained in this section is standard. The theme of representing lattices by suitable topologies—usually hull-kernel topologies—goes back to M.H. Stone's famous papers [1936],[1937] on the topological representation of Boolean algebras and distributive lattices. Of the authors who have pursued this theme we only quote Büchi [1952], Papert [1959], Bruns [1962], Thron [1962], and Drake and Thron [1965]. In these papers one finds the duality between sober spaces and complete lattices order generated by their prime elements, at least on the object level. Explicit formulations of this and other dualities in the language of category theory have been collected in the memoir of Hofmann and Keimel [1972].

5. DUALITY FOR CONTINUOUS HEYTING ALGEBRAS

In this section it is our aim to show that there is a one-one correspondence between distributive continuous lattices (that is, continuous Heyting algebras) and locally quasicompact sober spaces in the sense of a duality of categories. It will take some developement, however, to specify the maps of the desired categories precisely.

If X is a locally quasicompact space (meaning that every point in X has a neighborhood basis of quasicompact sets), it is easy to see that the lattice $\mathcal{O}(X)$ of open subsets is a distributive continuous lattice (see I-1.7(5)). Conversely, if L is a distributive continuous lattice, then the space Spec L of all prime elements $p \neq 1$ endowed with the hull-kernel topology is sober by 4.4; we have to prove that Spec L is locally quasicompact, too. For this we first need a general criterion for quasicompactness of sets in Spec L.

5.1. LEMMA. *Let* L *be a complete lattice. A subset Q of* Spec L *is quasicompact for the hull-kernel topology iff $\downarrow Q$ is Scott closed in* L.

Proof. Suppose that $Q \subseteq \operatorname{Spec} L$ is quasicompact, and let D be a directed set in $\downarrow Q$. Then $\{\nabla_L(d) \cap Q : d \in D\}$ is a filter base of (nonempty) closed subsets of Q. As Q is quasicompact, we have

$$\begin{aligned}\bigcap\{\nabla_L(d) \cap Q : d \in D\} &= \bigcap\{\nabla_L(d) : d \in D\} \cap Q \\ &= \nabla_L(\sup D) \cap Q \neq \emptyset.\end{aligned}$$

But this means that $\sup D \in \downarrow Q$. Thus, $\downarrow Q$ is Scott closed.

Suppose conversely that $\downarrow Q$ is Scott closed. In order to show that Q is quasicompact, let $\mathfrak{F}$ be a filter base of closed subsets F of Spec L with $F \cap Q \neq \emptyset$. Then $F = \nabla_L(\inf F)$ for all $F \in \mathfrak{F}$, and the set $\{\inf F : F \in \mathfrak{F}\}$ is directed and contained in $\downarrow Q$ as $F \cap Q \neq \emptyset$. As $\downarrow Q$ is Scott closed, we see that $\sup\{\inf F : F \in \mathfrak{F}\} \in \downarrow Q$. Thus, an element $q \in Q$ exists with we have $q \geq \sup\{\inf F : F \in \mathfrak{F}\}$; that is, $q \in \nabla_L(\inf F) = F$ for all $F \in \mathfrak{F}$. □

For a more specific criterion of quasicompactness we need the following notion:

5.2. DEFINITION. A subset Q of a topological space X is called *saturated* if $\{x\}^- \cap Q \neq \emptyset$ always implies $x \in Q$. □

Using the specialization order on X, where $x \leq y$ iff $x \in \{y\}^-$ (see II-3.6), we can say that Q is saturated in X iff $q \in Q$ and $q \leq x \in X$ always imply $x \in Q$ iff $Q = \uparrow Q$ in ΩX (cf. II-3.6). For more information on saturating, see Exercise 5.17 below. In the case X = Spec L with the hull-kernel topology, where L is a complete lattice, we have $\{x\}^- = \nabla_L(x)$; thus, Q is saturated in Spec L iff $q \in Q$ and $q \geq x \in \operatorname{Spec} L$ always imply $x \in Q$ iff Q is a lower subset of Spec L. The reader should note, by the way, that on Spec L the specialization order is ***opposite*** to the order induced from L.

5.3. LEMMA. *Let* L *be a complete lattice. A subset* Q *of* Spec L *is saturated and quasicompact for the hull-kernel topology iff there is a Scott-open filter* $F \subseteq$ L *such that* $Q =$ Spec L $\setminus F$ *and* $\downarrow Q =$ L$\setminus F$.

Proof. First let Q be saturated and quasicompact. Then $\downarrow Q$ is Scott closed by 5.1. Thus $F =$ L$\setminus\downarrow Q$ is Scott open. As L$\setminus\downarrow Q = \bigcap\{L\setminus\downarrow p : p \in Q\}$ and as all $p \in Q$ are prime, L$\setminus\downarrow Q$ is a filter. Clearly, because Q is saturated, we have $\downarrow Q =$ L$\setminus F$ and $Q =$ Spec L $\setminus F$.

Conversely, if there is a Scott-open filter F with both $Q =$ Spec L $\setminus F$ and $\downarrow Q =$ L$\setminus F$, then Q is saturated and quasicompact by 5.1. □

5.4. COROLLARY. *Let* L *be a distributive complete lattice. Then a subset* Q *of* Spec L *is saturated and quasicompact for the hull-kernel topology iff there is a Scott-open filter* $F \subseteq$ L *such that* $Q =$ Spec L$\setminus F$.

Proof. By 5.3 we only have to show that if F is a Scott-open filter and if $Q =$ Spec L $\setminus F$, then $\downarrow Q =$ L$\setminus F$. Indeed, if F is Scott open, then L$\setminus F$ is Scott closed. Thus, every element in L$\setminus F$ is dominated by a maximal element of that set. As F is a filter, these maximal elements are irreducible (cf. I-3.6), and consequently they are prime by the distributivity of L. Hence, they belong to Q, and so $\downarrow Q =$ L$\setminus F$. □

Now we prove the two principal results of this section:

5.5. THEOREM. *Let* L *be a distributive continuous lattice* L. *The space* Spec L *is sober and locally quasicompact and* $\Delta_L : L \to \mathcal{O}(\text{Spec } L)$ *is an isomorphism.*

Remark. As a consequence of I-3.43(10), all these spaces are Baire spaces. If L is infinite, then ω(Spec L) $= \omega$(L) by 4.7 and III-4.6(i).

Proof. By 4.4, Spec L is sober. Let U be a neighborhood of a point p in Spec L. We want to find a quasicompact neighborhood Q of p contained in U. We may suppose that $U = \Delta_L(a) =$ Spec L$\setminus\uparrow a$ for some a in L. As L is continuous, there is a an element $b \ll a$ with $b \not\leq p$. By I-3.3 there is a Scott-open filter F with $a \in F \subseteq \uparrow b$. Let $Q =$ Spec L $\setminus F$. By 5.4 Q is quasicompact. Further, $a \in F \subseteq \uparrow b$ implies $\Delta_L(b) \subseteq Q \subseteq \Delta_L(a) = U$. As $b \not\leq p$, we have $p \in \Delta_L(b)$, and Q is a quasicompact neighborhood of p contained in U. Finally, 4.7 and I-3.14 show that Δ_L is an isomorphism. □

5.6. THEOREM. (i) *For a sober space* X, *the lattice* $\mathcal{O}$(X) *of open subsets is continuous iff* X *is locally quasicompact;*

(ii) $U \ll V$ *for open subsets* U *and* V *of* X *if there is a quasicompact set* Q *with* $U \subseteq Q \subseteq V$. *The converse holds if* X *is locally quasicompact.*

Proof. The characterization of $U \ll V$ has been given in I-1.4(ii). In I-1.7(5) it has been shown that $\mathcal{O}$(X) is continuous for every locally quasicompact space. Finally, let X be sober. Then X is homeomorphic to Spec $\mathcal{O}$(X) by the remarks before 4.7. If, in addition, $\mathcal{O}$(X) is continuous, then Spec $\mathcal{O}$(X), and hence X, is locally quasicompact by 5.5. □

As every Hausdorff space is sober, Theorem 5.6 yields:

5.7. COROLLARY. *For a Hausdorff space* X, *the lattice* $\mathcal{O}(X)$ *of open subsets is continuous iff* X *is locally compact.* □

In general, that is, for nonsober spaces X, we do not have any characterization of the continuity of $\mathcal{O}(X)$ that is as satisfactory as Theorem 5.6. Of course, we can say that $\mathcal{O}(X)$ is continuous if and only if the sobrification X^s of X is locally quasicompact. For T_0-spaces this can be made more specific:

5.8. DEFINITION. An embedding $i : X \to Y$ of topological spaces is called *strict*, if the map $U \mapsto i^{-1}(U) : \mathcal{O}(Y) \to \mathcal{O}(X)$ is bijective and, hence, an isomorphism of lattices. □

We observe that the image of a strict embedding is always dense. The sobrification map $\xi_X : X \to X^s = \operatorname{Spec} \mathcal{O}(X)$ is a strict embedding if X is a T_0-space by the results of Section 4.

5.9. LEMMA. *Let* L *be a complete lattice in which* Spec L *is order generating. For a subset* $\Sigma \subseteq \operatorname{Spec} L$ *the inclusion map* $\Sigma \to \operatorname{Spec} L$ *is a strict embedding iff* Σ *is also order generating in* L.

Proof. The inclusion $\Sigma \to \operatorname{Spec} L$ is a strict embedding iff

$$\triangledown_L(s) \cap \Sigma = \triangledown_L(t) \cap \Sigma \text{ implies } \triangledown_L(s) = \triangledown_L(t)$$

for all s,t in L. As Spec L is order generating, this is equivalent to saying that $\triangledown_L(s) \cap \Sigma = \triangledown_L(t) \cap \Sigma$ implies $s = t$. As in any case

$$\triangledown_L(s) \cap \Sigma = \triangledown_L(\inf(\triangledown_L(s) \cap \Sigma)) \cap \Sigma,$$

the foregoing statement implies that $s = \inf(\triangledown_L(s) \cap \Sigma)$ for all s in L; that is, Σ is order generating in L. Clearly, if Σ is order generating, then we find that $\triangledown_L(s) \cap \Sigma = \triangledown_L(t) \cap \Sigma$ implies $s = \inf(\triangledown_L(s) \cap \Sigma) = \inf(\triangledown_L(t) \cap \Sigma) = t$. □

5.10. PROPOSITION. *For a* T_0*-space* X, *the following statements are equivalent:*

(1) $\mathcal{O}(X)$ *is a continuous lattice;*

(2) X *allows a strict embedding into a locally quasicompact space;*

(3) X *allows a strict embedding into a locally quasicompact sober space;*

(4) X *is homeomorphic to an order generating subspace of* Spec L *for some distributive continuous lattice* L;

(5) *The sobrification* X^s *of* X *is locally quasicompact.*

Proof. (1) iff (5): This follows from Theorem 5.6 and from the fact that $\mathcal{O}(X) \simeq \mathcal{O}(\operatorname{Spec}(\mathcal{O}(X))) = \mathcal{O}(X^s)$ by 4.7.

(5) implies (4): As X is a T_0-space, the sobrification map $\xi_X : X \to X^s = \operatorname{Spec}(\mathcal{O}(X))$ is a strict embedding. Hence $\xi_X(X)$ is order generating in $\mathcal{O}(X)$ by 5.9, and $\mathcal{O}(X) \simeq \mathcal{O}(X^s)$ is distributive and continuous.

(4) implies (3): By 5.9.
(3) implies (2) implies (1): Clear. □

The reader should recall that the spaces characterized by the equivalent conditions in the preceding proposition have been considered in Section II-4 because of their good behavior with respect to *function spaces.*

We have seen that the hull-kernel topology of Spec L is induced from the lower topology $\omega(L)$ of L. We now show that the topology induced from the Lawson topology $\lambda(L)$ of L on Spec L can be characterized—in the case of a continuous Heyting algebra L—in terms of the hull-kernel topology of Spec L. The procedure we are using associates with every topological space a refinement of its topology called the patch topology.

5.11. DEFINITION. Let X be a topological space and $X^1 = X \cup \{1\}$ the set obtained from X by adjoining a new element 1. On X^1 we define the *patch topology* to be that generated by $\mathcal{O}(X)$ together with all sets of the form $X^1 \setminus Q$, where Q is a quasicompact saturated subset of X. □

We now consider a complete lattice L and the patch topology for X = Spec L with the hull-kernel topology. In this case, the new element adjoined to X can be taken to be simply the greatest element 1 of L. We have X^1 = Spec $L \cup \{1\}$ = PRIME L.

We wish next to compare the patch topology on PRIME L with the topology induced from the Lawson topology on L. Recall from III-1.5 that the Lawson topology is generated by the lower topology together with the Scott topology. But the lower topology induces on Spec L exactly the hull-kernel topology. By 5.3, every set of the form PRIME L $\setminus$ Q, for a quasicompact saturated subset Q of Spec L, can be written as $F \cap$PRIME L for some Scott-open filter F of L. Thus, the patch topology on PRIME L is coarser than the topology induced from the Lawson topology on L.

If L is distributive, then by 5.4 the sets of the form PRIME L $\setminus$ Q with Q quasicompact and saturated are exactly the sets of the form $F \cap$PRIME L with F a Scott-open filter on L. Since on a continuous lattice the Scott topology is generated by the Scott-open filters (II-1.14), we have:

5.12. PROPOSITION. (i) *In an arbitrary complete lattice* L, *the patch topology on* PRIME L = (Spec L)^1 *is coarser than the toplogy induced by the Lawson topology on* L.

(ii) *In a distributive continuous lattice* L, *the two topologies on* PRIME L *agree.* □

The criteria in 3.7 for PRIME L to be closed with respect to the Lawson topology together with 5.12 yield the following:

5.13. COROLLARY. (i) *In a distributive continuous lattice* L, *the patch topology on* PRIME L *is compact iff the relation* $\ll$ on L *is multiplicative.*

(ii) *In a distributive algebraic lattice* L, *the patch topology on* PRIME L *is compact iff* L *is arithmetic.* □

Let us turn now to the functorial aspects of the correspondence between distributive continuous lattices and locally quasicompact sober spaces. Recall the duality between the categories ***SOB*** and ***HEYT***$_0$ of the last section. This duality restricts, by what we have just shown, to a duality between the full subcategory of ***SOB*** consisting of the locally quasicompact sober spaces and the full subcategory of ***HEYT***$\cap$***CONT*** consisting of the distributive continuous lattices, or, equivalently, the continuous Heyting algebras. Indeed, $\mathcal{O}(X)$ is a continuous distributive lattice for every locally quasicompact sober space, and Spec L is a locally quasicompact sober space for every distributive continuous lattice L by 5.5.

Unfortunately, the morphisms in ***SUP***$^\wedge$ (that is, the maps preserving arbitrary sups and finite infs) are of no particular significance for continuous lattices. What we want to do is to restrict attention to maps which, in addition, preserve the relation $\ll$. This is motivated by the fact that a map $\varphi : \mathrm{L} \to \mathrm{M}$ between continuous lattices preserves arbitrary sups and $\ll$ if and only if its upper adjoint $\tau : \mathrm{M} \to \mathrm{L}$ preserves arbitrary infs and directed sups. (That is, τ is a $\wedge$-semilattice homomorphism which is continuous for the respective Lawson topologies; see IV-1.4.) Now by 4.5, the upper adjoint maps Spec M into Spec L, and the restriction and corestriction Spec φ : Spec M$\to$Spec L is continuous for the respective hull-kernel topologies. What is new in the discussion is that we can verify an additional property for these maps, as we show in the next two lemmas.

5.14. LEMMA. *Let* L *and* M *be distributive continuous lattices and* $\varphi : \mathrm{L} \to \mathrm{M}$ *a map preserving arbitrary sups, finite infs and the relation* $\ll$. *Then for every saturated quasicompact subset* Q *of* Spec L, *the preimage* (Spec $\varphi)^{-1}(Q)$ *is also saturated and quasicompact.*

Proof. Let Q be saturated and quasicompact in Spec L. By 5.4, we can write Q = Spec L $\setminus F$ for some Scott-open filter F of L. Since the upper adjoint $\tau : \mathrm{M} \to \mathrm{L}$ is a $\wedge$-homomorphism which is continuous for the respective Lawson topologies, $\tau^{-1}(F)$ is a Scott-open filter of M. As Spec φ is the restriction of τ, we conclude from 5.4 that the set

$$(\mathrm{Spec}\ \varphi)^{-1}(Q) = \tau^{-1}(\mathrm{Spec\ L} \setminus F) = \mathrm{Spec\ M} \setminus \tau^{-1}(F)$$

is a saturated quasicompact subset of Spec M. □

5.15. LEMMA. *Let* X *and* Y *be locally quasicompact spaces and* $f : \mathrm{X} \to \mathrm{Y}$ *a continuous map with the property that* $f^{-1}(Q)$ *is quasicompact in* X *for every saturated quasicompact subset* Q *of* Y. *Then the map* $\mathcal{O}(f) : \mathcal{O}(\mathrm{Y}) \to \mathcal{O}(\mathrm{X})$ *preserves the relation* $\ll$.

Proof. Let $U \ll V$ in $\mathcal{O}(Y)$. We want to show that $f^{-1}(U) \ll f^{-1}(V)$. By I-1.4(ii) there is a quasicompact set Q with $U \subseteq Q \subseteq V$. Let P be the saturation of Q, that is, $P = \{y \in Y : \{y\}^- \cap Q \neq \emptyset\}$. Since every open set is saturated, the open coverings of Q and of P are the same. Thus P is also quasicompact, and we still have $U \subseteq P \subseteq V$. By hypothesis, $f^{-1}(P)$ is also quasicompact. Because $f^{-1}(U) \subseteq f^{-1}(P) \subseteq f^{-1}(V)$, we conclude that $f^{-1}(U) \ll f^{-1}(V)$. □

Consider now the following two categories: ***LQSOB*** which has as objects the locally quasicompact sober spaces and as its maps the continuous functions $f: X \to Y$ with the property that $f^{-1}(Q)$ is quasicompact for every saturated quasicompact subset Q of Y, and $\boldsymbol{CL}^{\mathrm{op}} \cap \boldsymbol{HEYT}$ which has as objects continuous Heyting algebras and as maps the morphisms $\varphi : L \to M$ preserving arbitrary sups, finite infs and the relation $\ll$. Moreover, let ***DL*** be the category of continuous distributive lattices and ***CL***-maps preserving spectra. Then ***DL*** and $\boldsymbol{CL}^{\mathrm{op}} \cap \boldsymbol{HEYT}$ are dual by IV-1.28.

5.16. PROPOSITION. (i) *A dual equivalence of categories is given by the functors*

$$\mathrm{Spec} : \boldsymbol{CL}^{\mathrm{op}} \cap \boldsymbol{HEYT} \to \boldsymbol{LQSOB} \text{ and } \mathcal{O} : \boldsymbol{LQSOB} \to \boldsymbol{CL}^{\mathrm{op}} \cap \boldsymbol{HEYT}.$$

(ii) *The categories **DL** and **LQSOB** are equivalent.* □

EXERCISES

5.17. EXERCISE. Let X be a topological space. For any subset $Y \subseteq X$ let

$$\mathrm{sat}\, Y = \{x \in X : \{x\}^- \cap Y \neq \emptyset\}.$$

It is called the *saturation* of Y. Show that:

(i) sat Y is the smallest saturated subset of X containing Y.
(ii) sat $U = U$ for every open subset U of X.
(iii) sat Y is the intersection of all open sets containing Y.
(iv) A subset $Q \subseteq X$ is quasicompact iff its saturation sat Q is quasicompact. □

5.18. EXERCISE. (i) Let L be a distributive algebraic lattice. Show that $\Delta_L(k)$ is quasicompact for the hull-kernel topology whenever k is a compact element of L, and that conversely every quasicompact open subset of Spec L is of this form. Show that, moreover, Spec L has a basis of quasicompact open sets (cf. also I-4.25).

(ii) Using 5.15 show that the functors Spec and $\mathcal{O}$ establish a dual equivalence between the following categories:

$\boldsymbol{AL}^{\mathrm{op}} \cap \boldsymbol{HEYT}$, whose objects are the distributive algebraic lattices and whose maps are the morphisms $\varphi : L \to M$ which preserve arbitrary sups, finite infs and which map compact elements of L onto compact elements of M;

BQSOB, whose objects are the sober spaces having a basis of quasicompact open sets and whose maps are the continuous functions $f: X \to Y$ such that $f^{-1}(Q)$ is quasicompact for every quasicompact open subset Q of Y. □

5.19. EXERCISE. Consider the following categories:

The full subcategory ***DAR*** of $\boldsymbol{ArL}^{op} \cap \boldsymbol{HEYT}$ the objects of which are the distributive arithmetic lattices with 1 a compact element.

The full subcategory ***CQSOB*** of ***BQSOB*** the objects of which are the quasicompact sober spaces with a basis of quasicompact open sets closed under finite intersections.

The category ***DLat*** of distributive lattices with 0 and 1 and all 0 and 1 preserving lattice homomorphisms.

Show that the categories ***DAR*** and ***CQSOB*** are dually equivalent under the functors Spec and $\mathcal{O}$, and that similarly, the categories ***DLat*** and ***CQSOB*** are dually equivalent.

(HINT: The categories ***DLat*** and ***DAR*** are equivalent. One may use the functor Id : ***DLat***$\to$***DAR*** which associates with every distributive lattice its ideal lattice, and the functor K : ***DAR***$\to$***DLat*** which associates with each arithmetic distributive lattice its lattice of compact elements (cf. IV-1.15).) □

5.20. EXERCISE. (i) Let X be a sober space and L a continuous Heyting algebra. Then Spec $[X,\Sigma L] \simeq X \times$ Spec L (see II-4.17, 18, 19).

(HINT: We know from II-4.17 that $[X,\Sigma L]$ is a complete Heyting algebra if L is one. For a sober space X and an arbitrary continuous lattice L we constructed a bijection $\beta : X \times (\mathrm{IRR}\, L \setminus \{1\}) \to \mathrm{IRR}\, F \setminus \{1\}$, $F = [X,\Sigma L]$, which was given by $\beta(x,p) = \chi_{X\setminus\{x\}^-} \wedge \mathrm{const}_p$.

We show that β is a homeomorphism with respect to the lower topologies induced on $\mathrm{IRR}\, L \setminus \{1\}$ and $\mathrm{IRR}\, F \setminus \{1\}$, respectively. The generic closed sets of $S = \mathrm{IRR}\, F \setminus \{1\}$ are of the form $\uparrow a \cap S$, $a \in F$. Now $\beta^{-1}(\uparrow a \cap S) = \{(x,p) : x \in X,\ p \in \mathrm{IRR}\, L,\ p \neq 1,\ a(x) \leq p\}$. We claim that the complement of this set is open in $X \times (\mathrm{IRR}\, L \setminus \{1\})$. Indeed suppose $a(x) \notin \downarrow p$. Pick an $s \in L$ with $s \notin \downarrow p$ and $s \ll a(x)$. Then $U = a^{-1}(\twoheaduparrow s)$ is an open neighborhood of x in X, and $V = (\mathrm{IRR}\, L) \setminus \uparrow s$ is an open neighborhood of p in $\mathrm{IRR}\, L \setminus \{1\}$. If $u \in U$ and $v \in V$, then $a(u) \not\leq v$, since otherwise $s \ll a(u) \leq v$ implies $v \in \uparrow s$. This proves the claim. Conversely, let A be closed in X and $s < 1$ in L so that $\uparrow s \cap (\mathrm{IRR}\, L \setminus \{1\})$ is a generic closed set of $T = \mathrm{IRR}\, L \setminus \{1\}$. Define $a: X \to L$ by

$$\begin{aligned} a(x) &= s, \text{ if } x \in A \\ &= 1, \text{ otherwise.} \end{aligned}$$

Then a is continuous, that is, $a \in F$. Moreover, $a \leq \beta(x,p)$ iff $a(x) \leq p$ iff $s \leq p$ for $x \in A$ and $1 \leq p$ for $x \notin A$; but $p<1$, hence $a \leq \beta(x,p)$ iff $x \in A$ and $p \in \uparrow s$. Thus $A \times (\uparrow s \cap \mathrm{T}) = \beta(\uparrow a \cap \mathrm{S})$ is the image of a generic closed set in IRR $F \setminus \{1\}$.)

Remark. We have in fact proved a stronger statement, since in the proof we did not use the distributivity of L.

(ii) With $\mathrm{L} = \mathcal{O}(\mathrm{Y})$ for a locally quasicompact sober space, and in view of the duality between these spaces and continuous Heyting algebras, retrieve the following corollary (cf. II-4.10): For two locally quasicompact sober spaces X and Y, one has $[\mathrm{X}, \Sigma\mathcal{O}(\mathrm{Y})] = \mathcal{O}(\mathrm{X}\times\mathrm{Y})$. □

5.21. EXERCISE. Let L_1 and L_2 be two continuous Heyting algebras.

(i) Then $\mathrm{Spec}\,(\mathrm{L}_1 \otimes \mathrm{L}_2) \simeq \mathrm{Spec}\,\mathrm{L}_1 \times \mathrm{Spec}\,\mathrm{L}_2$ (cf. IV-1.44 ff.).

(ii) Dually, for two locally quasicompact sober spaces X and Y, one has $\mathcal{O}(\mathrm{X}\times\mathrm{Y}) = \mathcal{O}(\mathrm{X}) \otimes \mathcal{O}(\mathrm{Y})$. (See Bandelt [1979].) □

The following exercise gives an indication of an example of a T_0-space X for which $\mathcal{O}(\mathrm{X})$ is a continuous lattice but which fails to be locally quasicompact so badly that every quasicompact subset of X has ***empty*** interior.

5.22. EXERCISE. Let $\mathbf{I} = [0,1]$ be the unit interval and L the continuous lattice $[\mathbf{I} \to \mathbf{I}] = \mathrm{LSC}(\mathbf{I},\mathbf{I})$ of lower semicontinuous functions from $\mathbf{I}$ into $\mathbf{I}$. Then $\mathrm{Y} = \mathrm{Spec}\,\mathrm{L}$ may be identified with $\mathbf{I} \times (\mathbf{I} \setminus \{1\})$ via 5.20 above, where we transport the hull-kernel topology of the spectrum to the topless unit square. By heavy use of the axiom of choice we pick a dense subset A of $\mathbf{I} \setminus \{1\}$ such that $A \cap U$ is not a Borel set (or, if one prefers, not even Lebesgue measurable) for every nonempty open subset U of the unit interval. Let Q be the set of rational points of $\mathbf{I}$. Now define

$$\mathrm{X} = (A \times Q \setminus \{1\}) \cup (\mathbf{I} \setminus A) \times (\mathbf{I} \setminus Q),$$

and give X the topology inherited from the spectrum of L. We have:

(i) X is a T_0-space for which $\mathcal{O}(\mathrm{X}) = \mathrm{L}$; in particular, $\mathcal{O}(\mathrm{X})$ is a continuous lattice;

(ii) Every quasicompact subset of X has empty interior. (For details see Hofmann and Lawson [1978].) □

5.23. EXERCISE. Let X be a sober topological space. Then the following statements are equivalent (see II-3.6):

(1) $\mathcal{O}(\mathrm{X})$ is completely distributive;

(2) X is locally quasicompact, $\sigma(\Omega\mathrm{X}) \subseteq \mathcal{O}(\mathrm{X})$ and $\Omega\mathrm{X}$ is a continuous poset.

Moreover, if these conditions are satisfied, then $\sigma(\Omega\mathrm{X}) = \mathcal{O}(\mathrm{X})$.

(HINT: (1) implies (2): If $\mathcal{O}(\mathrm{X})$ is completely distributive, we may identify X with Spec $\mathcal{O}(\mathrm{X})$ and $\Omega\mathrm{X}$ with (Spec $\mathcal{O}(\mathrm{X}), \supseteq$). Then 1.10 shows that $\Omega\mathrm{X}$ is a continuous poset.

(2) implies (1): We identify X with Spec L for a lattice $L \simeq \mathcal{O}(X)$, and ΩX with $(X, \geq)$. Since X is locally quasicompact, L is continuous (5.6), and the given topology on X is identified with the hull-kernel topology on X. Thus $U \in \mathcal{O}(X)$ means $U = X \setminus \uparrow x$ for some $x \in L$. But then clearly $U \in \sigma(\Omega X)$. Hence $\mathcal{O}(X) \subseteq \sigma(\Omega X)$. Thus $\mathcal{O}(X) = \sigma(\Omega X)$. But $\sigma(\Omega X)$ is completely distributive by II-1.22.) □

We utilize this information to give a detailed analysis of the subcategories of the category ***HEYT*** of complete Heyting algebras with morphisms preserving arbitrary sups and finite infs and their duals. This is best depicted in diagram form.

5.24. EXERCISE. (i) Consider the categories shown in the two tables at the end of this section. Show that the corresponding categories are dual under the duality

$$\boldsymbol{HEYT} \underset{\mathcal{O}}{\overset{\text{Spec}}{\rightleftarrows}} \boldsymbol{SOB}$$

(ii) The category ***HEYT*** has no dual based on topological spaces. [This has given rise to the study of "pointless topology" by passing to the formal dual (with objects sometimes called "frames" or "locales").] Show that by Theorem IV-1.26, $\boldsymbol{HEYT}^{\mathrm{op}}$, however, can be realized as a concrete category: namely the category of all complete Heyting algebras together with all maps g preserving arbitrary infs and prime ideals (that is, $\downarrow g(P)$ is prime for any prime ideal P). □

5.25. EXERCISE. Let $\boldsymbol{CL}_d$ be the full subcategory in ***CL*** of all distributive continuous lattices. For a ***CL***-morphism $f : L \to M$ between distributive continuous lattices L and M, let Spec f be the multivalued function Spec L $\to$ Spec M which associates with a $p \in$ Spec L the subset $\nabla_L(f(p)) \subseteq$ Spec M.

(i) Spec fg = Spec $f \circ$ Spec g (with composition of binary relations on the right hand side).

(ii) Spec f is a multivalued function $F : X \to Y$ between locally quasicompact sober spaces satisfying the following properties:

(a) The image of a point is closed.

(b) $F(A^-) = F(A)^-$ for all $A \subseteq X$.

(c) $F^{-1}(Q)$ is quasicompact saturated whenever Q is quasicompact and saturated in Y.

(iii) Spec is a functor from the category CL_d to the category of all multivalued maps between locally quasicompact sober spaces satisfying (a), (b) and (c) above, and this functor gives an equivalence of categories.

(iv) This theorem generalizes 5.16.

(HINT: (i): Use 1.13.

(ii): (a) is clear from the definition. For (b), $F(A)^- \subseteq F(A^-)$, use 1.13. For (c) recall 5.4(iii). Find an inverse functor for Spec by associating with a space X the lattice $\Gamma(X)^{op}$ (O-2.7(3)) and with a multivalued map $F : X \to Y$ sastisfying (a, b and c) the function $A \to F(A) : \Gamma(X)^{op} \to \Gamma(Y)^{op}$.) □

5.26. EXERCISE. Let X be a locally quasicompact sober space. The topology generated by the set of all quasicompact saturated sets as a subbasis for the closed sets is precisely the one making the function $x \mapsto X \setminus \{x\}^- : X \to \Sigma\mathcal{O}(X)$ an embedding.

(HINT: Consider X as Spec L for L = $\mathcal{O}(X)$ and modify the proof of 5.12(ii).) □

5.27. EXERCISE. (i) Let X be a T_0-space such that $\mathcal{O}(X)$ is a continuous lattice (cf. 5.10 below) and that its sobrification X^s (see 4.7 (Remark) and 4.9) is first countable. Then the following conditions are equivalent:

(1) X is sober;

(2) All closed subspaces of X are Baire spaces (cf. I-3.43.9);

(3) All closed irreducible subspaces of X are Baire spaces.

(ii) If X is a second countable T_0-space such that $\mathcal{O}(X)$ is continuous, then X is sober iff all closed subspaces are Baire spaces.

Remark. It is known that the primitive ideal spectrum Prim $\mathcal{A}$ of a $\mathbb{C}^*$-algebra $\mathcal{A}$ is a locally quasicompact Baire space and that it is second countable if $\mathcal{A}$ is separable. Every closed subset is a primitive ideal spectrum. Hence (ii) above shows that Prim $\mathcal{A}$ is sober in this case. This means that Prim $\mathcal{A}$ = Spec Id $\mathcal{A}$ in the case of a separable $\mathbb{C}^*$-algebra (cf. I-3.37 ff.)

(HINT: (ii) is a consequence of (i). In order to prove (i) we consider X as an order generating subset of Spec L for a continuous Heyting algebra L (see 5.10). First show that if $\mathcal{O} \in (\text{Spec } L) \setminus X$ and $\omega(L)$ has a countable basis at $\mathcal{O}$, then X is not a Baire space. For this purpose it is useful to argue that for $s \neq \mathcal{O}$, the set $\uparrow s \cap X$ is nowhere dense in X.

Next show that if each $p \in \text{Spec}\, L$ has a countable $\omega(L)$-neighborhood basis in $\uparrow p$, then for each $p \in (\text{Spec}\, L) \setminus X$ the set $\uparrow p \cap X$ is a closed irreducible subset of X which is not a Baire space.

In order to complete the proof, observe that (1) is equivalent to X = Spec L (see discussion of ξ_X preceding 4.7). Since closed subspaces of sober spaces are sober, (1) implies (2) implies (3) is clear, and not (1) implies not (3) follows now from the statement in the last paragraph.) □

TABLE OF COMPLETE HEYTING ALGEBRAS

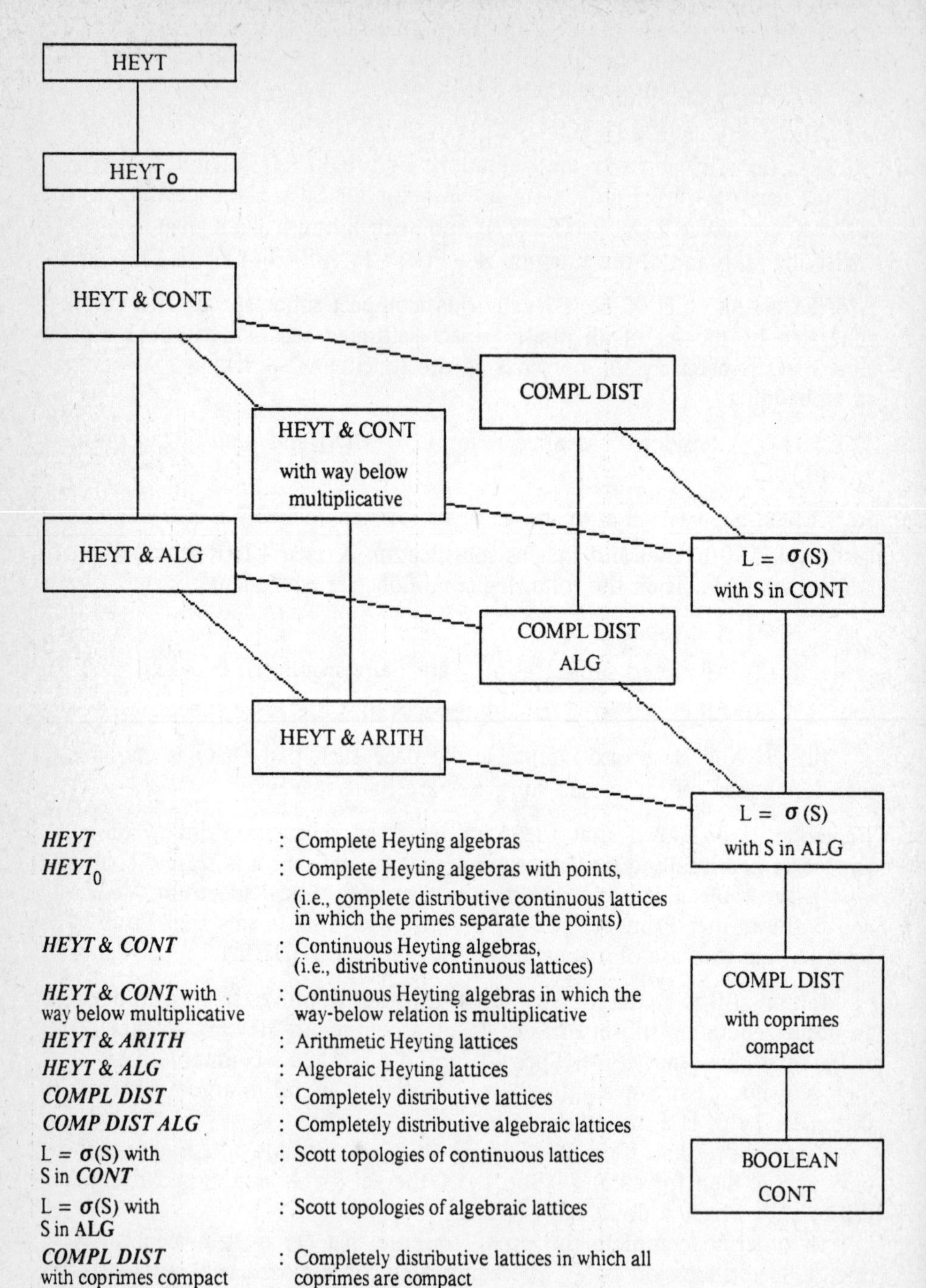

HEYT	: Complete Heyting algebras
$HEYT_0$	: Complete Heyting algebras with points, (i.e., complete distributive continuous lattices in which the primes separate the points)
HEYT & CONT	: Continuous Heyting algebras, (i.e., distributive continuous lattices)
HEYT & CONT with way below multiplicative	: Continuous Heyting algebras in which the way-below relation is multiplicative
HEYT & ARITH	: Arithmetic Heyting lattices
HEYT & ALG	: Algebraic Heyting lattices
COMPL DIST	: Completely distributive lattices
COMP DIST ALG	: Completely distributive algebraic lattices
L = σ(S) with S in *CONT*	: Scott topologies of continuous lattices
L = σ(S) with S in ALG	: Scott topologies of algebraic lattices
COMPL DIST with coprimes compact	: Completely distributive lattices in which all coprimes are compact
BOOLEAN CONT	: Continuous Boolean algebras, (i.e., complete and atomic Boolean algebras)

All morphisms preserve arbitrary sups and finite infs.

All subcategories are full.

TABLE OF SOBER SPACES

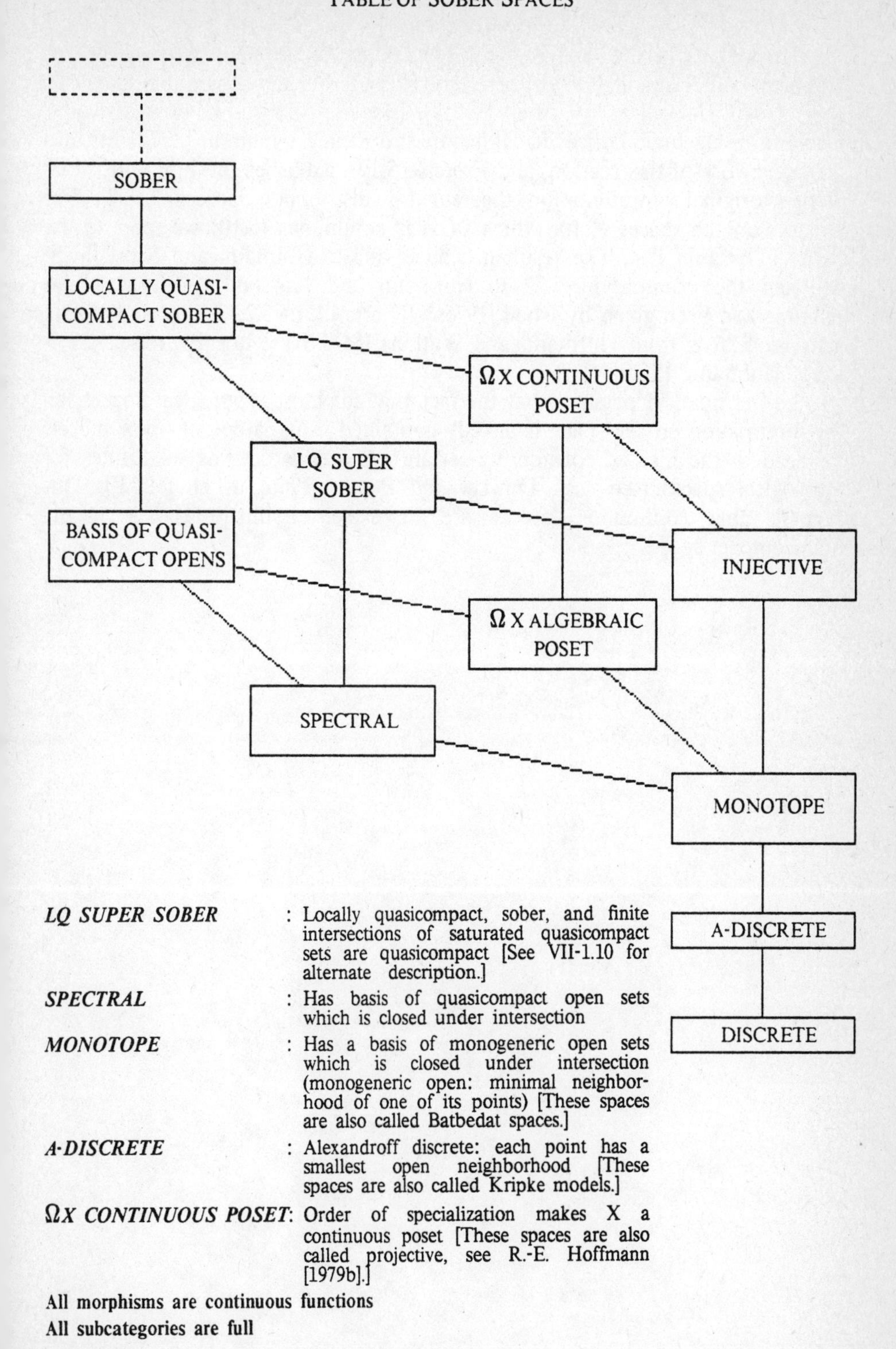

LQ SUPER SOBER : Locally quasicompact, sober, and finite intersections of saturated quasicompact sets are quasicompact [See VII-1.10 for alternate description.]

SPECTRAL : Has basis of quasicompact open sets which is closed under intersection

MONOTOPE : Has a basis of monogeneric open sets which is closed under intersection (monogeneric open: minimal neighborhood of one of its points) [These spaces are also called Batbedat spaces.]

A-DISCRETE : Alexandroff discrete: each point has a smallest open neighborhood [These spaces are also called Kripke models.]

ΩX CONTINUOUS POSET: Order of specialization makes X a continuous poset [These spaces are also called projective, see R.-E. Hoffmann [1979b].]

All morphisms are continuous functions

All subcategories are full

NOTES

The results in the body of this section are to be found in a paper of Hofmann and Lawson [1979]. Corollary 5.7 was already found by Day and Kelly [1970] as well as by Isbell [1975a]. Exercise 5.18 shows how a duality theorem on algebraic lattices in Hofmann and Keimel's memoir [1972] fits into the framework of this section, and Exercise 5.19 establishes the link with M.H. Stone's original representation theorem for distributive lattices [1937]. For remarks on the spaces X for which $\mathcal{O}(X)$ is continuous (5.10) we refer to the Notes of Section II-4. The result in 5.20 is due to Hofmann and Scott [SCS-42], and the example in 5.22 to Hofmann and Lawson [1979]; a similar example had been given by Isbell [1975a]. The result of 5.23 is due to Lawson. Exercise 5.26 is from Hofmann and Watkins [SCS-51], while Exercise 5.27 is from Hofmann [SCS-44].

Let us note, in passing, that the fact that for most topological spaces, all the information on the space is already contained in its lattice of open subsets has lead to the idea of considering certain types of lattices as substitutes for topological spaces (see, e.g., Dowker and Papert [1966], Isbell [1972]). This suggests that continuous lattices are in a sense substitutes for locally quasicompact spaces.

CHAPTER VI

Compact Posets and Semilattices

As the title of the chapter indicates, we now turn our attention from the principally algebraic properties of continuous lattices to the position these lattices hold in topological algebra as certain compact semilattices. Indeed, as the Fundamental Theorem 3.4 shows, continuous lattices are exactly the compact semilattices with small semilattices in the Lawson topology. Thus, continuous lattices not only comprise an intrinsically important subcategory of the category of compact semilattices but also form the most well-understood category of compact semilattices. In fact, there are only two known examples of compact semilattices which are ***not*** continuous lattices; these are presented in Section 4. The paucity of such examples attests to the unknown nature of compact semilattices in general.

We begin the chapter with some background remarks on compact pospaces and topological semilattices. This is followed by a lattice-theoretic description of the topology of a compact semilattice in Section 2. While this section is not in the mainstream of our development of continuous lattice theory, Theorems 2.7, 2.8 and their Corollaries 2.9 and 2.10 will prove invaluable in our further developments in Chapter VII. Section 3, the principal section of the chapter, contains the Fundamental Theorem together with numerous other useful results about continuous lattices as compact semilattices. Section 4 is devoted to the examples alluded to above, and the concluding Section 5 considers chains and order-arcs in compact semilattices.

1. POSPACES AND TOPOLOGICAL SEMILATTICES

There are several ways of interrelating a topology and a partial order. Our first definition singles out some of the topological properties of a relation that we often meet.

1.1. DEFINITION. Let X be a topological space. A partial order $\leq$ is said to be *lower semicontinuous* if $\downarrow x$ is closed for each $x \in X$; *upper semicontinuous* if each $\uparrow x$ is closed; *semicontinuous* if it is both lower and upper semicontinuous. The relation $\leq$ is said to be *closed* or have a *closed graph* if the relation $\leq$ is a closed subset of $X \times X$ in the product topology. In that case $(X,\leq)$ is called a *pospace.* □

Note that the concepts of semicontinuity and pospaces are symmetric with respect to the partial order; hence, each of the theorems and properties concerning them have order duals which are also valid. Note also that all pospaces are semicontinuous.

Recall from O-1.2 that a net $(x_j)_{j\in J}$ in a poset is *directed* if for $i,j\in J$ there exists $k\in J$ such that if $k\leq m$, then $x_i\leq x_m$ and $x_j\leq x_m$. If the net is directed and the set $\{x_j : j\in J\}$ has a supremum x, then x is called the *directed sup of the net.* Filtered nets and filtered infs of nets are defined dually. The notions of directed sups and filtered infs give an "algebraic" notion of convergence in a poset—in general we desire that such algebraic convergence imply topological convergence.

1.2. DEFINITION. Let $(X,\leq)$ be a poset equipped with a topology. The topology is said to be *compatible* if whenever x is the directed sup or filtered inf of a net $(x_j)_{j\in J}$, then the net converges to x topologically. □

1.3. PROPOSITION. *Let* $(X,\leq)$ *be a poset equipped with a topology. Let* $(x_j)_{j\in J}$ *be a directed net in* X.

(i) *If* $\leq$ *is upper (lower) semicontinuous,* x *is the directed sup of the net, and if the net clusters to* y *in the topological space* X, *then* $x\leq y$ (resp., $y\leq x$). *Hence, if the relation is semicontinuous, then* $x = y$.

(ii) *If* X *is quasicompact and* $\leq$ *is semicontinuous, then* $(x_j)_{j\in J}$ *has a directed sup to which it converges topologically. Hence, in this case the topology is compatible.*

Proof. (i): Suppose $\leq$ is upper semicontinuous. Then for each j, $\uparrow x_j$ is closed. Since the net is directed, there exists an index k_0 such that $x_k\in\uparrow x_j$ for $k\geq k_0$. Hence, $y\in\uparrow x_j$; that is, $x_j\leq y$ for each $j\in J$. Thus $x = \sup x_j\leq y$.

Suppose now that $\leq$ is lower semicontinuous. Then $\downarrow x$ is closed, and of course it contains $(x_j)_{j\in J}$. Since the net clusters to y, we have $y\in\downarrow x$; that is, $y\leq x$.

(ii): If X is quasicompact, then the net has a cluster point x. Since each $\uparrow x_j$ is closed and the net is eventually in this set, we have $x_j\leq x$ for each j. Thus, x is an upper bound. Suppose y is also an upper bound. Then $\downarrow y$ is closed and contains the net. Hence $x\in\downarrow y$, that is $x\leq y$. Therefore, x is a least upper bound. Since, as we have just seen, *any* cluster point is the directed sup of the net, there is a *unique* cluster point. But, as X is quasicompact, this implies that the net converges to x. By what we have just proved and its dual, it follows that the topology on X is compatible. □

The next obvious proposition gives a straightforward equivalent form of the definition of a pospace in terms of open sets.

1.4. PROPOSITION. *Let $(X,\leq)$ be a poset with a topology. The relation $\leq$ is closed iff whenever $a\not\leq b$, there exist open sets U and V with $a\in U$ and $b\in V$ such that if $x\in U$ and $y\in V$, then $x\not\leq y$. Hence, a pospace is Hausdorff.* □

1.5. DEFINITION. A subset A of a poset X is *order convex* (or simply *convex*) if $p\leq q\leq r$ and $p,r\in A$ always imply $q\in A$. For an arbitrary set $A\subseteq X$, the *order-convex hull* $[A]$ of A is defined to be $\uparrow A\cap\downarrow A$; it is the smallest convex set containing A. A pospace X is *locally order-convex* if X has a basis of open sets each of which is order-convex. □

1.6. PROPOSITION. (i) *Let X be a topological space with an upper semicontinuous partial order. If A is a quasicompact subset of X, then $\downarrow A$ is Scott closed.*

(ii) *Let X be a pospace. If A is a compact subset, then $\downarrow A$, $\uparrow A$, and $[A]$ are closed subsets of X. Hence, in particular, $\leq$ is semicontinuous.*

Proof. (i): Let D be a directed subset of $\downarrow A$. For each $d\in D$, $\uparrow d\cap A$ is nonempty and closed in A. Since A is quasicompact, there exists an $a\in\bigcap\{\uparrow d\cap A : d\in D\}$. Thus, a is an upper bound for D, and hence sup $D\in\downarrow A$.

(ii): Let π_1 denote projection into the first coordinate from $X\times A$ into X. Since $\downarrow A = \pi_1((\text{graph} \leq)\cap(X\times A))$, and since projection into the noncompact factor is a closed mapping, A is closed. Dually $\uparrow A$ is closed, and hence $[A] = \downarrow A\cap\uparrow A$ is closed. □

1.7. DEFINITION. A pospace X is said to be *monotone normal* if given two closed sets $A = \downarrow A$ and $B = \uparrow B$ such that $A\cap B = \emptyset$, then there exist open sets $U = \downarrow U$ and $V = \uparrow V$ such that $A\subseteq U$, $B\subseteq V$, and $U\cap V = \emptyset$. □

1.8. PROPOSITION. *If X is a compact pospace, then X is monotone normal.*

Proof. Let $A = \downarrow A$ and $B = \uparrow B$ be disjoint closed sets. Since X is compact Hausdorff, there exist open sets P and Q such that $A\subseteq P$, $B\subseteq Q$, and $P\cap Q = \emptyset$. Let $U = X\setminus\uparrow(X\setminus P)$ and $V = X\setminus\downarrow(X\setminus Q)$. By 1.6 U and V are open. Since $A = \downarrow A$, we have $A\subseteq U$; similarly $B\subseteq V$. Also $U\subseteq P$ and $V\subseteq Q$. Hence $U\cap V = \emptyset$. □

1.9. COROLLARY. *A compact pospace has a subbasis of open increasing and open decreasing sets. Hence, it is locally convex.*

Proof. Let X be a compact pospace with topology $\mathcal{U}$. Let $\mathcal{V}$ be the topology generated by the open increasing and open decreasing sets. By definition $\mathcal{U}$ is finer than $\mathcal{V}$. Hence, if $\mathcal{V}$ is Hausdorff, then $\mathcal{U} = \mathcal{V}$.

Let $x,y\in X$, $x\neq y$. Then $x\not\leq y$ or $y\not\leq x$. Assume $x\not\leq y$. Then $\uparrow x\cap\downarrow y = \emptyset$. By 1.8 there exist open sets $U = \uparrow U$, $V = \downarrow V$ such that $x\in U$, $y\in V$ and also $U\cap V = \emptyset$. Thus $(X,\mathcal{V})$ is Hausdorff.

Since open increasing and open decreasing sets are order-convex and the intersection of order-convex sets is order-convex, the space is locally order convex. □

1.10. PROPOSITION. *Let X be a compact pospace. Then X has a basis of order-convex compact neighborhoods at each point.*

Proof. Let $p\in X$ and let U be an open neighborhood of p. By 1.9 there exists an open order-convex set V such that $p\in V\subseteq U$. Since X is compact Hausdorff, there exists a compact neighborhood A of p such that $p\in A\subseteq V$. Then, by 1.6, the set $[A]$ is compact, and since it is the order-convex hull of A, we find $[A]\subseteq V$. Thus, $[A]$ is a compact order-convex neighborhood of p contained in U. □

We now specialize our considerations from pospaces to topological semilattices, but first we recall the definition.

1.11. DEFINITION. Let S be a semilattice endowed with a topology. The meet operation is said to be *separately continuous* on S if for each $y\in$S, the function $x \mapsto xy$ from S to S is continuous. In this case S is called a *semitopological semilattice.* The meet operation is said to be *jointly continuous* (or *continuous*) if the function $(x,y) \mapsto xy$ is continuous from S$\times$S into S, and in this case S is called a *topological semilattice.* If S is a lattice and the join operation is also continuous, then S is called a *topological lattice.* Topological semilattices are not necessarily Hausdorff; however, we use the adjective "compact" only for Hausdorff spaces.

1.12. REMARK. (i) *The meet operation is separately continuous iff given $x,y\in$S and an open set U containing xy, there exists an open set V containing x such that $Vy\subseteq U$.*

(ii) *The meet operation is jointly continuous if and only if given $x,y\in$S and an open set U containing xy, there exist open sets V and W such that $x\in V$, $y\in W$, and $VW\subseteq U$.* □

Dual definitions and remarks can be made for the join operation in a sup-semilattice or lattice. The next proposition lists several of the elementary properties of semitopological semilattices. Some of these results have appeared earlier, but we collect them here for convenient reference.

1.13. PROPOSITION. *Let S be a Hausdorff semitopological semilattice.*

(i) *For each $x\in$S, $\downarrow x =$ Sx is a retract of S and hence closed;*
(ii) *The relation $\leq$ is semicontinuous;*
(iii) *If U is open, then $\uparrow U$ is open (this holds without Hausdorffness).*

If further S *is compact, then*

(iv) *The topology of* S *is compatible*;
(v) S *is a complete semilattice, and hence has a least element* 0. *Also if* S *has a* 1, *it is a complete lattice*;
(vi) *For all* $A \subseteq S$, *we have* $\downarrow A^- \subseteq (\downarrow A)^-$;
(vii) *The semilattice* S *is meet continuous.*

Proof. (i): The mapping $y \mapsto xy$ is a retraction of S onto $\downarrow x$. A retract of a Hausdorff space is closed.

(ii): Let $\lambda_x : S \to S$ be defined by $\lambda_x(y) = xy$. Then $\uparrow x = (\lambda_x)^{-1}(\{x\})$, and hence is closed. Thus $\leq$ is semicontinuous.

(iii): For an open set U, we have $\uparrow U = \cup\{(\lambda_x)^{-1}(U) : x \in U\}$, and hence $\uparrow U$ is open.

(iv) and (v): By Proposition 1.3 filtered and directed sets converge to their greatest lower bounds and least upper bounds, respectively. Hence, by O-2.15 , S is a complete semilattice. A complete semilattice with a 1 is a complete lattice (O-2.12).

(vi): Let $y \leq x$ for some $x \in A^-$. Then there exists a net $(x_j) \subseteq A$ converging to x. Then yx_j converges to $yx = y$ and $yx_j \in \downarrow A$ for each j. Hence, $y \in (\downarrow A)^-$.

(vii): This was shown in O-4.4. □

1.14. PROPOSITION. *Let* S *be a Hausdorff topological semilattice. The partial order* $\leq$ *is closed, and hence* S *is a pospace.*

Proof. Define $f : S \times S \to S \times S$ by $f(x,y) = (x,xy)$. Then graph $\leq$ = $f^{-1}(\Delta)$, where $\Delta = \{(x,x) : x \in S\}$. But Δ is closed since S is Hausdorff. □

Note that the proof of 1.14 was used in proving III-2.4.

EXERCISES

1.15. EXERCISE. (THE URYSOHN-NACHBIN LEMMA). Let X be a monotone normal pospace. If A is a closed increasing set, B is a closed decreasing set, and $A \cap B = \emptyset$, then there exists a continuous order-preserving function $f: X \to \mathbb{I}$ such that $f(B) = 0, f(A) = 1$.

(HINT: Construct inductively a collection of open sets U_r , where r a dyadic rational between 0 and 1, with $U_1 = X$ and such that each U_r is decreasing and $r < s$ always implies $B \subseteq U_r \subseteq U_r^- \subseteq U_s \subseteq X \setminus A$ (by a process analogous to that employed in Urysohn's Lemma).

Define $f: X \to \mathbb{I}$ by $f(x) = \inf\{r \in \mathbb{I} : x \in U_r\}$. As in Urysohn's Lemma, f is continuous. It follows easily that f is order preserving and that we have $f(A) = 1$ and $f(B) = 0$.) □

1.16. DEFINITION. A metric $\boldsymbol{p}$ on a poset X is *radially convex* if $x \leq y \leq z$ implies that $\boldsymbol{p}(x,y) + \boldsymbol{p}(y,z) = \boldsymbol{p}(x,z)$.

1.17. EXERCISE. (THE URYSOHN-CARRUTH METRIZATION THEOREM). Let X be a compact metrizable pospace. Then there exists a radially convex metric giving rise to the original topology.

(HINT: If g is a continuous order-preserving function from X into $\mathsf{I} = [0,1]$, then $W_g = \{(x,y) \in X \times X : g(x) < ½ < g(y)\}$ is an open subset of $X \times X$. By Exercise 1.15 if $y \not\leq x$, there exists a continuous order-preserving function $g : X \to \mathsf{I}$ such that $g(x) = 0$, $g(y) = 1$. Hence $(x,y) \in W_g$. Since X is a compact metric space, there exists a sequence $\{g_n\}_{n \in \mathbf{N}}$ of continuous order-preserving functions such that if $y \not\leq x$, then $(x,y) \in W_{g_n}$, for some n. Then the mapping $g = (x \mapsto ((g_n(x))_{n \in \mathbf{N}}) : X \to \mathsf{I}^{\mathbf{N}}$ is a topological and order isomorphism. The metric $p(x,y) = \Sigma_n |x_n - y_n|/2^n$ is a radially convex metric on $\mathsf{I}^{\mathbf{N}}$; when restricted to $g(X)$, it gives a radially convex metric on X.) □

NOTES

The notion of a pospace has proved useful in topological algebra and in some aspects of functional analysis, and, in particular, in the study of topological semilattices. Although there were certainly important forerunners to his work, apparently L. Nachbin was the first to explicitly define and to investigate the notion of a pospace (Nachbin [1965], originally published in 1960). Most of the important results of this section are due to him (Propositions 1.6(ii), 1.8, Definition 1.7 and Corollary 1.9). J.H. Carruth's work on metrization appears in Carruth [1968], where a slightly weaker version of Exercise 1.17 appears.

Topological semilattices appear as early as Nachbin [1965]. They were also studied by workers in topological algebra as a special and important class of topological semigroups (see, e.g., Anderson and Ward [1961b], Koch [1959], Brown [1965] for some of the earlier work on topological semilattices). This section gives some of the most basic results concerning topological semilattices from the folklore of the subject.

2. COMPACT TOPOLOGICAL SEMILATTICES

In this section we extend some of the theory of continuous lattices to compact semilattices. Since this development goes beyond the main scope of the book, the reader may wish only to skim this material at first reading. Although such generalizations are usually more difficult and tedious than the corresponding results for continuous lattices (as we saw, for example, in the presentation of generalized continuous lattices in the exercises of Chapter III), we obtain such important results as an algebraic characterization of the topology of a compact semilattice (2.6) and its consequences (2.7, 2.8, 2.9).

2.1. DEFINITION. Let S be a semilattice. A pseudometric $\boldsymbol{d}$ is said to be *subinvariant* if $\boldsymbol{d}(ax,ay) \leq \boldsymbol{d}(x,y)$ for all $a,x,y \in S$. □

2.2. REMARK. (i) *If* $\boldsymbol{d}$ *is a subinvariant pseudometric on* S, *then*

$$\boldsymbol{d}(ax,by) \leq \boldsymbol{d}(ax,ay) + \boldsymbol{d}(ay,by) \leq \boldsymbol{d}(x,y) + \boldsymbol{d}(a,b).$$

(ii) *Hence, by induction,* $\boldsymbol{d}(a_1 \ldots a_n, b) = \boldsymbol{d}(a_1 \ldots a_n, b^n) \leq \boldsymbol{d}(a_1,b) + \ldots + \boldsymbol{d}(a_n,b)$. □

The topology of any compact semigroup is defined by a set of subinvariant pseudometrics. This fact is one of the early theorems about compact semigroups, first proved by S. Eilenberg around 1938. Alternate proofs have been given by Hofmann and Mostert [1966], Hofmann [1970] and Friedberg [1972]. The proof is deferred until the exercises.

2.3. PROPOSITION. *Let* S *be a compact topological semilattice, and let* $\mathbf{P}$ *be the set of continuous subinvariant pseudometrics on* S. *If* $\mathfrak{C}$ *is an open cover of* S, *then there exists* $\boldsymbol{d} \in \mathbf{P}$ *and an* $\varepsilon > 0$ *such that the set of all neighborhoods* $N_\varepsilon(x) = \{y : \boldsymbol{d}(x,y) < \varepsilon\}$ *for* $x \in S$ *refines* $\mathfrak{C}$. □

We next give a lattice-theoretical characterization of convergence in a compact semilattice.

2.4. LEMMA. *Let* S *be a compact semilattice and* $\boldsymbol{d}$ *a continuous subinvariant pseudometric on* S. *Let* $(x_j)_{j \in J}$ *be a net in* S *with limit* x. *Then for each sequence* $f : \mathbf{N} \to J$, *there exists a (monotone) sequence* $f_0 : \mathbf{N} \to J$ *with* $f \leq f_0$ *such that whenever* $f_0 \leq g$ *then* $\boldsymbol{d}(\underline{\lim}\, x_{g(n)}, x) = 0$.

Remark. Recall that $\underline{\lim}\, x_{g(n)} = \sup_n \inf_{m \geq n} x_{g(m)}$.

Proof. Let $n \in \mathbf{N}$. Since $x = \lim x_j$, there is an index $f_0(n) \geq f(n)$ (and also $f_0(n) \geq f_0(n-1)$) such that $d(x_j, x) \leq ½^{n+1}$ for all $j \geq f_0(n)$. Let $g \geq f_0$. For each n, let $y_n = \inf_{m \geq n} x_{g(m)}$. Since a filtered net converges to its infimum (by the dual of Proposition 1.3), we have

$$y_n = \lim_k x_{g(n)} x_{g(n+1)} \dots x_{g(n+k)}.$$

Employing the earlier Remark 2.2, we calculate,

$$d(x_{g(n)} \dots x_{g(n+k)}, x) \leq \Sigma_{0 \leq i \leq k} d(x_{g(n+i)}, x) \leq \Sigma_{0 \leq i \leq k+1} \tfrac{1}{2}^{n+i} \leq \tfrac{1}{2}^n$$

for all $k \in \mathbf{N}$. Thus, we conclude $d(y_n, x) \leq \frac{1}{2}^n$ for all n. Since, by definition, $\underline{\lim}\ x_{g(n)} = \sup y_n$, and since this is a directed supremum, we find that $\underline{\lim}\ x_{g(n)} = \lim y_n$. Because y_n converges to both x and $\underline{\lim}\ x_{g(n)}$, we have $d(\underline{\lim}\ x_{g(n)}, x) = 0$. □

2.5. PROPOSITION. *If $(x_j)_{j \in J}$ is a net in a compact semilattice S converging to x, then $x = \underline{\lim}_f \underline{\lim}_n x_{f(n)}$, where f ranges in $J^{\mathbf{N}}$ and that set has the pointwise ordering.*

Remark. The proposition remains valid if only monotone f are considered.

Proof. Let d be an arbitrary subinvariant pseudometric on S. For each $g \in J^{\mathbf{N}}$, set $y_g = \underline{\lim}\ x_{g(n)}$. By Lemma 2.4 above, for every $f \in J^{\mathbf{N}}$ there exists an f_0 such that $d(y_g, x) = 0$ for all $g \geq f_0$. If F is any finite set in $\{g \in J^{\mathbf{N}} : g \geq f_0\}$, then $d(\inf_{g \in F} y_g, x) = 0$ by Remark 2.2. Since

$$\inf_{g \geq f_0} y_g = \inf_{F \text{ finite}} \inf_{g \in F} y_g = \lim_{F \text{ finite}} \inf_{g \in F} y_g,$$

because the first infimum is filtered, we conclude that $d(\inf_{g \geq f_0} y_g, x) = 0$. Once again, since

$$\underline{\lim}\ y_f = \sup_f \inf_{g \geq f} y_g = \lim_f \inf_{g \geq f} y_g,$$

because the supremum is directed, we conclude that $d(\underline{\lim}_f y_f, x) = 0$. Now this holds for any continuous subinvariant pseudometric d, and these generate the topology, thus we conclude that $x = \underline{\lim}_f y_f = \underline{\lim}_f \underline{\lim}_n x_{f(n)}$. □

2.6. THEOREM. *Let $(x_j)_{j \in J}$ be a net in a compact semilattice. Then the following are equivalent:*

(1) $x = \lim x_j$,

(2) $x = \underline{\lim}_f \underline{\lim}_n y_{f(n)}$ *where f ranges in $K^{\mathbf{N}}$ for all subnets $(y_k)_{k \in K}$ of the given net.*

Proof. If $x = \lim x_j$, then the same is true for any subnet. Hence (2) follows from Proposition 2.5.

Conversely suppose (2) holds. If x is not equal to $\lim x_j$, then there exists a subnet converging to some $z \neq x$ since S is compact. Again by Proposition 2.5 we have $z = \underline{\lim}_f \underline{\lim}_n y_{f(n)}$ for this subnet. But, by hypothesis, this double limit must be x, a contradiction. Thus, $x = \lim x_j$. □

2.7. THEOREM. *Let $f : S \to T$ be a semilattice morphism between compact semilattices. Then the following are equivalent:*

(1) *f is continuous;*

(2) *f preserves arbitrary meets and directed sups;*

(3) *f preserves lim infs of nets.*

If, in addition, T is a continuous lattice, then these conditions are equivalent to

(4) *The lower adjoint $d : T \to S$ preserves $\ll$.*

Proof. The equivalence of (2) and (3) follows from the proof of Theorem 1.8 of Chapter III. The equivalence of (1) and (3) is a straightforward consequence of Theorem 2.6. Now suppose that T is a continuous lattice. Then (4) is equivalent to (2) by IV-1.4. □

Note that in view of Theorems III-1.8 and 2.7 a semilattice morphism between compact unital semilattices is continuous iff it is Lawson-continuous.

EXERCISES

2.8. EXERCISE. Let S be a complete semilattice and T a subsemilattice. We denote by $\bigwedge(T)$ the set of infima of nonempty subsets of T. Let $\nearrow(T)$ denote the set of all suprema of ***directed*** subsets of T. We let $\bigwedge_\omega(T)$ and $\nearrow_\omega(T)$ denote the corresponding notions employing only ***countable*** subsets of T.

(i) Prove that if S is a compact topological semilattice, then the topological closure can be written as $T^- = \nearrow\bigwedge\nearrow_\omega\bigwedge_\omega(T)$.

(ii) If $T = \downarrow T$, then $T^- = \nearrow\nearrow_\omega(T)$.

(HINT: Since T is a subsemilattice, $\bigwedge_\omega(T)$ consists of meets of down-directed sequences in T. Since S is compact, these sequences converge to their infima. Hence $\bigwedge_\omega(T) \subseteq T^-$. It is easily verified $\bigwedge_\omega(T)$ is again a subsemilattice. Again since S is compact, upward directed sequences in $\bigwedge_\omega(T)$ converge to their suprema. Thus $\nearrow_\omega\bigwedge_\omega(T) \subseteq T^-$. Continuity of the meet operation implies $\nearrow_\omega\bigwedge_\omega(T)$ is a subsemilattice. By an argument which is essentially a repetition of what we have just done, one concludes that $\nearrow\bigwedge\nearrow_\omega\bigwedge_\omega(T) \subseteq T^-$.

Conversely if $x \in T^-$, then there exists a net in T converging to x. Now employ Theorem 2.6.) □

2.9. EXERCISE. Let T be a subsemilattice of a compact semilattice S. The following statements are equivalent:

(1) T is closed (topologically);

(2) T is a complete subsemilattice (that is, T is closed with respect to arbitrary meets and directed sups; see O-2.11).

(3) T is closed with respect to taking lim infs. □

2.10. EXERCISE. Let $A = \downarrow A$ be a subset of a compact semilattice S. The following statements are equivalent:

(1) A is closed (topologically);

(2) A is Scott closed (that is, $A = \nearrow(A)$).

Thus the closed lower sets are precisely the Scott-closed sets. □

2.11. EXERCISE. Prove Proposition 2.3.

(HINT: By Urysohn's Lemma S can be topologically embedded in a product of intervals, $\mathbf{I}^A$. For each finite subset $F \subseteq A$ one defines a pseudometric on the product (and hence on S) by using the Euclidean metric $\boldsymbol{d}_F$ on the coordinates of F. Define a new pseudometric $\boldsymbol{p}_F$ in terms of $\boldsymbol{d}_F$ by

$$\boldsymbol{p}_F(u,v) = \sup\{\boldsymbol{d}_F(xu,xv) : x \in \mathrm{S} \cup \{1\}\}.$$

Continuity from $(\mathrm{S},\boldsymbol{p}_F)$ to $(\mathrm{S},\boldsymbol{d}_F)$ is clear. Using the compactness of S and continuity of the meet operation, one obtains continuity in the other direction. Hence $\boldsymbol{p}_F$ and $\boldsymbol{d}_F$ give rise to the same topology. It is easily verified that $\boldsymbol{p}_F$ is subinvariant. All such pseudometrics generate the topology of S; the compactness of S allows one to complete the proof.) □

2.12. EXERCISE. Let S be a compact topological semilattice. Show that if $k \in \mathrm{S}$ is a compact element, then $\uparrow k$ is open in S.

(HINT: If a net (x_j) in $\mathrm{S} \setminus \uparrow k$ converges to $x \in \uparrow k$, then Proposition 2.5 and the compactness of k imply $x_j \in \uparrow k$ for some index j.) □

NOTES

The ideas behind the results of this section originate in the work of Lawson [1973a] and [1973b]. The ideas are only implicit there although Theorem 2.7 and slightly weaker versions of Theorems 2.8, 2.9 and 2.10 do appear explicitly. H. Bauer and G. Gierz pointed toward the explicit characterization of convergence in compact semilattices given here; the pattern of proof was suggested by Gierz and Hofmann [1978] (replacing a somewhat more technical version given by Lawson).

3. THE FUNDAMENTAL THEOREM OF COMPACT SEMILATTICES

The class of topological semilattices which, at each point, possess a basis of neighborhoods which are subsemilattices was early singled out as an extremely important class of semilattices—both because of its widespread occurrence and its greater theoretical tractibility. In their study of the algebraic properties of these semilattices, Hofmann and Stralka[1976] discovered that the ***compact*** members of this class are precisely the continuous lattices, in a sense to be made explicit shortly. (This identification was actually only implicit in the paper and explicitly pointed out shortly thereafter by Stralka.) The consequences of this realization have been far reaching for both the theory of topological semilattices and that of continuous lattices.

We repeat, for the sake of easy reference, Definition III-2.10:

3.1. DEFINITION. A topological semilattice S is said to have *small semilattices at x* if the point x has a basis of neighborhoods which are subsemilattices of S. The semilattice S has *small semilattices* iff it has small semilattices at every point. ☐

Note that if S is regular and has small semilattices at x, then x has a basis of ***closed*** neighborhoods which are subsemilattices—because the closure of a subsemilattice is a subsemilattice. The next proposition is an easy consequence of the definition.

3.2. PROPOSITION. (i) *Let* S *be a topological semilattice with small semilattices, and let* T *be a subsemilattice (equipped with the relative topology). Then* T *has small semilattices.*

(ii) *Let* $\{S_j : j \in J\}$ *be a collection of topological semilattices with small semilattices. Then* $\mathsf{X}_j S_j$ *endowed with coordinatewise operations and the product topology has small semilattices.* ☐

There are some rather useful reformulations of the property of having small semilattices at a point. (Compare II-1.14 and III-2.11, III-2.13).

3.3. PROPOSITION. *Let* S *be a locally compact Hausdorff topological semilattice. For* $x \in S$ *the following statements are equivalent:*

(1) S *has small semilattices at* x;

(2) *The semilattice* $\downarrow x$ *has small semilattices at* x;

(3) *If* $U = \uparrow U$ *is open and* $x \in U$, *then there exists a filter* F *such that* $x \in \mathrm{int}(F) \subseteq F \subseteq U$;

(4) *If* V *is open and* $x \in V$, *then there exists a* $y \in V$ *such that* $x \in \mathrm{int}(\uparrow y)$.

Proof. (1) implies (2): Straightforward.

(2) implies (4): Let V be open, $x \in V$. By regularity there exists an open set U such that $x \in U$ and the closure of U is compact and a subset of V. Let N be a neighborhood of x in $\downarrow x$ which is both a subsemilattice and a subset of $U \cap \downarrow x$. Then N^- is a compact semilattice and, hence, has a least element y. Since N is a neighborhood of x in $\downarrow x$, and since translation by x from S to $\downarrow x$ is continuous, we have $\{w \in S : xw \in N\}$ is a neighborhood of x in S. Since this set is contained in $\uparrow y$, then $\uparrow y$ is a neighborhood of x. Clearly $y \in V$.

(4) implies (3): Let $x \in U = \uparrow U$. Then there exists an element $y \in U$ such that $x \in \text{int}(\uparrow y)$. Let $F = \uparrow y$.

(3) implies (1): Let U be an open set, $x \in U$. Without loss of generality we may assume that U^- is compact.

Since as a partially ordered space U^- is locally order convex, there exist open sets $V, W \subseteq U$ such that $x \in W$, $WW \subseteq V$, $VV \subseteq U$, and V is order convex in U^-. Choose a filter F such that $x \in \text{int}(F) \subseteq F \subseteq \uparrow W$. Let $N = V \cap F$. Then $x \in \text{int}(N) \subseteq N \subseteq U$. If $p, q \in N$, then there exist $u, v \in W$ such that $u \leq p$, $v \leq q$ (since $N \subseteq \uparrow W$). Then $uv \in WW \subseteq V$, $uv \leq pq \leq p \in V$, and $pq \in VV \subseteq W$ imply $pq \in V$ (since V is order-convex in W). Also $pq \in F$, and thus $pq \in N$. Hence, N is a subsemilattice. □

We come now to the main theorem of this section and chapter.

3.4. THEOREM. (THE FUNDAMENTAL THEOREM OF COMPACT SEMILATTICES).

(i) *Let* L *be a continuous lattice. Then with respect to the Lawson topology* L *is a compact unital topological semilattice with small semilattices.*

(ii) *Conversely, if* S *is a compact unital topological semilattice with small semilattices, then with respect to its semilattice structure* S *is a continuous lattice. Furthermore, the topology of* S *is the Lawson topology.*

(iii) *Let* L *and* M *be compact unital topological semilattices with small semilattices, and let* $f: L \to M$ *be a semilattice homomorphism. The following are equivalent:*

(1) *f is continuous;*

(2) *f preserves directed sups and arbitrary nonempty infs.*

If, moreover, $f(1) = 1$, *then* (1) *and* (2) *are equivalent to*

(3) *The lower adjoint g of f exists and preserves the relation* $\ll$.

Remark. The functor Λ which assigns to a continuous lattice the lattice endowed with its Lawson topology and is the identity on homomorphisms is an isomorphism from the category ***CL*** (see IV-1.9) to the category of compact unital topological semilattices with small semilattices and continuous (semilattice) homomorphisms preserving units. One may therefore identify the two categories.

Proof. (i): By Theorem 2.13 of Chapter III.

(ii): Conversely suppose S is a compact unital topological semilattice with 1 and with small semilattices. By Proposition 1.13 S is a complete lattice.

Let $x \in S$. By 3.3(4), $x = \sup\{y \in S : x \in \operatorname{int}(\uparrow y)\}$, since if $x \not\leq w$, $S \setminus \downarrow w$ is an open set around x. Suppose $x \in \operatorname{int}(\uparrow y)$, that is, there exists an open set U such that $x \in U \subseteq \uparrow y$. Then $x \in \uparrow U \subseteq \uparrow y$ and $\uparrow U$ is open by 1.13(iii). If D is directed and $\sup D \geq x$, then since D converges to $\sup D \in \uparrow U$, there exists $d \in \uparrow U \subseteq \uparrow y$. Hence $y \ll x$. Thus S is a continuous lattice.

To complete the proof, we must argue that the topology of S is the Lawson topology. It follows from Proposition 1.13 that each set of the form $S \setminus \downarrow x$ is open in S. Let V be a Scott-open set, $x \in V$. By the preceding paragraph we have that $x = \sup\{z \in S : x \in \operatorname{int}(\uparrow z)\}$. This set is easily verified to be directed. Thus, there exists $z \in V$ such that $x \in \operatorname{int}(\uparrow z) \subseteq \uparrow z \subseteq V$. Hence, the identity function on S is continuous from S with the given topology to S with the Lawson topology. Since the given topology is compact and the Lawson topology is Hausdorff (see III-1.10), we conclude that they agree.

(iii): This part is a consequence of part (i), III-1.8 (or 2.7 above) and IV-1.4 (where, if necessary, discrete (isolated) identities are adjoined to L and M and f is extended by sending the new identity of L to that of M). □

Theorem 3.4 is a powerful tool for the study of continuous lattices and compact semilattices. It allows an algebraic treatment of topological problems and vice-versa. We illustrate with the following important proposition.

3.5. PROPOSITION. *Let* S *and* T *be compact topological semilattices and let* f *be a continuous homomorphism from* S *onto* T. *If* S *has small semilattices, so also does* T.

Proof. By adjoining discrete identities to S and T and extending f to be identity preserving, we may assume without loss of generality that S and T have identities and f is identity preserving. The proposition now becomes a corollary to Theorem 3.4 and I-2.7(iii). □

We close this section with three important examples. In general these examples seem better suited to a free-flowing exposition rather than a formal series of propositions and proofs, and hence are presented in this fashion (although many of the properties displayed actually could easily be presented as propositions).

3.6. EXAMPLE. (THE ROLE OF THE UNIT INTERVAL). Let $\mathbf{I} = [0,1]$ be the unit interval equipped with its usual topology and order. Then $\mathbf{I}$ is a compact connected topological semilattice (in fact a topological lattice) which has small subsemilattices. Furthermore any topologically closed subsemilattice of a product of copies of $\mathbf{I}$ with the relative topology is again a compact semilattice. Furthermore, this example is exhaustive in a way the following proposition makes precise.

3.7. PROPOSITION. *Let* S *be a compact topological semilattice. The following statements are equivalent*:

(1) S^1 (S *with an identity attached, if necessary*) *is a continuous lattice with respect to its lattice structure*;

(2) S *has small semilattices*;

(3) Hom(S,**I**) *separates points, where* Hom(S,**I**) *denotes the set of all continuous semilattice homomorphisms*;

(4) S *is topologically isomorphic to a closed subsemilattice of a product of copies of* **I**.

Proof. The equivalence of (1) and (2) follows easily from the Fundamental Theorem, and that of (1) and (3) was established in IV-2.19 (where we use the characterization in 2.7 of continuous homomorphisms as those preserving arbitrary meets and directed joins). To see that (3) implies (4), consider $T = \mathbf{I}^{\mathrm{Hom}(S,\mathbf{I})}$. Define $F : S \to T$ by $\pi_f(F(x)) = f(x)$. Then F is a continuous isomorphism; since S is compact, F is a homeomorphism. Finally, (4) implies (2) by Proposition 3.2 (or (4) implies (1) by IV-2.19). □

3.8. EXAMPLE. (THE VIETORIS TOPOLOGY). (i) $\Gamma(\cdot)^{\mathrm{op}}$ *as a functor:* Let X be a compact Hausdorff space. Let $\Gamma(X)$ denote the set of closed subsets ordered by inclusion. The lattice $\Gamma(X)^{\mathrm{op}}$ is isomorphic (via complementation) to the lattice of open sets and hence is a continuous lattice (see Example 1.7(5) of Chapter I). The set $\Gamma(X)$ is an object of great interest to topologists, and is standardly endowed with the Vietoris topology in order to make it a topological space. The Vietoris topology has as subbasis sets of the form

$$N(U) = \{A \in \Gamma(X) : A \subseteq U\} \text{ and } D(V) = \{A \in \Gamma(X) : V \cap A \neq \emptyset\},$$

where U and V are open sets in X. Note that sets of the form $N(U)$ are open filters in the lattice $\Gamma(X)^{\mathrm{op}}$, and hence they generate the Scott topology since $\Gamma(X)^{\mathrm{op}}$ is a continuous lattice. Also $D(V)$ is the complement of the principal filter generated by $X \setminus V$; hence, these sets generate the lower topology on $\Gamma(X)^{\mathrm{op}}$. ***Thus, the Vietoris topology is precisely the Lawson topology of the lattice*** $\Gamma(X)^{\mathrm{op}}$ ***and so is compact.*** Since in a continuous lattice convergence in the Lawson topology is lim-inf convergence, it follows that if a net of closed sets A_j converges to A, then A is indeed the limit (in the technical topological sense) of the A_j.

The assignment of $\Gamma(X)^{\mathrm{op}}$ to X extends to a functor from the category of compact Hausdorff spaces to the category ***CL*** of continuous lattices and Lawson-continuous identity-preserving homomorphisms. If $f : X \to Y$ is a continuous function between compact Hausdorff spaces, define

$$\Gamma(f) = (A \mapsto f(A)) : \Gamma(X)^{\mathrm{op}} \to \Gamma(Y)^{\mathrm{op}}.$$

It is easily verified that

$$[f^{-1}] = (B \mapsto f^{-1}(B)) : \Gamma(Y)^{op} \to \Gamma(X)^{op}$$

is a lower adjoint for $\Gamma(f)$. From Proposition I-1.4(ii), it follows that $A \ll B$ if and only if $B \subseteq \operatorname{int}(A)$. Thus $[f^{-1}]$ preserves $\ll$, and therefore $\Gamma(f)$ is a ***CL***-morphism.

(ii) *The free continuous lattice generated by a compact Hausdorff space:* We next observe that Γ is the "free" functor on the category of compact Hausdorff spaces to the category ***CL*** in the sense that if $f : X \to L$ is a continuous mapping from a compact Hausdorff space X into a continuous lattice L (equipped with the Lawson topology), then there exists a unique, continuous, identity-preserving homomorphism F from $\Gamma(X)^{op} \to L$ such that $F \circ i = f$ where $i = (x \mapsto \{x\}) : X \to \Gamma(X)^{op}$. By standard categorical arguments, this is equivalent to saying that the functor Γ and the "forgetful" functor from ***CL*** to the category of compact Hausdorff spaces are adjoint functors.

To prove the assertion, let $f : X \to L$; then we have a continuous homomorphism $\Gamma(f) : \Gamma(X)^{op} \to \Gamma(L)^{op}$. By applying Proposition 3.9 below, the mapping $(A \mapsto \inf A) : \Gamma(L)^{op} \to L$ is a continuous homomorphism. Hence, we may let $F : \Gamma(X)^{op} \to L$ be the composition, which is continuous, identity-preserving, and a semilattice homomorphism. Moreover

$$F \circ i(x) = F(\{x\}) = \inf\{f(x)\} = f(x).$$

The uniqueness of F follows from the fact $i(X)$ order generates $\Gamma(X)^{op}$. □

3.9. PROPOSITION. *Let S be a compact unital topological semilattice. The function* $A \mapsto \inf A : \Gamma(S)^{op} \to S$ *is continuous iff S is a continuous lattice.*

Proof. Since $\{x\}$ goes to x, the function is onto. Hence, by Proposition 3.5, S is a continuous lattice if the function is continuous.

Conversely suppose S is a continuous lattice. It is easily verified that $x \mapsto \uparrow x : S \to \Gamma(S)^{op}$ is a lower adjoint. If $x \ll y$, then $\uparrow y \subseteq \operatorname{int}(\uparrow x)$. Whence, $\uparrow x \ll \uparrow y$ in $\Gamma(S)^{op}$. Thus, the way-below relation is preserved; and therefore we have shown that $A \mapsto \inf A$ is continuous. □

The preceding discussion has given us a free continuous lattice over any compact Hausdorff space. But in I-4.17 we have shown that the free continuous lattice over a ***set*** X is the lattice of filters on X. We may rederive this result now as follows: recall that the free compact Hausdorff space over the set X is the Stone-Cech compactification $\beta(X)$ of X as a discrete space. The free continuous lattice over $\beta(X)$ is $\Gamma(\beta(X))^{op}$. Since the composition of free functors is a free functor, it follows that $\Gamma(\beta(X))^{op}$ is the free continuous lattice over the set X. This latter turns out to be isomorphic to the lattice of filters on X. We suggest a detailed verification in the exercises (see 3.25) which can also be regarded as an alternate proof of I-4.17. □

3.10. EXAMPLE. (ANOTHER FREE CONSTRUCTION). Let X be a compact pospace. Let $\Xi(X)$ denote the closed ***upper*** subsets of X ordered by inclusion. Then $\Xi(X)^{op}$ is a continuous lattice.

By way of proof remark that if $A,B\in\Xi(X)^{op}$, and $A\subseteq \mathrm{int}(B)$, then $B\ll A$. Since X is monotone normal (by 1.8), we have $A = \sup \Downarrow A$. Hence, $\Xi(X)^{op}$ is a continuous lattice.

The Lawson topology on $\Xi(X)^{op}$ is a modified version of the Vietoris topology on $\Gamma(X)^{op}$. Take for a subbase of open sets, sets of the form $N(U)$ and $D(V)$ where U is open in X and V is open in X and satisfies $V = \downarrow V$. In $\Xi(X)$ arbitrary meets are just intersections, finite joins are unions, and the join of $\{A_i : i\in I\}$ is given by $\uparrow A$ where $A = (\bigcup_{i\in I} A_i)^-$.

The lattice $\Xi(X)^{op}$ is the "free" continuous lattice on the compact pospace X. Alternately Ξ is the adjoint to the forgetful functor from the category ***CL*** to the category of compact pospaces and continuous order-preserving mappings. The arguments are similar to those presented in Example 3.8 and are deferred to the exercises (see 3.22, 3.23). □

Remark. Since the opposite of a pospace is a pospace, the lattice $\Upsilon(X)$ of closed ***lower*** subsets is also a continuous lattice; however, it is not the "free" continuous lattice on X but on X^{op}.

Example 3.10 and the following Proposition 3.11 should be viewed in the context of Sections VII-3 and 4. The hypothesis of 3.11 is slightly different from that of 3.10.

3.11. PROPOSITION. *Let* S *be a compact semilattice.*

(i) $\Upsilon(S)$, *the lattice of closed lower sets, is the lattice of Scott-closed sets;*

(ii) $\Upsilon(S)^{op}$ *forms a continuous lattice;*

(iii) $\Upsilon(S)^{op}$ *is a closed subspace of* $\Gamma(S)^{op}$ *with the Lawson topology being the relative Vietoris topology.*

Proof. By 2.10 the Scott-closed sets are precisely the topologically closed sets A such that $A = \downarrow A$. This proves (i). The remarks in 3.10 establish (ii).

(iii): Let $A = \downarrow A$, let $x\in A^-$, and let $y\leq x$. We show $y\in A^-$. Let (x_j) be a net in A converging to x; by continuity $x_j\, y$ converges to $xy = y$. Since we have $A = \downarrow A$, $x_j\, y\in A$ for all j, and so $y\in A^-$. Thus $A^- = \downarrow A^-$.

If $(A_j)_{j\in J}$ is a collection of Scott-closed sets, then in $\Gamma(S)$ the meet of the collection is $\bigcap_{j\in J} A_j$ and the join is $(\bigcup_{j\in J} A_j)^-$. Since both of these are again closed lower sets, $\Upsilon(S)$ is closed in $\Gamma(S)$ by Exercise 2.9. It follows from the Fundamental Theorem that the relative Vietoris topology must be the Lawson topology, since the latter is the only one making $\Upsilon(S)$ into a compact topological semilattice with small semilattices. □

In earlier chapters we have seen that algebraic lattices form a basic subcategory of the category of continuous lattices. We close this section with that version of the Fundamental Theorem which applies to them.

3.12. LEMMA. *Let S be a compact topological semilattice. If the component of x in S is contained in $\uparrow x$, then S has small semilattices at x. Furthermore, x is the sup of compact elements.*

Proof. Let U be an open convex set in $\downarrow x$ containing x. Then $\uparrow U$ is open containing $\uparrow x$. Since every component in a compact Hausdorff space is the directed intersection of open and closed (clopen) sets, there exists a clopen set V such that $C \subseteq V \subseteq \uparrow U$, where C is the component of x in S. Let $Q = V \cap \downarrow x$. Then Q is clopen in $\downarrow x$. Let $W = \{y \in Q : yQ \subseteq Q\}$. By continuity of the meet operation W is clopen in $\downarrow x$. Clearly $x \in Q$ since x is an identity for $\downarrow x$. If $y, z \in W$, then $yzQ \subseteq yQ \subseteq Q$ and $yz = yzx \in yzQ \subseteq Q$. Thus $yz \in W$, that is, W is a subsemilattice. Now $W \cap \downarrow x \subseteq V \cap \downarrow x \subseteq \uparrow U \cap \downarrow x \subseteq U$ as U is convex.

By 1.10, the arguments of the preceding paragraph show that $\downarrow x$ has small semilattices at x, and hence by Proposition 3.3 so also does S.

Now W is a compact semilattice and by 1.13(v) has a least element z. Let D be a directed set with $z = \sup D$. Then by 1.13(iv) D converges to z. Since $D \subseteq \downarrow x$ and W is open in $\downarrow x$, there exists $d \in D$ such that $d \in W$. Hence $z \leq d$. Since S is meet-continuous, z is a compact element. As U was an arbitrary open convex set around x and $z \in U$, we conclude x is the supremum of the compact elements below it. □

3.13. THEOREM. (THE FUNDAMENTAL THEOREM FOR COMPACT TOTALLY DISCONNECTED SEMILATTICES).

(i) *Let* L *be an algebraic lattice. Then with respect to the Lawson topology* L *is a compact totally disconnected topological semilattice with unit.*

(ii) *Conversely if* L *is a compact totally disconnected topological semilattice with unit, then* L *has small semilattices and with respect to its semilattice structure is an algebraic lattice. Furthermore the topology of* L *is the Lawson topology.*

Proof. The proof follows from III-2.16, the Fundamental Theorem, and 3.12. □

EXERCISES

3.14. EXERCISE. (AN ALTERNATE VERSION OF THE FUNDAMENTAL THEOREM).

(i) Let S be a complete continuous semilattice. Then with respect to the Lawson topology S is a compact topological semilattice with small semilattices.

(ii) Conversely if S is a compact topological semilattice with small semilattices, then with respect to its semilattice structure S is a complete continuous semilattice. Furthermore the topology of S is the Lawson topology.

(iii) Let S and T be compact semilattices with small semilattices. Let f: S→T be a semilattice homomorphism. The following are equivalent:

(1) f is continuous;

(2) f preserves directed sups and arbitrary infs.

If, moreover, $\downarrow f(S) = T$, then (1) and (2) are equivalent to

(3) The lower adjoint g of f exists and preserves the relation $\ll$.

(HINT: The proof follows from the Fundamental Theorem by adjoining discrete identities where needed.)

3.15. EXERCISE. (i) Formulate and prove in the context of 3.13 the proper analogue to 3.4(iii).

(ii) Establish the analogue of 3.14 for the totally disconnected case. □

3.16. DEFINITION. A metric ρ on a semilattice S is called an *ultrametric* if

$$\rho(ax,by) \leq \max\{\rho(a,b),\rho(x,y)\}$$

holds for all $a,b,x,y \in S$. □

3.17. EXERCISE. (i) If ρ is an ultrametric, then prove that each open and each closed ε-ball around a point is a semilattice.

(ii) Establish in addition the formula:

$$\rho(xy,p) \leq \max\{\rho(x,p),\ \rho(y,p)\}.\ \square$$

3.18. EXERCISE. Let S be a compact metric topological semilattice. The following are equivalent:

(1) S has small semilattices.

(2) The topology of S is given by an ultrametric.

(HINT: (2) implies (1) by the 3.17. Conversely by 3.7 Hom(S,**I**) separates points. Since S is compact metric, countably many members of Hom(S,**I**) separate points. Embed S in $\mathbf{I}^{\mathbf{N}}$ with these homomorphisms. The metric $\rho((x_i),(y_i)) = \max_i \{|x_i - y_i|/2^i\}$ is an ultrametric on $\mathbf{I}^{\mathbf{N}}$, and, hence, is also one when restricted to the image of S.) □

The next exercise is a restatement of Example 3.8 for complete continuous semilattices.

3.19. EXERCISE. Let X be a compact Hausdorff space and let $\Gamma_0(X)$ denote the set of ***nonempty*** closed subsets ordered by inclusion.

(i) The Vietoris topology on $\Gamma_0(X)^{op}$ is the Lawson topology; with respect to this topology $\Gamma_0(X)^{op}$ is a compact topological semilattice with small semilattices.

(ii) Furthermore if $f: X \to S$ is a continuous mapping into a compact topological semilattice with small semilattices, then there exists an unique continuous homomorphism $F: \Gamma_0(X)^{op} \to S$ such that $F \circ i = f$ where we define $i: X \to \Gamma_0(X)^{op}$ by $i(x) = \{x\}$.

(HINT: The proof follows from Example 3.8 by throwing in empty sets and identities where needed.) □

3.20. EXERCISE. Let S be a compact topological semilattice. The function $A \mapsto \inf A : \Gamma_0(S)^{op} \to S$ is continuous iff S has small semilattices.

(HINT: Adjoin identities and use 3.9.) □

If S is a compact topological semilattice with small semilattices, then the mapping $A \mapsto \inf A$ is a retraction of $\Gamma_0(S)^{op}$ onto S. It is an old result that if S is a Peano continuum, then $\Gamma_0(S)$ is an absolute retract; indeed recently it has been proved that $\Gamma_0(S)$ is topologically the Hilbert cube; that is, it is a countable product of intervals (see Wojdyslawski [1939], Curtis and Schori [1976]). Thus, in this case S is also an absolute retract.

3.22. EXERCISE. Let X be a compact pospace. The mapping from X to $\Xi(X)^{op}$, the lattice of closed increasing subsets endowed with the Lawson topology, which sends x to $\uparrow x$ is continuous.

(HINT: Suppose $B \in \Xi(X)$, $B \ll \uparrow x$. Then $\uparrow x \subseteq \text{int}(B)$. If (x_α) is a net converging to x, then eventually $x_\alpha \in B$. Since $B = \uparrow B$, eventually $\uparrow x_\alpha \subseteq B$, that is, $B \leq \uparrow x_\alpha$. Suppose $D \in \Xi(X)$, $\uparrow x \not\subseteq D$. Since $D = \uparrow D$, $x \notin D$. Since D is closed, eventually $x_\alpha \notin D$. Thus for sets of the form $\Uparrow B$ and $\Xi(X) \setminus \uparrow D$ which contain $\uparrow x$ we have eventually that $\uparrow x_\alpha$ belongs to such sets. Since such sets form a subbase for the Lawson topology for continuous lattices, we conclude that the injection of X into $\Xi(X)^{op}$ is continuous.) □

3.23. EXERCISE. Let P be a compact poset and let $i : P \to \Xi(P)^{op}$ be the embedding $x \mapsto \uparrow x$. If $f : P \to L$ is a continuous order-preserving function from P into a compact topological semilattice with small semilattices, then there exists a unique continuous homomorphism $F : \Xi(P)^{op} \to L$ such that $F \circ i = f$.

Remark. There are actually two versions of this exercise depending on whether one includes the empty set as a closed descending set or not. If it is included then L must have an identity, that is, be a continuous lattice; the empty set is then mapped to this identity.

(HINT: Define $F : \Xi(P)^{op} \to L$ by $F(A) = \inf f(A)$ for a closed increasing set A. Let $M = \downarrow(f(P))$; note that $f(P)$ is compact and hence also is M. Also we note that $F(\Xi(P)) \subseteq M$. Define $G : M \to \Xi(P)^{op}$ by $G(y) = f^{-1}(\uparrow y)$. Since f is continuous and order-preserving $G(y) \in \Xi(P)$ (and is nonempty). A straightforward calculation gives that G is a lower adjoint for F (with codomain M). If $z \ll y$ in M, then by continuity of f, $f^{-1}(\uparrow y) \subseteq \text{int}\,(f^{-1}(\uparrow z))$. Thus G preserves $\ll$. Therefore F is continuous from $\Xi(P)^{op}$ to M and hence also to L.) □

3.24. EXERCISE. Let S be compact topological semilattice. If $A, B \in \Upsilon(S)$, then $B \ll A$ in $\Upsilon(S)^{op}$ iff $A \subseteq \text{int}\, B$.

(HINT: Suppose $A \subseteq \text{int } B$. If $\mathfrak{D}$ is a descending family of closed lower sets and $\bigcap \mathfrak{D} \subseteq A$, then by compactness $D \subseteq B$ for some $D \in \mathfrak{D}$. Thus $B \ll A$ in $\Upsilon(S)^{op}$.

Conversely suppose $B \ll A$. Define $\mathfrak{D}$ by

$$\mathfrak{D} = \{D \subseteq S : D \text{ is closed}, \downarrow D = D, A \subseteq \text{int } D\}.$$

Then $\mathfrak{D}$ is descending and using the monotone normality of S, one sees that in fact $A = \bigcap \mathfrak{D}$. Since $B \ll A$ in $\Upsilon(S)^{op}$, there exists $D \in \mathfrak{D}$ such that $D \subseteq B$. Hence $A \subseteq \text{int } B$.) □

3.25. EXERCISE. (i) Let X be a set and let $(\beta(X), i)$ be the Stone-Cech compactification of the discrete X. Then $(\Gamma(\beta(X))^{op}, j)$ is the free continuous lattice over X where $j(x) = \{i(x)\}$ for $x \in X$; that is, if $f : X \to L$ is a function into a continuous lattice L, then there exists a unique continuous homomorphism $F : \Gamma(\beta(X))^{op} \to L$ such that $F \circ j = f$.

(ii) Alternately let Filt 2^X denote the lattice of all set-theoretic filters on X ordered by inclusion (with the power set of X included as the largest element of Filt 2^X) and define $j : X \to \text{Filt } 2^X$ by $j(x) = \{A \subseteq X : x \in A\}$. Then (Filt $2^X, j$) forms the free continuous lattice over X. (See I-4.17)

(iii) Therefore, the two constructions are isomorphic.

(HINT: (i): Let $f : X \to L$ where L is a continuous lattice. If L is equipped with the Lawson topology, then by the universal properties of $\beta(X)$, there exists an unique $f^- : \beta(X) \to L$ such that $f^- \circ i = f$. By Example 3.8 there exists a unique continuous homomorphism $F : \Gamma(\beta(X))^{op} \to L$ such that $F \circ k = f$ where $k : \beta(X) \to \Gamma(\beta(X))^{op}$ is defined by $k(y) = \{y\}$. Then $F \circ j = F \circ k \circ i = f \circ i = f$. The uniqueness follows from the uniquenss of f^- and F.

(ii): Since F = Filt 2^X is the set of all filters of the lattice of all subsets of X, we know that it is an algebraic lattice. If $f : X \to L$ is a function into a continuous lattice L, define $G : L \to F$ by

$$G(y) = \{A \subseteq X : f^{-1}(\uparrow z) \subseteq A \text{ for some } z \ll y\}.$$

Since $\{z : z \ll y\}$ is directed, it follows that $G(y) \in F$ for each $y \in L$.

We claim G is a lower adjoint for the function $F(\mathfrak{F}) = \underline{\lim}\ \mathfrak{F}$. Indeed suppose $F(\mathfrak{F}) \geq y$. If $A \in G(y)$ then $f^{-1}(\uparrow z) \subseteq A$ for some $z \ll y$. By definition of F, there exists $B \in \mathfrak{F}$ such that $z \leq \inf B$; that is, $B \subseteq f^{-1}(\uparrow z)$. Thus, $A \in \mathfrak{F}$ since $\mathfrak{F}$ is a filter. Hence $\mathfrak{F} \supseteq G(y)$. Conversely, suppose $G(y) \subseteq \mathfrak{F}$. Then $f^{-1}(\uparrow z) \in \mathfrak{F}$ for all $z \ll y$. Hence, $\underline{\lim}\ \mathfrak{F} \geq \underline{\lim}\ \{z : z \ll y\} = y$, that is, $F(\mathfrak{F}) \geq y$.

To show F is continuous it suffices by IV-1.4 to show that G preserves $\ll$. Let $z \ll y$. Then there exists $w \in L$ such that $z \ll w \ll y$. Then $\mathfrak{F} = \{A \subseteq X : f^{-1}(\uparrow w) \subseteq A\}$ is a principal filter on the lattice of all subsets of X, and hence is a compact element of F. Since $G(z) \subseteq \mathfrak{F} \subseteq G(y)$, we conclude $G(z) \ll G(y)$. Thus F is a continuous homomorphism.

Clearly $F \circ j = f$. Now every principal filter is the intersection of principal ultrafilters, every filter is the directed union of principal filters, and $j(X)$ is the set of principal ultrafilters. Hence, there exists at most one F such that $F \circ j = f$. Since we have seen that one does exist, it is unique. Therefore (F, j) is free over X.

(iii): The equivalence between the two preceding constructions for the free object is obviously obtained by using (i) and (ii) to get maps in both directions whose compositions are both identity functions. However, the conclusion can also be obtained by assigning to a filter all the ultrafilters containing it (and identifying $\beta(X)$ with the set of all ultrafilters on X). It turns out this association defines a lattice isomorphism from Filt 2^X to $\Gamma(\beta(X))^{op}$. □

If we denote with ***CS*** the category of compact semilattices with identities and continuous semilattice morphisms preserving identities, and if we consider the category ***CL*** as a full subcategory of ***CS*** according to 3.4 and 2.7, then we can reinterpret Proposition 3.7 by saying that ***CL*** is the full subcategory in ***CS*** cogenerated by $\mathbf{I} = [0,1]$. This allows us to apply Freyd's existence theorem to obtain a left reflection of ***CS*** into ***CL***. However, the construction of the reflection of a compact semilattice S with identity into the category ***CL*** can easily be given explicitly.

3.26. EXERCISE. Let S be a compact semilattice with identity. Then there is a universal continuous lattice quotient $q : S \to T$ such that all compact semilattice morphisms $S \to T$ into a continuous lattice L factor through q in a unique fashion.

(HINT: Consider $H = \mathrm{Hom}(S, \mathbf{I})$ as in 3.7(3). Let $q^* : S \to \mathbf{I}^H$ be the evaluation map. Then $T = q^*(S)$ and the corestriction q of q^* to its image satisfy the requirements.) □

3.27. EXERCISE. Let $q : S \to T$ be as in Exercise 3.26. If s and t are in distinct components of S, then $q(s) \neq q(t)$.

(HINT: Consider the morphism $S \to S/R$ where R is the connectivity relation. Then S/R is a compact zero dimensional compact semilattice which is a continuous lattice by 3.13. Apply 3.26.) □

NOTES

The notion of a topological semilattice with small semilattices was introduced and studied in the 1967 University of Tennessee dissertation of J. Lawson. The major results appeared in Lawson [1969]. The same idea appeared independently and simultaneously in a paper of M. McWaters [1969].

The problem of whether every compact topological semilattice has small semilattices attracted attention to this class of semilattices. In Lawson [1973a] it was shown that the topology of any compact semilattice was an "intrinsic" topology, one that can be defined from the semilattice structure. This result indicated that these semilattices were some class of semilattices that could be defined in terms of the semilattice structure. Hofmann and Stralka [1976] addressed themselves to this problem and showed that a complete lattice L admitted a topology for which it was a compact topological semilattice with small semilattices (what they called a "Lawson semilattice") if and only if for every $x \in L$ there existed a smallest ideal I with sup $I \geq x$. (Of course by I-2.1 this is equivalent to L being a continuous lattice.)

The first explicit version of the Fundamental Theorem to appear in print was given by J. Lea [1976/77].

Example 3.8 is a composite from numerous sources. Lawson observed in his dissertation that $\Gamma(X)^{op}$ forms a topological semilattice with small semilattices for a compact Hausdorff space X. Hofmann (unpublished notes) recognized the "freeness" of the construction. A detailed treatment from a categorical viewpoint has been given by Wyler [1976].

A thorough treatment of algebraic lattices and compact totally disconnected topological semilattices appears in Hofmann, Mislove and Stralka [1973]. Additional results related to Example 3.10 may be found in Gierz and Keimel [1977].

For a treatment of ultrametrics and an alternate approach to Exercise 3.18, see Hofmann [1970].

Several of the results of this section were discovered much earlier than the Fundamental Theorem and hence were proved without the machinery of continuous lattices. For example it was shown that a compact semilattice with small semilattices could be embedded in a product of intervals by techniques similar to those employed in the proof of Urysohn's Lemma. This type of approach appears in Lawson [1969].

4. SOME IMPORTANT EXAMPLES

We give in this section two examples of unital compact topological semilattices which have no basis of subsemilattices and are thus ***not*** continuous lattices. The first example is topologically contained in the product of the unit interval and the Cantor set and, hence, is one dimensional and metric. The second example is constructed in terms of a space of closed convex subsets of a topological vector space and its topological structure, therefore, is not quite so immediate.

For the first example we develop a general method of construction and then apply it to a specific situation.

4.1. PROPOSITION. *Let* $(\mathrm{T},\wedge)$ *and* $(\mathrm{S},\cdot)$ *be semilattices, and let* $f: \mathrm{S}\to\mathrm{T}$ *be any order-preserving function. Set* $\mathrm{W} = \{(t,s)\in\mathrm{T}\times\mathrm{S} : t\leq f(s)\}$. *Then* W *is a semilattice with respect to the operation*

$$(t,s)\wedge(u,v) = (t\wedge u\wedge f(sv),sv).$$

Proof. The set W is a partially ordered set with respect to the order inherited from $\mathrm{T}\times\mathrm{S}$. Clearly W is closed under this product, and a product is a lower bound to any pair of arguments. To show that it is the greatest lower bound, suppose that $(p,q)\leq(t,s)$ and $(p,q)\leq(u,v)$ hold for three pairs in W. Then $q\leq sv$, and hence $p\leq f(q)\leq f(sv)$. Thus, it follows that $p\leq t\wedge u\wedge f(sv)$. Therefore, $(p,q)\leq(t,s)\wedge(u,v)$, as we wished to show. □

Another way to regard this construction is to think of $(t,s) \mapsto (t\wedge f(s),s)$ as a kernel operator k on $\mathrm{T}\times\mathrm{S}$, which is a semilattice under the pointwise operation. The set W is just the range of k, which is easily proved to be a semilattice under the operation of composing k with the product in $\mathrm{T}\times\mathrm{S}$.

Let now T be a *continuous* lattice and $\{\mathrm{S}_j : j\in J\}$ be a collection of *compact topological* semilattices. For each j let $f_j : \mathrm{S}_j\to\mathrm{T}$ be a given *continuous* order-preserving function. Let $\mathrm{S} = \mathsf{X}_{j\in J}\, \mathrm{S}_j$, and let $\pi_j : \mathrm{S}\to\mathrm{S}_j$ be the projection onto the jth coordinate. Define $f: \mathrm{S}\to\mathrm{T}$ by $f(s) = \bigwedge_{j\in J} f_j\pi_j(s)$. This function is order preserving. Define the semilattice W as in 4.1. In view of the topological assumptions, T, S and $\mathrm{T}\times\mathrm{S}$ are compact pospaces, and, since the maps $f_j\pi_j$ are continuous, W is closed and therefore compact—because we can write $\mathrm{W} = \{(t,s)\in\mathrm{T}\times\mathrm{S} : t\leq f_j\,\pi_j(s) \text{ for all } j\in J\}$. To prove that W is a topological semilattice with respect to the operation of 4.1 and the relative topology, we require a further assumption.

4.2. PROPOSITION. W *is a compact topological semilattice provided that for all* $x,y\in\mathrm{T}$ *with* $x\ll y$, *there exists a finite* $F\subseteq J$ *such that* $y\leq f(u)\wedge f(v)$ *always implies* $x\leq f_j\,\pi_j(uv)$ *for all* $j\notin F$.

Proof. All that is really required to show is that the semilattice operation in W is continuous. To this end let (a_α,r_α) and (b_α,s_α) be two nets in W converging to (a,r) and (b,s) respectively. Clearly $r_\alpha s_\alpha$ converges to rs by hypothesis on S, so what remains to show is that $z_\alpha = a_\alpha \wedge b_\alpha \wedge f(r_\alpha s_\alpha)$ converges to $z = a\wedge b \wedge f(rs)$. (This would be easy if f were continuous, but that is not quite our assumption.) We wish to show that for any subbasic neighborhood U of z, eventually $z_\alpha \in U$.

Let $x \ll z$ and pick y such that $x \ll y \ll z$. Let F be the finite set promised by assumption. For an arbitrary i, we have $f_i\pi_i(r_\alpha s_\alpha)$ converges to $f_i\pi_i(rs)$. Since $z \leq f(rs) \leq f_i\pi_i(rs)$ and $y \ll z$, we have eventually that the net $f_i\pi_i(r_\alpha s_\alpha)$ is in $\uparrow y \subseteq \uparrow x$, because $\uparrow y$ is a neighborhood of z. Thus, because F is finite, there is a β_0 such that $x \leq f_i\pi_i(r_\alpha s_\alpha)$ for all $\alpha \geq \beta_0$ and $i \in F$.

Since $z \leq a\wedge b$, there exists a β_1 such that $\alpha \geq \beta_1$ implies $a_\alpha, b_\alpha \in \uparrow y$, by the same style of argument. By reference to the definition of W and since the nets are in W, we conclude that also $f(r_\alpha)$, $f(s_\alpha) \in \uparrow y$ for $\alpha \geq \beta_1$. Hence, by our special assumption on F, we see that $x \leq f_j\pi_j(r_\alpha s_\alpha)$ for all $\alpha \geq \beta_1$, $j \notin F$.

Putting the two cases together, we have shown that eventually for ***all*** j ***simultaneously*** we have $x \leq f_j\pi_j(r_\alpha s_\alpha)$. Hence, eventually it is the case that $x \leq z_\alpha = a_\alpha \wedge b_\alpha \wedge f(r_\alpha s_\alpha)$. Since given any subbasic open set of the form $U = \{p : q \ll p\}$ for some $q \ll z$, we can find $x \ll z$ such that $q \ll x$, it follows from this argument that the net z_α is eventually in U.

The other type of subbasic open set is one of the form $T \setminus \uparrow d$ where $z \notin \uparrow d$. Hence either $a \notin \uparrow d$, $b \notin \uparrow d$ or $f(rs) \notin \uparrow d$. If $a \notin \uparrow d$ or $b \notin \uparrow d$, then eventually $a_\alpha \notin \uparrow d$ or $b_\alpha \notin \uparrow d$ resp. Thus $a_\alpha \wedge b_\alpha \wedge f(r_\alpha s_\alpha) \notin \uparrow d$ eventually. If $f(rs) \notin \uparrow d$, then $f_j\pi_j(rs) \notin \uparrow d$ for some j. Since $f_j\pi_j$ is continuous, eventually $f_j\pi_j(r_\alpha s_\alpha) \notin \uparrow d$. Therefore eventually $a_\alpha \wedge b_\alpha \wedge f(r_\alpha s_\alpha) \notin \uparrow d$. Either way we eventually have $z_\alpha \in T \setminus \uparrow d$. □

We now specialize further by letting T denote $[0,\infty]$, the extended nonnegative reals, which is a topological lattice with respect to its natural order. For each positive integer i, we will choose an integer $s(i) \geq 2$ and set $S_i = \{0,1\}^{s(i)}$, a finite lattice with respect to the coordinatewise order. We require a lemma to give us a suitably divergent series governing the choice of the $s(i)$.

For each positive integer n larger than 1, we set $\alpha_n = 1/m2^{m-1}$ where $2^{m-1} < n \leq 2^m$. The series $\Sigma_{n\geq 2}\, \alpha_n$ may be thought of as the result of dividing the mth term of the harmonic series into 2^{m-1} parts. Hence this series is divergent. The rate of growth is slow, however:

4.3. LEMMA. *For any $\varepsilon > 0$, there is an integer $p \geq 1$ such that if $k \geq p$, then*

$$\sum_{n=2}^{k} \alpha_n + \varepsilon > \sum_{n=2}^{2k} \alpha_n;$$

Proof. We first note that $\Sigma_{n\in A}\, \alpha_n = 1/m$ if $A = \{n : 2^{m-1} < n \leq 2^m\}$. Choose q and p such that $2/\varepsilon < q$ and $2^{q-1} < p$. If $k \geq p$, there exists an unique m such that $2^{m-1} < k \leq 2^m$. Then

$$\sum_{n=2}^{2k}\alpha_n \leq \sum_{n=2}^{2^{m+1}} \alpha_n = \sum_{n=2}^{2^{m-1}} \alpha_n + (1/m + 1/(m+1)) \leq \sum_{n=2}^{k} \alpha_n + 2/m.$$

Since $m \geq q$, we have $2/m \leq 2/q < \varepsilon$; this completes the proof. □

For each positive integer i, let $s(i)$ be the least integer where

$$i \leq \sum_{n=2}^{s(i)} \alpha_n.$$

such an integer exists since $\Sigma\alpha_n$ is divergent. For $x \in S_i$, let $\theta(x) \leq s(i)$ denote the number of zero entries of x. We define $f_i : S_i \to T$ by

$$\begin{aligned} f_i(x) &= \infty && \text{if } \theta(x) = 0; \\ &= i && \text{if } \theta(x) = 1; \\ &= 0 && \text{if } \theta(x) = s(i); \\ &= i - \sum_{n=2}^{\theta(x)} \alpha_n && \text{for all other cases.} \end{aligned}$$

4.4. LEMMA. (i) *Each f_i is a continuous order-preserving function from S_i into T;*

(ii) *If $\tau > \varepsilon > 0$, there exists a postive integer q such that, for all $i \geq q$ and all $u, v \in S_i$, if $f_i(u) > \tau$ and $f_i(v) > \tau$, then $f_i(uv) > \tau - \varepsilon$;*

(iii) *Consequently, f satisfies the assumption of 4.2.*

Proof. (i): That each f_i is order preserving is a straightforward consequence of its definition. Continuity is trivial since the lattice S_i is finite.

(ii): Assume that $\tau > \varepsilon > 0$. Choose the p guaranteed by Lemma 4.3 which corresponds to ε. Choose q larger than $\tau + \Sigma_{2 \leq n \leq 2p} \alpha_n$.

We suppose that $i \geq q$, $u, v \in S_i$, $f_i(u) > \tau$ and $f_i(v) > \tau$. If $z = uv$, then either $\theta(z) \leq 2\theta(u)$ or $\theta(z) \leq 2\theta(v)$ obtains; we arbitrarily assume $\theta(z) \leq 2\theta(u)$. (The reason one of the inequalities prevails is that uv can have at most twice as many zero entries as one of u or v.) We note from the definition of f_i that in all cases

$$f_i(z) \geq i - \sum_{n=2}^{\theta(z)} \alpha_n,$$

if the summation is interpreted to be 0 for $\theta(z)$ equal to 0 or 1. If $\theta(u) \leq p$, then

$$f_i(z) \geq i - \sum_{n=2}^{\theta(z)} \alpha_n \geq q - \sum_{n=2}^{2\theta(u)} \alpha_n \geq q - \sum_{n=2}^{2p} \alpha_n \geq \tau;$$

The last inequality follows from the choice of q. Hence $f_i(z) > \tau - \varepsilon$ if $\theta(u) \leq p$.

If $p<\theta(u)$, then

$$f_i(z)\geq i-\sum_{n=2}^{\theta(z)}\alpha_n\geq i-\sum_{n=2}^{2\theta(u)}\alpha_n\geq i-(\sum_{n=2}^{\theta(u)}\alpha_n+\varepsilon)=f_i(u)-\varepsilon>\tau-\varepsilon.$$

Hence $f_i(z)>\tau-\varepsilon$ for both cases.

(iii): Suppose now that $x\ll y$ holds in T. The case $x = 0$ is trivial, so we suppose that the element is positive. We then interpolate τ and ε so that $x<\tau-\varepsilon<\tau<y$. We choose q as in (ii) and let F be the set of indices below q. If we then had $y\leq f(u)\wedge f(v)$ for $u,v\in S$, this would imply $f_i\pi_i(u)>\tau$ and $f_i\pi_i(v)>\tau$ for all indices i. But then by (ii), we would have $x\leq f_j\ \pi_j(uv)$ for all $j\notin F$ as desired. □

4.5. THEOREM. W *is a unital compact topological semilattice without a basis of subsemilattices.*

Proof. Note that $1 = (\infty,(u_i))$ where each u_i has entries all 1, and that $1\in W$. (For simplicity we are using the subscript notation rather than the projection notation on S.) All that remains to check is the basis assertion; in fact, we show that if A is a subsemilattice and $1\in\text{int}(A)$, then $A\cap(0\times S)\neq\varnothing$.

There exists at 1 a basis of open sets of the form $U = \{(t,(u_i))\in W : n<t,$ u_i has entries all 1 for $i\leq n\}$, where n is a positive integer. We assume n is chosen so that $U\subseteq\text{int}(A)$. Let B be the set of all elements of the form $(n+1,(u_i))$ such that u_i has entries all 1 for $i\neq n+1$ and u_{n+1} has one zero entry. Then B has $s(n+1)$ elements. For each element of B, $\inf\{f_i(u_i) : 1\leq i\} = f_{n+1}(u_{n+1}) = n+1$; hence $B\subseteq W$ and thus $B\subseteq U$. Let $(t,(z_i))$ be the greatest lower bound in W of B. Then $(t,(z_i))\in A$, since A is a subsemilattice. As $(t,(z_i))\in W$, $t\leq f_{n+1}(z_{n+1}) = 0$, since z_{n+1} has entries all 0. Hence $t=0$. This completes the proof. □

The reader should consult Exercise VII-2.12 for an important further development of this example.

For the second example let V be any Hausdorff topological vector space and let K be a compact, convex subset of V. Let Con(K) denote the set of closed convex subsets of K (including the empty set). Define the function $f\colon K\times K\times[0,1]\to K$ by $f(x,y,t) = tx+(1-t)y$. Then f is continuous since V is a topological vector space. The function f induces a continuous mapping $F\colon\Gamma(K)\times\Gamma(K)\times\Gamma([0,1])\to\Gamma(K)$ by

$$F(A,B,M) = \{ta+(1-t)b : a\in A,\ b\in B,\ t\in M\}.$$

If $A,B\in\text{Con}(K)$, then $F(A,B,[0,1])$ is the closed convex hull of A and B. Hence $(A,B)\mapsto$ closed convex hull of $A\cup B$ is a continuous function on Con(K), and with respect to this operation $\text{Con}(K)^{\text{op}}$ is a compact topological semilattice.

Now if V is a locally convex space, then a closed convex subset of K has a basis of neighborhoods in K which are closed and convex. It is easily verified that if $K_1 \subseteq \mathrm{int}(K_2)$, then $K_2 \ll K_1$. Hence in this case $\mathrm{Con}(K)^{\mathrm{op}}$ is a continuous lattice (cf. I-1.22). Conversely, it is shown in Lawson [1976b] that if $\mathrm{Con}\,(K)^{\mathrm{op}}$ is a continuous lattice, then K can be embedded in a locally convex separated topological vector space by an affine homomorphism. J.W. Roberts [1977] has obtained examples of compact convex sets which have no extreme points and hence admit no such embedding; thus for such a K, $\mathrm{Con}(K)^{\mathrm{op}}$ is not a continuous lattice, although it is a compact unital topological semilattice. However, Roberts' constructions are at least as complicated as the one given for 4.5.

NOTES

After the notion of a compact topological semilattice with small semilattices was introduced, it remained an open question for several years whether every compact topological semilattice had small semilattices. Lawson solved the problem in the negative with the first counter-example of this section which appeared in [1970a].

An interesting topological question is to find topological properties which insure that a compact topological semilattice will have small semilattices. Theorem 3.13 states that total disconnectedness is such a condition. Lawson [1969] showed that this conclusion remains true for finite-dimensional Peano continua. The most general class of spaces so far discovered appear in Lawson [1977]; this class includes spaces that locally are homeomorphic to a product of a totally disconnected space and a finite-dimensional Peano continuum.

5. CHAINS IN COMPACT POSPACES AND SEMILATTICES

In this section we investigate the nature of chains (totally ordered sets) in pospaces, topological semilattices and lattices. Maximal chains are a particularly useful tool, and we invoke freely the axiom—equivalent to the Axiom of Choice—that in a poset every chain is contained in a maximal chain, the well-known Hausdorff Maximality Principle. Theorems 5.11 and 5.15 employ chains to give a criterion for connectedness.

The first proposition is quite straightforward.

5.1. **PROPOSITION.** *If M is a maximal chain in a poset, then we have*

$$M = \bigcap\{\downarrow x \cup \uparrow x : x \in M\}.$$

Hence, if $\leq$ is semicontinuous, M is closed. □

5.2. **COROLLARY.** *If $\leq$ is semicontinuous, then the closure of a chain is a chain.*

Proof. Any chain is contained in a maximal chain, which is closed and contains the closure of the given chain. Thus, the closure being contained in a chain is itself a chain. □

The next proposition is due to A.D. Wallace [1945] and is one of the oldest results in the theory of topological ordered spaces.

5.3. **PROPOSITION.** *Consider a poset equipped with a quasicompact topology for which the order is lower semicontinuous. Then any element has a minimal element below it.*

Proof. Let M be a maximal chain containing a given element q. Then $\{\downarrow x : x \in M\}$ is a tower of closed sets whose intersection is nonempty since the space is quasicompact. Let p be in the intersection. Then $\{p\} \cup M$ is a chain, and hence $p = \inf M \in M$. Then p is minimal in the whole poset, for otherwise the chain M could be extended. □

5.4. **PROPOSITION.** *If $\leq$ is semicontinuous and C is a quasicompact chain, then the relative topology of C is the order topology. Moreover, if C is nonempty, then C is complete.*

Proof. Since $\{x \in C : a < x < b\} = C \setminus (\downarrow a \cup \uparrow b)$, the relative topology is finer than the order topology. Since the order topology is Hausdorff, the two agree.

In case C is nonempty, then by 5.3 it has a minimum and a maximum element; hence, in particular, every subset is bounded from below. If $S \subseteq C$ is nonempty, let L be the set of lower bounds of S. The family of closed intervals $[x,y]$ for $x \in L$ and $y \in S$, must have a nonempty intersection in C; it is easy to argue that the intersection is in fact $\{\inf S\}$. □

5.5. DEFINITION. Let X be a pospace. We say $A \subseteq X$ is an *arc-chain* iff A is a nontrivial, compact, connected chain. □

Since we have just seen that the relative topology on an arc-chain A is the order topology, it follows that, topologically, A is an arc, that is, a continuum with exactly two non-cutpoints.

5.6. PROPOSITION. *Let* X *be a pospace, and let* $A \subseteq X$.

(i) *If A is an order-dense compact chain, then A is an arc-chain;*

(ii) *If* X *is compact and order dense and A is a maximal chain, then A is either an arc-chain or a point;*

(iii) *If* X *has a* 0 *and* 1 *and A is a connected chain containing* 0 *and* 1, *then A is a maximal chain.*

Proof. (i): The proof that A is connected is analogous to the proof that the unit interval is connected and can be left to the reader.

(ii): It follows easily from hypothesis that every maximal chain is order dense; hence, the conclusion follows from part (i) and 5.1.

(iii): Suppose A is not maximal. Then there exists $p \in X \setminus A$ such that $A \subseteq \uparrow p \cup \downarrow p$. Then $A \cap \uparrow p$ and $A \cap \downarrow p$ are closed, nonempty, disjoint subsets of A, which contradicts the assumption that A is connected. □

5.7. PROPOSITION. *Let* X *be a compact pospace. Every convergent net of arc-chains in* X *converges in the space of closed subsets of* X *endowed with the Vietoris topology to an arc-chain or a point.*

Proof. Let A be the limit of such a net. It is well-known that the limit of continua is a continuum. Also a set K is a chain if and only if $K \times K \subseteq (\leq \cup \geq)$, and the latter is a closed set, since X is a pospace. Hence, in the Vietoris topology, the family of all closed subsets whose square is contained in $(\leq \cup \geq)$ is closed; thus, A is a chain. □

5.8. DEFINITION. Let X be a pospace. A point $p \in X$ is a *local minimum* iff there exists an open set U with $\downarrow p \cap U = \{p\}$; that is, $\{p\}$ is open in $\downarrow p$. □

The next result was the discovery of R.J. Koch [1959] and is one of the principal results in the theory of pospaces.

5.9. THEOREM. (KOCH'S ARC THEOREM). *Let U be an open subset with compact closure in a pospace* X. *If U contains no local minimum, then every point of U lies on an arc-chain which meets the boundary of U.* □

The proof of this theorem is rather lengthy and is deferred to the exercises. The idea of the proof is to employ the nonexistence of local minima to construct, for each neighborhood $\mathcal{U}$ of the diagonal $\Delta = \{(x,x) \in X \times X\}$, a chain in U^- from p to $X \setminus U$ such that the chain is $\mathcal{U}$-connected (that is, if the chain is written as the disjoint union of two nonempty sets P and Q, then there exists $a \in P$, $b \in Q$ such that $(a,b) \in \mathcal{U}$). One then takes a limit of these chains over all neighborhoods $\mathcal{U}$ of the diagonal in the compact space of closed subsets of U^-. This limit is the desired chain.

We turn now to the topic of the existence of arc-chains in topological semilattices.

5.10. LEMMA. *Let* S *be a semitopological semilattice.*

(i) *Then* $k \in S$ *is a local minimum iff* $\uparrow k$ *is open.*

(ii) *If* S *is a compact topological semilattice, then* $k \in S$ *is a local minimum iff* $k \in K(S)$.

Proof. (i): If $\uparrow k$ is open, clearly k is a local minimum. Conversely, if $\{k\}$ is open in $\downarrow k = Sk$, then $\uparrow k$ is open as the inverse image of $\{k\}$ under the continuous map $x \mapsto xk : S \to \downarrow k$.

(ii): If $\uparrow k$ is open, then $k \in K(S)$ as soon as the topology is compatible (1.2). Conversely if $k \in K(S)$, then $\uparrow k$ is open by Exercise 2.12. □

5.11. THEOREM. *Let* S *be a compact semilattice. The following statements are equivalent:*

(1) S *is connected;*

(2) 0 *is the only compact element of* S;

(3) 0 *is the only local minimum in* S;

(4) *Each point of* S *lies on an arc-chain containing* 0.

Proof. The equivalence of (2) and (3) follows from Lemma 5.10. That (3) implies (4) follows easily from Theorem 5.9 applied to the open set $U = S \setminus \{0\}$. If (4) holds, then S is arcwise connected and hence connected. Finally if S is connected, then for $x \in S$, we have $\downarrow x$ is connected. If x is a local minimum, then $\{x\}$ is open and closed in $\downarrow x$, so $\downarrow x = \{x\}$, whence $x = 0$. Hence, (1) implies (3). □

It is not necessarily the case that there exists an arc-chain between x and y whenever $x \leq y$ in a compact connected topological semilattice. It becomes of interest to consider those pairs which are so connected, and we write $x \dashv y$ for this relationship (including the case $x = y$ and always implying $x \leq y$).

5.12. PROPOSITION. *Let* S *be compact semilattice.*

(i) $\dashv$ *is a partial order having a closed graph;*

(ii) $x \dashv y$ *iff* $x \leq y$ *and* $[x,y]$ *is connected.*

Proof. (i): Everything is immediate except that $\dashv$ is closed. Let nets (x_α) converge to x, (y_α) converge to y, and $x_\alpha \dashv y_\alpha$. For each α, let A_α be an arc-chain between x_α and y_α. Then by Proposition 5.7 a subnet of the A_α's converges to an arc-chain A containing x and y. Hence, $x \dashv y$.

(ii): Suppose $x \dashv y$. Then there exists an arc-chain A between x and y. Then $[x,y] = \bigcup\{Az : x \leq z \leq y\}$. Since each Az is connected and contains x, their union is connected; hence, $[x,y]$ is connected.

Conversely, suppose $[x,y]$ is connected. Then $[x,y]$ is a compact connected topological semilattice, and, by Proposition 5.11, there exists an arc-chain between x and y. □

5.13. DEFINITION. A topological semilattice S is *order connected* iff $[x,y]$ is connected for all $x,y \in S$ with $x \leq y$. □

5.14. PROPOSITION. *The following statements are equivalent in a compact topological semilattice* S:

(1) S *is order connected*;

(2) *The relations* $\leq$ *and* $\dashv$ *agree*;

(3) $\uparrow x$ *is connected for all* $x \in S$;

(4) S *is order-dense.*

If S *is unital, the above are also equivalent to*:

(5) *For all* $x \in S$, *there exists an arc-chain from* 1 *to* x.

Proof. The equivalence of (1) and (2) follows from Proposition 5.11.

Since $\uparrow x = \bigcup_{x \leq y}[x,y]$, (1) implies (3). Conversely if $\uparrow x$ is connected, then $[x,y] = (\uparrow x)y$ is connected.

Clearly (2) implies (4). Conversely, if $x < y$, let A be a maximal chain between x and y. Then A is closed and order dense and hence an arc-chain (see Proposition 5.6).

To conclude the proof, note that if S has a 1, then (2) implies (5). Assume (5) and let $x,y \in S$, $x < y$. If A is an arc-chain from 1 to x, then yA is a connected set containing x and y. Hence, $\uparrow x$ is connected. Thus, (5) implies (3). □

We consider now the existence of arc-chains in topological lattices. The third condition gives an algebraic characterization.

5.15. PROPOSITION. *Let* L *be compact topological lattice. The following statements are equivalent*:

(1) L is *connected*;

(2) *If* $x < y$, *then there exists an arc-chain between* x *and* y;

(3) L *is order dense.*

Proof. (1) implies (2): Since L is compact and connected and since $[x,y] = y(L \vee x)$, we have that $[x,y]$ is a compact connected topological lattice. Hence by 5.11 (applied to $[x,y]$) there exists an arc-chain between x and y.

(2) implies (3): Trivial.

(3) implies (1): Let x and y be elements of L, and let A and B be maximal chains containing $\{x,xy\}$ and $\{y,xy\}$ respectively. By 5.6 A and B are arc-chains which both contain xy. Thus, x and y lie in the same component of L. Hence, L is connected. □

EXERCISES

5.16. EXERCISE. Let X be a compact pospace, let $x<y$, and let U and V be disjoint open sets containing x and y respectively. Define a new relation $\preceq$ by

$p\preceq q$ iff $p\leq q$ and either $(p,q)\notin U\times V$ or there exists $t\notin U\cup V$ such that $p\leq t$ and $t\leq q$.

(i) Show $\preceq$ is a closed partial order contained in the original order.

(ii) Show that X has the same set of local minima for both orders.

(HINT: (i): Reflexivity and antisymmetry of $\preceq$ are immediate. Suppose $x\preceq y$ and $y\preceq z$. Then $x\leq y\leq z$. If $(x,z)\notin U\times V$, then $x\preceq z$. If $x\in U$, $z\in V$ and $y\notin U\cup V$, then $x\preceq z$. If $y\in U$, then $y\preceq z$ implies there exists $w\notin U\cup V$ such that $y\leq w$ and $w\leq z$. Then $x\leq w$ and $w\leq z$ imply $x\preceq z$. Similarly if $y\in V$. Hence $\preceq$ is transitive.

Let (x_α,y_α) be a net in X×X converging to (x,y) such that $x_\alpha\preceq y_\alpha$ for each α. Then $x_\alpha\leq y_\alpha$ for each α, and hence $x\leq y$. If $(x,y)\notin U\times V$ then $x\preceq y$. Otherwise suppose $x\in U$, $y\in V$. Then eventually $x_\alpha\in U$ and $y_\alpha\in V$. Thus, eventually there exists $w_\alpha\notin U\cup V$ and then clearly $x_\alpha\leq w_\alpha\leq y_\alpha$. The net w_α clusters to some $w\notin U\cup V$ such that $x\leq w\leq y$. Hence $x\preceq y$, and $\preceq$ is a closed relation.

(ii): A local minimum for $\leq$ is clearly one for $\preceq$. Conversely let p be a local minimum for $\preceq$. If $p\notin V$, then the lower set of p is the same for $\leq$ and $\preceq$. Hence p is a local minimum for $\leq$. If $p\in V$, then since $U\cap V = \varnothing$ the lower set of p intersected with V is the same for $\preceq$ and $\leq$.) □

5.17. EXERCISE. Let X be a compact Hausdorff space and let U be an open subset. If $\mathfrak{P}$ is a descending family of closed partial orders on X each of which has no local minimum in U, then the intersection is also a closed partial order with no local minimum in U.

(HINT: The intersection is easily seen to be a closed partial order. Let $x\in U$ and let W be an open set containing x. Pick open sets N and V such that $x\in N\subseteq N^-\subseteq V\subseteq V^-\subseteq W\cap U$.

Let $\leq$ be a partial order in $\mathfrak{P}$. By Proposition 5.3 there exists an element y minimal (with respect to $\leq$) in N^- such that $y\leq x$. Since y is not a local minimum, there exists $z\in V$ such that $z<y$. Then we must have $z\in V\backslash N^-$. Hence for each partial order $(\leq)_\alpha\in\mathfrak{P}$, there exists $z_\alpha\in V\backslash N^-$ such that $z_\alpha<x$. Since each z_α lies in the compact set $V^-\backslash N$, they cluster to some $z\in V^-\backslash N$.

One argues that the pair (z,x) is in each partial order and hence in the intersection. Since $z<x$ and $z\in W$, we conclude x is not a local minimum.) □

5.18. EXERCISE Prove Theorem 5.9.

(HINT: We may assume that X itself is a compact pospace. By picking a maximal chain of closed partial orders containing the given order which have no local minimum in U and taking the intersection, one obtains (5.17) a minimal such partial order. In this new order pick a maximal chain M containing p, and let $q = \sup\,(M\cap\downarrow p)\backslash U$. Since M is closed, $q\in M\backslash U$. To complete the proof, we need only show that $A = [q,p]\cap M$ is an arc-chain.

If A is order dense, then by Proposition 5.6 it is an arc-chain. But if $x,y\in A$, $x<y$, and $[x,y]\cap A = \{x,y\}$, then $[x,y] = \{x,y\}$ (otherwise one extends the maximal chain M by adding some element between x and y). But then (5.16) one eliminates the pair (x,y) from the order, contradicting the fact the order is minimal. Hence, A is order dense.) □

NOTES

Koch's Arc Theorem was one of the major early advances in the theory of pospaces and topological semilattices and has continued to be an important tool. The proof given in Exercises 5.16 through 5.18 is Ward's [1965]. Koch's work led to a detailed study of the existence of arc-chains in topological semilattices and lattices (see, e.g., Anderson and Ward [1961a], Brown [1965], Lawson [1969]).

CHAPTER VII

Topological Algebra and Lattice Theory: Applications

Our last chapter is devoted to exploring further links between topological algebra and continuous lattices. This theme has already played an important role: the Fundamental Theorem of Compact Semilattices (VI-3.4) is just one example. In this chapter, however, the methods of topological algebra occupy a more central role, while the methods of continuous lattices are somewhat less prominent.

Section 1 is devoted to recent and somewhat technical results about certain non-Hausdorff topological semilattices; they are included primarily to facilitate the proof of the principal result of Section 2: a complete lattice is a compact topological lattice if and only if it admits a compact topology relative to which the meet operation is jointly continuous and the join operation is separately continuous.

Section 3 continues the theme of Chapter V by developing a duality between a category of compact pospaces and continuous maps, on the one hand, and a category of distributive continuous lattices whose set of prime elements is closed in the Lawson topology, on the other. Section 4 then characterizes those meet-continuous complete lattices which admit a compact semilattice topology as being exactly those lattices whose lattice of Scott-open sets forms a continuous lattice; this augments II-1.12, which shows that the continuous lattices are exactly those meet-continuous complete lattices whose Scott-open sets form a completely distributive lattice. The final part of Section 4 is devoted to a proof that a compact semitopological semilattice is in fact topological. This is a particularly appropriate note on which to end this treatise, since the proof we present utilizes those aspects of the theory of continuous lattices which we have sought to stress: namely, the algebraic theory and its utility in applications to related areas of mathematics.

1. ONE-SIDED TOPOLOGICAL SEMILATTICES

So far, our consideration of topological semilatices has assumed that they are Hausdorff; however, in II-4.13 we had occasion to consider those non-Hausdorff topological semilattices that were topological semilattices with respect to the Scott topology. The purpose of this section is to study non-Hausdorff semilattices in more detail; not only is this class of interest in itself, but also we shall find useful applications of the theory to the Hausdorff setting.

We introduce first some convenient notation and an easy lemma.

1.1. DEFINITION. Let S be a semilattice, $A \subseteq S$. If $x \in S$, we define

$$x^{[-1]}A = \{y \in S : xy \in A\}. \square$$

1.2. LEMMA. *A semilattice* S *is a semitopological semilattice under a topology* $\mathcal{U}$ *iff* $x \in S$ *and* $U \in \mathcal{U}$ *always imply* $x^{[-1]}U \in \mathcal{U}$. $\square$

Every semitopological semilattice gives rise to a "one-sided" one, in the sense of having a new topology where all open sets are upper sets as was the case in the Scott topology.

1.3. PROPOSITION. *Let* $(S, \mathcal{U})$ *be a semitopological semilattice. Set*

$$\mathcal{V} = \{U \in \mathcal{U} : U = \uparrow U\}.$$

Then $(S, \mathcal{V})$ *is a semitopological semilattice.*

Proof. Let $V \in \mathcal{V}$. Then $V \in \mathcal{U}$ and $V = \uparrow V$. If $x \in S$, then by Lemma 1.2 $x^{[-1]}V \in \mathcal{U}$. If $y \in x^{[-1]}V$ and $z \geq y$, then $xz \geq xy \in V$. Thus $xz \in V$, that is, $z \in x^{[-1]}V$. Thus $x^{[-1]}V$ is also an upper set and hence in $\mathcal{V}$. By Lemma 1.2 again, $(S, \mathcal{V})$ is a semitopological semilattice. $\square$

1.4. DEFINITION. For a topology $\mathcal{U}$ on a semilattice S, define $\mathcal{U}^{\wedge}$ to be the topology generated by the subbase

$$\{x^{[-1]}U : x \in S, U \in \mathcal{U}\} \cup \mathcal{U}. \square$$

1.5. PROPOSITION. *For any topology* $\mathcal{U}$ *on a semilattice* S *the topology* $\mathcal{U}^{\wedge}$ *is the weakest topology on* S *containing* $\mathcal{U}$ *for which* S *is a semitopological semilattice.*

Proof. To show $(S, \mathcal{U}^{\wedge})$ is a semitopological semilattice, we apply Lemma 1.2 to the subbasis of $\mathcal{U}^{\wedge}$. If $U \in \mathcal{U}$, clearly $x^{[-1]}U \in \mathcal{U}^{\wedge}$. If $x \in S$ and $y^{[-1]}U \in \mathcal{U}^{\wedge}$, then $x^{[-1]}(y^{[-1]}U) = (yx)^{[-1]}U \in \mathcal{U}^{\wedge}$.

If $(S, \mathcal{V})$ is a semitopological semilattice and $\mathcal{U} \subseteq \mathcal{V}$, then by Lemma 1.2 each $x^{[-1]}U$ is in $\mathcal{V}$; hence, $\mathcal{U}^{\wedge} \subseteq \mathcal{V}$. $\square$

Recall from the definition preceding Exercise II-1.17, that a topology $\mathcal{U}$ on an up-complete poset X is called *order consistent* if the lower set of a point is the closure of that point and if directed nets converge to their sups. Equivalently we could say that $\mathcal{U}$ contains the upper topology and is contained in the Scott topology. The next definition introduces some other connections between order and topology and is somewhat technical, but it includes notions which will prove quite useful in the developments of this section. We recall from Definition VI-1.2 that a topology on a poset is called *compatible* if directed nets converge to their sups, and dually filtered nets converge to their infs. For a subset $V \subseteq X$, we write $V^{\diamond}$ for the set of all such directed sups and filtered infs of nets in V, and we remark that in the compatible case that $V \subseteq V^{\diamond} \subseteq V^{-}$.

1.6. DEFINITION. A topology on a poset X is called *order regular* (or *o-regular*) if for every open neighborhood $U \in \mathcal{U}$ of a point there exists another neighborhood $V \in \mathcal{U}$ such that $V^{\diamond} \subseteq U$ and filtered sets in V have infs..

A point in X is called an *ω-point* if there exists a countable collection of open neighborhoods of the point $\{U_n : n \in \mathbf{N}\}$ such that $U_{n+1}{}^{\diamond} \subseteq U_n$, and the point is a minimal element of the set $\bigcap\{U_n : n \in \mathbf{N}\}$. □

1.7. PROPOSITION. *Let* X *be a poset equipped with a topology.*

(i) *If the relation* $\leq$ *is lower semicontinuous and the topology is compatible, then the set of open upper sets is an order-consistent topology on X.*

(ii) *If* X *is a compact pospace, then the set of open upper sets forms an o-regular order-consistent topology on* X.

(iii) *If the relation* $\leq$ *is lower semicontinuous and the topology is o-regular, then each point in* X *is the supremum of the* ω*-points below it.*

Proof. (i): It is immediate that the collection of open upper sets is closed under arbitrary unions, finite intersections, and contains X and $\emptyset$; hence it forms a topology.

For $x \in X$ with respect to this topology $\{x\}^{-}$ is a lower set since its complement is open; thus $\downarrow x \subseteq \{x\}^{-}$. Since the order is lower semicontinuous, $\downarrow x$ is closed in both topologies; thus $\{x\}^{-} \subseteq \downarrow x$. Hence, the two sets are equal, and the first condition for being order consistent is satisfied. The second condition follows from the compatibility of the topology.

(ii): By part (i) the topology is order consistent. Suppose $x \in X$ and U is an open upper set containing x. Then $\uparrow x$ and $X \setminus U$ are disjoint, $\uparrow x$ is a closed upper set and $X \setminus U$ is a closed lower set. Since X is monotone normal, (see VI-1.8) there exist an open upper set P and an open lower set Q such that $\uparrow x \subseteq P$, $X \setminus U \subseteq Q$, and $P \cap Q = \emptyset$. Let F be a filtered set in P. Since X has a compatible topology, $\inf F \in P^{-} \subseteq X \setminus Q \subseteq U$. Clearly, since $P = \uparrow P$, if D is a directed subset of P then $\sup D \in P \subseteq U$. Thus the open upper sets form an o-regular topology.

(iii): Let X be a lower semicontinuous space equipped with an o-regular topology, and let $x \in X$. Suppose y is an upper bound for all ω-points below x. We show then that $x \leq y$ and hence x is the supremum of all ω-points below it.

Suppose on the contrary that $x \not\leq y$. Then $U_1 = X \setminus \downarrow y$ is an open neighborhood of x. Pick U_2 such that U_2 is open, $x \in U_2$, and $U_2^{\diamond} \subseteq U_1$. Continue this procedure inductively. Let $P = \bigcap\{U_n : n \in \mathbf{N}\}$.

Let M be a maximal chain in P containing x. Then since $M \subseteq U_{n+1}$, inf M exists and is a member of U_n for all n. Hence inf $M \in P$. Clearly inf $M \leq x$. Also inf M is minimal in P (otherwise the maximal chain M could be extended). Thus inf M is an ω-point. Since inf $M \in P \subseteq U_1 \subseteq X \setminus \downarrow y$, we have inf $M \not\leq y$, a contradiction. Thus $x \leq y$. □

We come now to a key (and difficult) lemma.

1.8. LEMMA. *Let* L *be a complete lattice with a topology* $\mathcal{U}$ *and* $a_1, \ldots, a_n \in$ L *be a finite set of* ω*-points. If an ultrafilter on* L *converges to each of the points* $a_1, \ldots, a_n$ *in the topology* $\mathcal{U}^{\wedge}$, *then it converges to* $s = a_1 \vee \ldots \vee a_n$ *in the Scott topology.*

Proof. The proof for any finite number of points is essentially the same as that for two. We restrict our attention to the latter case in order to simplify the bookkeeping. Suppose then that a and b are ω-points and the ultrafilter $\mathcal{F}$ converges to both a and b in the $\mathcal{U}^{\wedge}$ topology. Let M be a Scott-open set around $s = a \vee b$. Let $\{U_n\}$ and $\{V_n\}$ be the sequences of open neighborhoods of a and b guaranteed by the definition of ω-points. Let $F \in \mathcal{F}$.

We proceed recursively to obtain three sequences $\{a_n\}, \{b_n\}$, and $\{x_n\}$ such that for all n and all $j < n$ we have:

(i) $a_n = ax_n \in U_n$ and $b_n = bx_n \in V_n$;

(ii) $a_j a_{j+1} \cdots a_n = a_j a_{j+1} \cdots x_n \in U_j$ and $b_j b_{j+1} \cdots b_n = b_j b_{j+1} \cdots x_n \in V_j$;

(iii) $x_n \in F$.

Since $a \in a^{[-1]} U_1 \in \mathcal{U}^{\wedge}$ and $b \in b^{[-1]} V_1 \in \mathcal{U}^{\wedge}$, we have $a^{[-1]} U_1 \in \mathcal{F}$ and $b^{[-1]} V_1 \in \mathcal{F}$; thus $a^{[-1]} U_1 \cap b^{[-1]} V_1 \cap F \in \mathcal{F}$. Let x_1 be a point in this intersection. Then $a_1 = ax_1 \in U_1$ and $a_1 \leq a$; similarly $b_1 = bx_1 \in V_1$ and $b_1 \leq b$.

Next $a \in a^{[-1]} U_2$ and $a \in a_1^{[-1]} U_1$ because $aa_1 = a_1$. Similarly we find $b \in b^{[-1]} V_2 \cap b_1^{[-1]} V_1 \in \mathcal{U}^{\wedge}$. Thus there exists an element

$$x_2 \in a^{[-1]} U_2 \cap a_1^{[-1]} U_1 \cap b^{[-1]} V_2 \cap b_1^{[-1]} V_1 \cap F.$$

Let $a_2 = ax_2$ and $b_2 = bx_2$. Then $a_2 \leq a$, $a_2 \in U_2$, and $a_1 a_2 = a_1 x_2 \in U_1$; analogous statements hold for b_2. The recursive procedure should now be clear.

We set $c_n = \bigwedge\{a_j : j \geq n\}$, $d_n = \bigwedge\{b_j : j \geq n\}$, and $y_n = \bigwedge\{x_j : j \geq n\}$. Then the sequences $\{c_n\}$, $\{d_n\}$, and $\{y_n\}$ are all directed. Let $y = \bigvee_n y_n$. By part (ii) of the recursive definition, for each k, $\bigwedge_{i \leq k} a_{n+i} \in U_n$. Since c_n is the filtered inf of $\{\bigwedge_{i \leq k} a_{n+i} : k \geq 1\}$, we have $c_n \in U_{n-1}$ (from the definition of ω-point). Similarly $d_n \in V_{n-1}$ for all $n > 1$.

Let $c = \bigvee\{c_n : n \geq 1\}$. Since the set $\{U_n\}$ is towered, the preceding paragraph implies $c_m \in U_{n+1}$ for $m > n+1$. Since $\{c_n\}$ is a directed set, $c \in U_n$ (from the definition of an ω-point). Since n was arbitrary, $c \in \bigcap_n U_n$. Similarly $d \in \bigcap_n V_n$. For each n, $c_n \leq a_n \leq a$; hence $c \leq a$. Since a is minimal in $\bigcap_n U_n$, we conclude $c = a$. Similarly $d = b$.

From the definitions of c_n, d_n and y_n we conclude $c_n \leq y_n$ and $d_n \leq y_n$. Hence $c \leq y$ and $d \leq y$; that is, $s = a \vee b = c \vee d \leq y$. Since we began with a Scott-open set M around s and $s \leq y$, we conclude $y \in M$. Since y is the sup of a directed set, there exists n such that $y_n \in M$. Since $y_n \leq x_n$, we conclude $x_n \in M$. Finally, $x_n \in F$ (part (iii) of the recursive definition) implies $F \cap M \neq \emptyset$.

Now F was an arbitrary member of $\mathfrak{F}$, so we conclude $F \cap M \neq \emptyset$ for every $F \in \mathfrak{F}$. As $\mathfrak{F}$ is an ultrafilter, this implies $M \in \mathfrak{F}$. Since M was an arbitrary Scott-open set around s, we conclude that $\mathfrak{F}$ converges to s in the Scott topology. □

We come now to a major theorem.

1.9. THEOREM. *Let* L *be a complete lattice with a topology* $\mathfrak{U}$ *in which every open set is an upper set.*

(i) *If each point is a supremum of* ω*-points, then* $\mathfrak{U}^\wedge$ *contains the Scott topology.*

(ii) *If* $\mathfrak{U}$ *is order consistent and o-regular and* L *is meet continuous, then* $\mathfrak{U}^\wedge$ *is the Scott topology.*

(iii) *If* $\mathfrak{U}$ *is order consistent and o-regular and if* L *is a semitopological semilattice with respect to the meet operation, then* $\mathfrak{U}$ *is the Scott topology.*

Proof. (i): Showing $\mathfrak{U}^\wedge$ contains the Scott topology is equivalent to showing the identity function is continuous from $(L,\mathfrak{U}^\wedge)$ to $(L,\sigma(L))$. For this it suffices to show that if an ultrafilter $\mathfrak{F}$ converges to x in $(L,\mathfrak{U}^\wedge)$, then the same obtains in $(L,\sigma(L))$.

Let M be a Scott-open set containing x. Since M is Scott open and x is the sup of all ω-points below it, we conclude there exist ω-points $x_1,\ldots,x_n$ such that $y = x_1 \vee \ldots \vee x_n \in M$ and $y \leq x$. Since each open set in $\mathfrak{U}$, and hence in $\mathfrak{U}^\wedge$ is an upper set, the ultrafilter $\mathfrak{F}$ also converges to each of the points $x_1,\ldots,x_n$ in $(L,\mathfrak{U}^\wedge)$. By Lemma 1.8 $\mathfrak{F}$ converges to y in the Scott topology. Since $y \in M$, we conclude $M \in \mathfrak{F}$. Since M was an arbitrary Scott-open set containing x, we conclude $\mathfrak{F}$ converges to x in $\sigma(L)$.

(ii): Since L is meet-continuous, $(L,\sigma(L))$ is a semitopological semilattice. Hence by Proposition 1.5 $\mathfrak{U}^\wedge \subseteq \sigma(L)$.

Conversely by Proposition 1.7(iii), each point in L is a supremum of ω-points. Hence by part (i), we have $\sigma(L) \subseteq \mathfrak{U}^\wedge$. Thus $\sigma(L) = \mathfrak{U}^\wedge$.

(iii): Since $\mathcal{U}$ is order consistent, $\mathcal{U}\subseteq\sigma(L)$. Conversely by 1.7(iii) each point is a supremum of ω-points, and hence $\sigma(L)\subseteq\mathcal{U}^\wedge$ by part (i). But since $(L,\mathcal{U})$ is a semitopological semilattice, we have $\mathcal{U}^\wedge = \mathcal{U}$. □

Lattices of the types appearing in Theorem 1.9 possess other interesting properties; in the following we make a rather brief allusion to some of these. First of all we introduce an additional property closely related to that of sobriety.

1.10. DEFINITION. A topological space X is called *super sober* if the set of limit points of every ultrafilter is either empty or the closure of a (unique) point. □

The next proposition lends some credence to this terminology.

1.11. PROPOSITION. *A super-sober space is sober.*

Proof. Let A be an irreducible closed set in a super sober space X. Then every subset of A which is open in A is dense in A; hence, the open subsets of A form a filter base. Extend this filter base to an ultrafilter $\mathcal{F}$ on X. Since A is closed and in $\mathcal{F}$, the set of points of convergence of $\mathcal{F}$ is a subset of A. Conversely let $x\in A$, and let U be an open neighborhood of x. Since $U\cap A\in\mathcal{F}$ by definition of $\mathcal{F}$, we conclude $\mathcal{F}$ converges to x. We see then that A is precisely the set of limit points of $\mathcal{F}$. Thus, A is the closure of a point, and we have proved that X is sober. □

1.12. THEOREM. *Let* L *be a meet-continuous lattice with an order-consistent o-regular topology* $\mathcal{U}$. *Then* $(L,\sigma(L))$ *is super sober; that is, every ultrafilter in* L *has a largest point of convergence in the Scott topology.*

Proof. Let $\mathcal{F}$ be an ultrafilter which converges to points x and y in the Scott topology. Let $A = \{z : z \text{ is an } \omega\text{-point}, z\in\downarrow x\cup\downarrow y\}$. Since x and y are each the supremum of the ω-points below them by 1.7(iii), we conclude $\sup A = x\vee y$.

Since $\mathcal{F}$ converges to each of x and y, we have $\mathcal{F}$ converges to every member of A in the Scott topology. By Theorem 1.9 $\mathcal{U}^\wedge = \sigma(L)$; thus, by Lemma 1.8, the ultrafilter $\mathcal{F}$ Scott converges to $\sup F$ for every finite set $F\subseteq A$. Since the set $\{\sup F : F \text{ finite}, F\subseteq A\}$ is directed, $\mathcal{F}$ converges to its supremum, $x\vee y$.

Since x and y were arbitrary points of convergence for $\mathcal{F}$, we conclude the set of convergence points of $\mathcal{F}$ is directed. Then $\mathcal{F}$ will also converge to the supremum of the convergence points, since the set of all convergence points $\mathcal{F}$ must be Scott-closed. □

1.13. PROPOSITION. *Let* L *be a meet-continuous lattice such that the Scott topology on* $L\times L$ *is super sober.*

(i) *The Scott topology on* $L \times L$ *is the square of the Scott topology on* L;
(ii) $(L, \sigma(L))$ *is a topological lattice.*

Proof. (i): Since the product of two Scott-open sets is Scott open, we have always that $\sigma(L) \times \sigma(L) \subseteq \sigma(L \times L)$. To show containment in the other direction, we must show that any ultrafilter $\mathfrak{F}$ which converges to (x,y) in the product topology also converges to the same point in the Scott topology on $L \times L$.

Since each open set in the product topology is an upper set, $\mathfrak{F}$ also converges to $(x,0)$ in the product topology. Let U be a Scott-open set around $(x,0)$. Let $A = \{p \in L : (p,0) \in U\}$. Since U is Scott-open, it follows that A is Scott-open in L. Since U is an upper set, $A \times L \subseteq U$. But $A \times L$ is open in the product topology; hence $A \times L \in \mathfrak{F}$ and thus $U \in \mathfrak{F}$. Since U was arbitrary, $\mathfrak{F}$ converges to $(x,0)$ in the Scott topology. Similarly $\mathfrak{F}$ converges to $(0,y)$ in the Scott topology. Since $\mathfrak{F}$ has a largest point of convergence, we conclude that $\mathfrak{F}$ converges to (x,y) in the Scott topology.

(ii): It is easily verified that the inverse image of a Scott-closed set in L is Scott-closed in $L \times L$ for the mapping $(x,y) \rightarrow x \vee y$; a similar conclusion holds for $(x,y) \rightarrow xy$ if L is meet continuous. Since by part (i) the Scott topology is the product topology we conclude both mappings are continuous. □

1.14. COROLLARY. *Let* L *be a semitopological semilattice with an order-consistent o-regular topology. Then the topology is the Scott topology, and* L *is a super-sober topological lattice.*

Proof. Use 1.9, 1.12 and 1.13. (To apply 1.13 note that the product topology on $L \times L$ is order consistent and o-regular and that $L \times L$ is also meet-continuous since L is; thus, by Theorem 1.12, $L \times L$ is super sober with respect to the Scott topology.) □

We close this section with a result which will be of use in the final section of this chapter.

1.15. PROPOSITION. *Suppose that* X *is a poset with a compact Hausdorff topology such that the relation* $\leq$ *is semicontinuous and the family of open upper sets is the Scott topology on* X. *Then* X *is a pospace.*

Proof. We must show that the relation $\leq$ is closed in $X \times X$. Let $x, y \in X$ with $x \not\leq y$. Since the relation is semi-continuous, $\uparrow y$ is then a closed subset of X, and x is a point not in $\uparrow y$. Since X is compact Hausdorff, there is then a compact neighborhood N of x with $N \cap \uparrow y = \emptyset$. Hence $\downarrow N \cap \uparrow y = \emptyset$. By VI-1.6(i) we have that $\downarrow N$ is Scott closed.. Now, $x \in \operatorname{int}(\downarrow N)$ and $y \in X \setminus \downarrow N$, and these sets are open ($X \setminus \downarrow N$ is Scott open implies $X \setminus \downarrow N$ is open since the open upper sets in X are exactly the Scott-open sets). Moreover, if $z \in \operatorname{int}(\downarrow N)$ and $w \in X \setminus \downarrow N$, then $(z,w) \notin \geq$ since $\downarrow N$ is a lower set which is disjoint from its complement. Thus, $\leq$ is closed. □

EXERCISES

1.16. DEFINITION. Let X be a topological space. Define the *cocompact topology* on X to have as a subbase complements of saturated quasicompact sets (quasicompact sets which are intersections of open sets). Define the *patch topology* on X to be the common refinement of the original topology and the cocompact topology. □

1.17. EXERCISE. Let X be a poset with an order consistent topology.

(i) The saturated sets are precisely the upper sets;
(ii) Each set of the form $\uparrow x$ is a saturated quasicompact set;
(iii) Each saturated quasicompact set is Scott closed in X^{op};
(iv) The cocompact topology is order consistent on X^{op};
(v) If X is locally quasicompact, then under the patch topology X is a pospace (and hence Hausdorff).

(HINT: (i): Each open set is an upper set and hence the same is true for the intersection. Conversely if $A = \uparrow A$, then $A = \bigcap\{X\setminus\downarrow x : x \notin A\}$. Hence A is saturated.

(ii): Any open set containing x contains $\uparrow x$. Hence $\uparrow x$ is quasicompact.

(iii): Apply VI-1.6(i) to X^{op}.

(iv): It follows immediately from (ii) and (iii) that the cocompact topology contains the upper topology and is contained in the Scott topology on X^{op}. Hence it is order consistent.

(v): Let $x,y \in X$, $x \not\leq y$. Since $X\setminus\downarrow y$ is an open set containing x, there exists a quasicompact neighborhood Q of x contained in $X\setminus\downarrow y$. It is easily verified that $\uparrow Q$ is also quasicompact and $\uparrow Q \subseteq X\setminus\downarrow y$. The interior of Q is an open upper set containing x in the given and, hence, in the patch topology. The complement of $\uparrow Q$ is a lower set which is a neighborhood of y in the cocompact and, hence, in the patch topology. Clearly the two are disjoint. Therefore, we may apply VI-1.4.) □

1.18. EXERCISE. Let X be a compact pospace. Let $\mathcal{V}$ be the collection of open upper sets of X.

(i) $\mathcal{V}$ is an order- consistent topology on X;
(ii) $(X,\mathcal{V})$ is a locally quasicompact, super sober quasicompact space;
(iii) The cocompact topology with respect to $(X,\mathcal{V})$ consists of the open lower sets of X;
(iv) The patch topology with respect to $(X,\mathcal{V})$ is the original topology.

(HINT: (i): By VI-1.6(ii), $\downarrow x$ is closed for each $x \in X$. Hence, $\mathcal{V}$ contains the upper topology. That $\mathcal{V}$ is contained in the Scott topology follows from VI-1.3.

(ii): Let U be an open upper set and let $x \in U$. Since X is monotone normal (VI-1.8), there exists an open upper set V such that $x \in V \subseteq V^- \subseteq U$. Then V^- is compact in X and hence quasicompact in $(X,\mathcal{V})$.

Let $\mathfrak{F}$ be an ultrafilter in X. Since X is compact Hausdorff $\mathfrak{F}$ has a unique point of convergence, say p, in X. We show that the set of convergence points in $(X,\mathfrak{V})$ is $\downarrow p = \{p\}^-$. Since any open increasing set around a point in $\downarrow p$ contains p, we have that $\downarrow p$ is contained in the set of convergence points. If $q \not\leq p$, by the monotone normality, there exists an open increasing set U and an open decreasing set V such that $q \in U$, $p \in V$, $U \cap V = \emptyset$. Since $\mathfrak{F}$ converges to p in X, $V \in \mathfrak{F}$. Hence $U \notin \mathfrak{F}$, that is, $\mathfrak{F}$ does not converge to q in $(X,\mathfrak{V})$. Thus $\downarrow p$ is precisely the set of convergence points of $\mathfrak{F}$ in $(X,\mathfrak{V})$.

Since X is compact, we have that $(X,\mathfrak{V})$ is quasicompact.

(iii): This is equivalent to showing the quasicompact upper sets in $(X,\mathfrak{V})$ are precisely the closed upper sets. Since each closed upper set is compact, it is quasicompact in $(X,\mathfrak{V})$.

Conversely suppose $A = \uparrow A$ is quasicompact in $(X,\mathfrak{V})$. Suppose $x \notin A$. For each $a \in A$, pick an open upper set V_α and an open lower set U_α such that $a \in V_\alpha$, $x \in U_\alpha$, and $U_\alpha \cap V_\alpha = \emptyset$. Since A is quasicompact, finitely many of the U_α cover A. The intersection of the corresponding V_α is an open decreasing set around x missing A. Hence A is closed in X.

(iv): This follows from (iii) and VI-1.9.) □

1.19. EXERCISE. Let X be a poset with an order-consistent topology for which X is locally quasicompact, quasicompact and super sober. Then X endowed with the patch topology is a compact pospace.

(HINT: By Exercise 1.17 X is a pospace. Let $\mathfrak{F}$ be an ultrafilter on X and let the convergence points of $\mathfrak{F}$ be $\downarrow x$ (the set is nonempty since X is quasicompact). Let V be a set open in the cocompact topology containing x. Then $K = X \setminus V$ is a quasicompact upper set. Since $K \cap \downarrow x = \emptyset$, for each $w \in K$, there exists an open set U_w such that $U_w \notin \mathfrak{F}$. Hence each set $X \setminus U_w \in \mathfrak{F}$. Since finitely many of the U_w cover K, we have a finite intersection of the $X \setminus U_w$ contained in V. This implies $V \in \mathfrak{F}$. Hence every open set in $\mathfrak{U}$ containing x and every cocompact open set containing x is in $\mathfrak{F}$. Thus $\mathfrak{F}$ converges to x in the patch topology. Since every ultrafilter converges, we conclude that X is compact.) □

PROBLEM. Characterize those complete lattices for which the Scott topology is super sober. (By Exercise 1.18 this is so for continuous lattices.)

PROBLEM. Let L be a meet-continuous lattice. If L can be given some topology making it a compact pospace, is the Scott topology locally quasicompact?

1.20. EXERCISE. For each $\alpha \in A$, let L_α be a complete lattice with an order-consistent o-regular topology. Show the Scott-topology on $\times_\alpha L_\alpha$ is the product of the Scott topologies. □

1.21. EXERCISE. Let S be a compact topological semilattice. Then the topology on S is the patch topology associated with the Scott topology on S.

(HINT: The proof follows from Corollary 1.14 followed by Exercise 1.18.) □

The following exercise ties together some of the various formulations of the notion of a space being super sober.

1.22. EXERCISE. Let X be an up-complete poset, and let $\mathfrak{U}$ be an order-consistent topology on X.

(i) Consider the following statements:

(1) The topology $\mathfrak{U}$ constitutes the open upper sets for some topology on X making it a compact pospace.

(2) Every ultrafilter on X has a largest point of convergence.

(3) X is super sober and quasicompact.

(4) The intersection of finitely many quasicompact upper sets is quasicompact.

Prove that (1) implies (2) iff (3) implies (4).

(ii) If X is quasicompact, locally quasicompact, and sober, then (1)–(4) are equivalent.

(iii) Assume further that X is a sup-semilattice and consider the following condition:

(5) X is a topological semilattice.

Prove that (2) implies (5) implies (4). Hence, if X is also quasicompact, locally quasicompact, and sober, then (1)–(5) are equivalent.

(HINT: (i): That (1) implies (2) follows from Exercise 1.18. If every ultrafilter has a point of convergence, then X is quasicompact; if it has a largest point of convergence, then the set of points of convergence is just the down set of that point, which is its closure (since the topology is order consistent). Hence (2) implies (3). Conversely the fact that X is super sober and quasicompact implies the set of convergence points for an ultrafilter is the closure of a point, that is, the down set of a point. Hence there is a largest point of convergence.

Let A and B be quasicompact upper sets, and let $\mathfrak{F}$ be an ultrafilter in $A \cap B$. Then $\mathfrak{F}$ has a point of convergence in A and one in B. By (2) there exists a point of convergence for $\mathfrak{F}$ above both of these points, hence in $A \cap B$, since A and B are upper sets. Thus (2) implies (4).

(ii): We show (4) implies (1) under the assumptions of (ii). If $\mathfrak{F}$ is an ultrafilter on X, then the set A of convergence points is nonempty and closed. If $A = B \cup C$ where B and C are closed proper subsets, then there exists $p \in B \setminus C$ and $q \in C \setminus B$. Pick quasicompact neighborhoods U and V of p and q such that $U \subseteq X \setminus C$ and $V \subseteq X \setminus B$. Then $U, V \in \mathfrak{F}$ and by hypothesis $U \cap V$ is quasicompact. Since $U \cap V \in \mathfrak{F}$, $\mathfrak{F}$ has a point of convergence in $U \cap V$, a contradiction. Thus A is irreducible and hence the closure of a point (since X is sober). Thus, X is super sober.

By 1.19 the patch topology makes X a compact pospace. A member of $\mathcal{U}$ is certainly an open upper set in the patch topology. Conversely choose an ultrafilter $\mathcal{F}$ which converges to y for $\mathcal{U}$. Then $\mathcal{F}$ has a largest $\mathcal{U}$-convergence point x. By the proof of 1.19 $\mathcal{F}$ converges to x in the patch topology and hence in the topology of open upper sets. Thus $\mathcal{F}$ converges to y for the open upper sets.

(iii): Let $\mathcal{F}$ be an ultrafilter on $X \times X$ converging to (x,y) in the product topology. Let $\mathcal{G}$ be the ultrafilter generated by the image of $\mathcal{F}$ under the sup mapping. One verifies easily that if $\pi_1 : X \times X \to X$ is the projection into the first coordinate, then $\uparrow(\pi_1(F)) \in \mathcal{G}$ for each $F \in \mathcal{F}$. Since the topology on X is order consistent and since the first projection of $\mathcal{F}$ converges to x, it follows that $\mathcal{G}$ converges to x. Similarly $\mathcal{G}$ converges to y. By (2) $\mathcal{G}$ converges to something greater than or equal to $x \vee y$. Thus, the sup mapping is continuous.

Since the intersection of the two upper sets is just their image under the sup mapping, (5) easily implies (4).) □

NOTES

The techniques and results of this section constitute recent unpublished work of Gierz and Lawson. Traditionally topological algebraists have restricted their attention to the Hausdorff setting, but with the advent of non-Hausdorff topologies one is motivated to consider other assumptions. Certainly much remains to be done in this area.

The exercises also present new material. It is of interest to note that they show that compact pospaces can be viewed alternately as special kinds of T_0-spaces (where the order is simply the specialization order).

2. TOPOLOGICAL LATTICES

There are many examples of topological lattices. The unit interval $\mathbf{I} = [0,1]$, as everyone knows, is a topological lattice with respect to its usual order and topology. (In fact, more generally any chain under the order topology is a topological lattice.) Moreover, any cartesian product of the interval $\mathsf{X}_\alpha \mathbf{I}_\alpha$ of any number of copies of $\mathbf{I}$ is a distributive topological lattice with respect to the product topology and the coordinatewise order. We shall return to a characterization of certain sublattices of such products at the end of this section.

Examples of a different kind can be extracted from "hyperspaces" of closed subsets of suitable semilattices. Recall the discussion of the Vietoris topology in VI-3.8 and VI-3.10. Let S be a compact topological semilattice, hence, a compact pospace. In the notation of VI-3.10, $\Upsilon(S)$ is the lattice of all closed lower sets in S ordered by inclusion and equipped with the relative Vietoris topology. We have already remarked that $\Upsilon(S)^{op}$ is a continuous lattice (VI-3.11 and VI-3.24), but we can say more:

2.1. PROPOSITION. *Let* S *be a compact topological semilattice. Then* $\Upsilon(S)$ *is a distributive topological lattice. Furthermore, the embedding* $x \mapsto \downarrow x : S \to \Upsilon(S)$ *is a topological and semilattice monomorphism.*

Proof. Recall that $\Upsilon(S)$ is a closed subset of $\Gamma(S)$. It is standard fare in point-set topology that continuous mappings on compact spaces induce continuous mappings between their hyperspaces (see Example IV-3.8); thus, the mapping $(A,B) \mapsto AB = \{ab : a \in A, b \in B\}$ is continuous from $\Gamma(S) \times \Gamma(S) \to \Gamma(S)$ and also when restricted to $\Upsilon(S)$. Since the lattice operations are just union and intersection, $\Upsilon(S)$ is distributive. The mapping $x \mapsto \downarrow x$ is easily verified to be a semilattice monomorphism from S to $\Upsilon(S)$. It is also continuous (see Exercise VI-3.22). □

In the example just given the join operation is very quickly shown to be jointly continuous. But it is often enough to check separate continuity. In this regard the next proposition demonstrates the power of the theory developed in the preceding section.

2.2. PROPOSITION. *Let* L *be a complete lattice endowed with a topology making* L *into a compact topological semilattice. If the join operation is separately continuous, then* L *is a topological lattice.*

Proof. By VI-1.14 we know that L is a compact pospace. By 1.3 applied to L^{op}, we find L endowed with the topology of all open lower sets is a semitopological semilattice with respect to the join operation, and by 1.7(iii) this topology is order consistent and o-regular. Corollary 1.14 then implies L is a topological lattice with respect to this topology. A similar argument implies that L is a topological lattice when endowed with the topology of all open upper sets. Since VI-1.9 shows L has a subbasis of open increasing and open decreasing sets, we conclude that L with its original topology is a topological lattice. □

2.3. COROLLARY. *A compact topological lattice* L *has for its topology the topology generated by the Scott-open sets and their duals (the Scott-open subsets of* L^{op}*).*

Remark. The topology just mentioned will sometimes be referred to as the *bi-Scott topology.* See Exercise 2.11 for a general condition implying the Hausdorff property.

Proof. The proof follows immediately from the proof of 2.2 if one notes that by 1.14 the increasing open sets are the Scott-open sets and the decreasing open sets the dual Scott-open sets. □

2.4. COROLLARY. *Let* L *be a distributive continuous lattice (that is, a continuous Heyting algebra). The following statements are equivalent:*

(1) L *is join continuous;*

(2) L *is a topological lattice with respect to the Lawson topology.*

Proof. (2) implies (1): Use O-4.4.

(1) implies (2): We need only show the join operation is separately continuous. But this follows from the Fundamental Theorem VI-3.4 since the mapping $x \to x \vee y$ for a fixed $y \in L$ preserves arbitrary sups and meets. □

2.5. DEFINITION. A complete lattice L is a *bicontinuous* lattice if L is a continuous lattice with respect to both the meet and join operations (that is, both L and L^{op} are continuous). If further the two Lawson topologies on L and L^{op} agree, the lattice is called a *linked bicontinuous lattice.*

We now formulate a version of the Fundamental Theorem VI-3.4 appropriate to topological lattices.

2.6. PROPOSITION. (i) *Let* L *be a linked bicontinuous lattice. Then with respect to the Lawson topology* L *is a compact topological lattice which at each point has a basis of neighborhoods which are sublattices.*

(ii) *Conversely, given a compact topological lattice* L *which has small semilattices for both operations, then* L *is a linked bicontinuous lattice, and the topology on* L *is the Lawson topology.*

Proof. (i): Applying the Fundamental Theorem VI-3.4 to L endowed with the meet (resp., join operation) we conclude that L is a topological lattice. Let x be a member of an open set $U = \uparrow U$. Since L is continuous there exists an open filter F such that $x \in F \subseteq U$. Dually if $x \in V$, V is open, and $\downarrow V = V$, there exists an open ideal M such that $x \in M \subseteq V$. Since the intersection of open filters (resp., open ideals) is again an open filter (resp., open ideal) and since the intersection of an open filter and an open ideal is an open sublattice, it follows from VI-1.9 and VI-1.14 that L has a basis of open sublattices.

(ii): Conversely suppose L is a compact topological lattice which has small semilattices for both operations. Then by the Fundamental Theorem VI-3.4 L is a continuous lattice with respect to both the meet and join operations and the topology of L has to be the same as the Lawson topology and the dual Lawson topology. Thus, L is linked bicontinuous. □

2.7. DEFINITION. Let L be a complete lattice. The *interval topology* on L is the join of the lower topology and its order dual, the upper topology. Hence, the set of principal filters and principal ideals form a subbasis for the closed sets for the interval topology. □

The next proposition gives another important characterization of linked bicontinuous lattices.

2.8. PROPOSITION. *Let* L *be a complete lattice. The following statements are equivalent*:

(1) L *is a linked bicontinuous lattice*;

(2) L *is a meet-continuous and join-continuous lattice and the interval topology is Hausdorff.*

Proof. Suppose L is a linked bicontinuous lattice. Then with respect to the Lawson topology, the closed upper sets are precisely the sets which are closed in the lower topology (III-3.20(iv)). Similarly, since the dual Lawson topology is the Lawson topology, the closed lower sets are precisely the closed sets in the upper topology. Since by VI-1.9 these closed sets form a subbase for the topology of L, it follows that the topology of L is the interval topology. That L is meet and join continuous follows from I-1.12.

Conversely suppose L is both meet and join continuous and that the interval topology is Hausdorff. Since the Lawson topology is the join of the lower topology and the Scott topology, it follows that the Lawson topology is finer than the interval topology; hence, the Lawson topology is Hausdorff. By III-2.9, L is a continuous lattice and dually is also cocontinuous. Since the Lawson topology is compact and the interval topology Hausdorff, the two topologies must agree. Dually the dual Lawson topology is also the interval topology. Therefore, L is bicontinuous. □

Completely distributive lattices have already made their appearance in our study (see I-2.4, I-2.5, I-3.15 and II-1.14). The next proposition relates them to the considerations of this section.

2.9. PROPOSITION. *Let* L *be a distributive complete lattice. The following statements are equivalent*:

(1) L *is completely distributive*;

(2) L *is linked bicontinuous*;

(3) *The set of lattice homomorphisms into* **I** *preserving arbitrary meets and joins separates points*;

(4) L *admits a topology making it a compact topological lattice for which the set of continuous lattice homomorphisms into* **I** *separates points*;

(5) L *is bicontinuous.*

Remark. The conditions of this theorem therefore completely characterize those lattices representable as sublattices of direct powers of the unit interval closed under arbitrary inf and sup.

Proof. The equivalence of (1) and (5) follows from I-3.15, and the further equivalence with (3) follows from the note following it. (See also Exercise IV-2.30.)

(3) implies (4): Embed L in a product of intervals with lattice homomorphisms preserving arbitrary joins and meets. Then the image is closed under arbitrary joins and meets. Thus the image is closed under arbitrary infs and directed sups, and hence is closed in the product topology by III-1.11. Thus, with respect to the relative topology, it is a compact topological lattice. The projections into the coordinates show there are enough continuous lattice homomorphisms to separate points.

(4) implies (2): The hypothesis implies that L can be embedded in a product of intervals by a topological lattice isomorphism. Since a closed sublattice of a product of intervals is bicontinuous, (2) follows.

(2) implies (5): Immediate. □

Without assuming (1) implies (3), one can argue directly that (2) implies (4) and hence (3) by a Urysohn-type argument (employing Lemma 3.19 of Chapter I). These Urysohn-type arguments appear in Davies [1968] and Lawson [1967].

We have just seen that a distributive lattice for which L and L^{op} are both continuous lattices is linked bicontinuous. This conclusion can fail if the lattices are ***not*** assumed to be distributive. Indeed, let C be a countably infinite set with the trivial partial ordering (just the equality relation). Let $L = \{0,1\} \cup C$ be the resulting lattice obtained by adjoining a zero and a unit. Then L and L^{op} are (isomorphic) algebraic lattices. However, in the Lawson topology on L, 0 is the one-point compactification of C, while 1 is the one-point compactification of C in L^{op}. (That is to say, $\{0\} \cup C$ is closed but not open in the Lawson topology of L; whereas $\{1\} \cup C$ is open but not closed.) Thus, the two topologies are different.

EXERCISES

2.10. EXERCISE. Consider the following already established statements:

(i) Hom (L,I), the set of continuous lattice homomorphisms, separates points if L is a completely distributive complete lattice (Proposition 2.9).

(ii) Hom (S,I), the set of continuous semilattice homomorphisms, separates points if S is a continuous lattice with the Lawson topology (Proposition VI-3.7).

(iii) Hom (X,I), the set of continuous order-preserving functions, separates points if X is a compact pospace (Exercise VI-1.11).

Show that (i) implies (ii) implies (iii).

(HINT: (i) implies (ii): If S is as hypothesized, then $\Upsilon(S)$, the set of closed lower sets is bicontinuous (see I-3.15) and hence completely distributive. Furthermore S can be embedded in $\Upsilon(S)$ by a topological semilattice isomorphism by sending s to $\downarrow s$. Compose the members of Hom($\Upsilon(S)$,I) with the embedding.

(ii) implies (iii): $\Upsilon(X)$, the set of closed lower sets, is a continuous lattice (see Example VI-3.11). Embed X in $\Upsilon(X)$ by sending x to $\downarrow x$. Compose members of Hom ($\Upsilon(X)$,I) with the embedding.) □

2.11. EXERCISE. If L is a complete lattice admitting an o-regular order-consistent topology, then the bi-Scott topology is Hausdorff. □

It is known that the bi-Scott topology is not Hausdorff for the lattice of regular open sets on [0,1] (Floyd [1955]), and for the lattice of closed congruences on [0,1] (Clinkenbeard [1977]).

2.12. EXERCISE. Show that Corollary 2.4 cannot be sharpened to conclude that L^{op} is a continuous lattice.

(HINT: Let S be the compact unital topological semilattice constructed in Example VI-4.5 which is not a continuous lattice. $\Upsilon(S)$, the set of all (topologically) closed lower sets, is a compact distributive topological lattice, that is dually a continuous lattice with respect to the operation of union, and that the embedding $s \mapsto \downarrow s$ from S to $\Upsilon(S)$ is a topological and semilattice monomorphism. Hence, since S fails to have a basis of subsemilattices, the same must be true of $\Upsilon(S)$ with respect to the intersection operation. We have the peculiar phenomenon that a compact distributive topological lattice can be a continuous lattice with respect to one operation, but may fail to be one with respect to the other.) □

NOTES

Topological lattices have a rather long-standing history. The idea of a topological lattice is implicit in the work of G. Birkhoff on the order topology in the late thirties and a short time later in the work of O. Frink on the interval topology in a lattice. The theory of topological lattices was first explicitly studied by L.W. Anderson in his 1954 thesis directed by A.D. Wallace. The early work on topological lattices actually preceded the investigation of topological semilattices and was instrumental in shaping the direction of the latter research, although the study of semilattices has surpassed that of topological lattices in recent years.

The equivalence of (1) and (3) in Proposition 2.9 is an old result of Raney's [1953]. The construction of $\Upsilon(S)$ was early recognized by D.R. Brown and others as an important one in the investigation of topological semilattices. Proposition 2.2 is a result of Lawson's. The original proof employed the rather difficult theorem that a separately continuous compact Hausdorff topological semilattice is jointly continuous (a result we obtain in Section 4). The present proof is new and employs the machinery built up in Section 1 (rather than the intricate topological machinery involved in the other route). For further related details concerning intrinsic lattice topologies, topological lattices, and completely distributive lattices, see Strauss [1968], Lawson [1973], Gingras [1978] and Gierz and Lawson [198*].

3. COMPACT POSPACES AND CONTINUOUS HEYTING ALGEBRAS

In Section VI-3 we saw that the dual of the closed upper sets in a compact pospace is a continuous lattice. The dual of that lattice has a concrete representation by complements: the continuous lattice of open lower sets. It is this form of the construction that will be more appropriate to the considerations of this section.

Here we show that a distributive continuous lattice in which the set of prime elements is closed with respect to the Lawson topology can be represented as the lattice of all ***nonempty*** closed upper sets of the set of prime elements endowed with the topology induced from the Lawson topology. This topology can also be defined in terms of the hull-kernel topology on Spec L by a procedure that we have already considered quite generally in V-5.11, 5.12.

Now let L be a distributive continuous lattice in which the set PRIME L of prime elements is closed with respect to the Lawson topology. From V-3.7 we know that the latter is the case iff the way-below relation is multiplicative on L. From V-5.12 we know that the topology on PRIME L induced from the Lawson topology agrees with the patch topology. In the remainder of this section, PRIME L will always be endowed with this topology. Then PRIME L is a compact pospace in the sense of VI-1.1 with a greatest element 1.

Looking at this situation from the other side, let X be a compact pospace with a greatest element 1 and let D(X) denote the set of all ***proper*** *open lower sets* in X. Ordered by inclusion, D(X) is a complete distributive lattice (with the unit being X\{1}). We discuss the continuity in 3.3 below.

We now show that under the above hypotheses, L can be represented by the lattice D(PRIME L).

3.1. LEMMA. *Let* L *be a distributive continuous lattice in which* PRIME L *is closed with respect to the Lawson topology. A subset U of* PRIME L *is a proper open lower set iff $U = \Delta_L(a) =$* PRIME L$\setminus\uparrow a$ *for some* $a \in$L.

Proof. We show the equivalent assertion that a set A is a nonempty closed upper set in PRIME L iff $A = \uparrow a \cap$ PRIME L for some $a \in$L. Clearly $\uparrow a \cap$ PRIME L is a closed upper set for the topology on PRIME L induced from the Lawson topology on L. The set is nonempty because it contains 1. Conversely, let A be a closed upper set in PRIME L, and let $a =$ inf A. We claim that $A = \uparrow a \cap$ PRIME L. Indeed, if $p \in \uparrow a \cap$ PRIME L, then $p \geq a =$ inf A. By hypothesis A is compact so we conclude that $p \geq q$ for some $q \in A$ by V-1.5. As A is an upper set, we have $p \in A$; that is, the inclusion $\uparrow a \cap$PRIME L$\subseteq A$ holds. The converse inclusion is clear. □

3.2. PROPOSITION. *Let* L *be a distributive continuous lattice in which the set* PRIME L *of all prime elements is closed with respect to the Lawson topology. Then* PRIME L *is a compact pospace and the map Δ_L is an isomorphism of* L *onto the lattice* D(PRIME L).

Proof. By V-4.7, the map $a \mapsto \Delta_L(a)$ is an isomorphism of L onto the lattice of all hull-kernel open subsets of Spec L. Thus, the assertion follows from the preceding lemma. □

3.3. PROPOSITION. *Let* X *be a compact pospace with a greatest element* 1. *Then the lattice* D(X) *is distributive and continuous. The set of all prime elements in* D(X) *is closed with respect to the Lawson topology and the map* $\xi_X = (x \mapsto X\setminus\uparrow x)$ *is a topological and an order isomorphism from* X *onto* PRIME D(X).

Proof. It is easily checked that D(X) is continuous lattice, because $U \ll V$ holds iff $U^- \subseteq V$. If $U \ll V$ and $U \ll W$, then $U^- \subseteq V$ and $U^- \subseteq W$, hence $U^- \subseteq V \cap W$, whence $U \ll V \cap W$. Thus, the relation $\ll$ is multiplicative in D(X). By V-3.7 we conclude that PRIME D(X) is closed in D(X) with respect to the Lawson topology. Clearly $X\setminus\uparrow x$ is prime in D(X) for every $x \in X$, and the map $\xi_X = (x \mapsto X\setminus\uparrow x)$ from X into PRIME D(X) is continuous (see VI-3.22) and injective. Thus, the image of ξ_X is closed with respect to the Lawson topology on D(X). As PRIME D(X) is the smallest closed order-generating subset of D(X) by V-2.1, the surjectivity of ξ_X follows from the fact that the sets $X\setminus\uparrow x$ are order generating in D(X). Indeed, $U = \bigcap\{X\setminus\uparrow x : x \notin U\}$ for every lower set U in X. □

In Propositions 3.2 and 3.3 we have established a one-to-one correspondence between compact pospaces with a greatest element and distributive continuous lattices L for which PRIME L is closed with respect to the Lawson topology. We indicate how this correspondence can be extended to a duality of categories which we introduce as follows:

- ***DCL*** has as objects all distributive continuous lattices L for which PRIME L is closed with respect to the Lawson topology and as morphisms the maps $f : L \to M$ preserving arbitrary sups, finite infs and the relation $\ll$.
- ***CPO*** has as objects the compact pospaces with a greatest element 1 and as morphisms the continuous maps which preserve the order relation and the greatest element 1.

Notice that ***DCL*** is a full subcategory of the category $\boldsymbol{CL}^{op} \cap \boldsymbol{HEYT}$ used in proposition V-5.16.

We wish to show next that the constructions D and PRIME are actually functors. If $f : X \to Y$ is a morphism in ***CPO***, then $f^{-1}(U)$ is an open lower set of X for every open lower set U of Y. Thus, we obtain a map

$$D(f) = (U \mapsto f^{-1}(U)) : D(Y) \to D(X)$$

which clearly preserves arbitrary unions and finite intersections. If $U \ll V$ in $D(Y)$, then $U^- \subseteq V$; by the continuity of f,

$$f^{-1}(U)^{-}\subseteq f^{-1}(U^{-})\subseteq f^{-1}(V);$$

whence, $f^{-1}(U)\ll f^{-1}(V)$. Thus $D(f)$ also preserves the relation $\ll$ and is a morphism in the category ***DCL***. This shows in fact that D is a functor from $\boldsymbol{CPO}^{\mathrm{op}}$ to ***DCL***.

In the other direction, let $\varphi : L\to M$ be a morphism in the category ***DCL***. Denote by $\tau : M\to L$ its upper adjoint $x \mapsto \sup \varphi^{-1}(\downarrow x)$. Then $\tau(1) = 1$ and $\tau(\mathrm{Spec}\ M)\subseteq\mathrm{Spec}\ L$ by V-4.5. Thus, we obtain a function

$$\mathrm{PRIME}\ \varphi : \mathrm{PRIME}\ M\to\mathrm{PRIME}\ L$$

by restricting τ to PRIME M. As the upper adoint of φ, the map τ is a $\wedge$-homomorphism which is continuous with respect to the Lawson topologies on L and M (IV-1.10(ii) and III-1.8), its restriction PRIME φ is order preserving and continuous with respect to the restriction of the Lawson topologies to PRIME M and PRIME L. As in ***DCL***, we suppose that PRIME L is closed with respect to the Lawson topology. PRIME φ is indeed a morphism in ***CPO***. This shows in fact that PRIME is a functor from ***DCL*** to $\boldsymbol{CPO}^{\mathrm{op}}$.

Now Propositions 3.2 and 3.3 give us natural isomorphisms

$$\Delta_L : L\to D(\mathrm{PRIME}\ L) \text{ and } \xi_X : X\to\mathrm{PRIME}\ D(X).$$

Thus we have the following duality result which is a specialization of V-5.16.

3.4. PROPOSITION. *The categories **DCL** and **CPO** defined above are dually equivalent under the functors*

$$\mathrm{PRIME} : \boldsymbol{DCL}\to\boldsymbol{CPO}^{\mathrm{op}} \text{ and } D : \boldsymbol{CPO}^{\mathrm{op}}\to\boldsymbol{DCL}. \ \square$$

EXERCISES

3.5. EXERCISE. (i) Let L be a continuous lattice. Show that the patch topology on the set PRIME L is Hausdorff.

(ii) Let X be a locally quasicompact T_0-space. Show that the patch topology on X is Hausdorff. □

3.6. EXERCISE (FUNCTORIALITY OF THE PATCH TOPOLOGY). Let L and M be complete lattices and $\tau : M\to L$ a map preserving arbitrary infs and directed sups. Suppose, in addition, that the lower adjoint $\varphi : L\to M$ of τ preserves finite infs. Show that $\tau(\mathrm{PRIME}\ M)\subseteq\mathrm{PRIME}\ L$ and that the restriction of τ to PRIME M is continuous with respect to the patch topologies on PRIME M and PRIME L.

(HINT: Use V-4.5 and V-5.3.) □

3.7. EXERCISE. (i) Let L be a distributive continuous lattice in which the set Spec L = PRIME L\{1} is closed with respect to the Lawson topology. Show that L is isomorphic to the lattice $D_1(\mathrm{Spec}\ L)$ of ***all*** Lawson-open lower sets in Spec L.

(ii) For every compact pospace X, show that Spec $D_1(X)$ is Lawson-closed in the lattice $D_1(X)$ of all open lower sets of X, and that Spec $D_1(X)$ with the Lawson topology is isomorphic with X.

(iii) Establish a duality between the category of all compact pospaces and continuous order preserving maps and the full subcategory of ***DCL*** the objects of which are the distributive continuous lattices L for which Spec L is closed with respect to the Lawson topology. □

3.8. EXERCISE. (i) Let L be a distributive arithmetic lattice with compact 1. Show that Spec L is closed with respect to the Lawson topology; endowed with the topology induced from the Lawson topology on L, Spec L is a *totally order-disconnected compact pospace*, that is, a compact pospace such that for any two elements x and y with $x \not\leq y$ there is an open closed upper set containing x but not y; moreover, L is isomorphic to the lattice D_1(Spec L) of all open lower sets in Spec L; the compact elements of L correspond to the open closed lower sets in Spec L.

(ii) Let X be a totally order-disconnected compact pospace. Show that $D_1(X)$ is a distributive arithmetic lattice such that X is homeomorphic to the space Spec $D_1(X)$ endowed with the topology induced from the Lawson topology on $D_1(X)$.

(iii) Let M be a distributive lattice with 0 and 1. Let L be the lattice of all ideals of M. Show that L is a distributive arithmetic lattice and that M is isomorphic to the lattice $\mathcal{O}\Upsilon$(Spec L) of all clopen lower sets of Spec L endowed with the topology induced from the Lawson topology on L. □

3.9. EXERCISE. (PRIESTLEY DUALITY). Consider the following categories:

DAR has as objects all distributive arithmetic lattices with compact 1 and as morphisms all maps preserving arbitrary sups, finite infs and compact elements (cf. V-5.19);

DCPO has as objects all totally order disconnected pospaces and as morphisms all continuous order preserving maps;

DLat is the category of all distributive lattices with 0 and 1 and all 0 and 1 preserving lattice homomorphisms.

Show that the categories ***DAR*** and ***DCPO*** and likewise the categories ***DLat*** and ***DCPO*** are dually equivalent. □

NOTES

To the best of our knowledge, the first to consider the patch topology was Hochster [1969] who defined this topology for the spectrum of commutative rings. In a general setting, the patch topology has been introduced by Hofmann and Lawson [1979]. In Section 6 of that paper one finds most of the material treated here. The duality and representation theorems (3.2, 3.3, 3.4) are already contained in a paper of Gierz and Keimel [1977], although in the latter paper other morphisms are used. The exercises 3.8 and 3.9 show how our results are related to Priestley's representation theorems for distributive lattices by compact totally order-disconnected partially ordered spaces, and related duality theorems [1970], [1972]. The representation theorems for distributive lattices of Priestley are more appealing to the average mathematical intuition than the original ones of M.H. Stone, since the representing spaces are Hausdorff spaces.

4. LATTICES WITH CONTINUOUS SCOTT TOPOLOGY

In Chapter II we have provided much information on complete lattices L for which the lattice $\sigma(L)$ of Scott-open subsets is continuous (II-1.14, II-4.11, II-4.13 II-4.14, II-4.16). In this section, we give a characterization of these lattices. Remarkably, the fact that $\sigma(L)$ is a continuous lattice relates to the existence of a compact semilattice topology on L itself.

The key idea is the following: recall in V-4.7 the use made of the natural map $\xi_X = (x \mapsto X\backslash\{x\}^-) : X \to \mathcal{O}(X)$ for an arbitrary topological space X. For each $x \in X$, we have $\xi(x) \in \operatorname{Spec} \mathcal{O}(X)$, the set of all prime elements of $\mathcal{O}(X)$ other than X itself, and this map is continuous if $\operatorname{Spec} \mathcal{O}(X)$ is endowed with the hull-kernel topology.

We now assume that $X = (L, \sigma(L))$ where L is an arbitrary complete lattice with its Scott topology $\sigma(L)$. In this topology $\{x\}^- = \downarrow x$, so we have the natural map

$$\xi_L = (x \mapsto L\backslash\downarrow x) : L \to \operatorname{Spec} \sigma(L)$$

which is continuous on L under the Scott topology to $\operatorname{Spec} \sigma(L)$ with the hull-kernel topology. We also may consider $\operatorname{Spec} \sigma(L) \subseteq \sigma(L)$ as a set ordered by inclusion. One then has for every subset $A \subseteq L$ and for every directed $D \subseteq L$

$$\xi_L(\inf A) = \bigcup\{\xi_L(a) : a \in A\} \text{ and } \xi_L(\sup D) = \operatorname{int}(\bigcap\{\xi_L(d) : d \in D\}).$$

This means that the image of L under ξ_L is closed in $\sigma(L)$ with respect to arbitrary sups and filtered infs. Moreover, L^{op} is order isomorphic to its image.

Note that the image of ξ_L is *not* closed in $\sigma(L)$ under finite infs. Indeed, if x and y are incomparable elements of L, then $\xi_L(x) \cap \xi_L(y) = L\backslash(\downarrow x \cup \downarrow y) \neq L\backslash\downarrow z = \xi_L(z)$ for all $z \in L$.

Under the assumption that $\sigma(L)$ is continuous we can say rather more.

4.1. PROPOSITION. *Let* L *be a complete lattice such that* $\sigma(L)$ *is continuous.*

(i) $\xi_L : L \to \operatorname{Spec} \sigma(L)$ *is a homeomorphism, if* L *is endowed with the Scott topology and* $\operatorname{Spec} \sigma(L)$ *with the hull-kernel topology;*

(ii) $\xi_L : L \to \operatorname{Spec} \sigma(L)$ *is an order anti-isomorphism, and* $\operatorname{Spec} \sigma(L)$ *is closed in* $\sigma(L)$ *with respect to arbitrary sups and filtered infs;*

(iii) *With respect to the Scott topology,* L *is a locally quasicompact sober space;*

(iv) $\operatorname{Spec} \sigma(L)$ *is closed in* $\sigma(L)$ *with respect to the Lawson topology.*

Proof. (i): By II-4.14, L is sober with respect to the Scott topology. Thus ξ_L is a homeomorphism by the remarks preceding V-4.7.

(ii): Use (i) and the remarks preceding this proposition.

(iii): As $\sigma(L)$ is supposed to be a continuous lattice, $(L,\sigma(L))$ is locally quasicompact by V-5.6.

(iv): From I-1.4(ii) we see that two Scott-open sets U and V satisfy the relation $U \ll V$ iff there is a quasicompact set Q such that $U \subseteq Q \subseteq V$. As all Scott-open sets are upper sets, Q is quasicompact iff $\uparrow Q$ is quasicompact, and we may restrict our attention to quasicompact upper sets. Let U, V, W be Scott-open sets such that $U \ll V$ and $U \ll W$. There are quasicompact upper sets Q_1 and Q_2 such that $U \subseteq Q_1 \subseteq V$ and $U \subseteq Q_2 \subseteq W$. Then $U \subseteq Q_1 \cap Q_2 \subseteq V \cap W$, and $Q_1 \cap Q_2$ is also quasicompact; indeed, $Q_1 \cap Q_2 = Q_1 \vee Q_2$ as Q_1 and Q_2 are upper sets, and $Q_1 \vee Q_2$ is the image of $Q_1 \times Q_2$ under the map $(x,y) \mapsto x \vee y : (L,\sigma(L)) \times (L,\sigma(L)) \to (L,\sigma(L))$, which is continuous by II-4.13. Thus, $U \ll V \cap W$, that is, the relation $\ll$ on $\sigma(L)$ is multiplicative. By V-3.7, PRIME $\sigma(L)$ is then closed with respect to the Lawson topology, and as L (as an element of $\sigma(L)$) is isolated, we have the desired result that Spec $\sigma(L)$ = PRIME $\sigma(L) \setminus \{L\}$ is also closed. □

We can now characterize those lattices L for which $\sigma(L)$ is continuous.

4.2. THEOREM. *For a semilattice* L *with* 1, *the following properties are equivalent*:

(1) L *is a complete lattice for which the lattice* $\sigma(L)$ *of Scott-open subsets is continuous.*

(2) L *admits a compact topology* τ *finer than the Scott topology such that* (L,τ) *is a compact pospace.*

(3) L *admits a compact topology* τ *such that* (L,τ) *is a compact pospace and such that a lower subset of* L *is* τ*-closed iff it is Scott-closed.*

Proof. (1) implies (2): By 4.1(iv) Spec $\sigma(L)$ is Lawson-closed in the continuous lattice $\sigma(L)$. Thus, the restriction of the Lawson topology to Spec $\sigma(L)$ yields a compact pospace topology τ' on Spec $\sigma(L)$. This topology is finer than the hull-kernel topology on Spec $\sigma(L)$ (cf. remarks following V-4.3(1)). Let τ be the inverse image of the topology τ' under the map $\xi_L : L \to \mathrm{Spec}\ \sigma(L)$. As ξ_L is an order anti-isomorphism by 4.1(ii), τ is a compact pospace topology on L. From 4.1(i) we conclude that τ is finer than the Scott topology.

(2) implies (3): Indeed a closed lower set in a compact pospace is closed for directed sups, that is, Scott closed (see VI-1.3).

(3) implies (1): Under the assumption (3), the lattice $\sigma(L)$ of Scott-open sets in L is the opposite of the lattice of all τ-closed lower sets of the compact pospace (L,τ) and, hence, continuous by VI-3.10. Moreover, a compact pospace has filtered infs; if in addition it has finite infs and a greatest element, then it is a complete lattice. □

We insert at this point an interesting addendum to 4.2 concerning the existence of the compact topology.

4.3. PROPOSITION. *If a complete lattice* L *has a topology satisfying* (2) *or* (3) *in* 4.2 *then this topology is unique.*

Proof. Indeed, by 3.3—more exactly by the specialization of 3.3 indicated in 3.7—the map $x \mapsto L \setminus \downarrow x$ is a homomorphism of (L,τ) onto Spec $\mathcal{D}'(L,\tau)$, where $\mathcal{D}'(L,\tau)$ is the lattice of τ-open upper sets of L and Spec $\mathcal{D}'(L,\tau)$ is endowed with the topology induced from the Lawson topology on $\mathcal{D}'(L,\tau)$. By 4.2(3) $\mathcal{D}'(L,\tau) = \sigma(L)$ and we have proved that $\xi_L = (x \mapsto L \setminus \downarrow x)$ is a homomorphism from (L,τ) onto Spec $\sigma(L)$ endowed with the Lawson topology. This yields the unicity of τ. (Alternately the uniqueness of τ can be deduced from Exercise 1.18.) □

The following theorem characterizes those lattices that admit a compact semilattice topology.

4.4. THEOREM. *For a semilattice* L *with* 1, *the following properties are equivalent:*

(1) L *admits a compact* $\wedge$*-semilattice topology; that is, a compact topology such that the operation* $(x,y) \mapsto x \wedge y : L \times L \to L$ *is continuous.*

(2) L *is a meet-continuous lattice which admits a compact pospace topology finer than the Scott topology.*

(3) L *is a meet-continuous lattice such that* $\sigma(L)$ *is continuous.*

(4) L *is a complete lattice such that* $\sigma(L)$ *is continuous and join continuous.*

Proof. Let L be a compact semilattice with 1. Then L is meet-continuous by VI-1.13(vii); the topology of L is a pospace topology by VI-1.14 and finer than the Scott topology by VI-2.10. Thus, (1) implies (2). The implication (2) implies (3) follows from 4.2, and (3) implies (4) from II-4.15. Let us show finally that (4) implies (1). By 2.4 every continuous and join-Brouwerian lattice endowed with the Lawson topology is a compact lattice; in particular, the join operation is continuous on the lattice $\sigma(L)$ with respect to the Lawson topology on $\sigma(L)$ if (4) is fulfilled. By 4.1 L is order anti-isomorphic to a subset of $\sigma(L)$ which is closed with respect to joins and closed with respect to the Lawson topology. Thus, L can be endowed with a compact $\wedge$-semilattice topology by VI-2.9. □

In the following table we collect the information contained in this volume about the transfer of properties from a complete lattice L to its lattice $\sigma(L)$ of Scott-open sets and vice-versa:

L	σ(L)	REFERENCE
Meet-continuous	Join-continuous	II-4.15
Admits a (unique) compact pospace topology τ finer than the Scott topology	Continuous	VII-4.2
Admits a (unique) compact $\wedge$-semilattice topology	Continuous and join-continuous	VII-4.4
Continuous	Completely distributive	II-1.14
Algebraic	Algebraic and completely distributive	II-1.15

Our next goal is to show that the category of compact semilattices is (dually) equivalent to a certain subcategory of distributive continuous lattices (which itself is a subcategory of the category of compact semilattices). More precisely, we shall consider the following categories:

CS is the category of compact semilattices with 1 and all continuous semilattice homomorphisms.

H is the category whose objects are the lattices L with the following properties:

(1) L is distributive, continuous and join continuous;

(2) Spec L is closed in L with respect to arbitrary sups.

The morphisms are the maps $\varphi : L \to M$ which

(1) Preserve arbitrary sups, finite infs and the relation $\ll$;

(2) Have an adjoint $\tau : M \to L$ preserving finite sups.

Clearly, ***CS*** is a subcategory of the category ***CPO*** of compact pospaces considered in Section 3. The following lemma shows that ***H*** is a subcategory of the category ***DCL*** also introduced in Section 3.

4.5. LEMMA. *If* L *is in* ***H***, *then* Spec L *is closed in* L *with respect to the Lawson topology on* L *and also is a compact* $\vee$*-semilattice with respect to this topology.*

Proof. By definition, Spec L is closed in L with respect to arbitrary sups; as the inf of a filtered set of prime elements is also prime, Spec L is also closed with respect to filtered infs. As L is distributive continuous and join-continuous, L is a compact lattice with respect to the Lawson topology by 2.4 and in particular, a compact $\vee$-semilattice. From VI-2.9 we conclude that Spec L is closed with respect to the Lawson topology on L, and hence a compact $\vee$-subsemilattice of L. $\square$

In 3.3, more precisely in the specialization of 3.3 indicated in Exercise 3.7, we have seen that the category of compact pospaces is dually equivalent to the category of continuous distributive lattices L for which Spec L is Lawson-closed with appropriate morphisms; the duality is given by the functors Spec and D_1. We now show that, up to a reversal of the order, the restrictions of these functors establish a dual equivalence between the categories ***H*** and ***CS***.

4.6. PROPOSITION. *The categories* ***CS*** *and* ***H*** *are dually equivalent.*

Proof. With every object $L \in \boldsymbol{H}$ we associate Spec^{op} L, that is, the set Spec L with the topology induced from the Lawson topology on L and the order *opposite* to the order induced from the order on L. By 4.5 Spec^{op} L is a compact $\wedge$-semilattice. For every *H*-morphism $\varphi : L \to M$, $\mathrm{Spec}^{op}\ \varphi$ is the upper adjoint of φ restricted to Spec^{op} M; as in Section 3, $\mathrm{Spec}^{op}\ \varphi$ is continuous and order preserving. As, by hypothesis, τ also preserves finite sups, $\mathrm{Spec}^{op}\ \varphi$ preserves finite infs and is a *CS*-morphism. Thus Spec^{op} is indeed a functor from ***H*** to ***CS***.

Conversely, we associate with every compact semilattice S its Scott topology $\sigma(S)$ which is an *H*-object by 4.1 and 4.4. For every continuous semilattice homomorphism $f: S \to T$ we define $\sigma(f) : \sigma(T) \to \sigma(S)$ as usual by $U \mapsto f^{-1}(U)$. As in Section 3, $\sigma(f)$ preserves arbitrary sups, finite infs and the relation $\ll$. Let us show that the upper adjoint τ of $\sigma(f)$ preserves finite sups: a straightforward calculation shows that τ is given by $\tau(V) = T \setminus {\downarrow} f(S \setminus V)$ for all $V \in \sigma(S)$. We want to show that $\tau(V \cup W) = \tau(V) \cup \tau(W)$ for arbitrary $V, W \in \sigma(S)$. This is equivalent to saying that ${\downarrow} f(A \cap B) = {\downarrow} f(A) \cap {\downarrow} f(B)$ for arbitrary Scott-closed sets A and B. Thus equality holds, as $A \cap B = A \wedge B$ for all lower sets and as $f(A \wedge B) = f(A) \wedge f(B)$ by the hypotheses on f. Thus, σ is indeed a functor from $\boldsymbol{CS}$ to $\boldsymbol{H}$.

Now, Spec^{op} is nothing but the functor Spec from 3.7 up to a reversal of the order on the objects, and σ is nothing but the functor D_1 from 3.7 up to a reversal of the order, as the Scott-open upper sets are exactly the open upper set for the original topology on a compact semilattice (VI-2.10). Thus 3.4 in the special form of 3.7 yields the desired duality result. □

We note that the preceding duality result means in particular that the objects of H, that is, the distributive, continuous, and join continuous lattices L in which Spec L is closed with respect to arbitrary sups, are exactly the lattices which arise as Scott topologies of compact semilattices.

We close our work with an interesting application of the theory we have built up to compact semilattices. We first recall that a compact semitopological semilattice with identity is a meet-continuous lattice (VI-1.13(vii)). We next identify the open upper sets in S.

4.7. PROPOSITION. *The family of open upper sets in a compact semitopological semilattice is the family of Scott-open subsets of* S. *Hence, the graph of* $\leq$ *is closed in* $S \times S$.

Proof. We want to apply Theorem 1.9, and so we show that the family $\mathfrak{U}$ of open upper sets in S is an order consistent, o-regular topology and $(S, \mathfrak{U})$ is a semitopological semilattice.

If $x \in S$, then $xS = \downarrow x$ is closed in S by VI-1.13(i). Moreover, directed subsets of S converge to their sups by VI-1.13(iv), and so they do also in the topology $\mathfrak{U}$. Thus, $\mathfrak{U}$ is order consistent.

To show $\mathfrak{U}$ is o-regular, assume that $x \in U \in \mathfrak{U}$, and then choose an open subset V of S with $x \in V \subset V^- \subset U$. Then $\uparrow V \subseteq \uparrow V^- \subseteq U$ and $\uparrow V$ is open. By VI-1.13(ii) and the dual of VI-1.6(i) $\uparrow V^-$ is closed with respect to taking filtered infs.. This shows that $\mathfrak{U}$ is o-regular. Now $(S, \mathfrak{U})$ is semitopological, since S is; therefore, we have satisfied the hypotheses of Theorem 1.9(iii). We conclude that $\mathfrak{U} = \sigma(S)$, the Scott topology on S.

Proposition 1.15 now shows that the graph of $\leq$ is closed in $S \times S$. □

4.8. THEOREM. *A compact semitopological semilattice is in fact topological.*

Proof. By Proposition 4.7 $\sigma(S)$ is the family of open upper sets in S and S is a pospace. Hence by 4.2 $\sigma(S)$ is a continuous lattice.

Now, each compact semitopological semilattice is a meet-continuous lattice, and so Theorem 4.4 implies that S is a compact semilattice in the patch topology S inherits as the spectrum of $\sigma(S)$. By 4.3 the latter topology coincides with the original one. □

EXERCISES

4.9. EXERCISE. Show that the morphisms $\varphi : L \to M$ in the category $\boldsymbol{H}$ are characterized by the property that they are lower adjoints of maps $\tau : M \to L$ preserving arbitrary sups and infs, as well as primes. □

4.10. EXERCISE. Show that the category of continuous (algebraic) lattices is dually equivalent to a subcategory of the category of completely distributive lattices; more exactly, to the full subcategory of $\boldsymbol{H}$ whose objects are completely distributive (algebraic) lattices L for which the set Spec L is closed under arbitrary sups.

(HINT: Use 4.6, II-1.14 and II-1.15.) □

The next exercise is an alternate proof to Theorem 4.8.

4.11. EXERCISE Each compact semitopological semilattice is topological.

(HINT: Note that $S \times S$ in the product topology is also a compact semitopological semilattice, and so Proposition 4.7 applies to both S and $S \times S$ to show that the Scott topology on each is the family of open upper sets on each and that each is a pospace. Since the products of open upper sets in S form a base for the open upper sets in $S \times S$, it follows that the Scott topology on $S \times S$ is the product of the Scott topology on S with itself. Since S is a meet-continuous lattice, we easily conclude that the semilattice map $(x,y) \mapsto xy : S \times S \to S$ is continuous with respect to the Scott topologies.

Thus, the result is proved if we show that this map is also continuous when S and $S \times S$ are equipped with the topologies consisting of open lower sets, since both S and $S \times S$ are monotone normal by Nachbin's Lemma VI-1.8. Thus, we need to show that $xy \in U = {\downarrow}U$ and U open in S implies that there are open lower sets V and W containing x and y, respectively, with $VW \subseteq U$. But, $VW = V \cap W$ since V and W are lower sets. Finally, ${\downarrow}x$ and ${\downarrow}y$ are the intersection of those compact neighborhoods which are lower sets, and ${\downarrow}xy$ is then the intersection of these compact sets. Since ${\downarrow}xy \subseteq U$ and U is open, it follows that there is indeed some compact neighborhood V of ${\downarrow}x$ and some compact neighborhood of W of ${\downarrow}y$ with $V \cap W \subseteq U$. This proves the result.) □

The following exercise gives yet another equivalence between a class of lattices and its lattice of Scott-open sets (see Gierz and Lawson [1979]).

4.12. EXERCISE. Let L be a complete lattice. Show that L is a generalized continuous lattice iff the lattice of Scott-open sets $\sigma(L)$ is a hypercontinuous lattice.

(HINT: See Exercises III-1.17 and III-3.21 for definitions.) □

NOTES

The material of this section through 4.6 is due to Gierz and Hofmann [SCS-34]. Let us indicate the following consequence. There are compact distributive lattices that are not completely distributive or, equivalently, there are compact distributive lattices which are continuous, but the opposite of which is not continuous. Take any compact semilattice S with 1 which is not a continuous lattice (see Section VI-4.5 for examples). Then the lattice $\sigma(L)$ of all Scott-open subsets of S is distributive, continuous and join-continuous, hence a compact lattice by 2.4; but S is not completely distributive by II-1.14.

Theorem 4.8 was first shown by Lawson [1976a], but the proof contained therein utilizes some rather technical results from topology. An alternative proof was announced by Mislove [SCS-40], but a gap in the proof of what appears here as Corollary 2.4 was found by Harvey Carruth. The results of Section 1 serve to patch that gap, and also provide a much simpler proof of the Theorem. The alternate proof presented as Exercise 4.11 was first noticed by Gerhard Gierz.

BIBLIOGRAPHY

The notation in the first column refers to the following rough classification into subdisciplines:

APP	Applications	PO	Posets
CL	Continuous lattices	POTS	Partially ordered topological spaces
GT	General topology	REF	General references
GLT	General lattice theory	TSL	Topological semilattices
IT	Intrinsic topologies on semilattices and lattices	TL	Topological lattices

GLT **Abbot, J.C. [1969]**, *Sets, Lattices and Boolean Algebras*, Allyn and Bacon, 1969.

TL **Anderson, L.W. [1958]**, *Topological lattices and n-cells.* Duke Math. J., vol. 25 (1958), pp. 205-208.

TL **Anderson, L.W. [1959a]**, *On the breadth and co-dimension of a topological lattice*, Pac. J. Math., vol. 9 (1959), pp. 327-333.

TL **Anderson, L.W. [1959b]**, *On the distributivity and simple connectivity of plane topological lattices*, Trans. Amer. Math. Soc., vol. 91 (1959), pp. 102-112.

TL **Anderson, L.W. [1959c]**, *One dimensional topological lattices*, Proc. Amer. Math. Soc., vol. 10 (1959), pp. 715-720.

TL **Anderson, L.W. [1961]**, *Locally compact topological lattices*, Proc. Sympos. Pure Math. Amer. Math. Soc., vol. 2 (1961), pp. 195-197.

TL **Anderson, L.W. [1962]**, *The existence of continuous lattice homomorphisms*, J. London Math. Soc., vol. 37 (1962), pp. 60-62.

TL **Anderson, L.W. and Ward, L.E., Jr. [1961a]**, *A structure theorem for topological lattices*, Proc. Glasgow Math. Assn., vol. 5 (1961), pp. 1-3.

TSL **Anderson, L.W. and Ward, L.E., Jr. [1961b]**, *One-dimensional topological semilattices*, Ill. J. Math (1961), pp. 182-186.

IT **Atsumi, K. [1966]**, *On complete lattices having the Hausdorff interval topology*, Proc. Amer. Math. Soc., vol. 17 (1966), pp. 197-199.

GLT **Aumann, G. [1955]**, *Bemerkung über Galois-Verbindungen*, Bayer, Akad. Wiss. Math. -Nat. Kl. S. -B. (1955). pp. 281-284.

TSL **Austin, C.W. [1963]**, *Duality theorems for some commutative semigroups*, Trans. Amer. Math. Soc., vol. 109 (1963), pp. 245-256.

APP **Baartz, A. [1967]**, *The measure algebra of a locally compact semigroup*, Pac. J. Math, vol. 21 (1967), pp. 199-214.

TL **Baker, K.A. and Stralka, A.R. [1970]**, *Compact distributive lattices of finite breadth*, Pac. J. Math., vol. 34 (1970), pp. 311-320.

GLT **Balbes, R. and Dwinger, P. [1974]**, *Distributive Lattices*, Univ. Missouri Press, Columbia, 1974.

CL & GT **Banaschewski, B. [1973]**, *The filter space of a lattice, etc*, Proc. Lattice Theory Conf., Houston, 1973, pp. 147-155.

CL & GT **Banaschewski, B. [1977]**, *Essential extensions of* T_0*-spaces*, General Top. Appl., vol. 7 (1977), pp. 233-246.

CL **Banaschewski, B. [1978a]**, *Hulls, kernels, and continuous lattices*, Houston J. Math., vol. 4 (1978), pp. 577-525.

CL Banaschewski, B. [1978b], *The duality of distributive continuous lattices*, preprint, 1978, McMaster Univ.

CL Bandelt, H.J. [1979], *Complemented continuous lattices*, preprint, 1979.

CL Bandelt, H.J. [1980], *The tensor product of continuous lattices*, Math. Z. (to appear, 1980).

GT Birkhoff, G. [1937], *Moore-Smith convergence in general topology*, Annals of Math., vol. 38 (1937), pp. 39-56.

GLT & REF Birkhoff, G. [1967], *Lattice Theory*, 3rd ed., 1967, Amer. Math. Soc. Colloq. Publ., Providence, R.I.

GLT Birkhoff, G. and Frink, O. [1948], *Representations of lattices by sets*, Trans. Amer. Math. Soc., vol. 64 (1948), pp. 299-316.

GLT & REF Blyth, T.S. and Janowitz, M.F. [1972], *Residuation Theory*, Pergamon Press, Oxford, 1972.

TSL Borrego, J.T. [1970], *Continuity of the operation of a semilattice*, Coll. Math., vol. 21 (1970), pp. 49-52.

TSL Bowman, T.T. [1974], *Analogue of Pontryagin character theory for topological semigroups*, Proc. Amer. Math. Soc., vol. 46 (1974), pp. 95-105.

TSL Brown, D.R. [1965], *Topological semilattices on the 2-cell*, Pac. J. Math., vol. 15 (1965), pp. 35-46.

TSL Brown, D.R. and Stralka, A.R. [1973], *Problems on compact semilattices*, Semigroup Forum, vol. 6 (1973), pp. 265-270.

CL & TSL Brown, D.R. and Stralka, A.R. [1977], *Compact totally instable zero-dimensional semilattices*, Gen. Top. and Appl., vol. 7 (1977), pp. 151-159.

GLT Bruns, G. [1961], *Distributivität und subdirekte Zerlegbarkeit vollständiger Verbände*, Archiv d. Math, vol. 12 (1961), pp. 61-66.

GLT Bruns, G. [1961 & 1962], *Darstellungen und Erweiterungen geordneter Mengen I und II*, J.f.d. reine u. angew. Mathematik, vol. 209 (1961), pp. 167-220, and vol. 210 (1962), pp. 1-23.

GLT Bruns, G. [1967], *A lemma on directed sets and chains*, Arch d. Math., vol. 18 (1967), pp. 35-43.

GLT Büchi, J. [1952], *Representation of complete lattices by sets*, Port. Math, vol. 11 (1952), pp. 151-167.

GLT Bulman-Fleming, Fleischer, I. and Keimel, K. [1979], *The semilattices with distinguished endomorphisms which are equationally compact*, Proc. Amer. Math. Soc., vol. 73 (1979), pp. 7-10.

POTS Carruth, J.H. [1968], *A note on partially ordered compacta*, Pac. J. Math., vol. 24 (1968), pp. 229-231.

IT Choe, T.H. [1969a], *Intrinsic topologies in a topological lattice*, Pac. J. Math., vol. 28 (1969), pp. 49-52.

TL Choe, T.H. [1969b], *Notes on locally compact connected topological lattices*, Can. J. Math. (1969), pp. 1533-1536.

TL Choe, T.H. [1969c], *On compact topological lattices of finite dimension*, Trans. Amer. Math. Soc., vol. 140 (1969), pp. 223-237.

TL Choe, T.H. [1969d], *The breadth and dimension of a topological lattice*, Proc. Amer. Math. Soc., (1969), pp. 82-84.

TL Choe, T.H. [1971], *Locally compact lattices with small lattices*, vol. 18 (1971), pp. 81-85.

TL Choe, T.H. [1973a], *Injective compact distributive lattices*, Proc. Amer. Math. Soc., vol. 37 (1973), pp. 241-245.

TL Choe, T.H. [1973b], *Projective compact distributive topological lattices*, Proc. Amer. Math. Soc., vol. 39 (1973), pp. 606-608.

TL Choe, T.H. [1969], *Remarks on topological lattices*, Kyungpook Math. J., vol. 9 (1969), pp 59-62.

POTS & TL Choe, T.H. and Hong, U.H. [1976], *Extensions of completely regular ordered spaces*, Pac. J. Math., vol. 64 (1976).

TL Clark, C.E. and Eberhart, C. [1968], *A characterization of compact connected planar lattices*, Pac. J. Math., vol. 24 (1968), pp. 233-240.

TL Clinkenbeard, D. [1976], *Lattices of congruences on compact topological lattices*, Dissertation, Univ. of Calif. at Riverside, 1976.

TL Clinkenbeard, D. [1977], *Simple compact topological lattices*, preprint, 1977.

CL & TSL Crawley, C.W. [1976], *A note on epimorphisms of compact Lawson semilattices*, Semigroup Forum, vol. 13 (1976), pp. 92-94.

CL & TSL Crawley, C.W. [1977], *Amalgamation of compact Lawson semilattices*, Shippensburg State College, PA., preprint, 1977

APP Cunningham, F. and Roy, N.M. [1974], *Extreme functionals on an upper semicontinuous function space*, Proc. Amer. Math. Soc., vol. 42 (1974), pp. 461-465.

GT Curtis, D. and Schori, R. [1976], *2^X and $C(X)$ are homeomorphic to the Hilbert cube*, Bull. Amer. Math. Soc., vol. 80 (1976), pp. 927-931.

TL Davies, E.B. [1968], *The existence of characters on topological lattices*, J. London Math. Soc., vol. 43 (1968), pp. 219-220.

APP Day, A. [1975], *Filter monads, continuous lattices and closure systems*, Canad. J. Math., vol. 27 (1975), pp. 50-59.

CL Day, B.J. and Kelly, G.M. [1970], *On topological quotient maps preserved by pull-backs or products*, Proc. Cambridge Philos. Soc., vol. 67 (1970), pp. 553-558.

GLT Derderian, J.C. [1967], *Residuated mappings*, Pac. J. Math., vol. 20 (1967), pp. 35-43.

GLT Dilworth, R.P. and Crawley, P. [1960], *Decomposition theory for lattices without chain conditions*, Trans. Amer. Math. Soc., vol. 96 (1960), pp. 1-22.

REF Dilworth, R.P. and Crawley, P. [1973], *Algebraic Theory of Lattices*, Prentice Hall, Inc., 1973.

APP Dixmier, J. [1968], *Sur les espaces localement quasi-compacts*, Can. J. Math., vol. 20 (1968), pp. 1093-1100.

GT Dowker, C.H. and Papert, D. [1966], *Quotient frames and subspaces*, Proc. London Math. Soc., vol. 16 (1966), pp. 275-296.

GLT & GT Drake, D. and Thron, W.J. [1965], *On the representation of an abstract lattice and the family of closed subsets of a topological space*, Trans Amer. Math. Soc., vol. 120 (1965), pp. 57-71.

GTL Dubreil-Jacotin, M.L., Lesieur, L. and Croisot, R. [1953], *Leçons sur la théorie des treillis, etc.*, Gauthier-Villars, 1953.

TL Dyer, E. and Shields, A.S. [1959], *Connectivity of topological lattices*, Pac. J. Math., vol. 9 (1959), pp. 443-447.

TL Edmondson, D.E. [1956], *A non-modular compact connected topological lattice*, Proc. Amer. Math. Soc., vol. 7 (1956), p. 1157.

TL Edmondson, D.E. [1969a], *A modular topological lattice*, Pac. J. Math., vol. 29 (1969), pp. 271-297.

TL Edmondson, D.E. [1969b], *Modularity in topological lattices*, Proc. Amer. Math. Soc., vol. 21 (1969), pp. 81-82.

APP & CL Egli, H. [1973], *An analysis of Scott's λ-calculus models*, Cornell U. Tech. Report (TR 73-191), Dec. 1973.

APP & CL Egli, H. and Constable, R.L. [1976], *Computability concepts for programming language semantics*, Theoretical Computer Science, vol. 2 (1976), pp. 133-145.

REF Eilenberg, S. and Kelly, G.M. [1966]. *Closed categories*, Proceedings of the Conference on Categorical Algebra at La Jolla, 1965, Springer-Verlag, New York, 1966, pp. 421-562.

GLT Erné, M. [1979], *Order and Topology*, xi+675 pp., preprint.

IT Erné, M. and Weck, S. [1978], *Ordnungskonvergenz in Verbänden*, 1978.

APP & CL Ershov, Ju. L. [1972 & 1974], *Computable functionals of finite type*, Algebra i Logika, vol. 11 (1972), pp. 367-437. (Algebra and Logic, vol. 11 (1972), pp. 203-242 (1974).)

CL Ershov, Ju. L. [1972], *Continuous lattices and A-spaces*, Dokl. Akad. Nauk SSSR., vol. 207 (1972), pp. 523-526. (Soviet Math. Dokl., vol. 13 (1972), pp. 1551-1555.)

CL Ershov, Ju. L. [1973 & 1975], *The theory of A-spaces*, Algebra i Logika, vol. 12 (1973), pp. 369-416. (Algebra and Logic, vol. 12 (1973), pp. 209-232 (1975).)

GLT Everett, C.J. [1944], *Closure operators and Galois theory in lattices*, Trans. Amer. Math. Soc., vol. 55 (1944), pp. 514-525.

APP Fell, J.M.G. [1961], *The structure of algebras of operator fields*, Acta Math., vol. 106 (1961), pp. 233-280.

APP Fell, J.M.G. [1962], *A Hausdorff topology for the closed subsets of a locally compact non-Hausdorff space*, Proc. Amer. Math. Soc., vol. 13 (1962), pp. 472-476.

IT Floyd, E.E. [1955], *Boolean algebras with pathological order topologies*, Pac. J. Math., vol. 5 (1955), pp. 687-689.

REF Fourman, M.P., Mulvey, C.J., and Scott, D.S. [1979], *Applications of Sheaves: Proceedings, Durham, 1977*, Springer-Verlag Lecture Notes in Math., vol. 753, 1979.

TSL Friedberg, M. [1972], *Metrizable approximations of semigroups*, Colloq. Math., vol. 25 (1972), pp. 63-69, 164.

IT Frink, O. [1942], *Topology in lattices*, Trans. Amer. Math. Soc., vol. 5 (1942), pp. 569-582.

IT Frink, O. [1954], *Ideals in partially ordered sets*, Amer. Math. Monthly, vol. 6 (1954), pp. 223-234.

GLT Gaskill, H.S. [1973], *Classes of semilattices associated with an equational class of lattices*, Can. J. Math., vol. 25 (1973), pp. 361-365.

GLT Geissinger, L. and Graves, W. [1972], *The cateogry of complete algebraic lattices*, J. Combinatorial Theory, vol. 13 (1972), pp. 332-338.

CL Gierz, G. [1978], *Colimits of continuous lattices*, preprint 1978, Darmstadt.

CL Gierz, G. and Hofmann, K.H. [1978], *On a lattice-theoretical characterization of compact semilattices*, preprint, 1978.

CL Gierz, G. und Keimel, K. [1976], *Topologische Darstellung von Verbänden*, Math. Z., vol. 150 (1976), pp. 83-99.

APP & CL Gierz, G. und Keimel, K. [1977], *A lemma on primes appearing in algebra and analysis*, Houston J. Math., vol. 3 (1977). pp. 207-224.

CL Gierz, G. and Lawson, J.D. [1979], *Generalized continuous and hypercontinuous lattices*, preprint, 1979.

APP Giles, R. [1977], *Continuous lattices in the foundations of physics*, preprint, 1977.

CL & IT Gingras, A.R. [1976a], *Convergence lattices*, Rocky Mountain J. Math., vol. 6 (1976), pp. 85-104.

TL Gingras, A.R. [1976b], *Order convergence and order ideals*, Proc. Conf. Conv. Spaces, Univ. of Nevada, Reno, 1976, pp. 45-59.

IT Gingras, A.R. [1978], *Complete distributivity and order convergence*, preprint, 1978.

GLT & REF Grätzer, G. [1978], *General Lattice Theory*, Birkhäuser, 1978.

GLT & REF Halmos, P. [1963], *Lectures on Boolean Algebras*, Van Nostrand, Princeton, 1963. (Reprinted, Springer-Verlag 1974.)

GLT Hermes, H. [1967], *Einführung in die Verbandstheorie*, Springer-Verlag, Berlin, 1967.

APP Hochster, M. [1969], *Prime ideal structure in commutative rings*, Trans. Amer. Math. Soc., vol. 142 (1969), pp. 43-60.

CL Hoffmann, R.-E. [1979a], *Continuous posets and adjoint sequences*, Semigroup Forum (1979), to appear.

CL Hoffmann, R.-E. [1979b], *Projective sober spaces*, preprint, 1979.

CL Hoffmann, R.-E. [1979c], *Sobrification of partially ordered sets*, Semigroup Forum, vol. 17 (1979), pp 123-138.

CL Hoffmann, R.-E. [1979d], *Continuous posets and adjoint sequences*, Semigroup Forum, vol. 18 (1979), pp. 173-188.

GT Hoffmann, R.-E. [1979e], *Essentially complete* T_0*-spaces*, Manuscripta Math., vol. 27 (1979), pp. 401-432.

CL Hoffmann, R.-E. [1980], *Continuous posets, prime spectra of completely distributive complete lattices and Hausdorff compactificatons*, preprint, 47 pp..

GT Hofmann, K.H. [1970], *A general invariant metrization theorem for compact spaces*, Fund. Math, vol. 68 (1970), pp. 281-296.

CL Hofmann, K.H. [1977], *Continuous lattices, topology, and topological algebra*, Topology Proceedings, vol. 2 (1977), pp. 179-212.

GT & CL Hofmann, K.H. [1979], *A note on Baire spaces and continuous lattices*, preprint, 1979.

GLT Hofmann, K.H. and Keimel, K. [1972], *A general character theory for partially ordered sets and lattices*, Mem. Amer. Math. Soc., vol. 122 (1972), 121 pp.

CL Hofmann, K.H. and Lawson, J.D. [1976/77], *Irreducibility and generation in continuous lattices*, Semigroup Forum, vol. 13 (1976/77), pp. 307-353.

CL Hofmann, K.H. and Lawson, J.D. [1978], *The spectral theory of distributive continuous lattices*, Trans. Amer. Math. Soc., vol. 246 (1978) pp. 285-310.

CL & TSL Hofmann, K.H. and Mislove, M. [1973], *Lawson semilattices do have a Pontryagin duality*, Proc. Lattice Theory Conf., U. Houston, 1973, pp. 200-205.

CL & TSL Hofmann, K.H. and Mislove, M. [1975], *Epics of compact Lawson semilattices are surjective*, Arch. Math., vol. 26 (1975), pp. 337-345.

CL Hofmann, K.H. and Mislove, M. [1976], *Amalgamation in categories with concrete duals*, Algebra Universalis, vol. 6 (1976), pp. 327-347.

CL Hofmann, K.H. and Mislove, M. [1977], *The lattice of kernel operators and topological algebra*, Math. Z., vol. 154 (1977), pp. 175-188.

CL & TSL Hofmann, K.H., Mislove, M. and Stralka, A. [1973], *Dimension raising maps in topological algebra*, Math. Z., vol. 135 (1973), pp. 1-36.

CL & TSL Hofmann, K.H., Mislove, M. and Stralka, A.R. [1974], *The Pontryagin duality of compact 0-dimensional semilattices and its applications*, Springer-Verlag Lecture Notes in Math., vol. 396, 1974.

CL & TSL Hofmann, K.H., Mislove, M. and Stralka, A.R. [1975], *On the dimensional capacity of semilattices*, Houston J. Math., vol. 1 (1975), pp. 43-55.

TSL & REF Hofmann, K.H. and Mostert, P. [1966], *Elements of Compact Semigroups*, Merrill, Columbus, Ohio, 1966.

TSL Hofmann, K.H. and Stralka, A.R. [1973a], *Mapping cylinders and compact monoids*, Math. Ann., vol. 205 (1973), pp. 219-239.

TSL Hofmann, K.H. and Stralka, A.R. [1973b], *Push-outs and strict projective limits of semilattices*, Semigroup Forum, vol. 5 (1973), pp. 243-262.

CL & TSL Hofmann, K.H. and Stralka, A.R. [1976], *The algebraic theory of Lawson semilattices—Applications of Galois connections to compact semilattices*, Diss. Math., vol. 137 (1976), pp. 1-54.

APP Hofmann, K.H. and Thayer, F.J. [198*], *Almost finite dimensional C*-algebras*, Diss. Math., to appear.

GLT Horn, A. and Kimura, N. [1971], *The category of semilattices*, Algebra Universalis, vol. 1 (1971), pp. 26-38.

CL Hosono, Ch. and Sato, M. [1977], *The retracts in* R_ω *do not form a continuous lattice, etc.*, Theoretical Comp. Sci., vol. 4 (1977), pp. 137-142.

IT Insel, A.J. [1963], *A compact topology for a lattice*, Proc. Amer. Math Soc., vol. 14 (1963), pp. 382-385.

IT Insel, A.J. [1964], *A relationship between the complete topology and the order topology of a lattice*, Proc. Amer. Math. Soc., vol. 15 (1964), pp. 847-850.

GLT & GT Isbell, J.R. [1972], *Atomless parts of spaces*, Math. Scand., vol. 31 (1972), pp. 5-32.

CL Isbell, J.R. [1975a], *Function spaces and adjoints*, Math. Scand., vol. 36 (1975), pp. 317-339.

CL Isbell, J.R. [1975b], *Meet-continuous lattices*, Symp. Math., vol. 16 (1975), pp. 41-54.

APP Jamison, R.E. [1974]. *A general theory of convexity*, Dissertation, University of Washington, 1974.

APP Jónsson, B. [1967], *Algebras whose congruence lattices are distributive*, Math. Scand., vol. 9 (1967), pp. 110-121.

APP Kahn, G. [1978], *Concepts fondamentaux de théorie des modèles*, preprint, 1978.

CL Kamara, M. [1978], *Treillis continus et treillis complètement distributifs*, Semigroup Forum, vol. 16, no. 3 (1978), pp. 387-388.

GLT Keimel, K. [1972], *A unified theory of minimal prime ideals*, Acta Math. Acad. Sci. Hung., vol. 23 (1972), pp. 51-69.

GT & REF Kelley, J.L. [1955], *General Topology*, Van Nostrand, Princeton, 1955. (Reprinted, Springer-Verlag, 1975.)

IT Kent, D.C. [1966], *On the order topology in a lattice*, Illinois J. Math., vol. 10 (1966), pp. 90-96.

IT Kent, D.C. [1967], *The interval topology and order convergence as dual convergence structures*, Amer. Math. Monthly, vol. 74 (1967), pp. 426-427.

POTS Koch, R.J. [1959], *Arcs in partially ordered spaces*, Pac. J. Math., vol. 9 (1959), pp. 723-728.

POTS Koch, R.J. [1960] *Weak cutpoint ordering in hereditarily unicoherent continua*, Proc. Amer. Math. Soc., vol. 11 (1960), pp. 679-681.

POTS Koch, R.J. [1965], *Connected chains in quasi-ordered spaces*, Fund. Math., vol. 56 (1965), pp. 245-249.

IT Kolibiar, M. [1962], *Bemerkungen über Intervalltopologie in halbgeordneten Mengen*, General topology and its relations to modern analysis and algebra, Prague 1962, pp. 252-253.

POTS Krule, I.S. [1957], *Structs on the* 1-*sphere*, Duke Math. J., vol. 24 (1957), pp. 623-626.

TSL Lau, A.Y.W. [1972a], *Concerning costability of compact semigroups*, Duke Math. J., vol. 39 (1972), pp. 657-664.

TSL Lau, A.Y.W. [1972b], *Small semilattices*, Semigroup Forum, vol. 4 (1972), pp. 150-155.

TSL Lau, A.Y.W. [1973a], *Costability in SEM and TSL*, Semigroup Forum, vol. 5 (1973), pp. 370-372.

TSL Lau, A.Y.W. [1973b], *Coverable semigroups*, Proc. Amer. Math. Soc., vol. 38 (1973), pp. 661-664.

TSL Lau, A.Y.W. [1975], *The boundary of a semilattice on an n-cell*, Pac. J. Math, vol. 56 (1975), pp. 171-174.

TSL Lau, A.Y.W. [198*], *Existence of n-cells in Peano semilattices*, to appear.

APP Larsen, K.B. and Sinclair, A.M. [1975], *Lifting matrix units in C*-algebras, II*, Math. Scand., vol. 37 (1975), pp. 167-172.

TSL Lawson, J.D. [1967], *Vietoris mappings and embeddings of topological lattices*, Dissertation, Univ. of Tennessee, 1967.

TSL Lawson, J.D. [1969], *Topological semilattices with small semilattices*, J. London Math. Soc., vol. 2 (1969), pp. 719-724.

TL & TSL **Lawson, J.D. [1970a]**, *Lattices with no interval homomorphisms*, Pac. J. Math., vol. 32 (1970), pp. 459-465.

TSL **Lawson, J.D. [1970b]**, *The relation of breadth and codimension in topological semilattices*, Duke J. Math., vol. 37 (1970), pp. 207-212.

TSL **Lawson, J.D. [1971]**, *The relation of breadth and codimension in topological semilattices II*, Duke Math. J., vol. 38 (1971), pp. 555-559.

TSL **Lawson, J.D. [1972]**, *Dimensionally stable semilattices*, Semigroup Forum, vol. 5 (1972), pp. 181-185.

IT **Lawson, J.D. [1973a]**, *Intrinsic lattice and semilattice topologies*, Proc. Lattice Theory Conf., U. Houston, 1973, pp. 206-260.

IT **Lawson, J.D. [1973b]**, *Intrinsic topologies in topological lattices and semilattices*, Pac. J. Math., vol. 44 (1973), pp. 593-602.

TSL **Lawson, J.D. [1976a]**, *Additional notes on continuity in semitopological semigroups*, Semigroup Forum, vol. 12 (1976), pp. 265-280.

APP **Lawson, J.D. [1976b]**, *Embeddings of compact convex sets and locally compact cones*, Pac. J. Math., vol. 66, No. 2 (1976), pp. 443-453.

APP **Lawson, J.D. [1976c]**, *Applications of topological algebra to hyperspace problems*, Topology-Proc. Memphis State U. Conf., Dekker, 1976, pp. 201-206.

TSL **Lawson, J.D. [1977]**, *Compact semilattices which must have a basis of subsemilattices*, J. London Math. Soc. (2), vol. 16 (1977), pp. 369-371.

CL **Lawson, J.D [1979a]**, *The duality of continuous posets*, to appear, Houston J. Math, to appear.

CL **Lawson, J.D. [1979b]**, *Algebraic conditions leading to continuous lattices*, to appear, Proc. Amer. Math. Soc., to appear.

APP **Lawson, J.D., Liukkonen, J.R. and Mislove, M.W. [1977]**, *Measure algebras of semilattices with finite breadth*, Pac. J. Math., vol. 69 (1977), pp. 125-139.

TSL **Lawson, J.D. and Williams, W. [1970]**, *Semilattices and their underlying spaces*, Semigroup Forum, vol. 1 (1970), pp. 209-223.

TSL **Lea, J.W., Jr. [1972]**, *An embedding theorem for compact semilattices*, Proc. Amer. Math. Soc., vol. 34 (1972), pp. 325-331

TL **Lea, J.W., Jr. [1973]**, *The peripherality of irreducible elements of a lattice*, Pac. J. Math., vol. 45 (1973), pp. 555-560.

TL **Lea, J.W., Jr. [1974a]**, *The codimension of the boundary of a lattice ideal*, Proc. Amer. Math. Soc., vol. 43 (1974), pp. 36-38.

TL **Lea, J.W., Jr. [1974b]**, *Sublattices generated by chains in modular topological lattices*, Duke Math. J., vol. 41 (1974), pp. 241-246.

TL **Lea, J.W., Jr. [1976]**, *Breadth two topological lattices with connected sets of irreducibles*, Trans. Amer. Math. Soc., vol. 219 (1976), pp. 337-345.

CL **Lea, J.W., Jr. [1976/77]**, *Continuous lattices and compact Lawson semilattices*, Semigroup Forum, vol. 13 (1976/77), pp. 387-388.

TL **Lea, J.W., Jr. [1978]**, *Quasiplanar topological lattices*, Houston J. Math, vol. 4 (1978), pp. 85-90.

TSL **Lea, J.W., Jr. and Lau, A.Y.W. [1975]**, *Codimension of compact M-semilattices*, Proc. Amer. Math. Soc., vol. 52 (1975), pp. 406-408.

TL Liber, S.A. [1977], *On Z-free compact lattices (in Russ.)*, Issled. Algebre, vyp. 5, Saratov 1977, pp. 44-52.

TL Liber, S.A. [1978], *Free compact lattices*, Issled. Math. Notes, vol. 24 (1978), pp. 832-835.

CL Lystad, G. and Stralka, A. [198*], *Semilattices having bialgebraic congruence lattices*, to appear.

REF MacLane, S. [1971], *Categories for the working mathematician*, Springer-Verlag, New York, 1971, 262 pp.

APP Manna, Z. [1974], *Mathematical Theory of Computation*, McGraw-Hill, 1974.

PO Markowsky, G. [1976], *Chain-complete posets and directed sets with applications*, Alg. Universalis, vol. 6 (1976), pp. 53-68.

CL Markowsky, G. [1977a], *A motivation and generalization of Scott's notion of a continuous lattice*, preprint.

PO Markowsky, G. [1977b], *Categories of chain-complete posets*, Theoretical Computer Science, vol. 4 (1977), pp. 125-135.

CL Markowsky, G. [1979], *Free completely distributive lattices*, Proc. Amer. Math. Soc., vol. 74 (1979), pp. 227-234.

APP & CL Markowsky, G. and Rosen, B. [1976], *Bases for chain-complete posets*, IBM J. of Research and Development, vol. 20 (1976), pp. 138-147 .

IT Matsushima, Y. [1960], *Hausdorff interval topology on a partially ordered set*, Proc. Amer. Math. Soc., vol. 11 (1960), pp. 233-235.

POTS Maurice, M. [1964], *Compact ordered spaces*, Math. Centrum, 1964.

TSL McWaters, M. [1969], *A note on topological semilattices*, J. London Math. Soc., vol. 1 (1969), pp. 64-66.

GT Moore, E.H. and Smith, H.L. [1922], *A general theory of limits*, American Journal of Mathematics, vol. 44 (1922), pp. 102-121.

GLT Nachbin, L. [1949], *On characterizations of the lattice of all ideals of a ring*, Fund. Math., vol. 6 (1949), pp. 137-142.

POTS Nachbin, L. [1965], *Topology and Order*, Van Nostrand, 1965.

IT Naito, T. [1960], *On a problem of Wolk in interval topologies*, Proc. Amer. Math. Soc., vol. 11 (1960), pp. 156-158.

APP Newman, S.E. [1969], *Measure algebras on idempotent semigroups*, Pac. J. Math., vol. 31 (1969), pp. 161-169.

APP & IT Nickel, K. [1975], *Verbandstheoretische Grundlagen der Intervallmathematik*, Springer-Velag Lecture Notes in Computer Science, vol. 29 (1975), pp. 251-262.

TL Numakura, K. [1957], *Theorems on compact totally disconnected semigroups and lattices*, Proc. Amer. Math. Soc., vol. 8 (1957), pp. 623-626.

GLT Ore, O. [1944], *Galois connections*, Trans. Amer. Math. Soc., vol. 55 (1944), pp. 493-513.

CLT Papert, S. [1959], *Which distributive lattices are lattices of closed sets ?*, Proc. Cambr. Phil. Soc., vol. 55 (1959), pp. 172-176.

GLT Pickert, G. [1952], *Bemerkungen über Galois-Verbindungen*, Archiv d. Math., vol. 3 (1952), pp. 285-289.

APP&CL **Plotkin, G.D. [1976]**, *A powerdomain construction*, SIAM J. on Computing, vol. 5 (1976), pp. 452-487.

APP&CL **Plotkin, G.D. [1978]**, T^{ω} *as a universal domain*, J. Comp. Sys. Sci., vol. 17 (1978), pp. 209-236.

GT & GLT **Priestley, H.A. [1970]**, *Representation of distributive lattices by means of ordered Stone spaces*, Bull. Lond. Math. Soc., vol. 2 (1970), pp. 186-190.

POTS **Priestley, H.A. [1972]**, *Ordered topological spaces and the representation of distributive lattices*, Proc. London Math. Soc., vol. 24 (1972), pp. 507-530.

GLT **Raney, G.N. [1952]**, *Completely distributive complete lattices*, Proc. Amer. Math. Soc., vol. 3 (1952), pp. 667-680.

GLT **Raney, G.N. [1953]**, *A subdirect-union representation for completely distributive complete lattices*, Proc. Amer. Math. Soc., vol. 4 (1953), pp. 518-522.

GLT **Raney, G.N. [1960]**, *Tight Galois connections and complete distributivity*, Trans. Amer. Math. Soc., vol. 97 (1960), pp. 418-426.

IT **Rennie, B.C. [1950]**, *The Theory of Lattices*, Foister and Jag, 1950.

IT **Rennie, B.C. [1951]**, *Lattices*, Proc. London Math. Soc., vol. 52 (1951), pp. 386-400.

TL & TSL **Rhodes, J. [1973]**, *Decomposition semilattices with applications to topological lattices*, Pac. J. Math., vol. 44 (1973), pp. 299-307.

APP **Roberts, J.W. [1977]**, *A compact convex set with no extreme points*, Studia Math., vol. 60 (1977), pp. 255-266.

APP **Rogers, H.R. [1967]**, *Theory of Recursive Functions and Effective Computability*, McGraw-Hill, 1967.

GLT **Schmidt, J. [1973]**, *Each join completion of a partially ordered set is the solution of a universal problem*. J. Aust. Math. Soc., vol. 16 (1973).

TSL **Schneperman, L.B. [1968]**, *On the theory of characters of locally bicompact topological semigroups*, Math. Sbor., vol. 77 (1968), pp. 508-532. (Math. USSR sb., vol. 6 (1968), pp. 471-492.)

APP & CL **Scott, D.S. [1970]**, *Outline of a mathematical theory of computation*, Proc. 4th Ann. Princeton Conference on Information Science and Systems, 1970, pp. 169-176.

CL **Scott, D.S. [1972a]**, *Continuous lattices*, Springer-Verlag Lecture Notes in Math, vol. 274 (1972), pp. 97-136.

APP & CL **Scott, D.S. [1972b]**, *Lattice theory, data types, and semantics*, in Formal Semantics of Programming Languages, R. Rustin Ed., Courant Comp. Sc. Symp., vol. 2 (1972), pp. 65-106.

APP & CL **Scott, D.S. [1973]**, *Models for various type-free calculi*, in Logic, Methodology and Philosophy of Science IV, P. Suppes, *et al.*, eds., North-Holland Publ. Co., 1973, pp. 157-187.

APP **Scott, D.S. [1975a]**, *Combinators and classes*, Springer-Verlag Lecture Notes in Computer Science, vol. 37 (1975), pp. 1-26.

APP **Scott, D.S. [1975b]**, *Some philosophical issues concerning theories of combinators*, ibid. pp. 346-366.

APP & CL **Scott, D.S. [1976]**, *Data types as lattices*, SIAM J. Computing, vol. 5 (1976), pp. 522-587.

APP Scott, D.S. [1977], *Logic and programming languages*, Comm. Assoc. for Comp. Mach.,vol. 20 (1977), pp. 634-641.

TL Shirley, E.D. and Stralka, A.R. [1971], *Homomorphisms on connected topological lattices*, Duke J. Math. (1971), pp. 483-490.

GLT Shmuely, Z. [1974], *The structures of Galois connections*, Pac. J. Math., vol. 54 (1974), pp. 209-225.

REF Sikorski, R. [1964], *Boolean Algebras*, Springer-Verlag, New York, 1964. (Third edition, 1969.)

APP & CL Smyth, M.B. [1977], *Effectively given domains*, Theoretical Computer Science, vol. 5 (1977), pp. 257-274.

APP & CL Smyth, M.B. [1978], *Power domains*, J. Comp. and Sys. Sci., vol. 16 (1978), pp. 23-36.

APP & CL Smyth, M.B. and Plotkin, G.D. [1978], *The category-theoretic solution of recursive domain equations*, D.A.I. Research Report No. 60, Edinburgh, 1978.

TSL Stepp, J.W. [1971], *Semilattices which are embeddable in a product of min intervals*, Semigroup Forum, vol. 2 (1971), pp. 80-82.

TSL & TL Stepp, J.W. [1973], *Topological semilattices which are embeddable in topological lattices*, J. London Math. Soc., vol. 7 (1973), pp. 76-82.

TSL Stepp, J.W. [1975a], *Algebraic maximal semilattices*, Pac. J. Math., vol. 58 (1975), pp. 243-248.

TSL & TL Stepp, J.W. [1975b], *The free compact lattice generated by a topological semilattice*, J.f.d. reine u. angew. Mathematik, vol. 273 (1975), pp. 77-86.

GLT & GT Stone, M.H. [1936], *The theory of representations for Boolean algebras*, Trans. Amer. Math. Soc., vol. 40 (1936), pp. 37-111.

GLT & GT Stone, M.H. [1937], *Topological representation of distributive lattices and Browerian logics*, Cas. Mat. Fys., vol. 67 (1937), pp. 1-25.

APP Stoy, J.E. [1977], *Denotational Semantics - The Scott-Strachey Approach to Programming Language Theory*, MIT Press, Cambridge, Mass., 1977.

TL Stralka, A.R. [1970], *Locally convex topological lattices*, Trans. Amer. Math. Soc., vol. 151 (1970), pp. 629-640.

TL Stralka, A.R. [1971], *The congruence extension property for compact topological lattices*, Pac. J. Math, vol. 38 (1971), pp. 795-802.

TSL Stralka, A.R. [1972], *The lattice of ideals of a compact semilattice*, Proc. Amer. Math. Soc., vol. 33 (1972), pp. 175-180.

TL Stralka, A.R. [1973], *Distributive topological lattices*, Proc. Lattice Theory Conf., U. Houston, 1973, pp 269-276.

TL Stralka, A.R. [1974], *Imbedding locally convex lattices into compact lattices*, Coll. Math., vol. 29 (1974), pp. 144-150.

CL & TSL Stralka, A.R. [1977], *Congruence extension and amalgamation in CL*, Semigroup Forum, vol. 13 (1977), pp. 355-375.

CL Stralka, A.R. [1979], *Quotients of products of compact chains*, Bull. Lond. Math. Soc., vol. 11 (1979), pp. 1-4.

IT & TL Strauss, D.P. [1968], *Topological lattices*, Proc. London Math. Soc., vol. 18 (1968), pp. 217-230.

GLT & GT Thron, W.J. [1962], *Lattice equivalence of topological spaces*, Duke Math. J., vol. 29 (1962), pp. 671-680.

APP & CL Vuillemin, J. [1974], *Syntaxe, Sémantique et Axiomatique d'un Language de Programmation Simple*, Thèse d'état, University of Paris, Sept. 1974.

GT Wallace, A.D. [1945], *A fixed-point theorem*, Bull. Amer. Math. Soc., vol. 51 (1945), pp. 413-416.

POTS Wallace, A.D. [1954], *Partial order and indecomposability*, Proc. Amer. Math. Soc., vol. 5 (1954), pp. 780-781.

POTS Wallace, A.D. [1955], *Struct ideals*, Proc. Amer. Math. Soc., vol. 6 (1955), pp. 634-638.

TL Wallace, A.D. [1957a], *The center of a compact lattice is totally disconnected*, Pac. J. Math, vol. 7 (1957), pp. 1237-1238.

TL Wallace, A.D. [1957b], *The peripheral character of the central elements of a lattice*, Proc. Amer. Math. Soc., vol. 8 (1957), pp. 596-597.

TL Wallace, A.D. [1957c], *Two theorems on topological lattices*, Pac. J. Math, vol. 7 (1957), pp. 1239-1241.

TL Wallace, A.D. [1958], *Factoring a lattice*, Proc. Amer. Math. Soc., vol. 9 (1958), pp. 250-252.

TSL Wallace, A.D. [1961], *Acyclicity of compact connected semigroups*, Fund. Math., vol. 50 (1961), pp. 99-105.

IT Ward, A.S. [1955], *On relations between certain intrinsic topologies in certain partially ordered sets*, Proc. Cambridge Phil. Soc., vol. 51 (1955), pp. 254-261.

APP Ward, A.S. [1969], *Problem in "Topology and its Applications"* (Proceedings Herceg Novi 1968), Belgrade 1969, p. 352.

POTS Ward, L.E., Jr. [1954], *Partially ordered topological spaces*, Proc. Amer. Math. Soc., vol. 5 (1954), pp. 144-161.

TSL Ward, L.E., Jr. [1958], *Completeness in semilattices*, Canad. J. Math., (1958), pp. 578-582.

POTS Ward, L.E., Jr. [1965a], *Concerning Koch's theorem on the existence of arcs*, Pac. J. Math., vol. 15 (1965), pp. 347-355.

POTS Ward, L.E., Jr. [1965b], *On the conjecture of R.J. Koch*, Pacific J. Math., vol. 15 (1965), pp. 1429-1433.

TSL Williams, W.W. [1975], *Semilattices on Peano continua*, Proc. Amer. Math. Soc., vol. 49 (1975), pp. 495-500.

GT & POTS Wilson, R.L. [1978], *Intrinsic topologies on partially ordered sets and results on compact semigroups*, Dissertation, Univ. of Tennessee, 1978.

GT Wojdyslawski, M. [1939], *Rétracts absolus et hyperspaces des continus*, Fund. Math., vol. 32 (1939), pp. 184-192.

IT Wolk, E.S. [1958], *Topologies on a partially ordered set*, Proc. Amer. Math. Soc., vol. 9 (1958), pp. 524-529.

TL Wolk, E.S. [1961], *On order-convergence*, Proc. Amer. Math. Soc., vol. 12 (1961), pp. 379-384.

CL Wyler, O. [1977a], *On continuous lattices as topological algebras*, preprint.

GT Wyler, O. [1977b], *Injective spaces and essential extensions in **TOP***, General Topol. Appl., vol. 7 (1977), pp. 247-249.

APPENDIX

Chronological List of Memos Circulated in the Seminar on Continuity in Semilattices (SCS)

[1] 19 Jan 1976 Lawson, J.D.
More notes on spreads..

[2] 19 Jan 1976 Hofmann, K.H.
Notes on Memo [SCS-1].

[3] 29 Jan 1976 Keimel, K.
Equationally compact SENDOs are retracts of compact ones.

[4] 30 Mar 1976 Scott, D.S.
Notes on continuous lattices.

[5] 19 Apr 1976 Hofmann, K.H.
Notes on chains in CL-objects.

[6] 28 May 1976 Carruth, J.H., *et al.*
More notes on chains in CL-objects.

[7] 15 Jun 1976 Carruth, J.H., *et al.*
Still more notes on chains in CL-objects.

[8] 28 Jun 1976 Hofmann, K.H. and Mislove, M.
On the theorem of Lawson's that all compact locally connected finite dimensional semilattices are CL.

[9] 7 Jul 1976 Hofmann, K.H. and Mislove, M.
Commentary on Scott's function spaces.

[10] 12 Jul 1976 Lawson, J.D.
Points with small semilattices.

[11] 20 Jul 1976 Hofmann, K.H. and Mislove, M.
Errata and corrigenda to Memo [SCS-9].

[12] 1 Aug 1976 Gierz, G., Hofmann, K.H., Keimel, K. and Mislove, M.
Relations with the interpolation property and continuous lattices.

[13] 10 Aug 1976 Keimel, K.
Complements to relations with the interpolation property and continuous lattices.

[14] 18 Aug 1976 Mislove, M.
On Memo [SCS-10].

[15] 23 Aug 1976 Scott, D.S.
Continuous lattices and universal algebra.

[16] 1 Sep 1976 Hofmann, K.H. and Liukkonen, J.
The random unit interval (another example of a CL-object).

[17] 20 Sep 1976 Hofmann, K.H.
The space of lower semicontinuous functions into a CL-object, Applications (Part I): Copowers in CL.

[18] 21 Sep 1976 Day, A.
Continuous lattices and universal algebra.

[19] 30 Sep 1976 Keimel, K. and Mislove, M.
Several remarks:
1. The closed subsemilattices of a continuous lattice form a continuous lattice;
2. When do the prime elements of distributive lattice form a closed subset;

3. On lower semicontinuous function spaces;
4. On the continuity of the congruence lattice of a continuous lattice.

[20] 23 Oct 1976 Hofmann, K.H.
More on the coproduct. Errata and addenda.

[21] 10 Nov 1976 Carruth, J.D., Clark, C.E., Evans, E., Lea J.W. and Wilson, R.L.
$\leq(n)$.

[22] 10 Nov 1976 Gierz, G.
Representation of colimits in CL. Part I and II.

[23] 19 Nov 1976 Lawson, J.
Non-continuous lattices.

[24] 19 Nov 1976 Hofmann, K.H., Keimel, K.
Editorial.

[25] 23 Nov 1976 Hofmann, K.H.
Observations.

[26] 30 Nov 1976 Scott, D.
A reply to an editorial.

[27] 8 Dec 1976 Mislove, M.
Closure operators and kernel operators in CL.

[28] 15 Dec 1976 Keimel, K. and Mislove, M.
The lattice of open subsets of a topological space.

[29] 28 Dec 1976 Hofmann, K.H. and Wyler, O.
On the closedness of the set of primes in a continuous lattice.

[30] 4 Jan 1977 Lawson, J.
Continuous semilattices and duality.

[31] 13 Jan 1977 Hofmann, K.H.
The lattice of ideals of a C-algebra.*

[32] 8 Feb 1977 Hofmann, K.H. and Lawson, J.D.
The spectral theory of continuous lattices.

[33] 4 Mar 1977 Hofmann, K.H. and Lawson, J.D.
Complement to Memo [SCS-32].

[34] 8 Apr 1977 Gierz, G. and Hofmann, K.H.
On complete lattices for which $\mathcal{O}(L)$ *is continuous—A lattice theoretical characterization of CS.*

[35] 18 Apr 1977 Wyler, O.
Dedekind complete posets and Scott topologies.

[36] 16 May 1977 Scott, D.S.
Quotients of distributive continuous lattices: A result of S.A. Jalali.

[37] 20 May 1977 Wyler, O.
Comments on the spectral theory of continuous lattices.

[38] 1 Jul 1977 Kamara, M.
Treillis continus et treillis compléments distributifs.

[39] 15 Jul 1977 Stralka, A.R.
Quotients of cubes.

[40] 28 Jul 1977 Mislove, M.
A new approach to some results of Gierz, Hofmann and Lawson.

[41] 25 Sep 1977 Hofmann, K.H. and Scott, D.S.
An exercise on the spectrum of function spaces.

[42] 2 Nov 1977 Gierz, G. and Lawson, J.D.
Generalized continuous lattices.

[43] 18 Jan 1978 Hofmann, K.H.
Locally quasicompact sober spaces are Baire spaces.

[44] 9 Feb 1978 Bauer, H. and Keimel, K.
Remark on the Memo [SCS-43].

[45] 15 Apr 1978 Bauer, H.
Antichains and equational compactness.

[46] 19 May 1978 Gierz, G., Lawson, J.D. and Mislove, M.
A result about $\mathcal{O}(X)$.

[47] 28 May 1978 Hofmann, K.H.
Equivalence des espaces de Batbédat et des treillis algébriques.

[48] 29 Nov 1978 Hofmann, K.H. and Niño, J.
Projective limits in CL and Scott's construction.

[49] 30 Nov 1978 Hofmann, K.H. and Watkins, F.
A review of a theorem of Dixmier's.

[50] 30 May 1979 Hofmann, K.H. and Jones, L.W.
Scott continuous closure operators and modal operators. More self functors to which the Scott construction applies.

[51] 30 May 1979 Hofmann, K.H. and Watkins, F.
A new Lemma on primes and a topological characterisation of the category DCL of continuous Heyting algebras and CL-morphisms.

[52] 11 June 1979 Hofmann, K.H. and Keimel, K.
Bemerkungen zum "Neuen Lemma".

LIST OF SYMBOLS

a^*	Involution of the element a in a C*-algebra	I-1.20, *46*
$a \Rightarrow b$	Implication (in a Heyting algebra)	O-3.17, *25*
A ((B	A shields B	III-1.15, *148*
App(L)	Approximating auxiliary relations on L	I-1.11, *44*
Aux(L)	Auxiliary relations on L	I-1.9, *43*
$\mathfrak{B}_0$	Pederson ideal of the C*-algebra $\mathfrak{B}$	I-1.20, *46*
cBa	Complete Boolean algebra	O-2.6, *10*
(CD)	Complete distributivity law	I-2.4, *59*
cHa	Complete Heyting algebra	O-2.6, *10*
$\mathbb{C}$(L)	All subsets of L closed under all infs	Remarks preceding O-3.13, *23*
Con(K)	Compact convex subsets of K	I-1.22, *50*
Cong $\mathcal{A}$	Congruences on the abstract algebra $\mathcal{A}$	O-2.7(4), *11*
$\mathrm{Cong}^- \mathcal{A}$	Closed congruences on the compact algebra $\mathcal{A}$	O-2.7(6), *12*
const_p	Constant function with value p	II-4.18(2), *136*
COPRIME L	Coprimes of L	I-3.42, *81*
$C(X, \mathbb{R}^*)$	Continuous extended real-valued functions on X	O-2.10, *14*
d(L)	The density of the lattice L	III-4.16, *174*
D(X)	Proper open lower sets of X	Remarks preceding VII-3.1, *322*
diag : L→L×L	The diagonal map on L	O-3.27, *28*
f°	The corestriction of the map f	O-3.9, *21*
$f_\circ$	The inclusion induced by the map f	O-3.9, *21*
Filt L	Filters of L	O-1.3, *2*
$\mathrm{Filt}_0\mathrm{L}$	Filt L $\cup$ {Ø}	O-1.3, *2*
$\tilde{F}$L	Limit functor of F_n	Remarks preceding IV-4.1, *224*
Funct L	Function space functor on $INF^\uparrow$	Remarks following IV-4.9, *230*
$g^\wedge$	The lower adjoint of the map g	IV-3.4, *209*
(GA)	Condition for a GCL-lattice	III-1.17, *149*
G_f	Upper graph of the function f	I-1.21, *49*
Hom(L,$\mathbb{I}$)	Continuous lattice maps of L into $\mathbb{I}$	VII-2.10, *320*
Hom(S,$\mathbb{I}$)	Continuous semilattice maps of S into $\mathbb{I}$	VI-3.7, *284*
Hom(X,$\mathbb{I}$)	Continuous order-preserving maps of X into $\mathbb{I}$	VII-2.10, *320*
$\mathbb{I}$	The unit interval	O-2.7(9), *12*
I^+	Family of directed sups from I	O-4.13, *34*
Id L	The ideals of L	O-1.3, *2*
$\mathrm{Id}_0\mathrm{L}$	Id L $\cup$ {Ø}	O-1.3, *2*

Symbol	Meaning	Reference
$\mathrm{Id}^{-}\mathcal{A}$	The closed ideals of the Hausdorff ring $\mathcal{A}$	O-2.7(7), *12*
(INT)	Interpolation property	I-1.15, *46*
Irr L	The complete irreducibles of L	I-4.19, *92*
IRR L	The irreducible elements of L	I-3.5, *69*
(K)	Algebraic lattice axiom	I-4.4, *85*
K(L)	The compact elements of L	I-4.1, *85*
L^{op}	The opposite lattice of L	O-1.7, *4*
L_c	The range of p_c	O-3.11, *22*
L_k	The range of p_k	O-3.11, *22*
$\underline{\lim}\, x_i$	The lim inf of the $\{x_i\}$	II-1.1, *98*
L/R	The quotient of L by the relation R	I-2.11, *61*
LSC(X)	LSC(X,$\mathbb{R}^*$)	I-1.21, *49*
LSC(X,$\mathbb{R}^*$)	Lower semicontinuous functions from X to $\mathbb{R}^*$	O-2.10, *14*
$L_1 \otimes L_2$	The tensor product of L_1 and L_2	IV-1.44, *192*
$\mathbb{N}$	The natural numbers	O-2.7(10), *12*
$\mathbb{N}^*$	The extended natural numbers	O-2.7(10), *13*
$\mathcal{O}(\mathcal{L})$	Topology generated by all $\mathcal{L}$-convergent nets	II-1.1, *98*
$\mathcal{O}(X)$	The open sets in X	O-2.7(3), *11*
$\mathcal{O}$Filt X	The open filters of X	I-1.32, *54*
$\mathcal{O}\mathrm{Filt}_0$ X	$\mathcal{O}$Filt X $\cup \{\emptyset\}$	I-1.32, *54*
$\mathcal{O}_{reg}(X)$	The regular open sets in X	O-2.7(3), *11*
P	The power set functor on ***CL***	Remarks following IV-4.9, *230*
$\tilde{p}: F\tilde{F}L \to \tilde{F}L$	The map induced by $p: FL \to L$	IV-4.2(2), *225*
p_c	The closure operator induced by the projection p	O-3.11, *22*
$P(\mathbb{I})$	The random unit interval	I-2.17, *64*
p_k	The kernel operator induced by the projection p	O-3.11, *22*
p_L	The natural map from PL to L	IV-4.10(ii), *231*
$P(\mathbb{R})$	Distribution functions of real random variables on $\mathbb{R}$	I-2.17, *64*
PRIME L	The primes of L	I-3.11, *70*
$\mathbb{R}$	The real numbers	O-2.7(3), *11*
$\mathbb{R}^*$	The extended real numbers	O-2.7(9), *12*
R^{op}	The opposite relation of R	O-1.7, *4*
r_L	The sup map from Id L to L	IV-4.10(iii), *231*
(S)	Condition for a Scott-open set	Remarks preceding II-1.4, *100*
S^1	The poset S with an identity adjoined	O-2.12, *15*
$s_{\prec}$	The map from L to Id L induced by $\prec$	I-1.10, *43*
(SI)	Strong interpolation property	I-1.15, *46*
(SI_C)	Strong interpolation property for the chain C	IV-2.5(iii), *195*
Spec L	The spectrum of the lattice L	V-4.1, *251*

Symbol	Meaning	Reference
$\twoheaddownarrow x$	Way-below set of the point x	I-1.2, *39*
$\twoheaduparrow x$	Way-above set of the point x	I-1.2, *39*
$\twoheaduparrow \mathrm{M}$	Way-above set of M	II-1.28, *52*
$\lvert\mathrm{L}\rvert : \boldsymbol{I} \to \boldsymbol{C}$	Constant functor with value L	Remark following IV-3.1, *206*
$\top$	Top or identity of a poset	O-1.8, *4*
$\bot$	Bottom or zero of a poset	O-1.8, *4*
$\nabla_{\mathrm{L}}(a)$	$\uparrow a \cap \mathrm{Spec\,L}$	V-4.1, *251*
$\mathfrak{U}^{\wedge}$	Topology generated by all $x^{[-1]}A$ for $A \in \mathfrak{U}$	VII-1.4, *306*
$V^{\diamond}$	Set of direct sups and filtered infs from V	Remark preceding VII-1.6, *307*
$(\mathrm{S} \to \mathrm{T})$	Order-preserving maps from S to T	I-2.16, *63*
$[\mathrm{S} \to \mathrm{T}]$	Maps in $(\mathrm{S} \to \mathrm{T})$ preserving directed sups	I-2.16(iii), *64*
$\langle \mathrm{S} \to \mathrm{T} \rangle$	Cofinal maps in $\boldsymbol{CPoset}_0(\mathrm{S},\mathrm{T})$	IV-1.39, *191*
$\langle \mathrm{S} \rangle$	$\langle \mathrm{S} \to 2 \rangle$	IV-1.41, *191*
$\prec^{\sup}$	$\prec$ – sup relation on L	I-1.25, *51*
$\neg a$	Negation of the element a	O-3.17, *25*
2^X	Power set of X	O-2.7(1), *10*
$[\mathrm{X},\mathrm{Y}]$	$\boldsymbol{TOP}(\mathrm{X},\mathrm{Y})$ with the order from $\Omega\mathrm{Y}$	II-4.1, *128*
$[\mathrm{X},f]$	Function from $[\mathrm{X},\mathrm{Y}]$ to $[\mathrm{X},\mathrm{Z}]$ induced by $f : \mathrm{Y} \to \mathrm{Z}$	II-4.2, *128*
$\bigwedge : \boldsymbol{INF} \to \boldsymbol{SUP}^{\mathrm{op}}$	Functor taking an upper adjoint to its lower adjoint	IV-3.3, *209*
$\bigvee^{\mathrm{up}}$	Directed sup	Remark following I-2.3, *58*
$[A]$	The order-convex hull of A	VI-1.5, *273*
ζ-limit		II-1.1, *98*

LIST OF CATEGORIES

AL	Full subcategory of ***CL*** of algebraic lattices	IV-1.13, *183*
ALG	Full subcategory of ***CONT*** of algebraic lattices	II-2.2, *113*
ArL	Full subcategory of ***CL*** of arithmetic lattices	IV-1.13, *183*
BQSOB	Full subcategory of ***LQSOB*** of spaces having a basis of quasicompact open sets	V-5.18, *264*
CL	Full subcategory of $\boldsymbol{INF}^{\uparrow}$ of continuous lattices	IV-1.9, *182*
CONT	Full subcategory of ***UPS*** of continuous lattices	II-2.2, *113*
CPO	Compact pospaces with greatest element and continuous monotone maps preserving the unit	Remarks following VII-3.3, *323*
$\boldsymbol{CPoset}_0$	Continuous posets and Scott-continuous monotone maps	IV-1.39, *190*
CPoset	Continuous posets and cofinal maps	IV-1.39, *190*
CQSOB	Full subcategory of ***BQSOB*** of quasicompact spaces	V-5.19, *264*
CS	Compact semilattice monoids and continuous identity preserving semilattice morphisms	Remarks preceding V-3.26, *291*
$\boldsymbol{CSem}_0$	Continuous semilattices and Scott-continuous semilattice maps	IV-1.39, *190*
CSem	Continuous semilattices and cofinal semilattice maps	IV-1.39, *190*
DAR	Full subcategory of $\boldsymbol{ArL}^{\mathrm{op}} \cap \boldsymbol{HEYT}$ of distributive arithmetic lattices with compact identity element	V-5.19, *264*
DCL	Distributive continuous lattices with closed primes and sup-preserving lattice maps preserving way-below	VII-3.3, *323*
DCPO	Full subcategory of ***CPO*** of totally order-disconnected spaces	VII-3.9, p. 325
DL	Continuous distributive lattices and ***CL***-maps preserving spectra	Remarks preceding V-5.16, *263*
DLat	Distributive lattices with 0 and 1 and all 0 and 1 preserving lattice maps	V-5.19, *264*
GRAPH	Sup-semilattices and monotone relations	II-2.22, *120*
$\boldsymbol{HEYT}_0$	Full subcategory of ***HEYT*** of lattices where the primes order-generate	Remarks preceding V-4.6, *254*
HEYT	Complete Heyting algebras and sup-preserving lattice maps	Remarks preceding II-2.11, *118*
INF	Complete lattices and inf-preserving maps	IV-1.1, *179*
$\boldsymbol{INF}^{\uparrow}$	Complete lattices and inf- and directed sup-preserving maps	IV-1.9, *182*
LQSOB	Locally quasicompact sober spaces and "perfect" maps	Remarks preceding V-5.16, *263*
Lat	Unital lattices and lattice maps preserving the unit	IV-1.16, *185*
POSET	Posets and monotone maps	Remarks preceding II-3.6, *123*

INDEX

H. H. Scheafer

Banach Lattices and Positive Operators

1974. XI, 376 pages
(Grundlehren der mathematischen Wissenschaften, Band 215)
ISBN 3-540-06936-4

Contents:
Positive Matrices. – Banach Lattices. – Ideal and Operator Theory. – Lattices of Operators. – Applications.

Information:
This is a comprehensive account of Banach lattices and positive linear operators (including positive finite matrices) with main emphasis on their relevance for and applications to measure theory, mean ergotic theory, operators between L^p-spaces and spectral theory. Examples are given in each of the fifty sections, and the main text of each chapter is supplemented by bibliographical notes and by exercises.

Springer-Verlag
Berlin
Heidelberg
New York